Lecture Notes in Electrical Engineering

Volume 149

For further volumes:
http://www.springer.com/series/7818

Jean-Pierre Deschamps · Gustavo D. Sutter
Enrique Cantó

Guide to FPGA Implementation of Arithmetic Functions

Springer

Jean-Pierre Deschamps
University Rovira i Virgili
Tarragona
Spain

Enrique Cantó
University Rovira i Virgili
Tarragona
Spain

Gustavo D. Sutter
School of Computer Engineering
Universidad Autonoma de Madrid
Ctra. de Colmenar Km. 15
28049 Madrid
Spain

ISSN 1876-1100 e-ISSN 1876-1119
ISBN 978-94-007-2986-5 e-ISBN 978-94-007-2987-2
DOI 10.1007/978-94-007-2987-2
Springer Dordrecht Heidelberg New York London

Library of Congress Control Number: 2012935019

Printed on acid-free paper

Springer is part of Springer Science+Business Media (www.springer.com)

Preface

Field Programmable Gate Arrays constitute one of the technologies at hand for developing electronic systems. They form an attractive option for small production quantities as their non-recurrent engineering costs are much lower than those corresponding to ASIC's. They also offer flexibility and fast time-to-market. Furthermore, in order to reduce their size and, hence, the unit cost, an interesting possibility is to reconfigure them at run time so that the same programmable device can execute different predefined functions.

Complex systems, generally, are made up of processors executing programs, memories, buses, input-output interfaces, and other peripherals of different types. Those components are available under the form of Intellectual Property (IP) cores (synthesizable Hardware Description Language descriptions or even physical descriptions). Some systems also include specific components implementing algorithms whose execution on an instruction-set processor is too slow. Typical examples of such complex algorithms are: long-operand arithmetic operations, floating-point operations, encoding and processing of different types of signals, data ciphering, and many others. The way those specific components can be developed is the central topic of this book. So, it addresses to both, FPGA users interested in developing new specific components—generally for reducing execution times—and, IP core designers interested in extending their catalog of specific components.

This book distinguishes itself with the following aspects:

- The main topic is circuit synthesis. Given an algorithm executing some complex function, how can it be translated to a synthesizable circuit description? Which are the choices that the designer can make in order to reduce the circuit cost, latency, or power consumption? Thus, this is not a book on algorithms. It is a book on "how to efficiently translate an algorithm to a circuit" using, for that purpose, techniques such as parallelism, pipeline, loop unrolling, and others. In particular, this is not a new version of a previous book by two of the authors.[1]

[1] Deschamps JP, Bioul G, Sutter G (2006) Synthesis of Arithmetic Circuits. Wiley, New York.

It is a complement of the cited book; its aim is the generation of efficient implementations of some of the most useful arithmetic functions.
- All throughout this book, numerous examples of FPGA implementation are described. The circuits are modeled in VHDL. Complete and synthesizable source files are available at the authors' web site www.arithmetic-circuits.org.
- This is not a book on Hardware Description Languages, and the reader is assumed to have a basic knowledge of VHDL.
- It is not a book on Logic Circuits either, and the reader is assumed to be familiarized with the basic concepts of combinational and sequential circuits.

Overview

This book is divided into sixteen chapters. In the first chapter the basic building blocks of digital systems are briefly reviewed, and their VHDL descriptions are presented. It constitutes a bridge with previous courses, or books, on Hardware Description Languages and Logic Circuits.

Chapters 2–4 constitute a first part whose aim is the description of the basic principles and methods of algorithm implementation. Chapter 2 describes the breaking up of a circuit into Data Path and Control Unit, and tackles the scheduling and resource assignment problems. In Chaps. 3 and 4 some special topics of Data Path and Control Unit synthesis are presented.

Chapter 5 recalls important electronic concepts that must be taken into account for getting reliable circuits and Chap. 6 gives information about the main Electronic Design Automation (EDA) tools that are available for developing systems on FPGAs.

Chapters 7–13 are dedicated to the main arithmetic operations, namely addition (Chap. 7), multiplication (Chap. 8), division (Chap. 9), other operations such as square root, logarithm, exponentiation, trigonometric functions, base conversion (Chap. 10), decimal arithmetic (Chap. 11), floating-point arithmetic (Chap. 12), and finite-field arithmetic (Chap. 13). For every operation, several configurations are considered (combinational, sequential, pipelined, bit serial or parallel), and several generic models are available, thus, constituting a library of virtual components.

The development of Systems on Chip (SoC) is the topic of Chaps. 14–16. The main concepts are presented in Chap. 14: embedded processors, memories, buses, IP components, prototyping boards, and so on. Chapter 15 presents two case studies, both based on commercial EDA tools and prototyping boards. Chapter 16 is an introduction to dynamic reconfiguration, a technique that allows reducing the area by modifying the device configuration at run time.

Acknowledgments

The authors thank the people who have helped them in developing this book, especially Dr. Matthew Copley, for correcting the text. They are grateful to the following universities for providing them the means for carrying this work through to a successful conclusion: University Rovira i Virgili (Tarragona, Spain) and Autonomous University of Madrid (Spain).

Contents

Chapter 1
Basic Building Blocks

Digital circuits are no longer defined by logical schemes but by Hardware Description Language programs [1]. The translation of this kind of definition to an actual implementation is realized by Electronic Automation Design tools (Chap. 5). All along this book the chosen language is VHDL. In this chapter the most useful constructions are presented. For all of the proposed examples, the complete source code is available at the Authors' web page.

1.1 Combinational Components

1.1.1 Boolean Equations

Among the predefined operations of any Hardware Description Language are the basic Boolean operations. Boolean functions can easily be defined. Obviously, all the classical logic gates can be defined. Some of them are considered in the following example.

Example 1.1
The following VHDL instructions define the logic gates NOT, AND2, OR2, NAND2, NOR2, XOR2, XNOR2, NAND3, NOR3, XOR3, XNOR3 (Fig. 1.1).

```
b <= NOT(a);
c <= a AND b;
c <= a OR b;
c <= a NAND b;
```

J.-P. Deschamps et al., *Guide to FPGA Implementation of Arithmetic Functions*,
Lecture Notes in Electrical Engineering 149, DOI: 10.1007/978-94-007-2987-2_1,

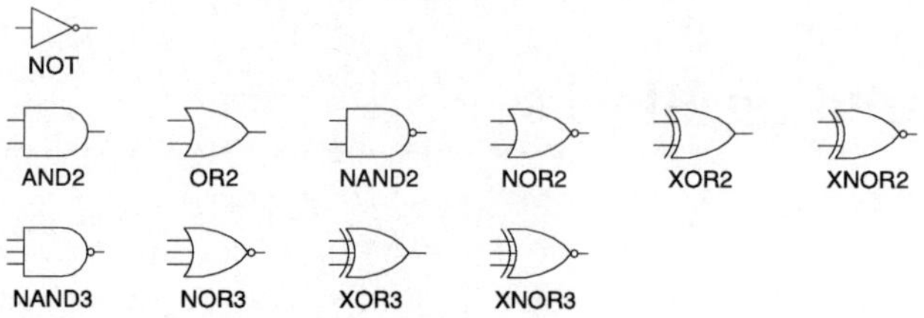

Fig. 1.1 Logic gates

```
c <= a NOR b;
c <= a XOR b;
c <= a XNOR b;
d <= NOT(a AND b AND c);
d <= NOT(a OR b OR c);
d <= a XOR b XOR c;
d <= NOT(a XOR b XOR c);
```

The same basic Boolean operations can be used to define sets of logic gates working in parallel. As an example, assume that signals $a = (a_{n-1}, a_{n-2},\ldots, a_0)$ and $b = (b_{n-1}, b_{n-2},\ldots, b_0)$ are n-bit vectors. The following assignation defines a set of n AND2 gates that compute $c = (a_{n-1}\cdot b_{n-1}, a_{n-2}\cdot b_{n-2},\ldots, a_0\cdot b_0)$:

```
c <= a AND b;
```

More complex Boolean functions can also be defined. It is worthwhile to indicate that within most Field Programmable Gate Arrays (FPGA) the basic combinational components are not 2-input logic gates but Look Up Tables (LUT) allowing the implementation of any Boolean function of a few numbers (4, 5, 6) of variables (Chap. 5). Hence, it makes sense to consider the possibility of defining small combinational components by the set of Boolean functions they implement.

Example 1.2
The following VHDL instruction defines an ANDORI gate (a 4-input 1-output component implementing the complement of $a{\cdot}b$ v $c{\cdot}d$, Fig. 1.2)

```
e <= NOT((a AND b) OR (c AND d));
```

and the two following instructions define a 1-bit full adder (a 3-input 2-output component implementing $a+b+c_{in}$ mod 2 and $a{\cdot}b$ v $a{\cdot}c_{in}$ v $b{\cdot}c_{in}$, Fig. 1.3).

```
s <= a XOR b XOR cin;
cout <= (a AND b) OR (a AND cin) OR (b AND cin);
```

Fig. 1.2 ANDORI gate

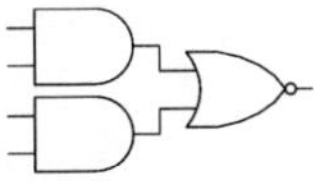

Fig. 1.3 One-bit full adder

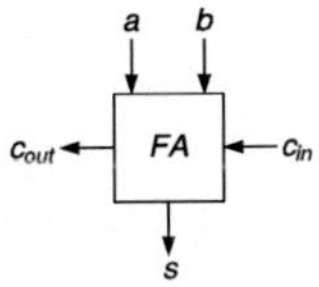

1.1.2 Tables

Combinational components can also be defined by their truth tables, without the necessity to translate their definition to Boolean equations. As a first example, consider a 1-digit to 7-segment decoder, that is a 4-input 7-output combinational circuit.

Example 1.3
The behavior of a 4-to-7 decoder (Fig. 1.4) can be described by a conditional assignment instruction

```
WITH digit SELECT segments <=
  "1110111" WHEN "0000",
  "0010010" WHEN "0001",
  "1011101" WHEN "0010",
  "1011011" WHEN "0011",
  "0111010" WHEN "0100",
  "1101011" WHEN "0101",
  "1101111" WHEN "0110",
  "1010010" WHEN "0111",
  "1111111" WHEN "1000",
  "1111011" WHEN "1001",
  "1111110" WHEN "1010",
  "0101111" WHEN "1011",
  "1100101" WHEN "1100",
  "0011111" WHEN "1101",
  "1101101" WHEN "1110",
  "1101100" WHEN OTHERS;
```

The last choice of a WITH… SELECT construction must be WHEN OTHERS in order to avoid the inference of an additional latch, and it is the same for other multiple choice instructions such as CASE.

As mentioned above, small Look Up Tables are basic components of most FPGA families. The following example is a generic 4-input 1-output LUT.

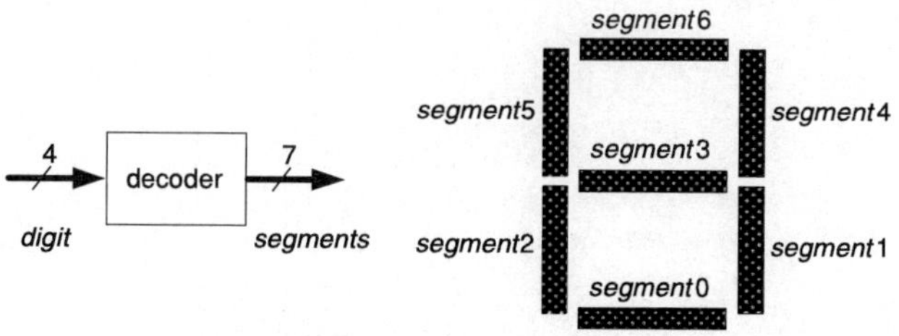

Fig. 1.4 Four-digit to seven-segment decoder

Fig. 1.5 Four-input Look Up Table

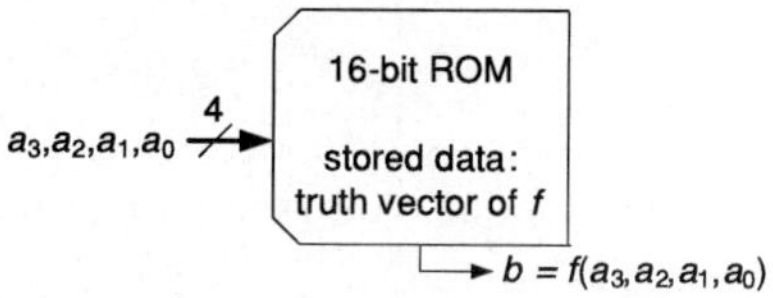

Example 1.4

The following entity defines a 4-input Boolean function whose truth vector is stored within a generic parameter (Fig. 1.5). Library declarations are omitted.

```
ENTITY lut4 IS
  GENERIC(truth_vector: STD_LOGIC_VECTOR(0 to 15));
PORT (
  a: IN STD_LOGIC_VECTOR (3 DOWNTO 0);
  b: OUT STD_LOGIC
);
END lut4;

ARCHITECTURE behavior OF lut4 IS
BEGIN
  PROCESS(a)
    VARIABLE c: NATURAL;
  BEGIN
    c := CONV_INTEGER(a);
    b <= truth_vector(c);
  END PROCESS;
END behavior;
```

Then the following component instantiation defines a 4-input XOR function.

```
xor4: lut4
GENERIC MAP(truth_vector => "0110100110010110")
PORT MAP(a => a, b => b);
```

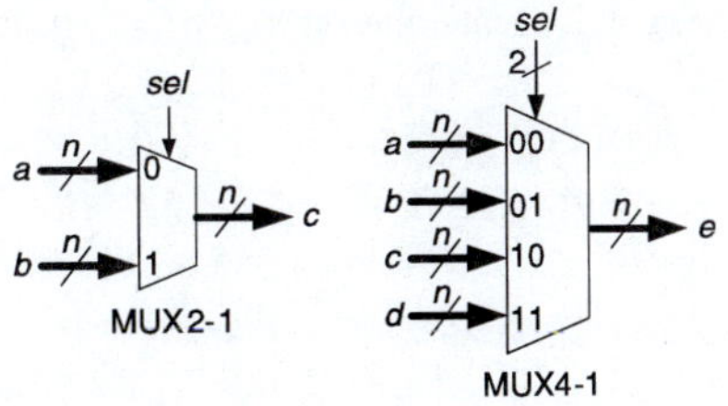

Fig. 1.6 Multiplexers

Comment 1.1
The VHDL model of the preceding Example 1.4 can be used for simulation purposes, independently of the chosen FPGA vendor. Nevertheless, in order to implement an actual circuit, the corresponding vendor's primitive component should be used instead (Chap. 5).

1.1.3 Controllable Connections

Multiplexers are the basic components for implementing controllable connections. Conditional assignments are used for defining them. Some of them are considered in the following example.

Example 1.5
The following conditional assignments define a 2-to-1 and a 4-to-1 multiplexer (Fig. 1.6). The signal types must be compatible with the conditional assignments: *a*, *b*, *c*, *d* and *e* are assumed to be *n*-bit vectors for some constant value *n*.

```
WITH sel SELECT c <= a WHEN '0', b WHEN OTHERS;

WITH sel SELECT e <=
  a WHEN "00",
  b WHEN "01",
  c WHEN "10",
  d WHEN OTHERS;
```

Demultiplexers, address decoders and tri-state buffers are other components frequently used for implementing controllable connections such as buses.

Example 1.6
The following equations define a 1-bit 1-to-2 demultiplexer (Fig. 1.7a)

```
b <= NOT(sel) AND a;
c <= sel AND a;
```

and the following conditional assignments a 1-to-4 demultiplexer (Fig. 1.7b)

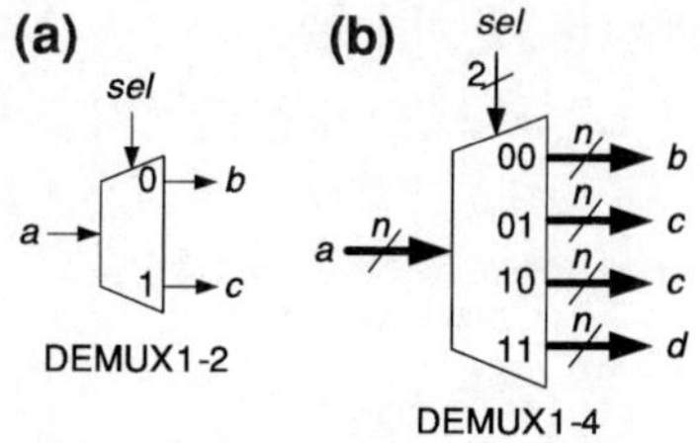

Fig. 1.7 Demultiplexers

```
WITH sel SELECT b <=
  a WHEN "00", (OTHERS => '0') WHEN OTHERS;
WITH sel SELECT c <=
  a WHEN "01", (OTHERS => '0') WHEN OTHERS;
WITH sel SELECT d <=
  a WHEN "10", (OTHERS => '0') WHEN OTHERS;
WITH sel SELECT e <=
  a WHEN "11", (OTHERS => '0') WHEN OTHERS;
```

In the second case the signal types must be compatible with the conditional assignments: a, b, c, d and e are assumed to be n-bit vectors for some constant value n.

An address decoder is a particular case of 1-bit demultiplexer whose input is 1.

Example 1.7
The following equations define a 3-input address decoder (Fig. 1.8).

```
rows(0) <=
  NOT(address(2)) AND NOT(address(1)) AND NOT(address(0));
rows(1) <=
  NOT(address(2)) AND NOT(address(1)) AND address(0);
rows(2) <=
  NOT(address(2)) AND address(1)      AND NOT(address(0));
rows(3) <=
  NOT(address(2)) AND address(1)      AND address(0);
rows(4) <=
  address(2)      AND NOT(address(1)) AND NOT(address(0));
rows(5) <=
  address(2)      AND NOT(address(1)) AND address(0);
rows(6) <=
  address(2)      AND address(1)      AND NOT(address(0));
rows(7) <=
  address(2)      AND address(1)      AND address(0);
```

Three-state buffers implement controllable switches.

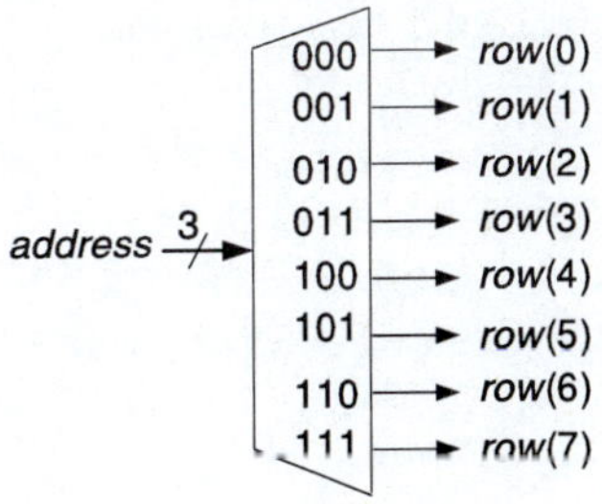

Fig. 1.8 Address decoder

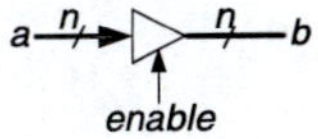

Fig. 1.9 Three-state buffer

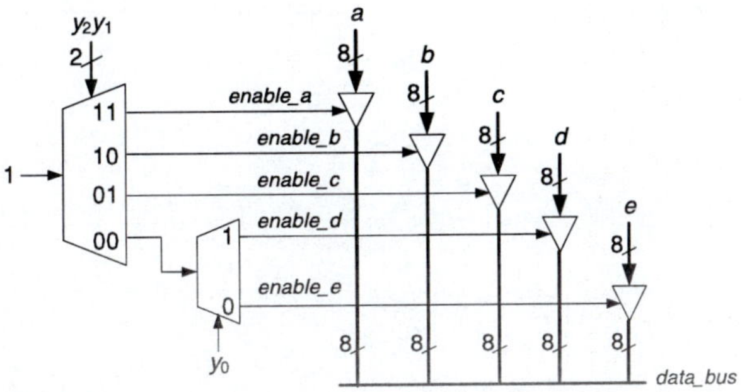

Fig. 1.10 Example of a data bus

Example 1.8
The following conditional assignment defines a tri-state buffer (Fig. 1.9): when the enable signal is equal to 1, output *b* is equal to input *a*, and when the enable signal is equal to 0, output *b* is disconnected (high impedance state). The signal types must be compatible with the conditional assignments: *a* and *b* are assumed to be *n*-bit vectors for some constant value *n*.

```
WITH enable SELECT b <=
  a WHEN '1',
  (OTHERS =>'Z') WHEN OTHERS;
```

An example of the use of demultiplexers and tri-state buffers, namely a data bus, is shown in Fig. 1.10.

Example 1.9
The circuit of Fig. 1.10, made up of demultiplexers and three-state buffers, can be described by Boolean equations (the multiplexers) and conditional assignments (the three-sate buffers).

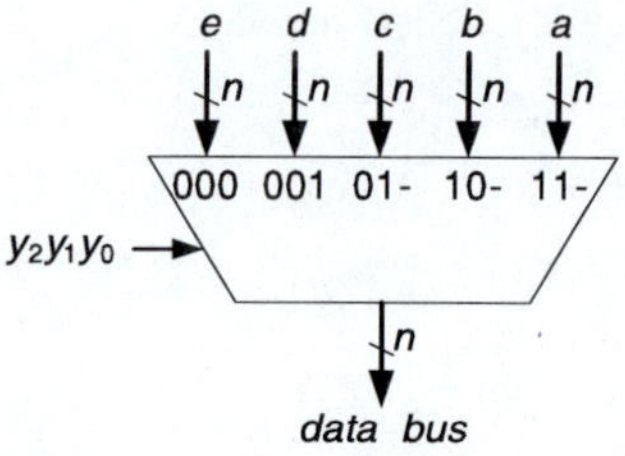

Fig. 1.11 Data bus, second version

```
enable_a <= y(2)       AND y(1);
enable_b <= y(2)       AND NOT(y(1));
enable_c <= NOT(y(2)) AND y(1);
enable_d <= NOT(y(2)) AND NOT(y(1))    AND y(0);
enable_e <= NOT(y(2)) AND NOT(y(1))    AND NOT(y(0));
WITH enable_a SELECT data_bus <=
  a WHEN '1', (OTHERS =>'Z') WHEN OTHERS;
WITH enable_b SELECT data_bus <=
  b WHEN '1', (OTHERS =>'Z') WHEN OTHERS;
WITH enable_c SELECT data_bus <=
  c WHEN '1', (OTHERS =>'Z') WHEN OTHERS;
WITH enable_d SELECT data_bus <=
  d WHEN '1', (OTHERS =>'Z') WHEN OTHERS;
WITH enable_e SELECT data_bus <=
  e WHEN '1', (OTHERS =>'Z') WHEN OTHERS;
```

Nevertheless, the same functionality can be implemented by an 8-bit 5-to-1 multiplexer (Fig. 1.11).

```
WITH y SELECT data_bus <=
  a WHEN "111" | "110",
  b WHEN "101" | "100",
  c WHEN "011" | "010",
  d WHEN "001",
  e WHEN OTHERS;
```

Generally, this second implementation is considered safer than the first one. In fact, tri-state buffers should not be used within the circuit core. They should only be used within I/O-port components (Sect. 1.4).

1.1.4 Arithmetic Circuits

Among the predefined operations of any Hardware Description Language there are also the basic arithmetic operations. The translation of this kind of description to

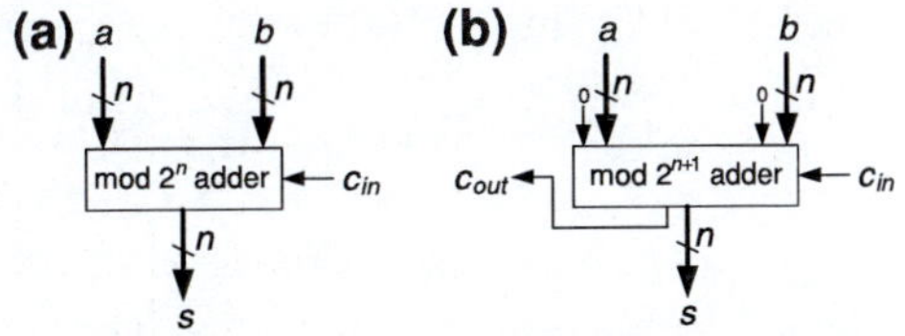

Fig. 1.12 *n*-bit adders

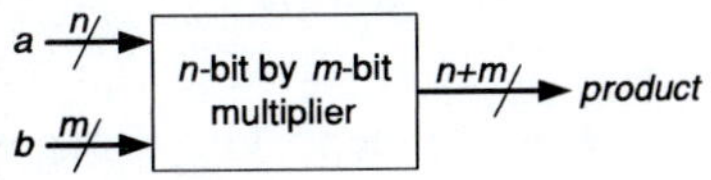

Fig. 1.13 *n*-bit by *m*-bit multiplier

actual implementations, using special purpose FPGA resources (carry logic, multiplier blocks), is realized by Electronic Automation Design tools (Chap. 5).

Example 1.10
The following arithmetic equation defines an adder mod 2^n, being *a*, *b* and *s* *n*-bit vectors and c_{IN} an input carry (Fig. 1.12a).

```
s <= a + b + c_in;
```

By adding a most significant bit 0 to one (or both) of the *n*-bit operands *a* and *b*, an *n*-bit adder with output carry c_{OUT} can be defined (Fig. 1.12b). The internal signal *sum* is an (*n*+1)-bit vector.

```
sum <= ('0'&a) + b + c_in;
s <= sum(n-1 DOWNTO 0);
c_out <= sum(n);
```

The following arithmetic equation defines an *n*-bit by *m*-bit multiplier where *a* is an *n*-bit vector, *b* an *m*-bit vector and *product* an (*n*+*m*)-bit vector (Fig. 1.13).

```
product <= a * b;
```

Comment 1.2
In most VHDL models available at the Authors' web page, the type *unsigned* has been used, so that bit-vectors can be treated as natural numbers. In some cases, it could be better to use the *signed* type, for example when bit-vectors are interpreted as 2's complement integers, and when magnitude comparisons or sign-bit extensions are performed.

Fig. 1.14 D flip-flops

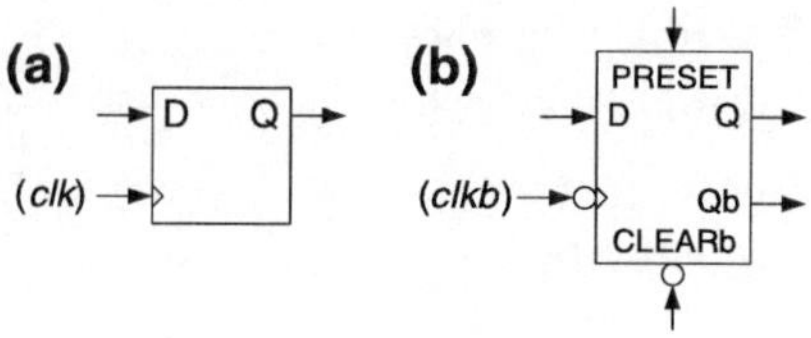

1.2 Sequential Components

1.2.1 Flip-Flops

The basic sequential component is the D-flip-flop. In fact, several types of D-flip-flop can be considered: positive edge triggered, negative edge triggered, with asynchronous input(s) and with complemented output.

Example 1.11
The following component is a D-flip-flop triggered by the positive edge of *clk* (Fig. 1.14a),

```
PROCESS(clk)
BEGIN
  IF clk'EVENT AND clk = '1' THEN q <= d; END IF;
END PROCESS;
```

while the following, which is controlled by the negative edge of *clkb*, has two asynchronous inputs *clearb* (active at low level) and *preset* (active at high level), having *clearb* priority, and has two complementary outputs *q* and *qb* (Fig. 1.14b).

```
PROCESS(clkb, preset, clearb)
BEGIN
  IF clearb = '0' THEN q <= '0'; qb <= '1';
  ELSIF preset = '1' THEN q <= '1'; qb <= '0';
  ELSIF clkb'EVENT AND clkb = '0' THEN q <= d; qb <= not(d);
  END IF;
END PROCESS;
```

Comment 1.3
The use of level-controlled, instead of edge-controlled, components is not recommendable. Nevertheless, if it were necessary, a D-latch could be modeled as follows:

```
IF en = '1' THEN q <= d; END IF;
```

where *en* is the enable signal and *d* the data input.

Fig. 1.15 Parallel register

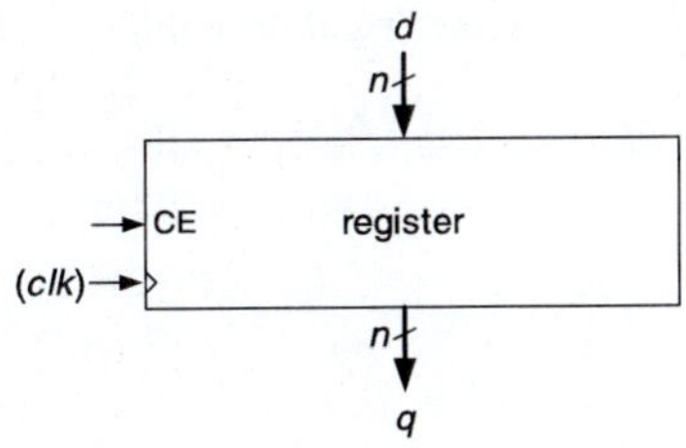

1.2.2 Registers

Registers are sets of D-flip-flops controlled by the same synchronization and control signals, and connected according to some regular scheme (parallel, left or right shift, bidirectional shift).

Example 1.12
The following component is a parallel register with *ce* (*clock enable*) input, triggered by the positive edge of *clk* (Fig. 1.15).

```
PROCESS(clk)
BEGIN
  IF clk'EVENT AND clk = '1' THEN
    IF ce = '1' THEN q <= d; END IF;
  END IF;
END PROCESS;
```

As a second example (Fig. 1.16), the next component is a right shift register with parallel input (controlled by *load*) and serial input (controlled by *shift*).

```
PROCESS(clk)
BEGIN
  IF clk'EVENT AND clk = '1' THEN
    IF load = '1' THEN q <= parallel_in;
    ELSIF shift = '1' THEN
      q <= serial_in & q(n-1 downto 1);
    END IF;
  END IF;
END PROCESS;
```

1.2.3 Counters

A combination of registers and arithmetic operations permits the definition of counters.

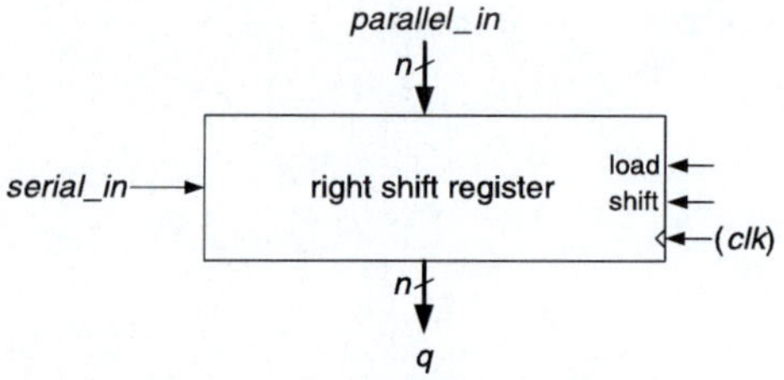

Fig. 1.16 Right shift register

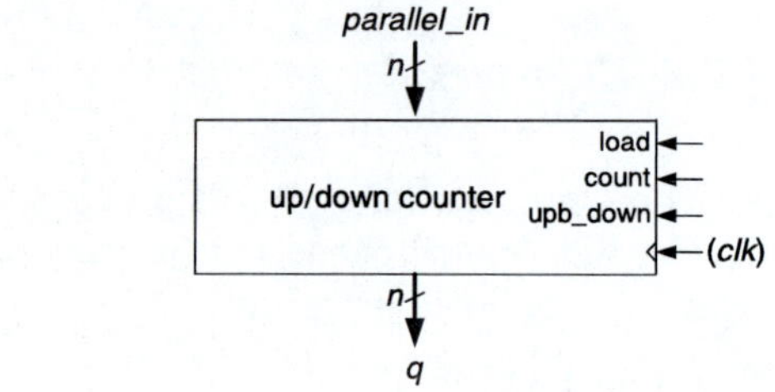

Fig. 1.17 Up/down counter

Example 1.13
This defines an up/down counter (Fig. 1.17) with control signals *load* (input *parallel_in*), *count* (update the state of the counter) and *upb_down* (0: count up, 1: count down).

```
PROCESS(clk)
BEGIN
  IF clk'EVENT AND clk = '1' THEN
    IF load = '1' THEN q <= parallel_in;
    ELSIF count = '1' AND upb_down = '0' THEN
      q <= q + 1;
    ELSIF count = '1' AND upb_down = '1' THEN
      q <= q - 1;
    END IF;
  END IF;
END PROCESS;
```

The following component is a down counter (Fig. 1.18) with control signals *load* (input *parallel_in*) and *count* (update the state of the counter). An additional binary output *equal_zero* is raised when the state of the counter is *zero* (all 0's vector).

```
PROCESS(clk)
BEGIN
  IF clk'EVENT AND clk = '1' THEN
    IF load = '1' THEN q <= parallel_in;
    ELSIF count = '1' THEN q <= q - 1;
    END IF;
  END IF;
END PROCESS;
equal_zero <= '1' WHEN q = zero ELSE '0';
```

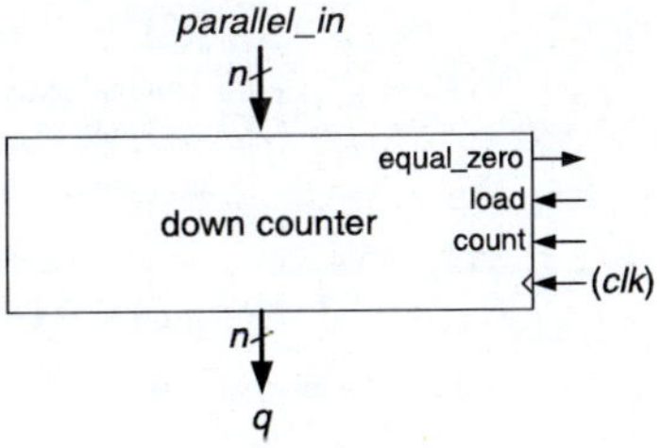

Fig. 1.18 Down counter

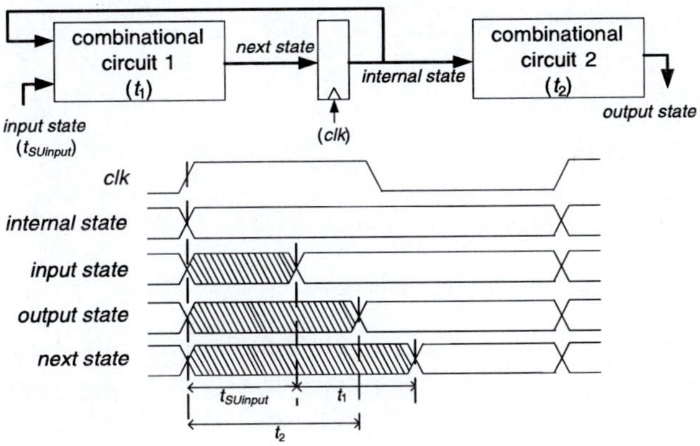

Fig. 1.19 Moore machine

1.2.4 Finite State Machines

Hardware Description Languages allow us to define finite state machines at an input/output behavioral level. The translation to an actual implementation including registers and combinational circuits—a classical problem of traditional switching theory—is realized by Electronic Automation Design tools (Chap. 5).

In a Moore machine, the output state only depends on the current internal state (Fig. 1.19) while in a Mealy machine the output state depends on both the input state and the current internal state (Fig. 1.20). Let $t_{SUinput}$ be the maximum set up time of the input state with respect to the positive clock edge, t_1 the maximum delay of the combinational block that computes the next internal state, and t_2 the maximum delay of the combinational block that computes the output state. Then, in the case of a Moore machine, the following conditions must hold

$$t_{SUinput} + t_1 < T_{CLK} \text{ and } t_2 < T_{CLK}, \tag{1.1}$$

and in the case of a Mealy machine

$$t_{SUinput} + t_1 < T_{CLK} \text{ and } t_{SUinput} + t_2 < T_{CLK}. \tag{1.2}$$

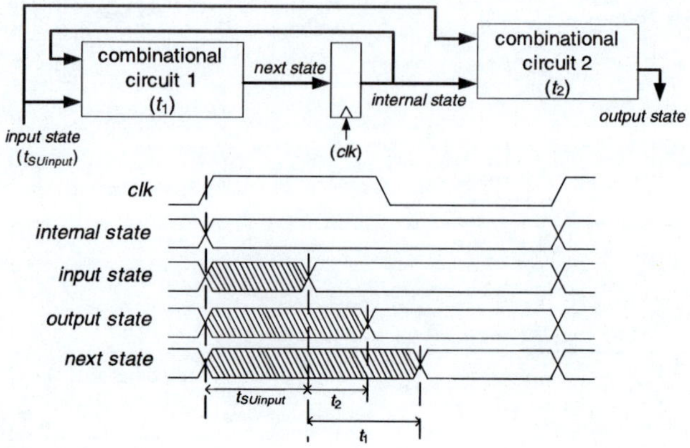

Fig. 1.20 Mealy machine

The set up and hold times of the register (Chap. 6) have not been taken into account.

Example 1.14

A Moore machine is shown in Fig. 1.21. It is the control unit of a programmable timer (Exercise 2.6.2). It has seven internal states, three binary input signals *start*, *zero* and *reference*, and two output signals *operation* (2 bits) and *done*. It can be described by the following processes.

```
next_state: PROCESS(reset, clk)
BEGIN
  IF reset = '1' THEN current_state <= 0;
  ELSIF clk'EVENT AND clk = '1' THEN
    CASE current_state IS
      WHEN 0 => IF start = '0'
                THEN current_state <= 1;
                END IF;
      WHEN 1 => IF start = '1'
                THEN current_state <= 2;
                END IF;
      WHEN 2 => current_state <= 3;
      WHEN 3 => IF zero = '1'
                THEN current_state <= 0;
                ELSE current_state <= 4;
                END IF;
      WHEN 4 => IF reference = '0'
                THEN current_state <= 5;
                END IF;
      WHEN 5 => IF reference = '1'
                THEN current_state <= 6;
```

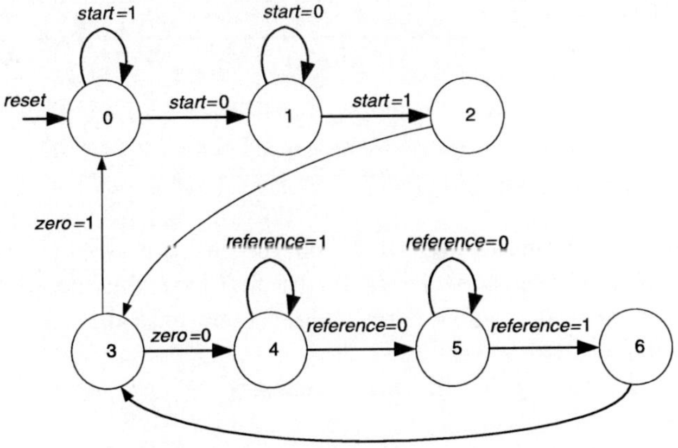

state	*operation*	*done*
0	00	1
1	00	1
2	11	0
3	00	0
4	00	0
5	00	0
6	01	0

Fig. 1.21 An example of a Moore machine

```
            END IF;
        WHEN 6 => current_state <= 3;
      END CASE;
    END IF;
END PROCESS next_state;

output_state: PROCESS(current_state)
BEGIN
  CASE current_state IS
    WHEN 0 TO 1 => operation <= "00"; done <= '1';
    WHEN 2 =>      operation <= "11"; done <= '0';
    WHEN 3 TO 5 => operation <= "00"; done <= '0';
    WHEN 6 =>      operation <= "01"; done <= '0';
  END CASE;
END PROCESS output_state;
```

Example 1.15
Consider the Mealy machine of Table 1.1. It has four internal states, two binary inputs x_1 and x_0, and one binary output z. Assume that x_0 and x_1 are

Table 1.1 A Mealy machine: next state/z

	X_1 x_0 : 00	01	10	11
A	A/0	B/0	A/1	D/1
B	B/1	B/0	A/1	C/0
C	B/1	C/1	D/0	C/0
D	A/0	C/1	D/0	D/1

periodic, but out of phase, signals. Then the machine detects if x_0 changes before x_1 or if x_1 changes before x_0. In the first case the sequence of internal states is A B C D A B… and $z = 0$. In the second case the sequence is D C B A D C… and $z = 1$.

It can be described by the following processes.

```
input_state <= x1&x0;
next_state: PROCESS (reset, clk)
BEGIN
  IF reset = '1' THEN current_state <= A;
  ELSIF clk'EVENT AND clk = '1' THEN
    CASE current_state IS
      WHEN A => IF input_state = "01" THEN
                  current_state <= B;
                ELSIF input_state = "11" THEN
                  current_state  <= D;
                END IF;
      WHEN B => IF input_state = "10" THEN
                  current_state <= A;
                ELSIF input_state = "11" THEN
                  current_state  <= C;
                END IF;

      WHEN C => IF input_state = "00" THEN
                  current_state <= B;
                ELSIF input_state = "10" THEN
                  current_state  <= D;
                END IF;
      WHEN D => IF input_state = "00" THEN
                  current_state <= A;
                ELSIF input_state = "01" THEN
                  current_state  <= C;
                END IF;
    END CASE;
  END IF;
END PROCESS;
```

```
output_state: PROCESS (current_state, input_state)
BEGIN
  CASE current_state IS
    WHEN A => z <= x1;
    WHEN B => z <= not(x0);
    WHEN C => z <= not(x1);
    WHEN D => z <= x0;
  END CASE;
END PROCESS output_state;
```

1.3 Memory Blocks

With regards to memory blocks, a previous comment similar to Comment 1.1 must be outlined: VHDL models can be generated for simulation purposes; nevertheless, in order to implement an actual circuit, the corresponding vendor's primitive component should be used instead (Chap. 5).

Example 1.16
The following entity defines a Read Only Memory storing 2^n m-bit words. The stored data is defined by a generic parameter (Fig. 1.22). Library declarations are omitted.

```
ENTITY rom IS
  GENERIC(n, m: NATURAL; stored_data: STD_LOGIC_VECTOR);
PORT (
  address: IN STD_LOGIC_VECTOR(n-1 DOWNTO 0);
  word: OUT STD_LOGIC_VECTOR(m-1 DOWNTO 0)
);
END rom;

ARCHITECTURE behavior OF rom IS
BEGIN
  PROCESS(address)
    VARIABLE c: NATURAL;
  BEGIN
    c := CONV_INTEGER(address)* m;
    word <= stored_data(c to c + m -1);
  END PROCESS;
END behavior;
```

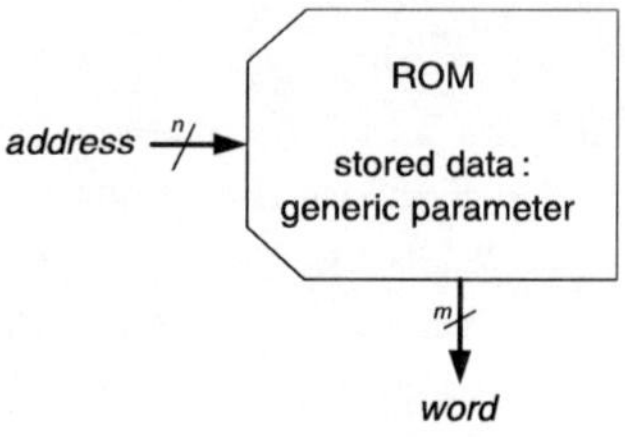

Fig. 1.22 Read only memory

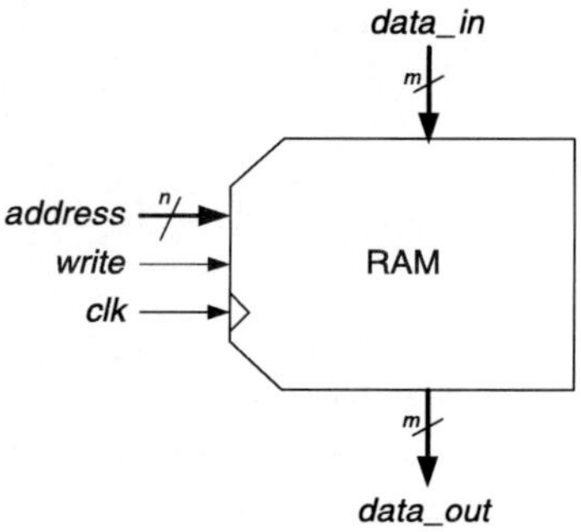

Fig. 1.23 Random access memory

Then the following component instantiation defines a ROM storing 16 4-bit words, namely

0100,1010,1011,1100,0111,1000,1001,0010,0001,1101,1101,1110,1111,0101,0100,0001

(from address 0000 to address 1111).

```
DUT: rom
GENERIC MAP(n => 4, m => 4, stored_data =>
X"4abc78921ddef541")
PORT MAP(address => address, word => word);
```

Example 1.17
The following entity defines a synchronous Random Access Memory storing 2^n m-bit words (Fig. 1.23). A *write* input enables the writing operation. Functionally, it is equivalent to a Register File made up of 2^n m-bit registers whose *clock enable* inputs are connected to *write*, plus an address decoder. Library declarations are omitted.

```
ENTITY ram IS
  GENERIC(n, m: NATURAL);
PORT (
  address: IN STD_LOGIC_VECTOR(n-1 DOWNTO 0);
  data_in: IN STD_LOGIC_VECTOR(m-1 DOWNTO 0);
  clk, write: IN STD_LOGIC;
  data_out: OUT STD_LOGIC_VECTOR(m-1 DOWNTO 0)
);
END ram;

ARCHITECTURE behavior OF ram IS
  TYPE memory IS ARRAY (0 TO 2**n-1) of
    STD_LOGIC_VECTOR(m-1 DOWNTO 0);
  SIGNAL stored_data: memory;
BEGIN
  PROCESS(clk)
  BEGIN
    IF clk'EVENT AND CLK = '1' THEN
      IF write = '1' THEN

      stored_data(CONV_INTEGER(address)) <= data_in;
    END IF;
  END IF;
END PROCESS;
  data_out <= stored_data(CONV_INTEGER(address));
END behavior;
```

The following component instantiation defines a synchronous RAM storing 16 4-bit words.

```
DUT: ram
GENERIC MAP(n => 4, m => 4)
PORT MAP(address => address, data_in => data_in,
  clk => clk, write => write, data_out => data_out);
```

1.4 IO-Port Components

Once again, a previous comment similar to Comment 1.1 must be outlined: VHDL models can be generated for simulating input and output amplifiers; nevertheless, in order to implement an actual circuit, the corresponding vendor's I/O component should be used instead (Chap. 5).

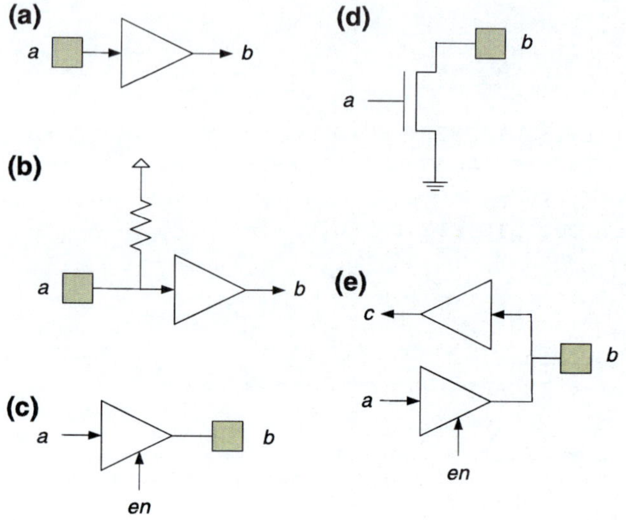

Fig. 1.24 I/O ports

In order to generate VHDL models of I/O amplifiers, it is convenient to understand the meaning of the STD_LOGIC type elements, that is

'X': forcing unknown, '0' : forcing 0, '1' : forcing 1,
'W': weak unknown, 'L' : weak 0, 'H' : weak 1,
'Z': high impedance

("uninitialized" and "don't care" states have no sense in the case of I/O amplifiers).

An amplifier generates low-impedance (forcing) signals when enabled and high impedance signals when disabled. Thus the possible outputs generated by an amplifier are 'X', '0', '1' and 'Z'. In the following example several types of input, output and bidirectional amplifiers are defined.

Example 1.18

The following conditional assignment defines an input buffer (Fig. 1.24a):

```
WITH a SELECT b <=
'0' WHEN '0' | 'L',
'1' WHEN '1' | 'H',
'X' WHEN OTHERS;
```

An input buffer with a pull-up resistor can be defined as follows (Fig. 1.24b).

```
WITH a SELECT b <=
'0' WHEN '0' | 'L',
'1' WHEN '1' | 'H' | 'Z',
'X' WHEN OTHERS;
```

The definition of a tri-state output buffer (Fig. 1.24c) is the same as in example 1.8, that is

```
WITH en SELECT b <= a WHEN '1', 'Z' WHEN OTHERS;
```

An open-drain output can be defined as follows (Fig. 1.24d).

```
WITH a SELECT b <= '0' WHEN '1' | 'H', 'Z' WHEN OTHERS;
```

As an example of hierarchical description, a bidirectional I/O buffer can be defined by instantiating an input buffer and a tri-state output buffer (Fig. 1.24e).

```
an_input_buffer: input_buffer
PORT MAP(a => b, b => c);
an_output_buffer: tri_state_output
PORT MAP(a => a, en => en, b => b) ;
```

1.5 VHDL Models

The following complete VHDL models are available at the Authors' web page www.arithmetic-circuits.org:

logic_gates.vhd (Sects. 1.1.1, 1.1.2 and 1.1.3),
arithmetic_blocks.vhd (Sects. 1.1.4, 1.2.1 and 1.2.2),
sequential_components.vhd (Sects. 1.2.1, 1.2.2 and 1.2.3),
fnite_state_machines.vhd (Sect. 1.2.4),
memories.vhd (Sect. 1.3),
input_output.vhd (Sect. 1.4).

1.6 Exercises

1. Generate the VHDL model of a circuit that computes $y = a{\cdot}x$ where a is a bit, and x and y are n-bit vectors, so that $y = (a{\cdot}x_{n-1}, a{\cdot}x_{n-2}, \ldots, a{\cdot}x_0)$.
2. Generate several models of a 1-bit full subtractor (Boolean equations, table, LUT instantiation).
3. Generate a generic model of an n-bit 8-to-1 multiplexer.

4. Generate a generic model of an n-input address decoder.
5. Design an n-bit magnitude comparator: given two n-bit naturals a and b, it generates a 1-bit output gt equal to 1 if $a \geq b$ and equal to 0 if $a < b$.
6. Design a 60-state up counter with *reset* and *count* control inputs.
7. Design a finite state machine with two binary inputs x and y and a binary output z defined as follows: if the input sequence is (x, t) = 00 01 11 10 00 01 11… then $z = 0$, and if the input sequence is (x, y) = 00 10 11 01 00 10 11… then $z = 1$.

Reference

1. Hamblen JO, Hall TS, Furman MD (2008) Rapid prototyping of digital systems. Springer, New York

Chapter 2
Architecture of Digital Circuits

This chapter describes the classical architecture of many digital circuits and presents, by means of several examples, the conventional techniques that digital circuit designers can use to translate an initial algorithmic description to an actual circuit. The main topics are the decomposition of a circuit into Data Path and Control Unit and the solution of two related problems, namely scheduling and resource assignment.

In fact, modern Electronic Design Automation tools have the capacity to directly generate circuits from algorithmic descriptions, with performances—latency, cost, consumption—comparable with those obtained using more traditional methods. Those development tools are one of the main topics of Chap. 5. So, it is possible that, in the future, the concepts and methods presented in this chapter will no longer be of interest to circuit designers, allowing them to concentrate on algorithmic innovative aspects rather than on scheduling and resource assignment optimization.

2.1 Introductory Example

As a first example, a "naive" method for computing the square root of a natural x is considered. The following algorithm sequentially computes all the pairs $[r, s = (r + 1)^2]$ with $r = 0, 1, 2$, etc.:

```
r := 0; s := 1;
loop
  s := s + 2*(r+1) + 1;
  r := r + 1;
end loop;
```

Initially $r = 0$ and thus $s = 1$. Then, at each step, the pair $[r + 1, (r + 2)^2]$ is computed in function of r and $s = (r + 1)^2$:

$$(r+2)^2 = ((r+1)+1)^2 = (r+1)^2 + 2\cdot(r+1) + 1 = s + 2\cdot(r+1) + 1.$$

J.-P. Deschamps et al., *Guide to FPGA Implementation of Arithmetic Functions*, Lecture Notes in Electrical Engineering 149, DOI: 10.1007/978-94-007-2987-2_2,

The same method can be used for computing the square root of x. For that, the loop execution is controlled by the condition $s \leq x$.

Algorithm 2.1: Square root

```
r := 0; s := 1;
while s <= x loop
  s := s + 2*(r+1) + 1;
  r := r + 1;
end loop;
root := r;
```

The loop is executed as long as $s \leq x$, that is $(r + 1)^2 \leq x$. Thus, at the end of the loop execution,

$$r^2 \leq x < (r + 1)^2.$$

Obviously, this is not a good algorithm as its computation time is proportional to the square root itself, so that for great values of x ($x \cong 2^n$) the number of steps is of the order of $2^{n/2}$. Efficient algorithms are described in Chap. 10.

In order to implement Algorithm 2.1, the list of operations executed at each clock cycle must be defined. In this case, each iteration step includes three operations: evaluation of the condition $s \leq x$, $s + 2{\cdot}(r + 1) + 1$ and $r + 1$. They can be executed in parallel. On the other hand, the successive values of r and s must be stored at each step. For that, two registers are used. Their initial values (0 and 1 respectively) are controlled by a common *load* signal, and their updating at the end of each step by a common *ce* (clock enable) signal. The circuit is shown in Fig. 2.1.

To complete the circuit, a control unit in charge of generating the *load* and *ce* signals must be added. It is a finite state machine with one input *greater* (detection of the loop execution end) and two outputs, *load* and *ce*. A *start* input and a *done* output are added in order to allow the communication with other circuits. The finite state machine is shown in Fig. 2.2.

The circuit of Fig. 2.1 is made up of five blocks whose VHDL models are the following:

- computation of *next_r*:

```
next_r <= r + 1;
```

- computation of *next_s*:

```
next_s <= s + (next_r(n-2 DOWNTO 0)&'0') + 1;
```

(multiplying by 2 is the same as shifting one position to the right)

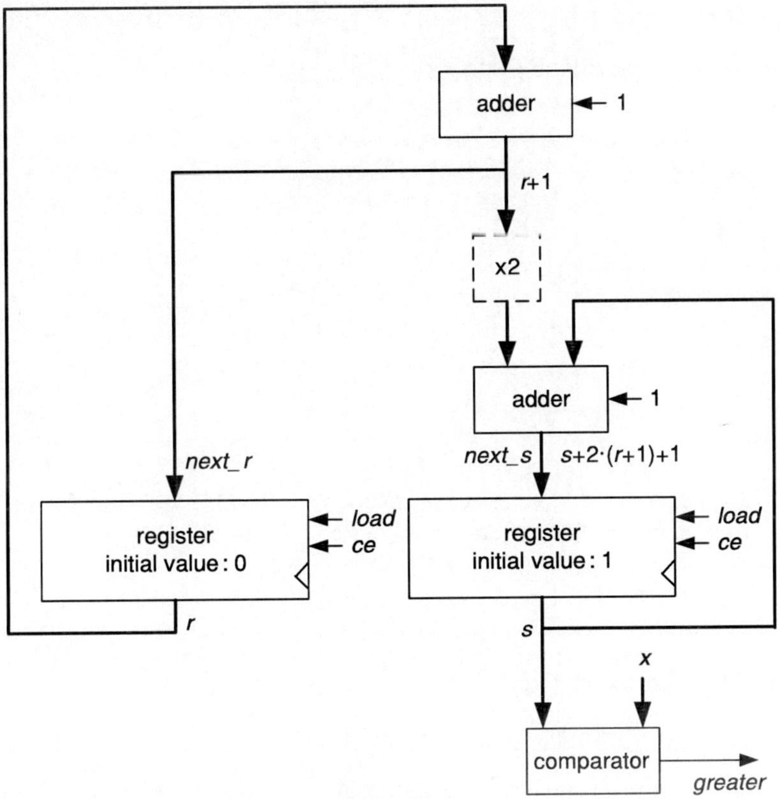

Fig. 2.1 Square root computation: data path

Fig. 2.2 Square root computation: control unit

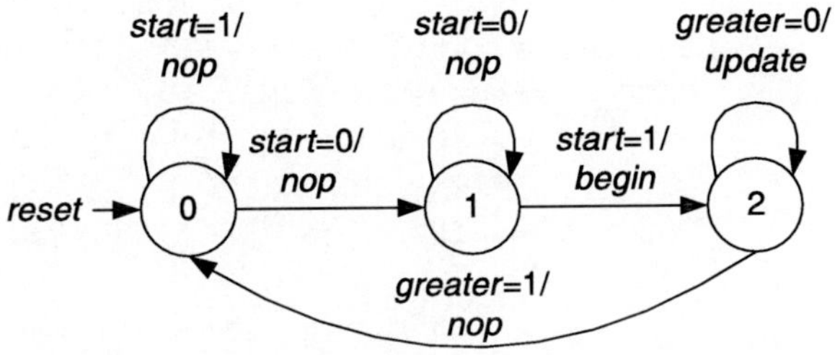

	ce	*load*	*done*
nop	0	0	1
begin	0	1	0
update	1	0	0

- register *r*:

```
register_r: PROCESS(clk)
BEGIN
  IF clk'EVENT AND clk = '1' THEN
    IF LOAD = '1' THEN r <= (OTHERS => '0');
    ELSIF ce = '1' THEN r <= next_r;
    END IF;
  END IF;
END PROCESS;
```

- register *s*:

```
register_s: PROCESS(clk)
BEGIN
  IF clk'EVENT AND clk = '1' THEN
    IF LOAD = '1' THEN s <= CONV_STD_LOGIC_VECTOR(1,8);
    ELSIF ce = '1' THEN s <= next_s;
    END IF;
  END IF;
END PROCESS;
```

- end of loop detection:

```
greater <= '1' WHEN s > x ELSE '0';
```

The control unit is a Mealy finite state machine that can be modeled as follows:

- next state computation:

```
control_unit_next_state: PROCESS(clk, reset)
BEGIN
  IF reset = '1' THEN current_state <= 0;
  ELSIF clk'event AND clk= '1' THEN
    CASE current_state IS
     WHEN 0 => IF start = '0' THEN current_state <= 1;
               END IF;
     WHEN 1 => IF start = '1' THEN current_state <= 2;
               END IF;
     WHEN 2 => IF greater = '1' THEN current_state <= 0;
               END IF;
    END CASE;
  END IF;
END PROCESS;
```

- output state computation:

```
control_unit_output: PROCESS(current_state, start, greater)
BEGIN
  CASE current_state IS
    WHEN 0 => ce <= '0'; load <= '0'; done <= '1';
    WHEN 1 => ce <= '0';
              IF start = '1' THEN load <= '1'; done <= '0';
              ELSE load <= '0'; done <= '1'; END IF;
    WHEN 2 => IF greater = '0' THEN ce <= '1';
              ELSE ce <= '0'; END IF;
              load <= '0'; done <= '0';
  END CASE;
END PROCESS;
```

The circuit of Fig. 2.1 includes three *n*-bit adders: a half adder for computing *next_r*, a full adder for computing *next_s* and another full adder (actually a subtractor) for detecting the condition $s > x$. Another option is to use one adder and to decompose each iteration step into three clock cycles. For that, Algorithm 2.1 is slightly modified.

Algorithm 2.2: Square root, version 2

```
r := 0; s := 1;
if s > x then greater := true; else greater := false; end if;
while not(greater) loop
  --cycle 1:
  r := r + 1;
  --cycle 2:
  s := s + 2*r + 1;
  --cycle 3:
  if s > x then greater := true; end if;
end loop;
```

A circuit able to execute the three operations, that is $r + 1$, $s + 2 \cdot r + 1$ and evaluation of the condition $s > x$ must be defined. The condition $s > x$ is equivalent to $s \geq x + 1$ or $s + 2^n - 1 - x \geq 2^n$. The binary representation of $2^n - 1 - x$ is obtained by complementing the bits of the binary representation of x. So, the condition $s > x$ is equivalent to $s + not(x) \geq 2^n$. Thus, the three operations amount to additions: $r + 1$, $s + 2 \cdot r + 1$ and $s + not(x)$. In the latter case, the output carry defines the value of *greater*. The corresponding circuit is shown in Fig. 2.3. It is an example of *programmable computation resource*: under the control of a 2-bit command *operation*, it can execute the three previously defined operations.
The corresponding VHDL description is the following:

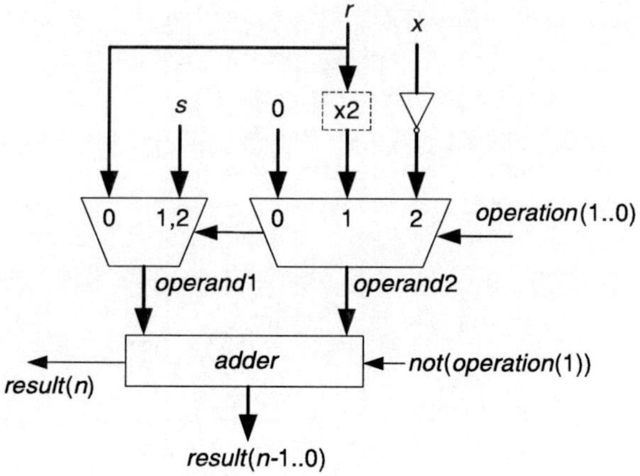

Fig. 2.3 Square root computation: programmable computation resource

```
shifted_r <= r(n-2 DOWNTO 0)&'0';
complemented_x <= NOT(x);
WITH operation SELECT operand_1 <=
  r WHEN "00", s WHEN OTHERS;
WITH operation SELECT operand_2 <= zero WHEN "00",
  shifted_r WHEN "01", complemented_x WHEN OTHERS;
result <= '0'&operand_1 + operand_2 + NOT(operation(1));
```

The complete circuit is shown in Fig. 2.4.

A control unit must be added. It is a finite state machine with one input *greater* and five outputs *load*, *ce_r*, *ce_s*, *ce_greater*. As before, a *start* input and a *done* output are added in order to allow the communication with other circuits. The finite state machine is shown in Fig. 2.5.

The building blocks of the circuit of Fig. 2.4 (apart from the programmable resource) are the following:

- register *r*:

```
register_r: PROCESS(clk)
BEGIN
  IF clk'EVENT AND clk = '1' THEN
    IF LOAD = '1' THEN r <= (OTHERS => '0');
    ELSIF ce_r = '1' THEN r <= result(n-1 DOWNTO 0);
    END IF;
  END IF;
END PROCESS;
```

- register *s*:

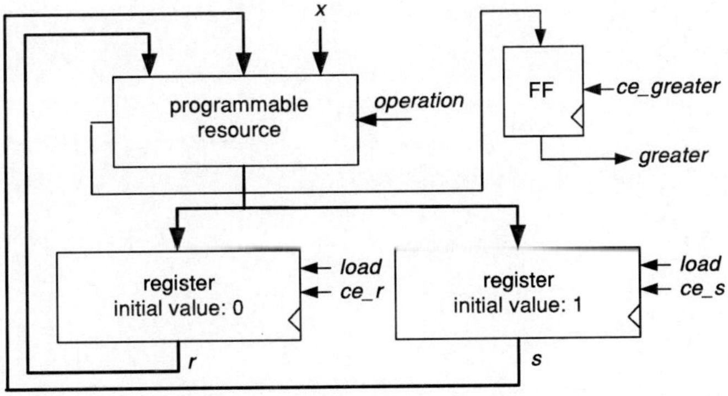

Fig. 2.4 Square root computation, second version: data path

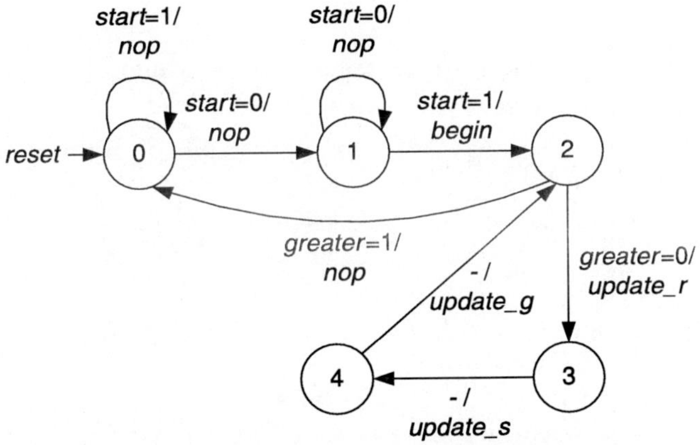

commands	*ce_r*,	*ce_s*,	*ce_greater*,	*load*,	*operation*,	*done*
nop	0	0	0	0	0	1
begin	0	0	0	1	0	0
update_r	1	0	0	0	0	0
update_s	0	1	0	0	1	0
update_greater	0	0	1	0	2	0

Fig. 2.5 Square root computation, second version: control unit

```
register_s: PROCESS(clk)
BEGIN
  IF clk'EVENT AND clk = '1' THEN
    IF LOAD = '1' THEN s <= CONV_STD_LOGIC_VECTOR(1,8);
    ELSIF ce_s = '1' THEN s <= result(n-1 DOWNTO 0);
    END IF;
  END IF;
END PROCESS;
```

- flip-flop *greater*

```
ff_greater: PROCESS(clk)
    BEGIN
      IF clk'EVENT AND clk = '1' THEN
        IF LOAD = '1' THEN greater <= '0';
        ELSIF ce_greater = '1' THEN greater <= result(n);
        END IF;
      END IF;
    END PROCESS;
```

The control unit is a Mealy finite state machine whose VHDL model is the following:

- next state computation:

```
control_unit_next_state: PROCESS(clk, reset)
BEGIN
  IF reset = '1' THEN current_state <= 0;
  ELSIF clk'event AND clk= '1' THEN
   CASE current_state IS
    WHEN 0 => IF start = '0' THEN
              current_state <= 1; END IF;
    WHEN 1 => IF start = '1' THEN
              current_state <= 2; END IF;
    WHEN 2 => IF greater = '1' THEN current_state <= 0;
              ELSE current_state <= 3; END IF;
    WHEN 3 => current_state <= 4;
    WHEN 4 => current_state <= 2;
   END CASE;
  END IF;
END PROCESS;
```

- output state computation:

```
control_unit_output: PROCESS(current_state, start, greater)
BEGIN
  CASE current_state IS
    WHEN 0 => ce_r <= '0'; ce_s <= '0'; ce_greater <= '0';
              load <= '0'; operation <= "00"; done <= '1';
    WHEN 1 => ce_r <= '0'; ce_s <= '0'; ce_greater <= '0';
              operation <= "00";
              IF start = '1' THEN load <= '1'; done <= '0';
              ELSE load <= '0'; done <= '1'; END IF;
    WHEN 2 => IF greater = '0' THEN ce_r <= '1';
              ELSE ce_r <= '0'; END IF;
              ce_s <= '0'; ce_greater <= '0'; load <= '0';
              operation <= "00"; done <= '0';
    WHEN 3 => ce_r <= '0'; ce_s <= '1'; ce_greater <= '0';
              load <= '0'; operation <= "01"; done <= '0';
    WHEN 4 => ce_r <= '0'; ce_s <= '0'; ce_greater <= '1';
              load <= '0'; operation <= "10"; done <= '0';
  END CASE;
END PROCESS;
```

Complete VHDL models (*square_root.vhd*) of both circuits (Figs. 2.1, 2.4) are available at the Authors' web page.

2.2 Data Path and Control Unit

The general structure of a digital circuit is shown in Fig. 2.6. It consists of a *data path* and a *control unit*. The data path (leftmost part of Fig. 2.6) includes computation resources executing the algorithm operations, registers storing the algorithm variables, and programmable connections (for example multiplexers, not represented in Fig. 2.6) between resource outputs and register inputs, and between register outputs and resource inputs. The control unit (rightmost part of Fig. 2.6) is a finite state machine. It controls the sequence of data path operations by means of a set of control signals (*commands*) such as clock enables of registers, programming of computation resources and multiplexers, and so on. It receives from the data path some feedback information (*conditions*) corresponding to the algorithm control statements (loop, if, case).

In fact, the data path could also be considered as being a finite state machine. Its internal states are all the possible register contents, the next-state computation is performed by the computation resources, and the output states are all the possible values of *conditions*. Nevertheless, the number of internal states is enormous and there is generally no sense in using a finite state machine model for the data path. However, it is interesting to observe that the data path of Fig. 2.6 is a Moore

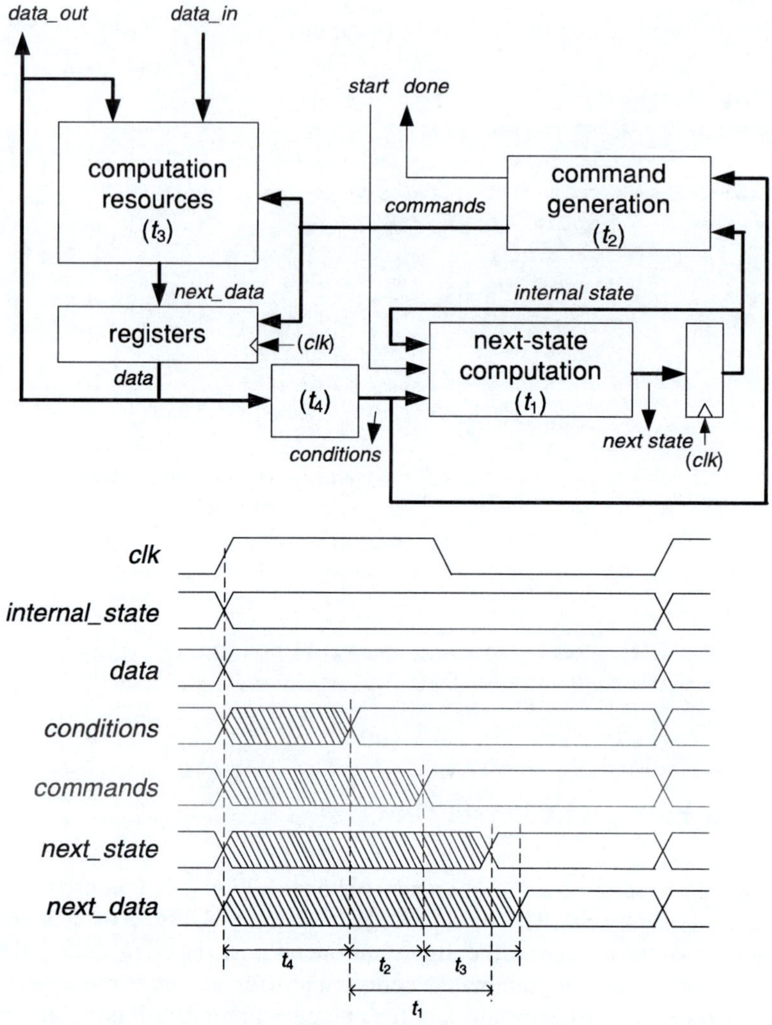

Fig. 2.6 Structure of a digital circuit: data path and control unit

machine (the output state only depends on the internal state) while the control unit could be a Moore or a Mealy machine. An important point is that, when two finite state machines are interconnected, one of them must be a Moore machine in order to avoid combinational loops.

According to the chronograms of Fig. 2.6, there are two critical paths: from the data registers to the internal state register, and from the data registers to the data registers. The corresponding delays are

$$T_{data-state} = t_4 + t_1 \tag{2.1}$$

and

$$T_{data-data} = t_4 + t_2 + t_3, \tag{2.2}$$

where t_1 is the computation time of the next internal state, t_2 the computation time of the commands, t_3 the maximum delay of the computation resources and t_4 the computation time of the conditions (the set up and hold times of the registers have not been taken into account).

The clock period must satisfy

$$T_{clk} > max\{t_4 + t_1, t_4 + t_2 + t_3\}. \tag{2.3}$$

If the control unit were a Moore machine, there would be no direct path from the data registers to the data registers, so that (2.2) and (2.3) should be replaced by

$$T_{state-data} = t_2 + t_3 \tag{2.4}$$

and

$$T_{clk} > max\{t_4 + t_1, t_2 + t_3\}. \tag{2.5}$$

In fact, it is always possible to use a Moore machine for the control unit. Generally it has more internal states than an equivalent Mealy machine and the algorithm execution needs more clock cycles. If the values of t_1 to t_4 do not substantially vary, the conclusion could be that the Moore approach needs more, but shorter, clock cycles. Many designers also consider that Moore machines are safer than Mealy machines.

In order to increase the maximum frequency, an interesting option is to insert a command register at the output of the command generation block. Then relation (2.2) is substituted by

$$T_{data-commands} = t_4 + t_2 \quad \text{and} \quad T_{commands-data} = t_3, \tag{2.6}$$

so that

$$T_{clk} > max\{t_4 + t_1, t_4 + t_2, t_3\}. \tag{2.7}$$

With this type of *registered Mealy machine*, the commands are available one cycle later than with a non-registered machine, so that additional cycles must be sometimes inserted in order that the data path and its control unit remain synchronized.

To summarize, the implementation of an algorithm is based upon a decomposition of the circuit into a data path and a control unit. The data path is in charge of the algorithm operations and can be roughly defined in the following way: associate registers to the algorithm variables, implement resources able to execute the algorithm operations, and insert programmable connections (multiplexers) between the register outputs (the operands) and the resource inputs, and between the resource outputs (the results) and the register inputs. The control unit is a finite state machine whose internal states roughly correspond to the algorithm steps, the input states are conditions (flags) generated by the data path, and the output states are commands transmitted to the data path.

In fact, the definition of a data path poses a series of optimization problems, some of them being dealt with in the next sections, for example: scheduling of the operations, assignment of computation resources to operations, and assignment of registers to variables. It is also important to notice that minor algorithm modifications sometimes yield major circuit optimizations.

2.3 Operation Scheduling

Operation scheduling consists in defining which particular operations are in the process of execution during every clock cycle. For that purpose, an important concept is that of *precedence relation*. It defines which of the operations must be completed before starting a new one: if some result r of an operation A is an initial operand of some operation B, the computation of r must be completed before the execution of B starts. So, the execution of A must be scheduled before the execution of B.

2.3.1 Introductory Example

A *carry-save adder* or 3-to-2 *counter* (Sect. 7.7) is a circuit with 3 inputs and 2 outputs. The inputs x_i and the outputs y_j are naturals. Its behavior is defined by the following relation:

$$x_1 + x_2 + x_3 = y_1 + y_2. \tag{2.8}$$

It is made up of 1-bit full adders working in parallel. An example where x_1, x_2 and x_3 are 4-bit numbers, and y_1 and y_2 are 5-bit numbers, is shown in Fig. 2.7.

The delay of a carry-save adder is equal to the delay T_{FA} of a 1-bit full adder, independently of the number of bits of the operands. Let CSA be the function associated to (2.8), that is

$$(y_1, y_2) = \mathrm{CSA}(x_1, x_2, x_3). \tag{2.9}$$

Using carry-save adders as computation resources, a 7-to-3 counter can be implemented. It allows expressing the sum of seven naturals under the form of the sum of three naturals, that is

$$x_1 + x_2 + x_3 + x_4 + x_5 + x_6 + x_7 = y_1 + y_2 + y_3.$$

In order to compute y_1, y_2 and y_3, the following operations are executed (op_1 to op_4 are labels):

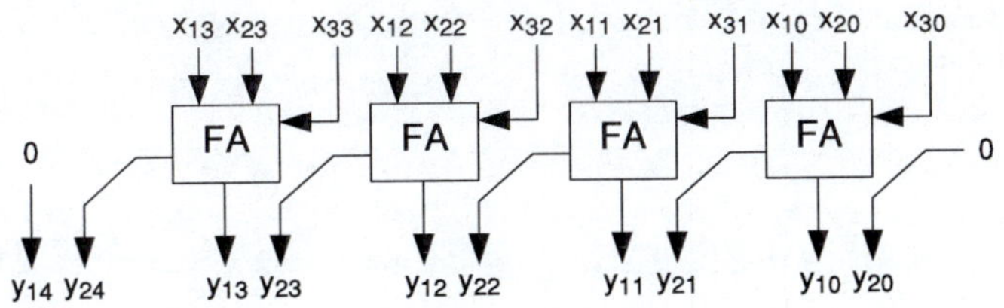

Fig. 2.7 Carry-save adder

$$
\begin{aligned}
op_1 &: (a_1, a_2) = \mathrm{CSA}(x_1, x_2, x_3),\\
op_2 &: (b_1, b_2) = \mathrm{CSA}(x_4, x_5, x_6),\\
op_3 &: (c_1, c_2) = \mathrm{CSA}(a_2, b_2, x_7),\\
op_4 &: (d_1, d_2) = \mathrm{CSA}(a_1, b_1, c_1).
\end{aligned} \tag{2.10}
$$

According to (2.10) and the definition of CSA

$$
\begin{aligned}
a_1 + a_2 &= x_1 + x_2 + x_3,\\
b_1 + b_2 &= x_4 + x_5 + x_6,\\
c_1 + c_2 &= a_2 + b_2 + x_7,\\
d_1 + d_2 &= a_1 + b_1 + c_1,
\end{aligned}
$$

so that

$$
\begin{aligned}
c_1 + c_2 + d_1 + d_2 &= a_2 + b_2 + x_7 + a_1 + b_1 + c_1\\
&= x_1 + x_2 + x_3 + x_4 + x_5 + x_6 + x_7 + c_1.
\end{aligned}
$$

Thus

$$c_2 + d_1 + d_2 = a_2 + b_2 + x_7 + a_1 + b_1 + c_1 = x_1 + x_2 + x_3 + x_4 + x_5 + x_6 + x_7$$

and y_1, y_2 and y_3 can be defined as follows:

$$y_1 = d_1,\ y_2 = d_2,\ y_3 = c_2.$$

The corresponding precedence relation is defined by the graph of Fig. 2.8, according to which op_1 and op_2 must be executed before op_3, and op_3 before op_4. Thus, the minimum computation time is equal to $3{\cdot}T_{FA}$.

For implementing (2.10) the following options could be considered:

1. A combinational circuit, made up of four carry-save adders, whose structure is the same as that of the graph of Fig. 2.8. Its computation time is equal to $3{\cdot}T_{FA}$ and its cost to $4{\cdot}C_{CSA}$, being C_{CSA} the cost of a carry-save adder. This is probably a bad solution because the cost is high (4 carry-save adders) and the delay is long (3 full-adders) so that the minimum clock cycle of a synchronous circuit including this 7-to-3 counter should be greater than $3{\cdot}T_{FA}$.
2. A data path including two carry-save adders and several registers (Sect. 2.5). The computation is executed in three cycles:

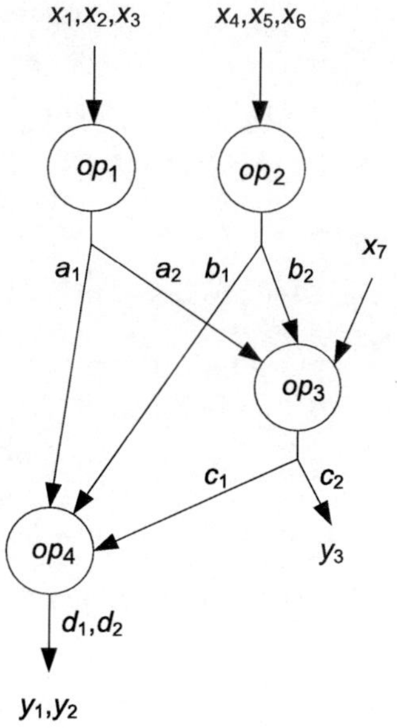

Fig. 2.8 Precedence relation of a 7-to-3 counter

```
1: (a₁, a₂, b₁, b₂) := (CSA(x₁, x₂, x₃), CSA(x₄, x₅, x₆));
2: (c₁, y₃) := CSA(a₂, b₂, x₇);
3: (y₁, y₂) := CSA(a₁, b₁, c₁);
```

The computation time is equal to $3 \cdot T_{clk}$, where $T_{clk} > T_{FA}$, and the cost equal to $2 \cdot C_{CSA}$, plus the cost of the additional registers, controllable connections and control unit.

3. A data path including one carry-save adder and several registers. The computation is executed in four cycles:

```
1: (a₁, a₂) := CSA(x₁, x₂, x₃);
2: (b₁, b₂) := CSA(x₄, x₅, x₆);
3: (c₁, y₃) := CSA(a₂, b₂, x₇);
4: (y₁, y₂) := CSA(a₁, b₁, c₁);
```

The computation time is equal to $4 \cdot T_{clk}$, where $T_{clk} > T_{FA}$, and the cost equal to C_{CSA}, plus the cost of the additional registers, controllable connections and control unit.

In conclusion, there are several implementations, with different costs and delays, corresponding to the set of operations in (2.10). In order to get an optimized circuit, according to some predefined criteria, the space for possible implementations must be explored. For that, optimization methods must be used.

2.3.2 Precedence Graph

Consider a *computation scheme*, that is to say, an algorithm without branches and loops. Formally it can be defined by a set of operations

$$op_J : (x_i, x_k, \ldots) = f(x_l, x_m, \ldots), \tag{2.11}$$

where x_i, x_k, x_l, x_m,... are variables of the algorithm and f one of the algorithm operation types (*computation primitives*). Then, the precedence graph (or *data flow graph*) is defined as follows:

- associate a vertex to each operation op_J,
- draw an arc between vertices op_J and op_M if one of the results generated by op_J is used by op_M.

An example was given in Sect. 2.3.1 (operations (2.10) and Fig. 2.8).

Assume that the computation times of all operations are known. Let t_{JM} be the computation time, expressed in number of clock cycles, of the result(s) generated by op_J and used by op_M. Then, a schedule of the algorithm is an application *Sch* from the set of vertices to the set of naturals that defines the number $Sch(op_J)$ of the cycle at the beginning of which the computation of op_J starts. A necessary condition is that

$$Sch(op_M) \geq Sch(op_J) + t_{JM} \tag{2.12}$$

if there is an arc from op_J to op_M.

As an example, if the clock period is greater than the delay of a full adder, then, in the computation scheme (2.10), all the delays are equal to 1 and two admissible schedules are

$$Sch(op_1) = 1,\ Sch(op_2) = 1,\ Sch(op_3) = 2,\ Sch(op_4) = 3, \tag{2.13}$$

$$Sch(op_1) = 1,\ Sch(op_2) = 2,\ Sch(op_3) = 3,\ Sch(op_4) = 4. \tag{2.14}$$

They correspond to the options 2 and 3 of Sect. 2.3.1.

The definition of an admissible schedule is an easy task. As an example, the following algorithm defines an ASAP (as soon as possible) schedule:

- initial step: $Sch(op_J) = 1$ for all initial (without antecessor) vertices op_J;
- step number $n + 1$: choose an unscheduled vertex op_M whose total amount of antecessors, say op_P, op_Q,... have already been scheduled, and define $Sch(op_M) = maximum\{Sch(op_P) + t_{PM}, Sch(op_Q) + t_{QM}, \ldots\}$.

Applied to (2.10) the ASAP algorithm gives (2.13). The corresponding data flow graph is shown in Fig. 2.9a.

An ALAP (as late as possible) schedule can also be defined. For that, assume that the latest admissible starting cycle for all the final vertices (without successor) has been previously specified:

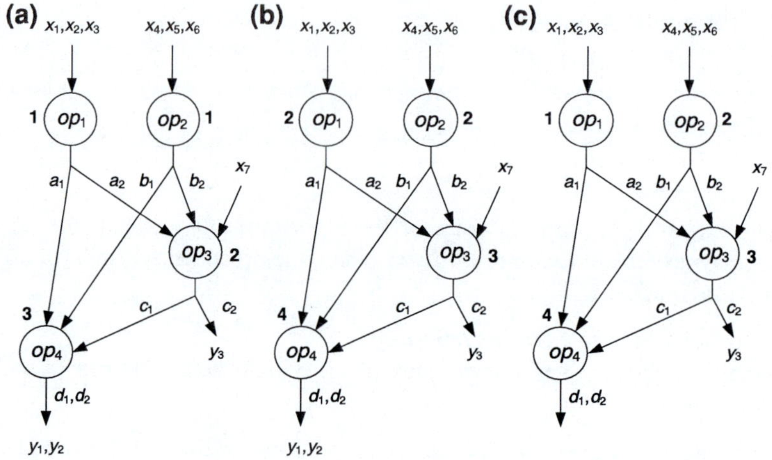

Fig. 2.9 7-to-3 counter: **a** ASAP schedule. **b** ALAP schedule. **c** Admissible schedule

- initial step: $Sch(op_M)$ = latest admissible starting cycle of op_M for all final vertices op_M;
- step number $n + 1$: choose an unscheduled vertex op_J whose all successors, say op_P, op_Q,... have already been scheduled, and define $Sch(op_J) = minimum\{Sch(op_P) - t_{JP}, Sch(op_Q) - t_{JQ},\ldots\}$.

Applied to (2.10), with $Sch(op_4) = 4$, the ALAP algorithm generates the data flow graph of Fig. 2.9b.

Let ASAP_*Sch* and ALAP_*Sch* be ASAP and ALAP schedules, respectively. Obviously, if op_M is a final operation, the previously specified value ALAP_-$Sch(op_M)$ must be greater than or equal to ASAP_$Sch(op_M)$. More generally, assuming that the latest admissible starting cycle for all the final operations has been previously specified, for any admissible schedule *Sch* the following relation holds:

$$\text{ASAP\_}Sch(op_J) \leq Sch(op_J) \leq \text{ALAP\_}Sch(op_J),\ \forall op_J. \tag{2.15}$$

Along with (2.12), relation (2.15) defines the admissible schedules.

An example of admissible schedule is defined by (2.14), to which corresponds the data flow graph of Fig. 2.9c.

A second, more realistic, example is now presented. It corresponds to part of an Elliptic Curve Cryptography algorithm.

Example 2.1

Given a point $P = (x_P, y_P)$ of an elliptic curve and a natural k, the scalar product $kP = P + P + \cdots + P$ can be defined [1, 2]. In the case of the curve $y^2 + xy = x^3 + ax + 1$ over the binary field, the following formal algorithm [3] computes kP. The initial data are the scalar $k = k_{m-1}\,k_{m-2}\ldots k_0$ and the x-coordinate x_P of P. All the algorithm variables are elements of the Galois field $GF(2^m)$, that is, polynomials of degree m over the binary field $GF(2)$ (Chap. 13).

Algorithm 2.3: Scalar product, projective coordinates

```
xA := 1; zA := 0; xB := xP; zB := 1;
for i in 1 .. m loop
  u := (xA·zB + xB·zA)^2;
  v := xP·u + xA·xB·zA·zB;
  if k(m-i) = 0 then
    xB := v;
    zB := u;
    y := xA^2·zA^2;
    xA := xA^4+zA^4;
    zA := y;
  else
    xA := v;
    zA := u;
    z := xB^2·zB^2;
    xB := xB^4+zB^4;
    zB := z;
  end if;
end loop;
```

In fact, the preceding algorithm computes the value of four variables x_A, z_A, x_B and z_B in function of k and x_P. A final, not included, step would be to compute the coordinates of kP in function of the coordinates of P (x_P and y_P) and of the final values of x_A, z_A, x_B and z_B.

Consider one step of the main iteration of Algorithm 2.3, and assume that $k_{m-i} = 0$. The following computation scheme computes the new values of x_A, z_A, x_B and z_B in function of their initial values and of x_P. The available computation primitives are the addition, multiplication and squaring in $GF(2^m)$ (Chap. 13).

```
op1: a = xA·zB,
op2: b = xB·zA,
op3: c = a + b,
op4: d = c^2,
op5: e = xP·d,
op6: f = a·b,
op7: g = e+f,
op8: h = xA·zA,
op9: i = h^2,
op10: j = xA+zA,
op11: k = j^2,
op12: l = k^2.
```

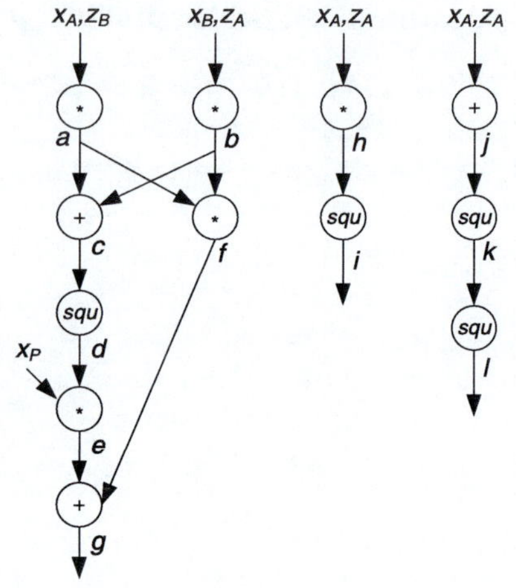

Fig. 2.10 Example 2.1: precedence graph

The updated values of x_A, z_A, x_B and z_B are $x_A = l$, $z_A = i$, $x_B = g$, $z_B = d$. The corresponding data flow graph is shown in Fig. 2.10. The operation type corresponding to every vertex is indicated (instead of the operation label). If $k_{m-i} = 1$ the computation scheme is the same but for the interchange of indexes A and B.

Addition and squaring in $GF(2^m)$ are relatively simple one-cycle operations, while multiplication is a much more complex operation whose maximum computation time is $t_m \gg 1$. In what follows it is assumed that $t_m = 300$ cycles. An ASAP schedule is shown in Fig. 2.11. The computation of g starts at the beginning of cycle 603 so that all the final results are available at the beginning of cycle 604. The corresponding circuit must include three multipliers as the computations of a, b and h start at the same time.

The computation scheme includes 5 multiplications. Thus, in order to execute the algorithm with only one multiplier, the minimum computation time is 1,500. More precisely, one of the multiplications e, f or h cannot start before cycle 1,201, so that the next operation (g or i) cannot start before cycle 1,501. An ALAP schedule, assuming that the computations of g and i start at the beginning of cycle 1,501, is shown in Fig. 2.12.

2.3.3 Optimization Problems

Assuming that the latest admissible starting cycle for all the final operations has been previously specified then any schedule, such that (2.12) and (2.15) hold true, can be chosen. This poses optimization problems. For example:

1. Assuming that the maximum computation time has been previously specified, look for a schedule that minimizes the number of computation resources of each type.

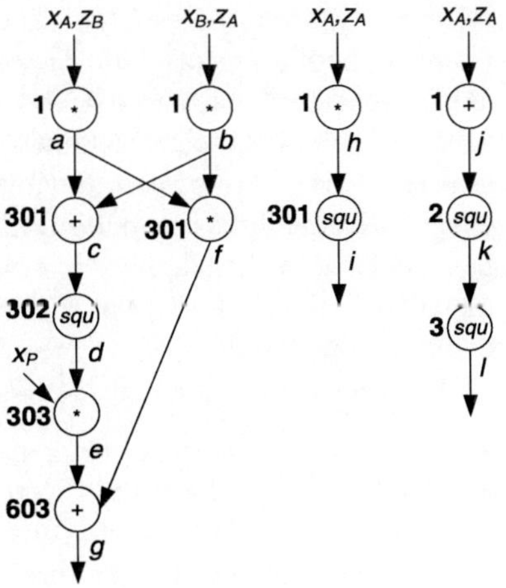

Fig. 2.11 Example 2.1: ASAP schedule

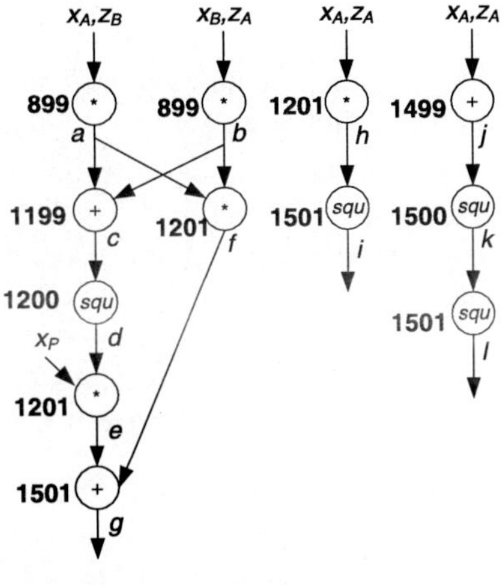

Fig. 2.12 Example 2.1: ALAP schedule with Sch(g) = 1501

2. Assuming that the number of available computation resources of each type has been previously specified, minimize the computation time.

An important concept is the *computation width* $w(f)$ with respect to the computation primitive (operation type) f. First define the *activity intervals* of f. Assume that f is the primitive corresponding to the operation op_J, that is

$$op_J : (x_i, x_k, \ldots) = f(x_l, x_m, \ldots).$$

Then

$$[Sch(op_J), Sch(op_J) + maximum\{t_{JM}\}]$$

is an activity interval of *f*. This means that a resource of type *f* must be available from the beginning of cycle $Sch(op_J)$ to the end of cycle $Sch(op_J) + t_{JM}$ for all M such that there is an arc from op_J to op_M. An incompatibility relation over the set of activity intervals of *f* can be defined: two intervals are incompatible if they overlap. If two intervals overlap, it is obvious that the corresponding operations cannot be executed by the same computation resource. Thus, a particular resource of type *f* must be associated to each activity interval of *f* in such a way that if two intervals overlap, then two distinct resources of the same type must be used. The minimum number of computation resources of type *f* is the *computation width w(f)*.

The following graphical method can be used for computing $w(f)$.

- Associate a vertex to every activity interval.
- Draw an edge between two vertices if the corresponding intervals overlap.
- Color the vertices in such a way that two vertices connected by an edge have different colors (a classical problem of graph theory).

Then, $w(f)$ is the number of different colors, and every color defines a particular resource assigned to all edges (activity intervals) with this color.

Example 2.2
Consider the scheduled precedence graph of Fig. 2.11. The activity intervals of the multiplication are

$$a : [1,300],\ b : [1,300],\ h : [1,300], f : [301,600],\ e : [303,602].$$

The corresponding incompatibility graph is shown in Fig. 2.13a. It can be colored with three colors (c_1, c_2 and c_3 in Fig. 2.13a). Thus, the computation width with respect to the multiplication is equal to 3.

If the scheduled precedence graph of Fig. 2.12 is considered, then the activity intervals of the multiplication are

$$a : [899,1198],\ b : [899,1198],\ h : [1201,1500], f : [1201,1500],\ e : [1201,1500].$$

The corresponding incompatibility graph is shown in Fig. 2.13b. It can be colored with three colors. Thus, the computation width with respect to the multiplication is still equal to 3.

Nevertheless, other schedules can be defined. According to (2.15) and Figs. 2.11 and 2.12, the time intervals during which the five multiplications can start are the following:

$$a : [1,899],\ b : [1,899],\ h : [1,1201], f : [301,1201],\ e : [303,1201].$$

As an example, consider the admissible schedule of Fig. 2.14. The activity intervals of the multiplication operation are

$$a : [1,300]\ , b : [301,600],\ h : [601,900], f : [901,1200],\ e : [1201,1500].$$

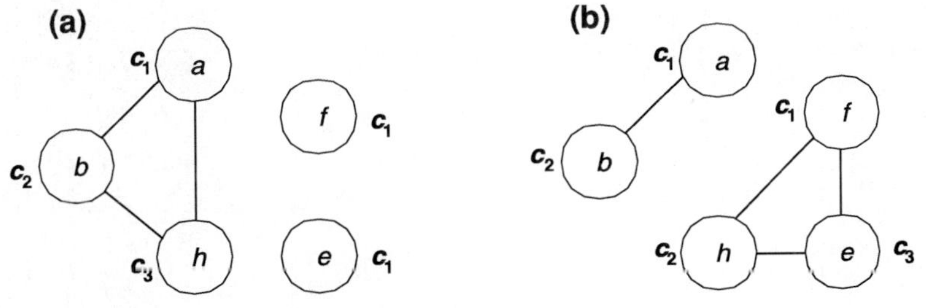

Fig. 2.13 ColoringComputation width: graph coloring

Fig. 2.14 Example 2.1: admissible schedule using only one multiplier

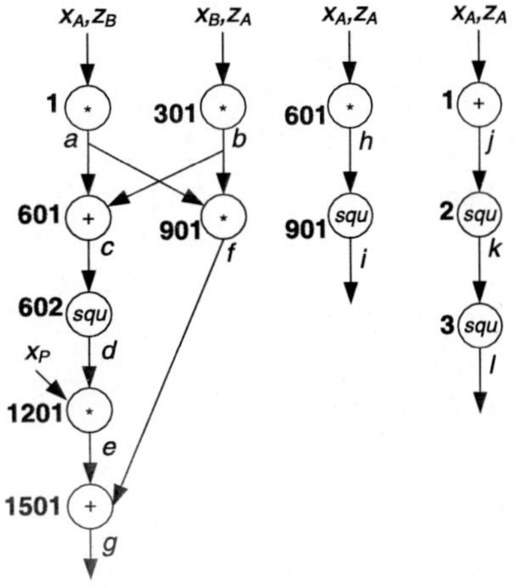

They do not overlap hence the incompatibility graph does not include any edge and can be colored with one color. The computation width with respect to the multiplication is equal to 1.

Thus, the two optimization problems mentioned above can be expressed in terms of computation widths:

1. Assuming that the maximum computation time has been previously specified, look for a schedule that minimizes some cost function

$$C = c_1 \cdot w(f^1) + c_2 \cdot w(f^2) + \cdots + c_m \cdot w(f^m) \tag{2.16}$$

   where $f^1, f^2, \ldots, f^m$ are the computation primitives and $c_1, c_2, \ldots, c_m$ their corresponding costs.
2. Assuming that the maximum computation width $w(f)$ with respect to every computation primitive f has been previously specified, look for a schedule that minimizes the computation time.

Both are classical problems of scheduling theory. They can be expressed in terms of integer linear programming problems whose variables are x_{It} for all

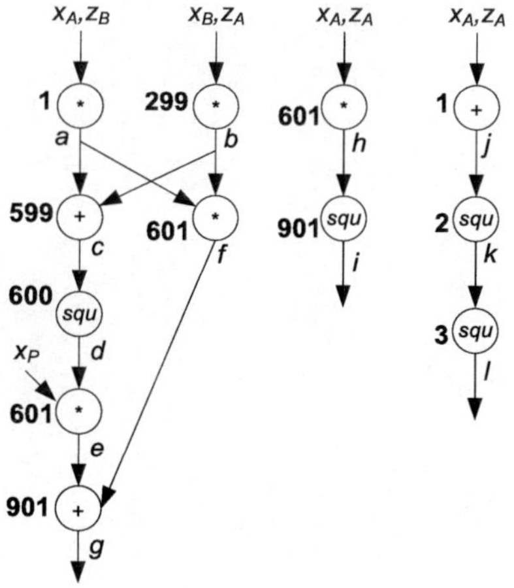

Fig. 2.15 Example 2.3: schedule corresponding to the first optimization problem

operation indices I and all possible cycle numbers t: $x_{It} = 1$ if $Sch(e_I) = t$, 0 otherwise. Nevertheless, except for small computation schemes—generally tractable by hand—the so obtained linear programs are intractable. Modern Electronic Design Automation tools execute several types of heuristic algorithms applied to different optimization problems (not only to schedule optimization). Some of the more common heuristic strategies are *list scheduling*, *simulated annealing* and *genetic algorithms*.

Example 2.3
The list scheduling algorithm, applied to the graph of Fig. 2.10, with $t_m = 300$ and assuming that the latest admissible starting cycle for all the final operations is cycle number 901 (first optimization problem), would generate the schedule of Fig. 2.15. The list scheduling algorithm, applied to the same graph of Fig. 2.10, with $t_m = 300$ and assuming that the computation width is equal to 1 for all operations (second optimization problem), would generate the schedule of Fig. 2.14.

2.4 Resource Assignment

Once the operation schedule has been defined, several decisions must be taken.

- The number $w(f)$ of resources of type f is known, but it remains to decide which particular computation resource executes each operation. Furthermore the definition of multifunctional programmable resources could also be considered.
- As regards the storing resources, a simple solution is to assign a particular register to every variable. Nevertheless, in some cases the same register can be used for storing different variables.

A key concept for assigning registers to variables is the *lifetime* $[t_I, t_J]$ of every variable: t_I is the number of the cycle during which its value is generated, and t_J is the number of the last cycle during which its value is used.

Example 2.4
Consider the computation scheme of Example 2.1 and the schedule of Fig. 2.14. The computation width is equal to 1 for all primitives (multiplication, addition and squaring). The computation is executed as follows:

```
initial data: xA, zA, xB, zB
cycle 1: j := xA+zA; start xA·zB;
cycle 2: k := j²;
cycle 3: l := k²;
cycle 300: a := xA·zB;
cycle 301: start xB·zA;
cycle 600: b := xB·zA;
cycle 601: c := a + b; start xA·zA;
cycle 602: d := c²;
cycle 900: h := xA·zA;
cycle 901: i := h²; start a·b;
cycle 1200: f := a·b;
cycle 1201: start xP·d;
cycle 1500: e := xP·d;
cycle 1501: g := e+f;
final results: (xA, zA, xB, zB) := (l, i, g, d);
```

In order to compute the variable lifetimes, it is assumed that the multiplier reads the values of the operands during some initial cycle, say number I, and generates the result during cycle number $I + t_m - 1$ (or sooner), so that this result can be stored at the end of cycle number $I + t_m - 1$ and is available for any operation beginning at cycle number $I + t_m$ (or later). As regards the variables x_A, z_A, x_B and z_B, in charge of passing values from one iteration step to the next (Algorithm 2.3), their initial values must be available from the first cycle up to the last cycle during which those values are used. At the end of the computation scheme execution they must be updated with their new values. The lifetime intervals are given in Table 2.1.

The definition of a minimum number of registers can be expressed as a graph coloring problem. For that, associate a vertex to every variable and draw an edge between two variables if their lifetime intervals are incompatible, which means that they have more than one common cycle. As an example, the lifetime intervals of j and k are compatible, while the lifetime intervals of b and d are not.

The following groups of variables have compatible lifetime intervals:

Table 2.1 Lifetime intervals

a	[300, 901]
j	[1, 2]
k	[2, 3]
l	[3, *final*]
b	[600, 901]
h	[900, 901]
c	[601, 602]
d	[602, *final*]
f	[1200, 1501]
i	[901, *final*]
e	[1500, 1501]
g	[1501, *final*]
x_A	[*initial*, 601]
z_A	[*initial*, 601]
x_B	[*initial*, 301]
z_B	[*initial*, 1]

$z_B(\textit{initial} \to 1),\ j(1 \to 2),\ k(2 \to 3),\ l(3 \to \textit{final});$
$x_B(\textit{initial} \to 301),\ b(600 \to 901),\ f(1200 \to 1501),\ g(1501 \to \textit{final});$
$z_A(\textit{initial} \to 601),\ c(601 \to 602),\ d(602 \to \textit{final});$
$x_A(\textit{initial} \to 601),\ h(900 \to 901),\ e(1500 \to 1501);$
$a(300 \to 901),\ i(901 \to \textit{final}).$

Thus, the computing scheme can be executed with five registers, namely x_A, z_A, x_B, z_B and R:

```
cycle 1: z_B := x_A+z_A; start x_A·z_B;
cycle 2: z_B := z_B^2;
cycle 3: z_B := z_B^2;
cycle 300: R := x_A·z_B;
cycle 301: start x_B·z_A;
cycle 600: x_B := x_B·z_A;
cycle 601: z_A := R + x_B; start x_A·z_A;
cycle 602: z_A := z_A^2;
cycle 900: x_A := x_A·z_A;
cycle 901: R := x_A^2; start R·x_B;
cycle 1200: x_B := R·x_B;
cycle 1201: start x_P·z_A;
cycle 1500: x_A := x_P·z_A;
cycle 1501: (x_A, z_A, x_B, z_B) := (Z_B, R, x_A+x_B, z_A) ;
```

2.5 Final Example

Each iteration step of Algorithm 2.3 consists of executing a computation scheme, either the preceding one when $k_{m-i} = 0$, or a similar one when $k_{m-i} = 1$. Thus, Algorithm 2.3 is equivalent to the following algorithm 2.4 in which sentences separated by commas are executed in parallel.

Algorithm 2.4: Scalar product, projective coordinates (scheduled version)

```
x_A := 1; z_A := 0; x_B := x_P; z_B := 1;
for i in 1 .. m loop
  if k_m-i = 0 then
    R := x_A·z_B, z_B := x_A+z_A;
    z_B := z_B^2;
    z_B := z_B^2;
    x_B := x_B·z_A;
    x_A := x_A·z_A, z_A := R + x_B;
    z_A := z_A^2;
    x_B := R·x_B, R := x_A^2;
    x_A := x_P·z_A;
    x_B := x_A+x_B, x_A := z_B, z_A := R, z_B := z_A ;
  else
    R := x_B·z_A, z_A := x_B+z_B;
    z_A := z_A^2;
    z_A := z_A^2;
    x_A := x_A·z_B;
    x_B := x_B·z_B, z_B = R + x_A;
    z_B := z_B^2;
    x_A := R·x_A, R := x_B^2;
    x_B := x_P·z_B;
    x_A := x_B+x_A, x_B := z_A, z_B := R, z_A := z_B ;
  end if;
end loop;
```

The data processed by Algorithm 2.4 are m-bit vectors (polynomials of degree m over the binary field $GF(2)$) and the computation resources are field multiplication, addition and squaring. Field addition amounts to bit-by-bit modulo 2 additions (XOR functions). On the other hand, VHDL models of computation resources executing field squaring and multiplication are available at the Authors' web page, namely *classic_squarer.vhd* and *interleaved_mult.vhd* (Chap. 13). The *classic_squarer* component is a combinational circuit. The *interleaved_mult* component reads and internally stores the input operands during the first cycle after detecting a positive edge on *start_mult* and raises an output flag *mult_done* when the multiplication result is available.

The operations executed by the multiplier are

$$x_A \cdot z_B, x_B \cdot z_A, x_A \cdot z_A, R \cdot x_B, x_P \cdot z_A, x_B \cdot z_B, R \cdot x_A, x_P \cdot z_B.$$

An incompatibility relation can be defined over the set of involved variables: two variables are incompatible if they are operands of a same operation. As an example, x_A and z_B are incompatible, as $x_A \cdot z_B$ is one of the operations. The corresponding graph can be colored with two colors corresponding to the sets

$$\{x_A, x_B, x_P\} \text{ and } \{z_A, z_B, R\}.$$

The first set of variables can be assigned to the leftmost multiplier input and the other to the rightmost input.

The operations executed by the adder are

$$x_A + z_A, R + x_B, x_A + x_B, x_B + z_B, R + x_A, x_B + x_A.$$

The incompatibility graph can be colored with three colors corresponding to the sets

$$\{x_A, z_B\}, \{x_B, z_A\} \text{ and } \{R\}.$$

The first one is assigned to the leftmost adder input, the second to the rightmost input, and R to both inputs.

Finally, the operations realized by the squaring primitive are

$$z_B^2, z_A^2, x_A^2, x_B^2.$$

The part of the data path corresponding to the computation resources and the multiplexers that select their input data is shown in Fig. 2.16. The corresponding VHDL model can easily be generated. As an example, the multiplier, with its input multiplexers, can be described as follows.

```
WITH sel_p1 SELECT
  mult1 <= xA WHEN "00", xB WHEN "01", xP WHEN OTHERS;
WITH sel_p2 SELECT
  mult2 <= zA WHEN "00", zB WHEN "01", R WHEN OTHERS;
a_mod_f_multiplier: interleaved_mult
  PORT map (A => mult1, B => mult2, clk => clk,
  reset => reset, start => start_mult, Z => product,
  done => mult_done);
```

Consider now the storing resources. Assuming that x_P and k remain available during the whole algorithm execution, there remain five variables that must be internally stored: x_A, x_B, z_A, z_B and R. The origin of the data stored in every register must be defined. For example, the operations that update x_A are

$x_A := 1;\ x_A := x_A \cdot z_A;\ x_A := x_P \cdot z_A;\ x_A := z_B;\ x_A := x_A \cdot z_B;$
$x_A := R \cdot x_A;\ x_A := x_B + x_A;$

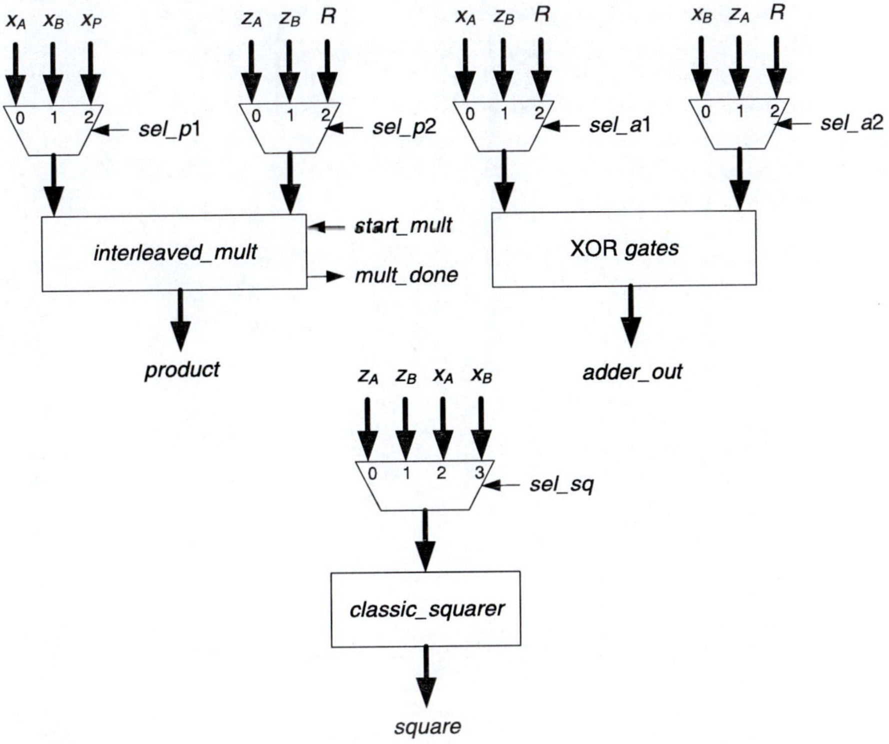

Fig. 2.16 Example 2.4: computation resources

So, the updated value can be 1 (initial value), *product*, *adder_out* or z_B. A similar analysis must be done for the other registers. Finally, the part of the data path corresponding to the registers and the multiplexers that select their input data is shown in Fig. 2.17. The corresponding VHDL model is easy to generate. As an example, the x_A register, with its input multiplexers, can be described as follows.

```
WITH sel_xA SELECT
  next_xA <=
  product WHEN "00", adder_out WHEN "01", zB WHEN OTHERS;
register_xA: PROCESS(clk)
BEGIN
  IF clk'EVENT AND clk = '1' THEN
    IF load = '1' THEN xA <= one;
    ELSIF en_xA = '1' THEN xA <= next_xA;
    END IF;
  END IF;
END PROCESS;
```

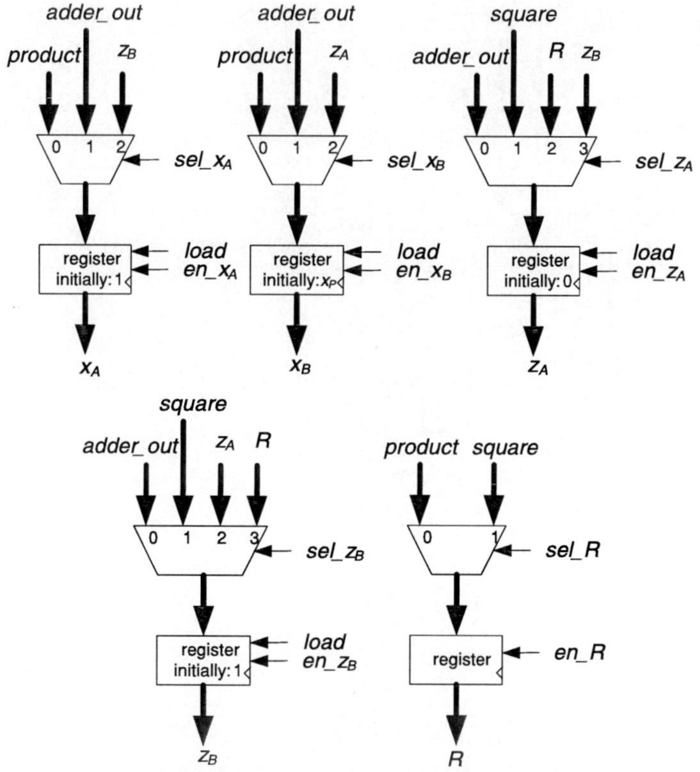

Fig. 2.17 Example 2.5: data registers

A complete model of the data path *scalar_product_data_path.vhd* is available at the Authors' web page.

The complete circuit is defined by the following entity.

```
ENTITY scalar_product IS
PORT (
    xP, k: IN STD_LOGIC_VECTOR(M-1 DOWNTO 0);
    clk, reset, start: IN STD_LOGIC;
    xA, zA, xB, zB: INOUT STD_LOGIC_VECTOR(M-1 DOWNTO 0);
    done: OUT STD_LOGIC
);
END scalar_product;
```

It is made up of

- the data path;
- a shift register allowing sequential reading of the values of k_{m-i};
- a counter for controlling the *loop* execution;

- a finite state machine in charge of generating all the control signals, that is *start_mult*, *load*, *shift*, *en_xA*, *en_xB*, *en_zA*, *en_zB*, *en_R*, *sel_p*1, *sel_p*2, *sel_a*1, *sel_a*2, *sel_sq*, *sel_xA*, *sel_xB*, *sel_zA*, *sel_zB* and *sel_R*. In particular, the control of the multiplier operations is performed as follows: the control unit generates a positive edge on the *start_mult* signal, along with the values of *sel_p*1 and *sel_p*2 that select the input operands; then, it enters a wait loop until the *mult_done* flag is raised (instead of waiting for a constant time, namely 300 cycles, as was done for scheduling purpose); during the wait loop the *start_mult* is lowered while the *sel_p*1 and *sel_p*2 values are maintained; finally, it generates the signals for updating the register that stores the result. As an example, assume that the execution of the fourth instruction of the main loop, that is $x_B := x_B \cdot z_A$, starts at state 6 and uses identifiers *start*4, *wait*4 and *end*4 for representing the corresponding commands. The corresponding part of the next-state function is

```
WHEN 6 => current_state <= 7;
WHEN 7 => IF mult_done = '1' THEN
          current_state <= 8; END IF;
```

  and the corresponding part of the output function is

```
WHEN 6  => command <= start4;
WHEN 7 => IF mult_done = '0' THEN command <= wait4;
          ELSE command <= end4; END IF;
```

- a command decoder (Chap. 4). Command identifiers have been used in the definition of the finite state machine output function, so that a command decoder must be used to generate the actual control signal values in function of the identifiers. For example, the command *start*4 initializes the execution of $x_B := x_B \cdot z_A$ and is decoded as follows:

```
WHEN start4 =>
  start_mult <= '1'; load <= '0'; shift <= '0';
  en_xA <= '0'; en_xB <= '0'; en_zA <= '0'; en_zB <= '0';
  en_R <= '0'; sel_p1 <= "01"; sel_p2 <= "00";
  sel_a1 <= "00"; sel_a2 <= "00"; sel_sq <= "00";
  sel_xA <= "00"; sel_xB <= "00"; sel_zA <= "00";
  sel_zB <= "00"; sel_R <= '0';
```

  In the case of operations such as the first of the main loop, that is $R := x_A \cdot z_B$, $z_B := x_A + z_A$, the 1-cycle operation $z_B := x_A + z_A$ is executed in parallel with the final cycle of $R := x_A \cdot z_B$ and not in parallel with the initial cycle. This makes the algorithm execution a few cycles (3) longer, but this is not significant as t_m is generally much greater than 3. Thus, the control signal values corresponding to the identifier *end*1 are:

```
WHEN end1 =>
   start_mult <= '0'; load <= '0'; shift <= '0';
   en_xA <= '0'; en_xB <= '0'; en_zA <= '0'; en_zB <= '1';
   en_R <= '1'; sel_p1 <= "00"; sel_p2 <= "00";
   sel_a1 <= "00"; sel_a2 <= "01"; sel_sq <= "00";
   sel_xA <= "00"; sel_xB <= "00"; sel_zA <= "00";
   sel_zB <= "00"; sel_R <= '0';
```

The control unit also detects the *start* signal and generates the *done* flag. A complete model *scalar_product.vhd* is available at the Authors' web page.

Comment 2.1

The *interleaved_mult* component is also made up of a data path and a control unit, while the *classic_squarer* component is a combinational circuit. An alternative solution is the definition of a data path able to execute all the operations, including those corresponding to the *interleaved_mult* and *classic_squarer* components. The so-obtained circuit could be more efficient than the proposed one as some computation resources could be shared between the three algorithms (field multiplication, squaring and scalar product). Nevertheless, the hierarchical approach consisting of using pre-existing components is probably safer and allows a reduction in the development times.

Instead of explicitly disassembling the circuit into a data path and a control unit, another option is to describe the operations that must be executed at each cycle, and to let the synthesis tool define all the details of the final circuit. A complete model *scalar_product_DF2.vhd* is available at the Authors' web page.

Comment 2.2

Algorithm 2.4 does not compute the scalar product *kP*. A final step is missing:

```
if zB = 0 then xA := xP; yA := xP + yP;
else
 xA := xA / zA;
 xB := xB / zB;
 yA := ((xA + xP)[(xA + xP)(xB + xP) + xP^2 + yP] / xP) +
yP;
end if;
xR := xA; yR := yR;
```

The design of a circuit that executes this final step is left as an exercise.

2.6 Exercises

1. Generate several VHDL models of a 7-to-3 counter. For that purpose use the three options proposed in Sect. 2.3.1.
2. Generate the VHDL model of a circuit executing the final step of the scalar product algorithm (Comment 2.2). For that purpose, the following entity, available at the Authors' web page, is used:

```
entity binary_algorithm_polynomials is
port(
  g, h: in std_logic_vector(M-1 downto 0);
  clk, reset, start: in std_logic;
  z: out std_logic_vector(M-1 downto 0);
  done: out std_logic
);
end binary_algorithm_polynomials;
```

   It computes $z = g{\cdot}h^{-1}$ over $GF(2^m)$. Several architectures can be considered.
3. Design a circuit to compute the greatest common divisor of two natural numbers, based on the following simplified Euclidean algorithm.

```
while a ≠ b loop
  if a > b then a := a - b;
  else b := b - a;
end loop;
```

4. Design a circuit for computing the greatest common divisor of two natural numbers, based on the following Euclidean algorithm.

```
while b > 0 loop
  (a, b) := (b, a mod b);
end loop;
gcd := a;
```

5. The distance d between two points (x_1, y_1) and (x_2, y_2) of the (x, y)-plane is equal to $d = ((x_1 - x_2)^2 + (y_1 - y_2)^2)^{0.5}$. Design a circuit that computes d with only one subtractor and one multiplier.
6. Design a circuit that, within a three-dimensional space, computes the distance between two points (x_1, y_1, z_1) and (x_2, y_2, z_2).
7. Given a point (x, y, z) of the three-dimensional space, design a circuit that computes the following transformation.

$$\begin{bmatrix} x_t \\ y_t \\ z_t \end{bmatrix} = \begin{bmatrix} a_{11} & a_{21} & a_{31} \\ a_{21} & a_{22} & a_{32} \\ a_{31} & a_{32} & a_{11} \end{bmatrix} \times \begin{bmatrix} x \\ y \\ z \end{bmatrix}$$

8. Design a circuit for computing $z = e^x$ using the formula

$$e^x = 1 + \frac{x}{1!} + \frac{x^2}{2!} + \frac{x^3}{3!} + \cdots$$

9. Design a circuit for computing x^n, where n is a natural, using the following relations: $x^0 = 1$; if n is even then $x^n = (x^{n/2})^2$, and if n is odd then $x^n = x \cdot (x^{(n-1)/2})^2$.
10. Algorithm 2.4 (scalar product) can be implemented using more than one *interleaved_multiplier*. How many multipliers can operate in parallel? Define the corresponding schedule.

References

1. Hankerson D, Menezes A, Vanstone S (2004) Guide to elliptic curve cryptography. Springer, New York
2. Deschamps JP, Imaña JL, Sutter G (2009) Hardware implementation of finite-field arithmetic. McGraw-Hill, New York
3. López J, Dahab R (1999) Improved algorithm for elliptic curve arithmetic in $GF(2^n)$. Lect Notes Comput Sci 1556:201–212

Chapter 3
Special Topics of Data Path Synthesis

Several important implementation techniques are presented in this chapter. The first one is pipelining, a very commonly used method in systems that process great volumes of data. Self-timing is the topic of the second section. To some extent it can be considered as an extension of the pipelining concept and is especially attractive in the case of very big circuits. The third section is an introduction to a circuit level, or even algorithm level, transformation known as "loop unrolling". It permits the exploration of different cost—performance tradeoffs, from combinational iterative circuits to completely sequential circuits. Finally, the last section tackles the problem of reducing the number of connection resources.

3.1 Pipeline

A very useful implementation technique, especially for signal processing circuits, is pipelining [1, 2]. It consists of inserting additional registers so that the maximum clock frequency and input data throughput are increased. Furthermore, in the case of FPGA implementations, the insertion of pipeline registers has a positive effect on the power consumption.

3.1.1 Introductory Example

Consider the introductory example of Sect. 2.3.1. The set of Eq. (2.10) can be implemented by a combinational circuit (option 1. of Sect. 2.3.1) made up of four carry-save adders, with a computation time equal to $3 \cdot T_{FA}$. That means that the minimum clock period of a synchronous circuit including this 7-to-3 counter should be greater than $3 \cdot T_{FA}$, and that the introduction interval between successive data inputs should also be greater than $3 \cdot T_{FA}$. The corresponding circuit is shown

J.-P. Deschamps et al., *Guide to FPGA Implementation of Arithmetic Functions*, Lecture Notes in Electrical Engineering 149, DOI: 10.1007/978-94-007-2987-2_3,

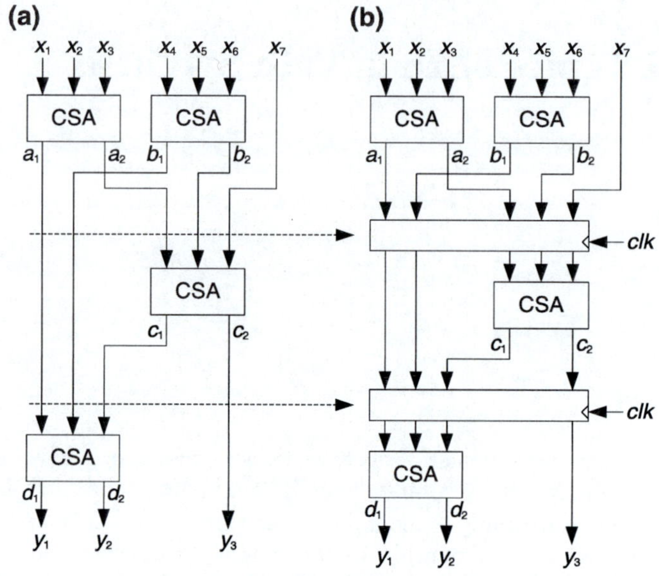

Fig. 3.1 **a** Combinational circuit. **b** Pipelined circuit

in Fig. 3.1a. As previously commented, this is probably a bad circuit because its cost is high and its maximum clock frequency is low.

Consider now the circuit of Fig. 3.1b in which registers have been inserted in such a way that operations scheduled in successive cycles, according to the ASAP schedule of Fig. 2.9a, are separated by a register. The circuit still includes four carry-save adders, but the minimum clock period of a synchronous circuit including this counter must be greater than T_{FA}, plus the set-up and hold times of the registers, instead of $3 \cdot T_{FA}$. Furthermore, the minimum data introduction interval is now equal to T_{clk}: as soon as a_1, a_2, b_1 and b_2 have been computed, their values are stored within the corresponding output register, and a new computation, with other input data, can start; at the same time, new computations of c_1 and c_2, and of d_1 and d_2 can also start. Thus, at time t, three operations are executed in parallel:

```
(a1(t), a2(t), b1(t), b2(t)) :=
(CSA(x1(t), x2(t), x3(t)), CSA(x4(t), x5(t), x6(t)));

(c1(t-1), y3(t-1)) := (CSA(a2(t-1), b2(t-1), x7(t-1)));

(y1(t-2), y2(t-2)) := (CSA(a1(t-2), b1(t-2), c1(t-2)));
```

To summarize, assuming that the set-up and hold times are negligible,

$$T_{clk} > T_{FA}, \; latency = 3 \cdot T_{clk}, \; r = T_{clk},$$

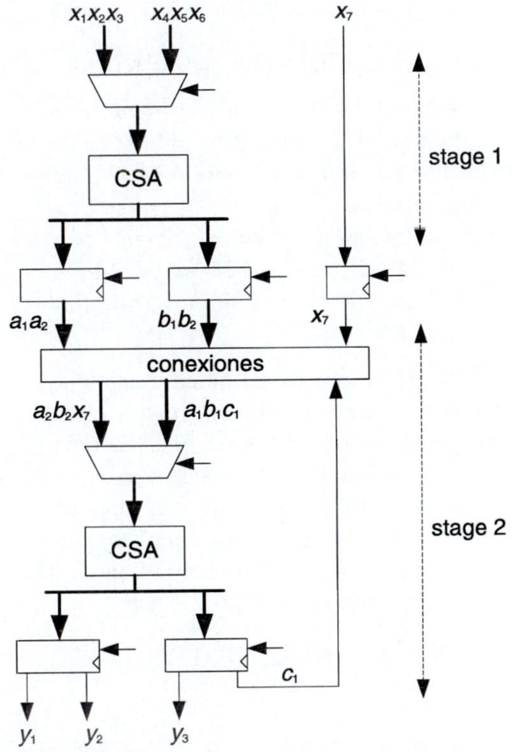

Fig. 3.2 Two-stage two-cycle implementation

where *latency* is the total computation time and r is the minimum data introduction interval.

Another implementation, based on the admissible schedule of Fig. 2.9c is show in Fig. 3.2. In this case the circuit is made up of two stages separated by a pipeline register. Within every stage the operations are executed in two cycles. During the first cycle the following operations are executed

```
stage 1:  (a1(t), a2(t))  := CSA(x1(t), x2(t), x3(t));

stage 2 :
  (c1(t-1), y3(t-1))  := (CSA(a2(t-1), b2(t-1), x7(t-1)));
```

and during the second cycle, the following ones are executed

```
stage 1:  (b1(t), b2(t))  := CSA(x4(t), x5(t), x6(t)) ;

stage 2 :
  (y1(t-1), y2(t-1))  := (CSA(a1(t-1), b1(t-1), c1(t-1)));
```

The circuit of Fig. 3.2 includes two carry-save adders instead of four, and its timing constraints are the following:

$$T_{clk} > T_{FA} + T_{multiplexor},\ latency = 4 \cdot T_{clk},\ r = 2 \cdot T_{clk}.$$

To summarize, the main parameters of a pipelined circuit are the following.

- *Latency* (also called *delay, response time*): total delay between the introduction of a new set of input data and the generation of the corresponding output results. It is equal to $n \cdot T_{clk}$ where n is the number of pipeline stages and T_{clk} the clock period.
- *Pipeline rate* (also called *pipeline period*): data introduction interval.
- *Throughput* (also called *speed, bandwidth, production*): number of input data processed per time unit. For great numbers of processed data, it is the inverse of the pipeline rate r.

Assuming that the combinational delay of stage number i is equal to t_i, and that the register set-up and hold times are negligible, the minimum clock period is the maximum of all t_i's. In the first example (Fig. 3.1a) $t_1 = t_2 = t_3 = T_{FA}$, while in the second example (Fig. 3.2) $t_1 = t_2 = T_{FA} + T_{multiplexor}$.

A very common situation is that of the first example (Fig. 3.1). An initial combinational circuit has a delay equal to C, so that it is able to process $1/C$ input data per time unit. It is divided up into n pipeline stages, all of them with the same delay C/n (balanced stages). Then

$$T_{clk} \cong C/n,\ r = 1/T_{clk} \cong n/C,\ latency = n \cdot T_{clk} \cong C,\ T(m) = n \cdot T_{clk} + (m-1) \cdot T_{clk},$$

where $T(m)$ is the time necessary to process m input data. Thus, the average number of input data processed per time unit is equal to $m/T(m) = m/(n + m - 1) \cdot T_{clk} \cong mn/(n + m - 1) \cdot C$. For great values of m, the number of input data processed per time unit is equal to n/C. Thus, with respect to the initial combinational circuit, the throughput has been multiplied by n.

The actual speedup factor is smaller if the connection and register delays are taken into account. Assume that those additional delays are equal to δ time units. Then, the minimum clock period is equal to $T_{clk} = C/n + \delta$, $r = 1/T_{clk} = n/(C + n\delta)$, $latency = n \cdot T_{clk} = C + n\delta$, $T(m) = (n + m - 1) \cdot T_{clk} = (n + m - 1) \cdot (C/n + \delta) \cong m \cdot (C/n + \delta)$, $m/T(m) \cong 1/(C/n + \delta) = n/(C + n\delta)$. Hence, the throughput increase factor is equal to $n \cdot C/(C + n\delta) = n/(1 + \alpha)$ where $\alpha = n\delta/C$.

3.1.2 Segmentation

Given a computation scheme and its precedence graph G, a *segmentation* of G is an ordered partition $\{S_1, S_2, \ldots, S_k\}$ of G. The segmentation is *admissible* if it respects the precedence relationship. This means that if there is an arc from $op_J \in S_i$ to op_M then either op_M belongs to the same segment S_i or it belongs to a different segment S_j with $j > i$. Two examples are shown in Fig. 3.3 in which the segments are separated by dotted lines.

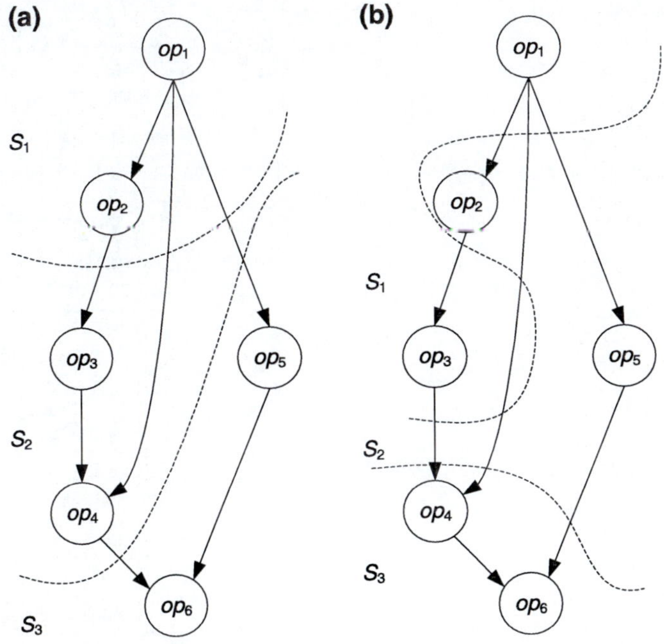

Fig. 3.3 **a** Admissible segmentation. **b** Non-admissible segmentation

The segmentation of Fig. 3.3a, that is $S_1 = \{op_1, op_2\}$, $S_2 = \{op_3, op_4\}$, $S_3 = \{op_5, op_6\}$, is admissible, while that of Fig. 3.3b, that is $S_1 = \{op_1, op_3\}$, $S_2 = \{op_2, op_5\}$, $S_3 = \{op_4, op_6\}$, is not (there is an arc $op_2 \rightarrow op_3$ from S_2 to S_1).

Once an admissible partition has been defined, every segment can be synthesized separately, using the same methods as before (scheduling, resource assignment). In order to assemble the complete circuit, additional registers are inserted: if an arc of the precedence graph crosses the line that separates segments i and $i + 1$, then a register must be inserted; it will store the output data generated by segment i that in turn are input data to segment $i + 1$. As an example, the structure of the circuit corresponding to Fig. 3.3a is shown in Fig. 3.4.

Assume that C_i and T_i are the cost and computation time of segment i. The cost and latency of the complete circuit are

$$C = C_1 + C_2 + \cdots + C_k + C_{registers} \text{ and } T = T_1 + T_2 + \cdots + T_k + T_{registers}$$

where $C_{registers}$ represents the total cost of the pipeline registers and $T_{registers}$ the additional delay they introduce. The time interval δ between successive data inputs is

$$\delta = max\{T_1, T_2, \ldots, T_k\} + T_{SU} + T_P \cong max\{T_1, T_2, \ldots, T_k\}$$

where T_{SU} and T_P are the set-up and propagation times of the used flip-flops.

Fig. 3.4 Pipelined circuit

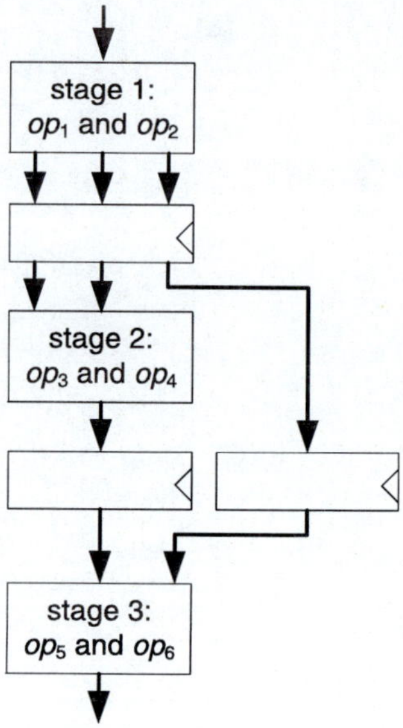

Fig. 3.5 An admissible segmentation

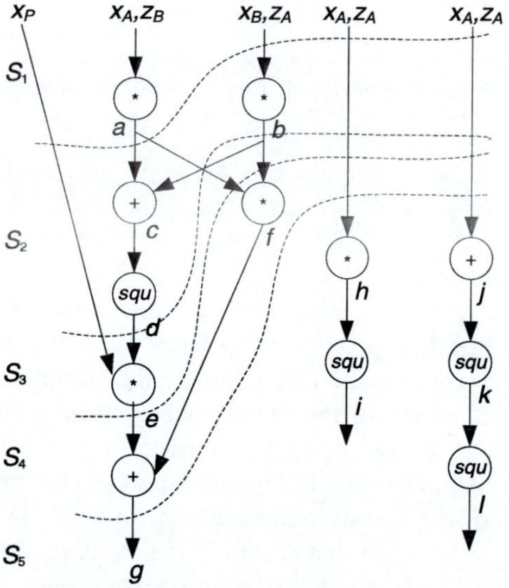

A second, more realistic, example is now presented. It corresponds to part of an Elliptic Curve Cryptography algorithm (Example 2.1).

Example 3.1
An admissible segmentation of the precedence graph of Fig. 2.10 is shown in Fig. 3.5.

The operations corresponding to each segment are the following.

- Segment 1:

```
a = xA·zB;
```

- Segment 2:

```
b = xB·zA,
c = a + b,
d = c2;
```

- Segment 3:

```
e = xP·d;
```

- Segment 4:

```
f = a·b,
g = e + f;
```

- Segment 5:

```
h = xA·zA,
i = h2,
j = xA+zA,
k = j2,
l = k2;
```

So, every segment includes a product over a finite field plus some additional 1-cycle operations (finite field additions and squares) in segments 2, 4 and 5. The corresponding pipelined circuit, in which it is assumed that the output results are g, d, l and i, is shown in Fig. 3.6.

A finite field product is a complex operation whose maximum computation time t_m, expressed in number of clock cycles, is much >1. Thus, the latency T and the time interval δ between successive data inputs of the complete circuit are

$$T \cong 5t_m \text{ and } \delta \cong t_m.$$

The cost of the circuit is very high. It includes five multipliers, three adders, four squarers and four pipeline registers. Furthermore, if used within the scalar product circuit of Sect. 2.5, the fact that the time interval between successive data inputs has been reduced ($\delta \cong t_m$) does not reduce the execution time of Algorithm 2.4 as the input variables x_A, z_A, x_B and z_B are updated at the end of every main loop execution.

As regards the control of the pipeline, several options can be considered. A simple one is to previously calculate the maximum multiplier computation time t_m and to choose $\delta > t_m + 2$ (computation time of segment 2). The control unit updates the pipeline registers and sends a *start* pulse to all multipliers every δ cycles. In the following VHDL process, *time_out* is used to enable the pipeline register clock every *delta* cycles and *sync* is a synchronization procedure (Sect. 2.5):

```
control_unit: PROCESS
BEGIN
  LOOP
    time_out <= '0'; start <= '0'; sync;
    IF reset = '1' THEN EXIT; END IF;
    start <= '1'; sync;
    FOR i IN 1 TO delta-3 LOOP
      sync;
    END LOOP;
    time_out <= '1'; sync;
  END LOOP;
END PROCESS;
```

In order to describe the complete circuit, five multipliers, three adders and four squarers are instantiated, and every pipeline register, for example the segment 1 output register, can be described as follows:

```
segment1: PROCESS
BEGIN
  WAIT UNTIL time_out = '1';
  xP1 <= xP; a1 <= a; xB1 <= xB; zA1 <= zA; xA1 <= xA;
END PROCESS;
```

A complete VHDL model *pipeline_DF2.vhd* is available at the Authors' web page.

Another interesting option could be a self-timed circuit (Example 3.4).

3.1.3 Combinational to Pipelined Transformation

A very common situation is the following: a combinational circuit made up of relatively small blocks, all of them with nearly equal delays, has been designed, and its computation time is equal to T seconds. If this combinational circuit is used as a computation resource of a synchronous circuit, then the clock cycle must be greater than T, and in some cases it could be an over extended time (a too low frequency). In order to increase the clock frequency, as well as to reduce the minimum time interval between successive data inputs, the solution is pipelining. As the combinational version already exists, it is no longer necessary to use the general method of Sect. 3.1.2. The combinational circuit can be directly segmented into stages.

Consider a generic example. The iterative circuit of Fig. 3.7 is made up twelve identical blocks, each of them with a maximum delay of t_{cell} seconds. The maximum propagation time of every connection is equal to $t_{connection}$ seconds. Thus, the computation time of this circuit is equal to $T = 6t_{cell} + 7t_{connection}$ (input and output connections included). Assume that this circuit is part of a synchronous

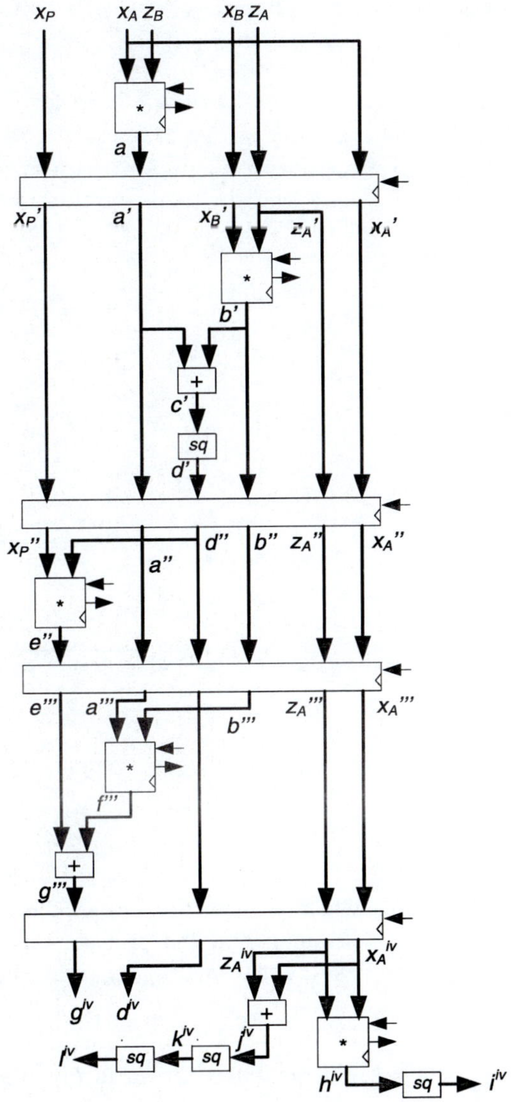

Fig. 3.6 Pipelined circuit ($\forall$ s: $s' = s(t-1)$, $s'' = s(t-2)$, $s''' = s(t-3)$, $s^{iv} = s(t-4)$)

circuit and that all inputs come from register outputs and all outputs go to register inputs. Then the minimum clock cycle T_{CLK} is defined by the following relation:

$$T_{CLK} > 6t_{cell} + 7t_{connection} + t_{SU} + t_P, \tag{3.1}$$

where t_{SU} and t_P are the minimum set-up and propagation times of the registers (Chap. 6).

If the period defined by condition (3.1) is too long, the circuit must be segmented. A 2-stage segmentation is shown in Fig. 3.8. Registers must be inserted in

Fig. 3.7 Combinational circuit

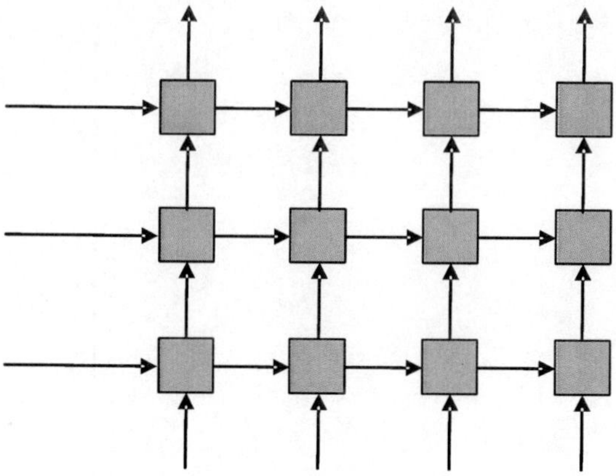

Fig. 3.8 2-stage segmentation

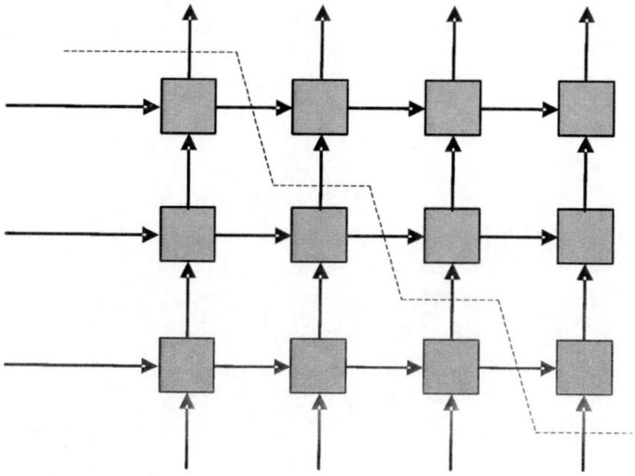

all positions where a connection crosses the dotted line. Thus, seven registers must be added. Assuming that the propagation time of every part of a segmented connection is still equal to $t_{connection}$, the following condition must hold:

$$T_{CLK} > 3t_{cell} + 4t_{connection} + t_{SU} + t_P. \quad (3.2)$$

A 5-stage segmentation is shown in Fig. 3.9. In this case, 32 registers must be added and the following condition must hold:

$$T_{CLK} > t_{cell} + 2t_{connection} + t_{SU} + t_P. \quad (3.3)$$

Consider a practical example.

Example 3.2
Implement a 128-bit adder made up of four 32-bit adders. A combinational implementation is described in Fig. 3.10. The computation time T of the circuit is equal to $4 \cdot T_{adder}$, where T_{adder} is the computation time of a 32-bit adder.

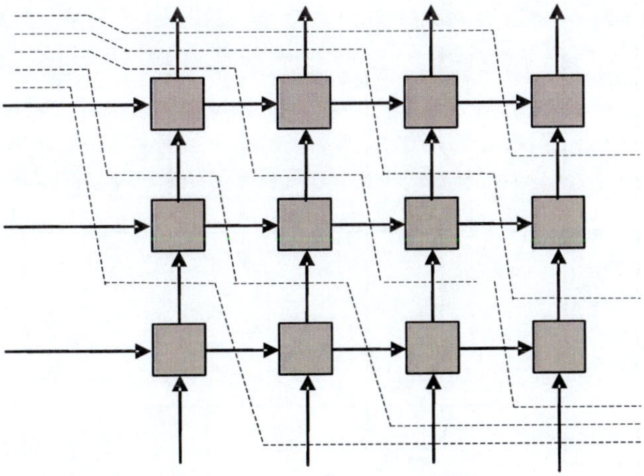

Fig. 3.9 5-stage segmentation

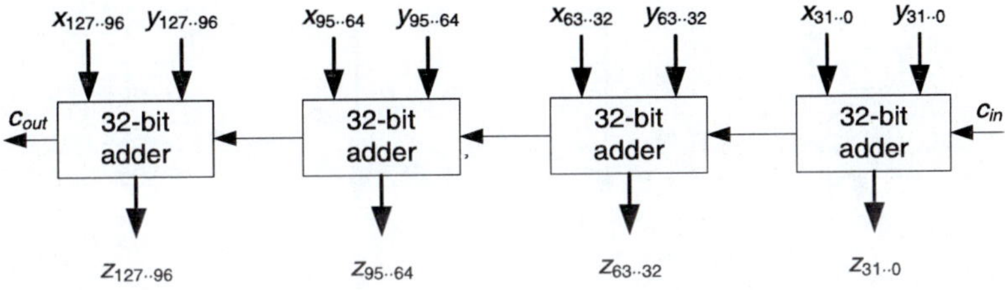

Fig. 3.10 128-bit adder

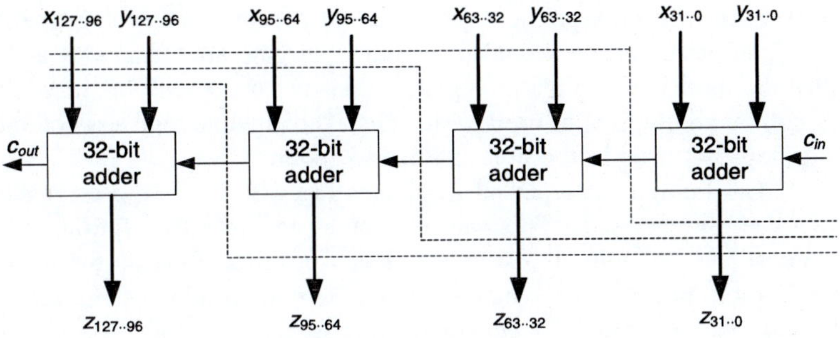

Fig. 3.11 4-stage segmentation

A 4-stage segmentation is shown in Fig. 3.11. Every stage includes one 32-bit adder so that the minimum clock cycle, as well as the minimum time interval between successive data inputs, is equal to T_{adder}. The corresponding circuit is shown in Fig. 3.12. In total, $(7 \cdot 32 + 1) + (6 \cdot 32 + 1) + (5 \cdot 32 + 1) = 579$ additional flip-flops are necessary in order to separate the pipeline stages.

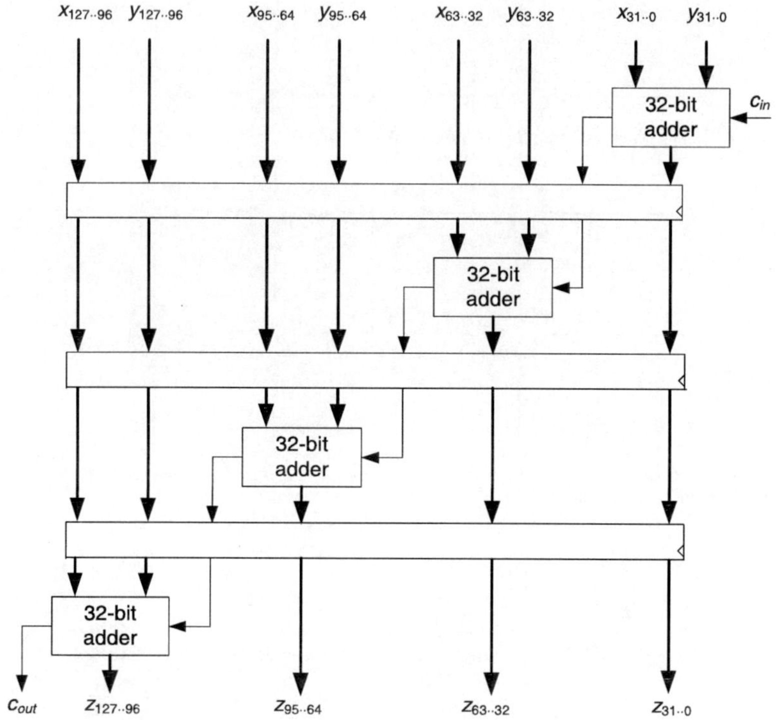

Fig. 3.12 Pipelined 128-bit adder

Comments 3.1

- The extra cost of the pipeline registers could appear to be prohibitive. Nevertheless, the basic cell of a field programmable gate array includes a flip-flop, so that the insertion of pipeline registers does not necessarily increase the total cost, computed in terms of used basic cells. The pipeline registers could consist of flip-flops not used in the non-pipelined version.
- Most FPGA families also permit implementing with LUTs those registers that do not need reset signals. This can be another cost-effective option.
- The insertion of pipeline registers also has a positive effect on the power consumption: the presence of synchronization barriers all along the circuit drastically reduces the number of generated spikes.

3.1.4 Interconnection of Pipelined Components

Assume that several pipelined circuits are used as computational resources for generating a new pipelined circuit. For example, consider a circuit that computes $g = a{\cdot}b$ and $f = (a{\cdot}b + c){\cdot}d$, and uses a 2-stage pipelined multiplier and a 3-stage

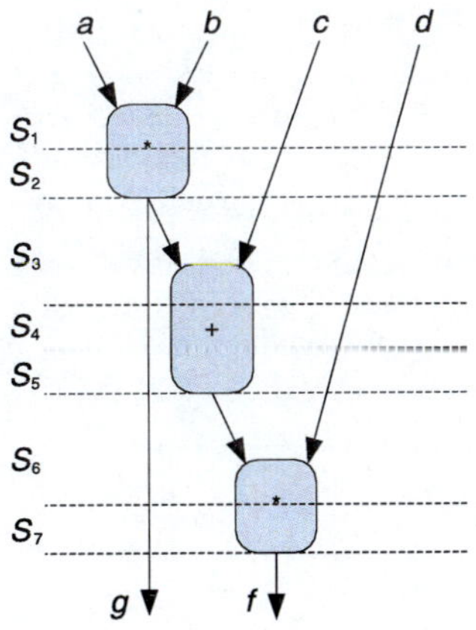

Fig. 3.13 Interconnection of pipelined components: scheduling

pipelined adder, both of them working at the same frequency $1/T_{clk}$ and with the same pipeline rate $r = 1$. The operations can be scheduled as shown in Fig. 3.13. Some inputs and outputs must be delayed: input c must be delayed 2 cycles, input d must be delayed 5 cycles, and output g must be delayed 5 cycles. The corresponding additional registers, which maintain the correct synchronization of the data, are sometimes called *skewing* (c and d) and *deskewing* (g) registers.

An alternative solution, especially in the case of large circuits, is *self-timing*.

As a generic example, consider the pipelined circuit of Fig. 3.14a. To each stage, for example number i, are associated a maximum delay $t_{MAX}(i)$ and an average delay $t_{AV}(i)$. The minimum time interval between successive data inputs is

$$\delta = max\{t_{MAX}(1), t_{MAX}(2), \ldots, t_{MAX}(n)\}, \tag{3.4}$$

and the minimum circuit latency T is

$$T = n \cdot max\{t_{MAX}(1), t_{MAX}(2), \ldots, t_{MAX}(n)\}. \tag{3.5}$$

A self-timed version of the same circuit is shown in Fig. 3.14b. The control is based on a *Request/Acknowledge* handshaking protocol:

- a *req_in* signal to stage 1 is raised by an external circuit; if stage 1 is free, the input data is registered ($ce = 1$), and an *ack_out* signal is issued;
- the *start* signal of stage 1 is raised; after some amount of time, the *done* signal of stage 1 elevates indicating the completion of the computation;
- a *req_out* signal to stage 2 is issued by stage 1; if stage 2 is free, the output of stage 1 is registered and an *ack_out* signal to stage 1 is issued; and so on.

If the probability distribution of the internal data were uniform, inequalities (3.4) and (3.5) would be substituted by the following:

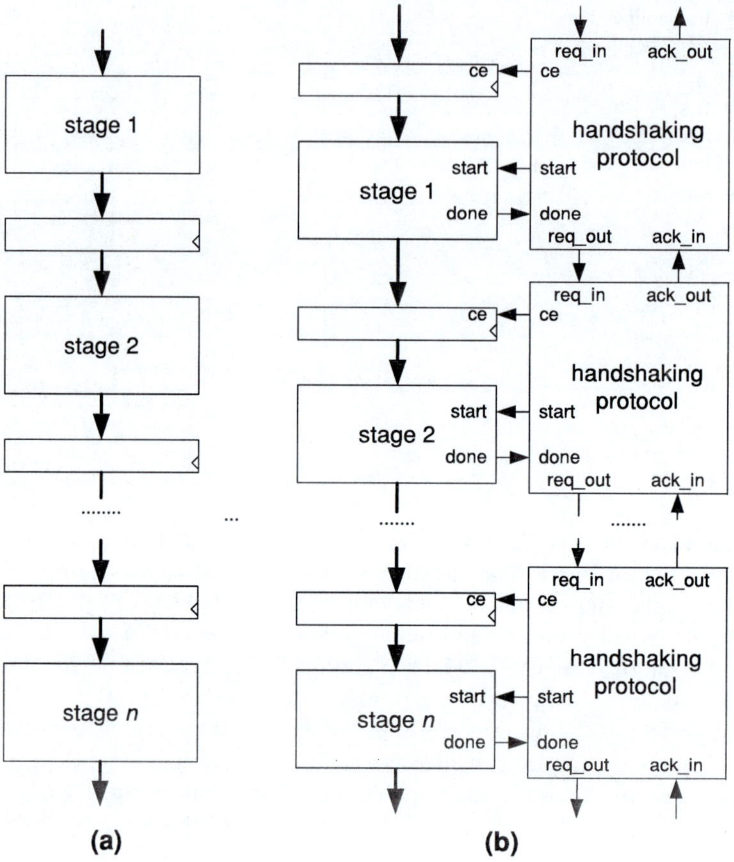

Fig. 3.14 **a** Pipelined circuit. **b** Self-timed pipelined circuit

$$\delta = max\{t_{AV}(1), t_{AV}(2), \ldots, t_{AV}(n)\}, \tag{3.6}$$

$$T = t_{AV}(1) + t_{AV}(2) + \cdots + t_{AV}(n). \tag{3.7}$$

Example 3.3
The following process describes a *handshaking protocol* component. As before, *sync* is a synchronization procedure (Sect. 2.5):

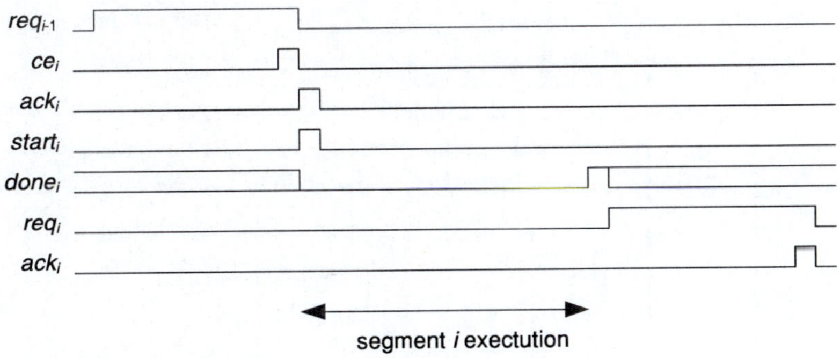

Fig. 3.15 Handshaking protocol

```
PROCESS
BEGIN
  ce <= '0'; start <= '0'; req_out <= '0'; ack_out <= '0';
  sync;
  LOOP
    IF reset = '1' THEN EXIT; END IF;
    IF req_in = '1'THEN
      ce <= '1'; sync;
      ce <= '0'; ack_out <= '1'; sync;
      ack_out <= '0'; start <= '1'; sync;
      start <= '0';
      WAIT UNTIL done = '1';
    req_out <= '1';
    WAIT UNTIL ack_in = '1';
    req_out <= '0'; sync;
  ELSE sync;
  END IF;
END LOOP;
END PROCESS;
```

The corresponding signals are shown in Fig. 3.15.

Example 3.4
Consider a self-timed version of the circuit of Example 3.1 (Fig. 3.6). In stages 1 and 3, the *done* signals are the corresponding *done* outputs of the multipliers. In stages 2, 4 and 5 an additional delay must be added. Stage 2 is shown in Fig. 3.16: a synchronous delay Δ, greater than the sum of the computation times of $c' = a' + b'$ and $d' = (c')^2$, has been added. A complete VHDL model *pipeline_ST.vhd* is available at the Authors' web page.

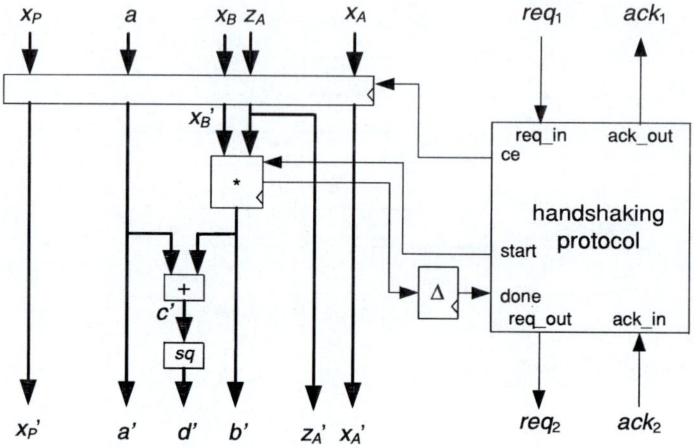

Fig. 3.16 Self-timed pipelined circuit: stage 2

Table 3.1 Redundant encoding

s	s_1	s_0
Reset or *in transition*	0	0
0	0	1
1	1	0

With regards to the generation of the *done* signal in the case of combinational components, an interesting method consists of using a redundant encoding of the binary signals (Sect. 10.4 of [3]: every signal s is represented by a pair (s_1, s_0) according to the definition of Table 3.1.

The circuit will be designed in such a way that during the initialization (*reset*), and as long as the value of s has not yet been computed, $(s_1, s_0) = (0, 0)$. Once the value of s is known $s_1 = s$ and $s_0 = not(s)$.

Assume that the circuit includes n signals $s_1, s_2, \ldots, s_n$. Every signal s_i is substituted by a pair (s_{i1}, s_{i0}). Then the *done* flag is computed as follows:

$$done = (s_{11} + s_{10}) \cdot (s_{21} + s_{20}) \ldots (s_{n1} + s_{n0}).$$

During the initialization (*reset*) and as long as at least one of the signals is *in transition*, the corresponding pair is equal to (0, 0), so that *done* = 0. The *done* flag will be raised only when all signals have a stable value.

In the following example, only the signals belonging to the critical path of the circuit are encoded.

Example 3.5

Generate an n-bit ripple-carry adder (Chap. 7) with end of computation detection. For this purpose, all signals belonging to the carry chain, that is $c_0, c_1, c_2, \ldots, c_{n-1}$, are represented by the form $(c_0, cb_0), (c_1, cb_1), (c_2, cb_2), \ldots, (c_{n-1}, cb_{n-1})$. During the initialization, all c_i and cb_i are equal to 0. When *reset* goes down,

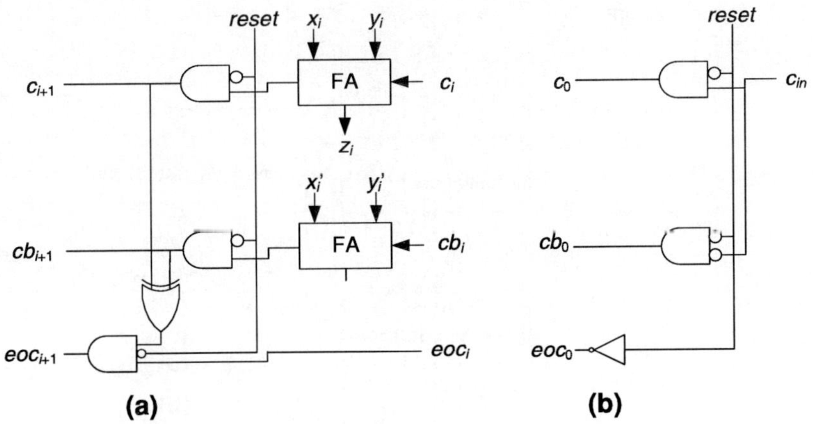

Fig. 3.17 **a** Iterative cell. **b** Initial cell

$$c_0 = c_{in}, \quad cb_0 = \overline{c_{in}},$$

$$c_{i+1} = x_i \cdot y_i + x_i \cdot c_i + y_i \cdot c_i, \quad cb_{i+1} = \overline{x_i} \cdot \overline{y_i} + \overline{x_i} \cdot cb_i + \overline{y_i} \cdot cb_i, \\ \forall i \in \{0, 1, \ldots n-1\}.$$

The end of computation is detected when

$$cb_i = \overline{c_i}, \quad \forall i \in \{0, 1, \ldots, n\}.$$

The following VHDL process describes the circuit:

```
  adder: PROCESS(reset, cy, cyb, eoc)
  BEGIN
  IF reset = '1' THEN cy <= (OTHERS => '0');
    cyb <= (OTHERS => '0'); eoc <= (OTHERS => '0');
  ELSE
    cy(0) <= c_in; cyb(0) <= NOT(c_in); eoc(0) <= '1';
    FOR i IN 0 TO n-1 LOOP
      cy(i+1) <= (x(i) AND y(i)) OR
        (x(i) AND cy(i)) OR (y(i) AND cy(i));
      cyb(i+1) <= (NOT(x(i)) AND NOT(y(i))) OR
        (NOT(x(i)) AND cyb(i)) OR (NOT(y(i)) AND cyb(i));
      eoc(i+1) <= eoc(i) AND (cy(i+1) XOR cyb(i+1));
    END LOOP;
  END IF;
END PROCESS;
z <= x XOR y XOR cy(n-1 DOWNTO 0);
c_out <= cy(n);
done <= eoc(n);
```

The corresponding circuit is shown in Fig. 3.17.

A complete model *adder_ST2.vhd* is available at the Authors' web page. In order to observe the carry chain delay, *after* clauses have been added (1 ns for

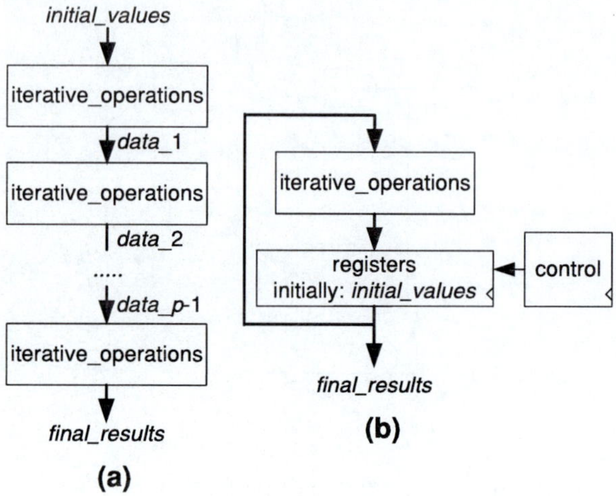

Fig. 3.18 Iterative algorithm implementation

c_{i+1} and cb_{i+1}, 0.2 ns for eoc_{i+1}). For synthesis purpose, they must be deleted.

3.2 Loop Unrolling and Digit-Serial Processing

Consider an iterative algorithm whose main operation consists of executing a procedure *iterative_operations*(*a*: *in*; *b*: *out*):

```
data0 := initial_values;
for i in 1 .. p loop
  iterative_operations(datai-1, datai);
end loop;
final_results := datap;
```

Assuming that a combinational component that implements *iterative_operations* has been previously developed, two straightforward implementations of the algorithm are shown in Fig. 3.18. The first one is an iterative combinational circuit whose cost and delay are

$$C_{combinational} = p \cdot C_{component},\ T_{combinational} < p \cdot T_{component}.$$

The second one is a sequential circuit whose main characteristics are

$$C_{sequential} = C_{component} + C_{registers} + C_{control},\ T_{clk} > T_{component},\ T_{sequentiall} = p \cdot T_{clk}.$$

An alternative option consists of a partial *unroll* of the "for" loop [1, 2]. Assume that $p = k{\cdot}s$. Then, s successive iteration steps are executed at each clock cycle.

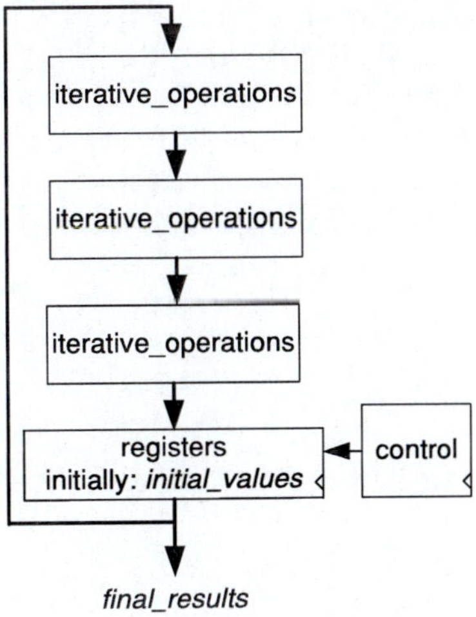

Fig. 3.19 Unrolled loop implementation ($s = 3$)

An example, with $s = 3$, is shown in Fig. 3.19. Obviously, the clock cycle, say T_{clk}', must be longer than in the sequential implementation of Fig. 3.18b (T_{clk}). Nevertheless, it will be generally shorter than $s{\cdot}T_{clk}$. On the one hand, the critical path length of s serially connected combinational circuits is generally shorter than the critical path length of a single circuit, multiplied by s. For example, the delay of an n-bit ripple-carry adder is proportional to n; nevertheless the delay of two serially connected adders, that compute $(a + b) + c$, is proportional to $n + 1$, and not to $2n$. On the other hand, the register delays are divided by s. Furthermore, when interconnecting several circuits, some additional logical simplifications can be performed by the synthesis tool. So,

$$C_{unrolled} = s \cdot C_{component} + C_{registers} + C_{control},\ T_{unrolled} = (p/s) \cdot T'_{clk},\ \text{where } T'_{clk} < s \cdot T_{clk}.$$

Example 3.6

Given two naturals x and y, with $x < y$, the following *restoring division algorithm* computes two fractional numbers $q = 0.q_{-1}\, q_{-2}\, \ldots\, q_{-p}$ and $r < y{\cdot}2^{-p}$ such that $x = q{\cdot}y + r$ and, therefore, $q \leq x/y < q + 2^{-p}$:

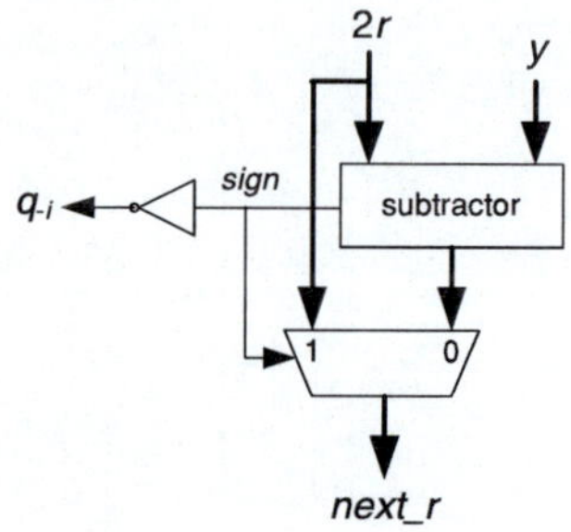

Fig. 3.20 Restoring algorithm (combinational component)

Algorithm 3.1: Restoring division algorithm

```
r0 := x;
for i in 1 .. p loop
  z := 2·ri-1 - y;
  if z < 0 then q-i := 0; ri := 2·ri-1;
  else q-i := 1; ri := z;
  end if;
end loop;
r := rp·2-p;
```

The corresponding circuit is made up of a combinational component (Fig. 3.20), a register that stores the successive remainders $r_0, r_1, \ldots, r_p$, a shift register that serially stores the quotient bits $q_{-1}, q_{-2}, \ldots, q_{-p}$, and a control unit. The combinational component can be defined as follows:

```
long_y <= '0'&y;
two_r <= r&'0';
dif <= two_r - long_y;
WITH dif(n) SELECT next_r <=
  dif(n-1 DOWNTO 0) WHEN '0',
  two_r(n-1 DOWNTO 0) WHEN OTHERS;
q_i <= NOT(dif(n));
```

A complete model *restoring.vhd* is available at the Authors' web page.

An unrolled version, with $s = 2$, is made up of a combinational component consisting of two serially connected copies of the component of Fig. 3.20, a register that stores the successive remainders $r_0, r_2, r_4, \ldots$, a shift register that stores the successive quotient bits $q_{-1}, q_{-3}, q_{-5}, \ldots$, another shift register that stores $q_{-2}, q_{-4}, q_{-6}, \ldots$, and a control unit. The combinational component can be defined as follows:

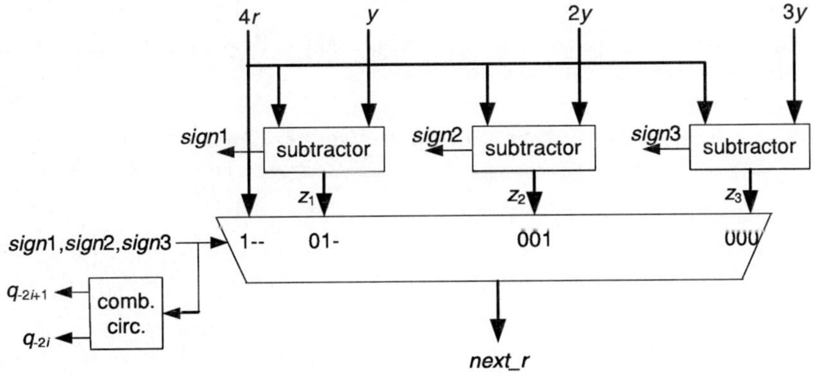

Fig. 3.21 Digit serial restoring divider ($D = 2$)

```
long_y <= '0'&y;
two_r_even <= r_even&'0';
dif_even <= two_r_even - long_y;
WITH dif_even(2*n) SELECT r_odd <=
  dif_even(2*n-1 DOWNTO 0) WHEN '0',
  two_r_even(2*n-1 DOWNTO 0) WHEN OTHERS;
two_r_odd <= r_odd&'0';
dif_odd <= two_r_odd - long_y;
WITH dif_odd(2*n) SELECT next_r <=
  dif_odd(2*n-1 DOWNTO 0) WHEN '0',
  two_r_odd(2*n-1 DOWNTO 0) WHEN OTHERS;
q_even <= NOT(dif_even(n)); q_odd <= NOT(dif_odd(n));
```

A complete model *unrolled_divider.vhd* is available at the Authors' web page.

The first implementation (*restoring.vhd*) of Example 3.6 generates one quotient bit at each step, while the second one (*unrolled_divider.vhd*) generates two quotient bits at each step. So, as regards to the quotient generation, the first implementation could be considered as *bit-serial* and the second one as *digit-serial*, defining in this case a digit as a 2-bit number. This is a common situation in arithmetic function implementation: an algorithm processes data, or part of them, in a bit-serial manner; a modified version of this initial algorithm permits the processing of several bits, described as *D*, concurrently. The second implementation is called *digital-serial* and *D* is the digit size. This technique is used in many examples during the course of this book, in order to explore cost—performance tradeoffs: small values of *D* generate cost-effective circuits, while high values of *D* yield fast circuits.

Example 3.7
Consider again a restoring divider (Example 3.6). Algorithm 3.1 is modified in order to generate two quotient bits ($D = 2$) at each step.

Algorithm 3.2: Base-4 restoring division algorithm (p even)

```
r0 := x;
for i in 1 .. p/2 loop
  z1 := 4·r2i-2 - y;
  z2 := 4·r2i-2 - 2·y;
  z3 := 4·r2i-2 - 3·y;
  if z1 < 0 then q-(2i-1) := 0; q-2i := 0; r2i := 4·r2i-2;
  elsif z2 < 0 then q-(2i-1) := 0; q-2i := 1; r2i := z1;
  elsif z3 < 0 then q-(2i-1) := 1; q-2i := 0; r2i := z2;
  else q-(2i-1) := 1; q-2i := 1; r2i := z3;
  end if;
end loop;
r := rp/2^p;
```

The corresponding circuit is made up of a combinational component (Fig. 3.21), a register that stores the successive remainders r_0, r_2, r_4,..., a shift register that stores the successive quotient bits q_{-1}, q_{-3}, q_{-5},..., another shift register that stores q_{-2}, q_{-4}, q_{-6},..., a circuit that computes $3y$, and a control unit. The combinational component can be defined as follows:

```
z1 <= four_r - long_y;
z2 <= four_r - two_y;
z3 <= four_r - three_y;
signs <= z1(n+2)&z2(n+2)&z3(n+2);
WITH signs SELECT next_r <=
  four_r(n-1 DOWNTO 0) WHEN "111",
  z1(n-1 DOWNTO 0) WHEN "011",
  z2(n-1 DOWNTO 0) WHEN "001",
  z3(n-1 DOWNTO 0) WHEN OTHERS;
  digits: PROCESS(signs)
  BEGIN
    IF signs(2) = '1' THEN q_even <= '0'; q_odd <= '0';
    ELSIF signs(1) = '1' THEN q_even <= '0'; q_odd <= '1';
    ELSIF signs(0) = '1' THEN q_even <= '1'; q_odd <= '0';
    ELSE q_even <= '1'; q_odd <= '1';
    END IF;
  END PROCESS;
```

A complete model *restoringDS.vhd* is available at the Authors' web page.

The components of Fig. 3.20 (initial restoring algorithm) and Fig. 3.21 (digit-serial restoring algorithm, with $D = 2$) have practically the same delay, namely the computation time of an n-bit subtractor, so that the minimum clock period of the corresponding dividers are practically the same. Nevertheless, the first divider needs p clock periods to perform a division while the second only needs $p/2$. So, in this example, the digit-serial approach practically divides by 2 the divider latency. On the other hand the second divider includes three subtractors instead of one.

Loop unrolling and digit-serial processing are techniques that allow the exploration of cost—performance tradeoffs, in searching for intermediate options between completely combinational (maximum cost, minimum latency) and completely sequential (minimum cost, maximum latency) circuits. Loop unrolling can be directly performed at circuit level, whatever the implemented algorithm, while digit-serial processing looks more like an algorithm transformation. Nevertheless it is not always so clear that they are different techniques.

3.3 Data Path Connectivity

In the data paths described in Chap. 2, multiplexers are associated with all the computation resource and register data inputs. With this structure, sets of operations such as $R_i := CR_j(\ldots)$ using different resources CR_j can be executed in parallel. In other words, this type of data path has *maximum connectivity*.

Assuming that the computation resources have at most p data inputs, another option is to add $p - 1$ registers $acc_1, acc_2, \ldots, acc_{p-1}$, and to realize all the data transfers with two multiplexers: the first one connects all the register outputs and external signals to the first data input of every resource as well as to every register acc_i; the second one connects the resource data outputs and the first multiplexer output to the register data inputs. With this structure, an operation such as

$$R_i := CR_j(R_0, R_1, \ldots, R_{p-1});$$

must be decomposed as follow:

$$acc_1 := R_1; acc_2 := R_2; \ldots acc_{p-1} := R_{p-1}; R_i := CR_j(R_0, acc_1, \ldots, acc_{p-1});$$

Obviously, it is no longer possible to execute several operations in parallel. On the contrary, every operation is divided up into (at most) p steps. So, this option only makes sense if one resource of each type is used.

Example 3.8
Figure 3.22 shows a *minimum connectivity* data path for the example of Sect. 2.5. The operations of the first branch ($k_{m-i} = 0$) of Algorithm 2.4 are executed as follows:

```
acc := zB;
R := xA·acc;
acc := zA;
zB := xA+acc;
zB := zB^2;
zB := zB^2;
xB := xB·acc;
xA := xA·acc ;
acc := xB;
zA := R + acc;
zA := zA^2;
acc := xB;
xB := R·acc;
R := xA^2;
acc := zA;
xA := xP·acc ;
acc := xB;
xB := xA+acc;
xA := zB;
zA := R;
zB := zA;
```

This 2-*multiplexer structure* includes ten multiplexer inputs instead of thirty-two in Figs. 2.18 and 2.19, but does not allow the concurrent execution of compatible operations. Nevertheless, in this case, the total computation time is defined by the only time consuming operations, which are the five products $R := x_A{\cdot}acc$, $x_B := x_B{\cdot}acc$, $x_A := x_A{\cdot}acc$, $x_B := R{\cdot}acc$ and $x_A := x_P{\cdot}acc$, so that the latency of the circuit is still of the order of $5t_m$, t_m being the delay of a multiplier. In conclusion, the circuit of Fig. 3.22 has practically the same computation time as that of Figs. 2.18 and 2.19, and uses less multiplexing resources.

3.4 Exercises

1. Generate VHDL models of different pipelined 128-bit adders.
2. Design different digit-serial restoring dividers ($D = 3$, $D = 4$, etc.).
3. The following algorithm computes the product of two natural numbers x and y:

```
r := y;
for i in 0 to n loop
    r := 2·r + xi·y;
end loop;
```

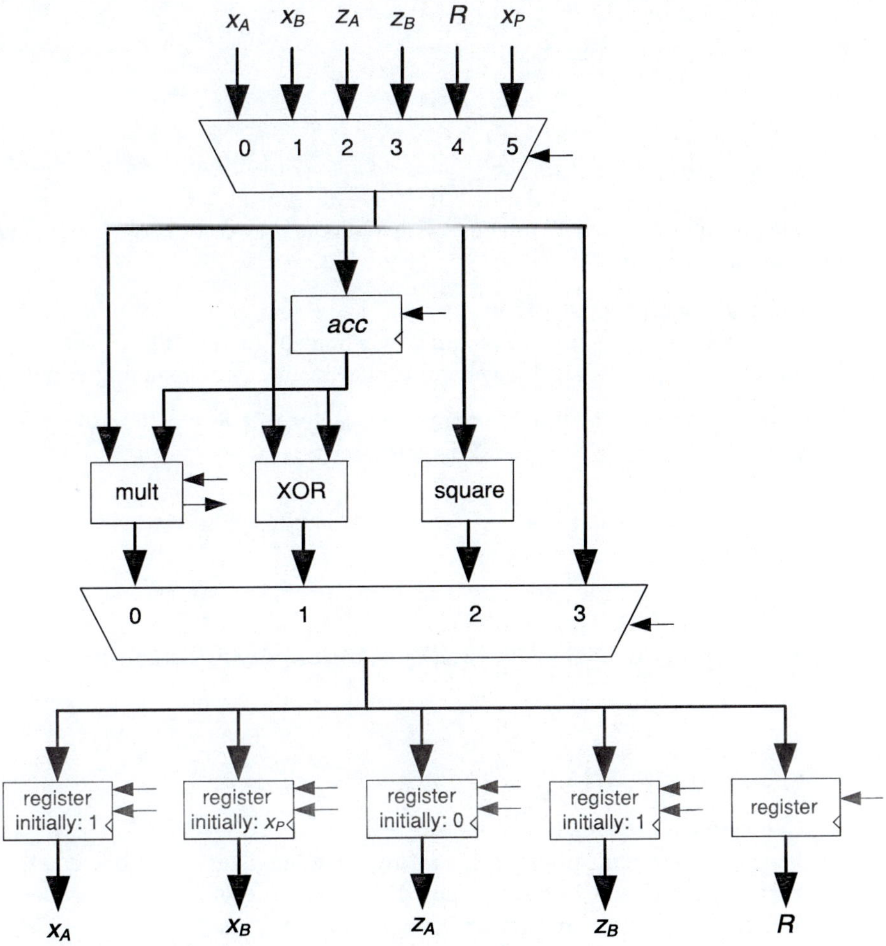

Fig. 3.22 Example of minimum connectivity data path

A computation resource *mult_and_add* that computes $2r + x_i{\cdot}y$, is available.

a. Define a combinational circuit using *mult_and_add* as a computation resource.
b. Define and compare several pipelined versions of the circuit.
c. Unroll the loop in several ways and synthesize the corresponding circuits.

4. The following algorithm divides an integer x by a natural y, where $-y \le x < y$, and generates a quotient $q = q_0{\cdot}q_1\ q_2 \ldots q_p \cong x/y$ (Chap. 9).

```
r0 := x;
for i in 1 .. p loop
  if ri-1 < 0  then ri := 2·ri-1 + y ; qi-1 := 0;
  else ri := 2·ri-1 - y ; qi-1 := 1; end if;
end loop;
q0 := 1 - q0; qp := 1; r := rp;
```

An adder-subtractor that computes $2r \pm y$, under the control of an *add/sub* variable, is available.

a. Define a combinational circuit.
b. Define and compare several pipelined versions of the circuit.
c. Unroll the loop in several ways and synthesize the corresponding circuits.

5. In the following combinational circuits, the delays of every cell and of every connection are equal to 5 ns and 2 ns, respectively.

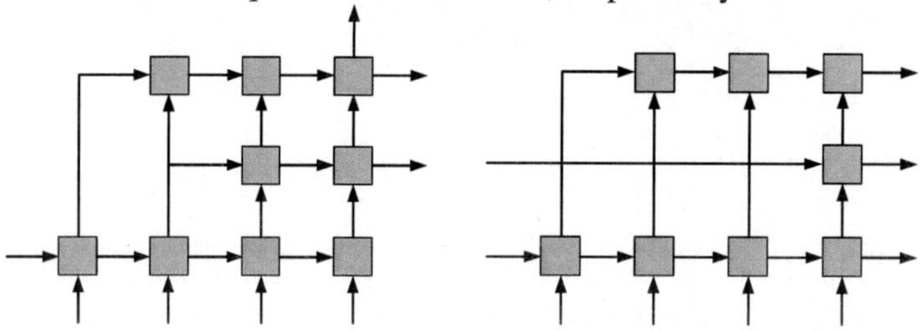

For each circuit:

a. Compute the combinational delay.
b. Segment the circuit in two stages. How many registers must be added?
c. Segment the circuit in three stages. How many registers must be added?
d. What is the maximum number of segmentation stages?
e. Assume that the cutting of a connection generates two new connections whose delays are still equal to 2 ns, and that the registers have a propagation delay of 1 ns and a setup time of 0.5 ns. Which is the maximum frequency of circuits b. and c. ?

6. The following pipelined floating-point components are available: *fpmul* computes the product in 2 cycles, *fpadd* computes the sum in 3 cycles, and *fpsqrt* computes the square root in 5 cycles, all of them with a rate $r = 1$.

a. Define the schedule of a circuit that computes the distance d between two points (x_1, y_1) and (x_2, y_2) of the (x, y)-plane.
b. Define the schedule of a circuit that computes the distance d between two points (x_1, y_1, z_1) and (x_2, y_2, z_2) of the three-dimensional space.
c. Define the schedule of a circuit that computes $d = a + ((a - b)\cdot c)^{0.5}$.
d. In every case, how many registers must be added?

References

1. Parhami B (2000) Computer arithmetic: algorithms and hardware design. Oxford University Press, New York
2. De Micheli G (1994) Synthesis and optimization of digital circuits. McGraw-Hill, New York
3. Rabaey JM, Chandrakasan A, Nikolic B (2003) Digital integrated circuits: a design perspective. Prentice Hall, Upper Saddle River

Chapter 4
Control Unit Synthesis

Modern Electronic Design Automation tools have the capacity to synthesize the control unit from a finite state machine description, or even to extract and synthesize the control unit from a functional description of the complete circuit (Chap. 5). Nevertheless, in some cases the digital circuit designer can himself be interested in performing part of the control unit synthesis. Two specific synthesis techniques are presented in this chapter: command encoding and hierarchical decomposition [1]. Both of them pursue a double objective. On the one hand they aim at reducing the circuit cost. On the other hand they can make the circuit easier to understand and to debug. The latter is probably the most important aspect.

The use of components whose latency is data-dependent has been implicitly dealt with in Sect. 2.5. Some additional comments about variable-latency operations are made in the last section of this chapter.

4.1 Command Encoding

Consider the control unit of Fig. 2.6 and assume that *commands* is an m-bit vector, *conditions* a p-bit vector and *internal_state* an n-bit vector. Thus, the command generation block generates $m + 1$ binary function of $p + n$ binary variables. Nevertheless, the number s of different commands is generally much smaller than 2^m. An alternative option is to encode the s commands with a t-bit vector, with $2^t \geq s$. The command generation block of Fig. 2.6 can be decomposed into two blocks as shown in Fig. 4.1: the first one generates $t + 1$ binary functions of $p + n$ variables, and the second one (the command decoder) m binary functions of t binary variables.

A generic circuit-complexity measure is the number of bits that a memory (ROM) must store in order to implement the same functions. Thus, the complexity of a circuit implementing $m + 1$ functions of $p + n$ variables is

J.-P. Deschamps et al., *Guide to FPGA Implementation of Arithmetic Functions*, Lecture Notes in Electrical Engineering 149, DOI: 10.1007/978-94-007-2987-2_4,

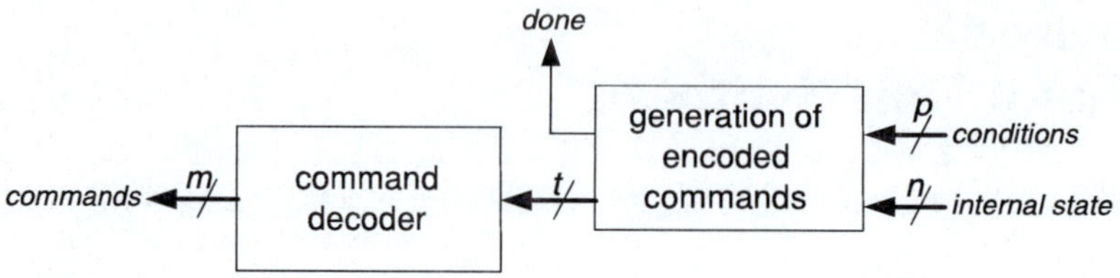

Fig. 4.1 Command encoding

$$(m + 1) \cdot 2^{p+n}\text{bits}, \tag{4.1}$$

and the total complexity of two circuits implementing $t + 1$ function of $p + n$ variables and m functions of t variables, respectively, is

$$(t + 1) \cdot 2^{p+n} + m \cdot 2^t\text{bits}. \tag{4.2}$$

Obviously, this complexity measure only takes into account the numbers of outputs and inputs of the combinational blocks, and not the functions they actually implement.

Another generic complexity measure is the minimum number of LUTs (Chap. 1) necessary to implement the functions, assuming that no LUT is shared by two or more functions. If k-input LUTs are used, the minimum number of LUTs for implementing a function of r variables is

$$\lceil (r - 1)/(k - 1) \rceil \text{LUTs},$$

and the minimum delay of the circuit is

$$\lceil log_k r \rceil \cdot T_{LUT}$$

being T_{LUT} the delay of a k-input LUT.

The complexities corresponding to the two previously described options are

$$(m + 1) \cdot \lceil (p + n - 1)/(k - 1) \rceil \text{LUTs} \tag{4.3}$$

and

$$(t + 1) \cdot \lceil (p + n - 1)/(k - 1) \rceil + m \cdot \lceil (t - 1)/(k - 1) \rceil \text{LUTs}, \tag{4.4}$$

and the delays

$$\lceil log_k(p + n) \rceil \cdot T_{LUT} \text{ and } (\lceil log_k(p + n) \rceil + \lceil log_k t \rceil) \cdot T \tag{4.5}$$

Example 4.1
Consider the circuit of Sect. 2.5 (*scalar_product.vhd*, available at the Authors' web page). The commands consist of 26 bits: eight one-bit signals

```
start_mult, load, en_xA, en_xB, en_zA, en_zB, en_R, sel_R
```

and nine two-bit signals

```
sel_p1, sel_p2, sel_a1, sel_a2, sel_sq, sel_xA, sel_xB,
sel_zA, sel_zB.
```

There are four binary conditions:

```
start, k(m-1), mult_done, count < m-1
```

and the finite-state machine has 40 states. Thus, $m = 26$, $p = 4$ and $n = 6$. Nevertheless, there are only $31 \ll 2^{26}$ different commands, namely

```
nop, first, update, start1, end1, second, third, start4,
end4, start5, end5, sixth, start7, end7, start8, end8, ninth,
start10, end10, eleventh, twelfth, start13, end13, start14,
end14, fifteenth, start16, end16, start17, end17, eighteenth,
```

that can be encoded with $t = 5$ bits.

Thus, the complexities in numbers of stored bits (4.1 and 4.2) to be compared are

$$(m+1) \cdot 2^{p+n} = 27 \cdot 2^{10} = 27{,}648 \text{ bits}, \tag{4.6}$$

$$(t+1) \cdot 2^{p+n} + m \cdot 2^{t} = 6 \cdot 2^{10} + 26 \cdot 2^{5} = 6{,}976 \text{ bits}, \tag{4.7}$$

and the complexities in numbers of LUTs (4.3 and 4.4), assuming that 4-input LUTs are used, are

$$(m+1) \cdot \lceil (p+n-1)/3 \rceil = 27 \cdot \lceil 9/3 \rceil = 81 \text{ LUTS}, \tag{4.8}$$

$$(t+1) \cdot \lceil (p+n-1)/3 \rceil + m \cdot \lceil (t-1)/3 \rceil = 6 \cdot \lceil 9/3 \rceil + 26 \cdot \lceil 4/3 \rceil = 70 \text{ LUTs}. \tag{4.9}$$

The corresponding minimum delays (4.7) are

$$\lceil log_k(p+n) \rceil = \lceil log_4 10 \rceil = 2T_{LUT}, \tag{4.10}$$

$$(\lceil log_k(p+n) \rceil + \lceil log_k t \rceil) \cdot T_{LUT} = \lceil log_4 10 \rceil + \lceil log_4 5 \rceil = 4T_{LUT}. \tag{4.11}$$

The second complexity measure (number of LUTs) is surely more accurate than the first one. Thus, according to (4.8–4.11), the encoding of the commands hardly reduces the cost and increases the delay. So, in this particular case, the main advantage is clarity, flexibility and ease of debugging, and not cost reduction.

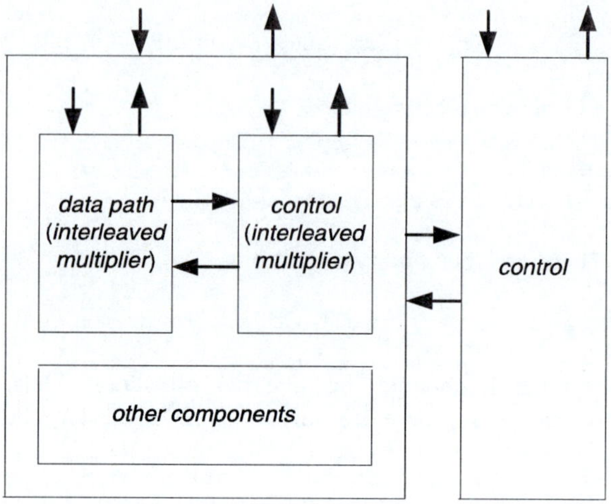

Fig. 4.2 Hierarchical circuit

4.2 Hierarchical Control Unit

Complex circuits are generally designed in a hierarchical way. As an example, the data path of the *scalar product* circuit of Sect. 2.5 (Fig. 2.18) includes a *polynomial adder* (XOR gates), a *classic squarer* and an *interleaved multiplier*, and the latter in turn consists of a data path and a control unit (Fig. 4.2). This is a common strategy in many fields of system engineering: hierarchy improves clarity, security, ease of debugging and maintenance, thus reducing development times.

Nevertheless, this type of hierarchy based on the use of previously defined components does not allow for the sharing of computation resources between several components. As an example, one of the components of the circuit of Sect. 2.5 is a polynomial adder, and the *interleaved multiplier* also includes a polynomial adder. A slight modification of the operation scheduling, avoiding executing field multiplications and additions at the same time, would allow to use the same polynomial adder for both operations. Then, instead of the architecture of Fig. 4.2, a conventional (flat) structure with a data path including only a polynomial adder could be considered. In order to maintain some type of hierarchy, the corresponding control unit could be divided up into a main control unit, in charge of controlling the execution of the main algorithm (*scalar product*) and a secondary control unit, in charge of controlling the execution of the interleaved multiplication.

Consider another, simpler, example.

Example 4.2
Design a circuit that computes

$$z = \sqrt{x^2 + y^2}.$$

The following algorithm computes z:

```
a := x^2;
b := y^2;
c := a + b;
z := square_root(c);
```

A first solution is to use three components: a squaring circuit, an adder and a square rooting circuit, for example that of Sect. 2.1. The corresponding circuit would include two adders, one for computing c, and the other within the *square_root* component (Fig. 2.3). Another option is to substitute, in the preceding algorithm, the call to *square_root* with the corresponding sequence of operations. After scheduling the operations and assigning registers to variables, the following algorithm is obtained:

```
1: r := x^2;
2: s := y^2;
3: c := r + s;
4: r := 0; s := 1;
5: r := r+1;
6: s := s + 2·r + 1;
7: greater := signb(s-c);
8: if greater = 0 then r := r+1; goto 6; else z := r; end if;
```

This algorithm can be executed by the data path of Fig. 4.3.

In order to distinguish between the main algorithm and the square root computation, the control unit can be divided up as shown in Fig. 4.4. A command decoder (Sect. 4.1) is used. There are eight different commands, namely

```
nop, r := x^2, s := y^2, c := r + s, (r, s) := (0, 1),
r := r+1, s := s + 2·r + 1, greater := signb(s - c);
```

that are encoded with three bits. The following process describes the command decoder:

```
command_decoder: PROCESS(command)
BEGIN
  CASE command IS
    WHEN "000" => sel_sq <= '0'; sel_a1 <= '0';
      sel_a2 <= "00"; cy_in <= '0'; sel_r <= '0';
      sel_s <= '0'; load <= '0'; en_r <= '0';
      en_s <= '0'; en_c <= '0'; en_signb <= '0';
    WHEN "001" => sel_sq <= '0'; sel_a1 <= '0';
      sel_a2 <= "00"; cy_in <= '0'; sel_r <= '0';
```

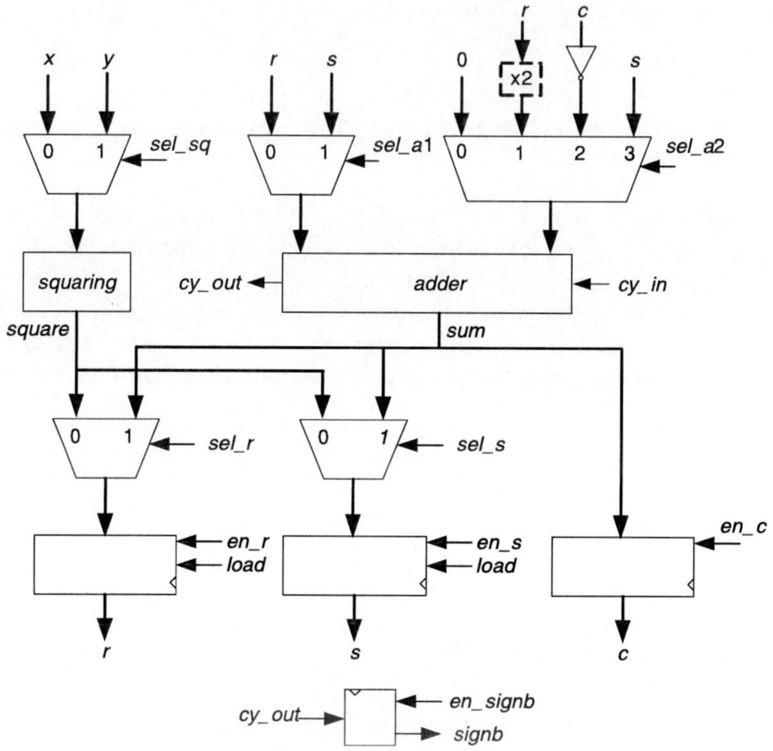

Fig. 4.3 Data path

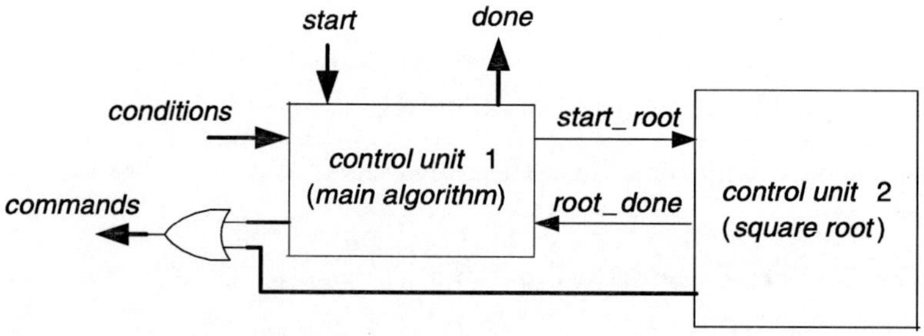

Fig. 4.4 Hierarchical control unit

```
      sel_s <= '0'; load <= '0'; en_r <= '1';
      en_s <= '0'; en_c <= '0'; en_signb <= '0';
    ...
  END CASE;
END PROCESS;
```

The two control units communicate through the *start_root* and *root_done* signals. The first control unit has six states corresponding to a "wait for start" loop, four steps of the main algorithm (operations 1, 2, 3, and the set of operations 4–8), and an "end of computation" detection. It can be described by the following process:

```
control_unit1:
PROCESS(clk, reset, current_state1, start, root_done)
BEGIN
  CASE current_state1 IS
    WHEN 0 =>
      command1 <= "000"; start_root <= '0'; done <= '1';
    WHEN 1 => IF start = '0' THEN command1 <= "000";
       ELSE command1 <= "001"; END IF;
       start_root <= '0'; done <= '1';
    WHEN 2 =>
      command1 <= "010"; start_root <= '0'; done <= '0';
    WHEN 3 =>
      command1 <= "011"; start_root <= '0'; done <= '0';
    WHEN 4 =>
      command1 <= "000"; start_root <= '1'; done <= '0';
    WHEN 5 =>
      command1 <= "000"; start_root <= '0'; done <= '0';
  IF reset = '1' THEN current_state1 <= 0;
  ELSIF clk'EVENT AND clk = '1' THEN
    CASE current_state1 IS
      WHEN 0 =>
        IF start = '0' THEN current_state1 <= 1; END IF;
      WHEN 1 =>
        IF start = '1' THEN current_state1 <= 2; END IF;
      WHEN 2 => current_state1 <= 3;
      WHEN 3 => current_state1 <= 4;
      WHEN 4 => current_state1 <= 5;
      WHEN 5 =>
        IF root_done = '1' THEN current_state1 <= 0; END IF;
    END CASE;
  END IF;
END PROCESS;
```

The second control unit has five states corresponding to operations 4, 5, 6, and 7, and "end of root computation" detection:

```
control_unit2: PROCESS(clk, reset, current_state2,
  start_root, signb)
BEGIN
  CASE current_state2 IS
    WHEN 0 => IF start_root = '0' THEN command2 <= "000";
      ELSE command2 <= "100"; END IF; root_done <= '0';
    WHEN 1 => command2 <= "101"; root_done <= '0';
    WHEN 2 => command2 <= "110"; root_done <= '0';
    WHEN 3 => command2 <= "111"; root_done <= '0';
    WHEN 4 =>
      IF signb = '0' THEN
        command2 <= "101"; root_done <= '0';
      ELSE command2 <= "000"; root_done <= '1'; END IF;
  END CASE;

  IF reset = '1' THEN current_state2 <= 0;
  ELSIF clk'EVENT AND clk = '1' THEN
    CASE current_state2 IS
      WHEN 0 => IF start_root = '1' THEN
          current_state2 <= 1; END IF;
      WHEN 1 => current_state2 <= 2;
      WHEN 2 => current_state2 <= 3;
      WHEN 3 => current_state2 <= 4;
      WHEN 4 => IF signb = '0' THEN current_state2 <= 2;
        ELSE current_state2 <= 0; END IF;
    END CASE;
  END IF;
END PROCESS;
```

The code corresponding to *nop* is 000, so that the actual command can be generated by ORing the commands generated by both control units:

```
command <= command1 OR command2;
```

A complete VHDL model *example*4_1*.vhd* is available at the Authors' web page.

Comments 4.1

- This technique is similar to the use of procedures and functions in software generation.
- In the former example, the dividing up of the control unit was not necessary. It was done only for didactic purposes. As in the case of software development, this method is useful when there are several calls to the same procedure or function.

- This type of approach to control unit synthesis is more a question of clarity (well structured control unit) and ease of debugging and maintenance, than of cost reduction (control units are not expensive).

4.3 Variable-Latency Operations

In Sect. 2.3, operation scheduling was performed assuming that the computation times t_{JM} of all operations were constant values. Nevertheless, in some cases the computation time is not a constant but a data-dependent value. As an example, the latency t_m of the field multiplier *interleaved_mult.vhd* of Sect. 2.5 is dependent on the particular operand values. In this case, the scheduling of the operations was done using an upper bound of t_m. So, an implementation based on this schedule should include "wait for t_m cycles" loops. Nevertheless, the proposed implementations (*scalar_product.vhd* and *scalar_product_DF2.vhd*) are slightly different: they use the *mult_done* flag generated by the multiplier. For example, in *scalar_product_DF2.vhd* (Sect. 2.5), there are several sentences, thus:

```
wait until mult_done = '1';
```

In an implementation that strictly respects the schedule of Fig. 2.14, these particular sentences should be substituted by constructions equivalent to

```
wait for tm cycles;
```

In fact, the pipelined circuit of Fig. 3.6 (*pipeline_DF2.vhd*) has been designed using such an upper bound of t_m. For that, a generic parameter *delta* was defined and a signal *time_out* generated by the control unit every *delta* cycles. On the other hand, the self-timed version of this same circuit (Example 3.4) used the *mult_done* flags generated by the multipliers.

Thus, in the case of variable-latency components, two options could be considered: a first one is to previously compute an upper bound of their computation times, if such a bound exists; another option is to use a *start-done* protocol: *done* is lowered on the *start* positive edge, and raised when the results are available. The second option is more general and generates circuits whose average latency is shorter. Nevertheless, in some cases, for example for pipelining purpose, the first option is better.

Comment 4.2
A typical case of data-dependent computation time corresponds to algorithms that include *while* loops: some iteration is executed as long as some condition holds true. Nevertheless, for unrolling purpose, the algorithm should be modified and the *while* loop substituted by a *for* loop including a fixed number of steps, such as *for i*

in 0 *to* $n - 1$ *loop*. Thus, in some cases it may be worthwhile to substitute a variable-latency slow component by a constant-latency fast one.

An example of a circuit including variable-latency components is presented.

Example 4.3

Consider again Algorithm 2.3, with the schedule of Fig. 2.17, so that two finite field multipliers are necessary. Assume that they generate output flags $done_1$ and $done_2$ when they complete their respective operations. The part of the algorithm corresponding to $k_{m-i} = 0$ can be executed as follows:

cycle 1: start $x_A \cdot z_B$; start $x_B \cdot z_A$; $z_B := x_A + z_A$;
cycle 2: $z_B := z_B^2$;
cycle 3: $z_B := z_B^2$;
wait until ($done_1$ and $done_2$) = 1;
cycle i: $R := x_A \cdot z_B$; $x_B := x_B \cdot z_A$;
cycle i+1: start $x_A \cdot z_A$; start $R \cdot x_B$; $z_A := R + x_B$;
wait until ($done_1$ and $done_2$) = 1;
cycle j: $x_A := x_A \cdot z_A$; $x_B := R \cdot x_B$; $z_A := z_A^2$;
cycle j+1: start $x_P \cdot z_A$; $R := x_A^2$;
wait until $done_1$ = 1;
cycle k: $x_A := x_P \cdot z_A$;
cycle k+1 : $x_A := z_B$; $z_A := R$; $x_B := x_A + x_B$; $z_B := z_A$;

The following VHDL model describes the circuit:

```
mod_f_multiplier1: interleaved_mult PORT MAP (
  A => mult1a, B => mult1b, clk => clk, reset => reset,
  start => start_mult1, Z => product1, done => done1);
mod_f_multiplier2: interleaved_mult PORT MAP (
  A => mult2a, B => mult2b, clk => clk, reset => reset,
  start => start_mult2, Z => product2, done => done2);
a_squarer: classic_squarer PORT MAP (
  a => square_in, c => square);
done12 <= done1 AND done2;
PROCESS
BEGIN
  LOOP
    WAIT UNTIL start = '0'; WAIT UNTIL start = '1';
    xA <= one; zA <= zero; xB <= xP; zB <= one;
    start_mult1 <= '0'; start_mult2 <= '0'; done <= '0';
    sync;
```

```
FOR i in 1 TO m LOOP
  IF reset = '1' THEN EXIT; END IF;
  IF k(m-i) = '0' THEN
    zB <= xA XOR zA; mult1a <= xA; mult1b <= zB;
    start_mult1 <= '1'; mult2a <= xB; mult2b <= zA;
    start_mult2 <= '1'; sync;
    start_mult1 <= '0'; start_mult2 <= '0'; sync;
    square_in <= zB; sync;
    zB <= square; sync;
    square_in <= zB; sync;
    zB <= square; sync;
    WAIT UNTIL done12 = '1';
    R <= product1; xB <= product2; sync;
    zA <= R XOR xB; mult1a <= xA; mult1b <= zA;
    mult2a <= R; mult2b <= xB; start_mult1 <= '1';
    start_mult2 <= '1'; sync;
    start_mult1 <= '0'; start_mult2 <= '0'; sync;
    WAIT UNTIL done12 = '1';
    square_in <= zA; xA <= product1; xB <= product2;
    sync;
    zA <= square; sync;
    square_in <= xA; mult1a <= xP; mult1b <= zA;
    start_mult1 <= '1'; sync;
    R <= square; start_mult1 <= '0'; sync;
    WAIT UNTIL done1 = '1';
    xA <= product1; sync;
    xB <= xA XOR xB; xA <= zB; zA <= R; zB <= zA; sync;
  ELSE

      --same operations interchanging A and B
      ....
    END IF;
  END LOOP;
  done <= '1'; sync;
  END LOOP;
END PROCESS;
```

A complete model *unbounded_DF.vhd* is available at the Authors' web page. Other implementations, using latency upper bounds and/or pipelining or self-timing, are proposed as exercises.

4.4 Exercises

1. Design a circuit that computes $z = (x_1 - x_2)^{1/2} + (y_1 - y_2)^{1/2}$ with a hierarchical control unit (separate square rooter control units, see Example 4.2).
2. Design a 2-step self-timed circuit that computes $z = (x_1 - x_2)^{1/4}$ using two square rooters controlled by a *start/done* protocol.
3. Design a 2-step pipelined circuit that computes $z = (x_1 - x_2)^{1/4}$ using two square rooters, with a *start* input, whose maximum latencies are known.
4. Consider several implementations of the scalar product circuit of Sect. 2.5, taking into account Comment 2.2. The following options could be considered:

- hierarchical control unit;
- with an upper bound of the multiplier latency;
- pipelined version;
- self-timed version.

Reference

1. De Micheli G (1994) Synthesis and optimization of digital circuits. McGraw-Hill, New York

Chapter 5
Electronic Aspects of Digital Design

This chapter is devoted to those electronics aspects important for digital circuit design. The digital devices are built with analog components, and then some considerations should be taken into account in order to obtain good and reliable designs.

Some important electronics aspects, related to the circuit design, the timing and synchronization aspects are discussed in this chapter. Most of those details are hidden in reprogrammable logic for simplicity, but these do not eliminate the consequences.

5.1 Basic Electronic Aspects of Digital Design

Digital devices represent signals by discrete bands of analog levels, rather than by a continuous range. All levels within a range represent the same signal state. Typically the number of these states is two, and they are represented by two voltage bands: one near zero volts (referenced also as ground or earth) and a higher level near the supply voltage, corresponding to the "false" ("0") and "true" ("1") values of the Boolean domain, respectively (Fig. 5.4).

5.1.1 Basic Concepts

Before we start looking at more specific ideas, we need to remember a few basic concepts.

J.-P. Deschamps et al., *Guide to FPGA Implementation of Arithmetic Functions*, Lecture Notes in Electrical Engineering 149, DOI: 10.1007/978-94-007-2987-2_5,

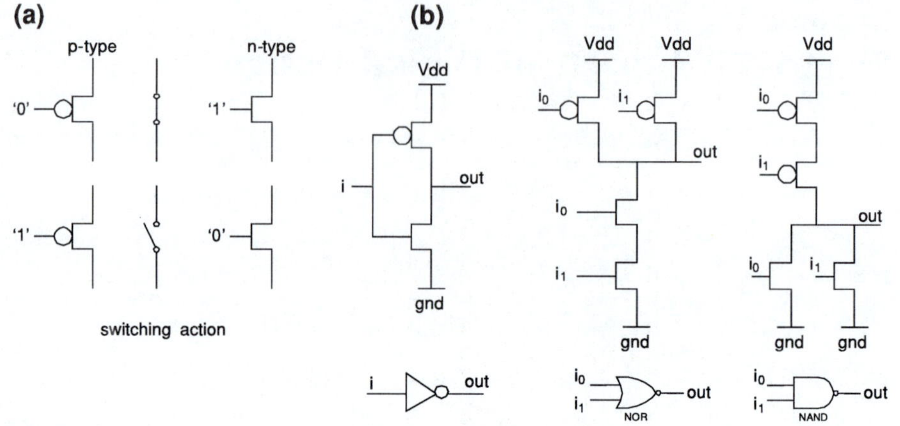

Fig. 5.1 CMOS transistors. **a** Switching action for p-type and n-type. **b** Some basic gates

5.1.1.1 CMOS Circuits

Complementary metal–oxide–semiconductor (CMOS) is the dominant technology for constructing digital integrated circuits since mid 1970s. The word "complementary" refers to the fact that the typical digital design style with CMOS uses complementary and symmetrical pairs of p-type and n-type metal oxide semiconductor field effect transistors (MOSFETs) for logic functions. The key characteristics of CMOS devices are high noise immunity and low static power consumption. Neglecting other technological details, the power is only drawn while the transistors in the CMOS device are switching between on and off states. The CMOS transistors allow creating the logic gates of Fig. 1.1 that are the basics of the digital design. The key idea is the use of transistors as switches as described in Fig. 5.1a. The p transistor conducts better the supply voltage meanwhile the n-type conducts to the ground. With this, the main principle behind CMOS circuits is the use of p-type and n-type transistors to create paths to the output from either the voltage source or ground. Figure 5.1b shows the "complementary-symmetric" interconnecttion of p-type and n-type transistors to build some basic gates. More technological details of these constructions are out of the scope of this book. One can find excellent surveys and material in [1, 2].

The FPGA technology is a CMOS (re)programmable Application Specific Integrated Circuit (ASIC). Hence, all the characteristics, peculiarities and consequences of an ASIC design are present.

5.1.1.2 Fan-in and Fan-out

In ASICs technologies and others the fan-in and fan-out are measured in capacitance units (in actual technologies in *picofarad* or *femtofarad*). So, the fan-in is the

Table 5.1 Fan-in, fan-out, internal and external delay of typical gates

	fan-in (pf)	fan-out (pf)	t_int_hl (ns)	t_int_lh (ns)	t_ext_hl (ns/pf)	t_ext_lh (ns/pf)
INV	0.003	0.337	0.079	0.151	2.710	4.891
BUF	0.004	0.425	0.265	0.056	1.334	2.399
AND2	0.003	0.334	0.105	0.144	4.470	4.271
AND3	0.007	0.673	0.211	0.131	1.362	2.376
NAND2	0.004	0.197	0.105	0.144	4.470	4.271
NAND3	0.003	0.140	0.071	0.192	6.088	7.212
NOR2	0.004	0.205	0.091	0.162	3.101	8.035
XOR2	0.008	0.645	0.279	0.331	1.435	2.560
CKBUF	0.006	1.160	0.355	0.350	0.782	1.782
CKBUF_N	0.006	1.460	0.183	0.537	0.628	1.881
DFF	0.003	0.703	0.402	0.354	1.256	2.360

capacitance that an input of a gate has. Thus, the fan-out is the maximum capacitance controllable by a gate, while providing voltage levels in the guaranteed range. The fan-out really depends on the amount of electric current a gate can source or sink while driving other gates. Table 5.1 shows examples of fan-in and fan-out for gates in 0.25 μm technologies (a ten years old technology). Observe, for example, that a NAND2 gate can drive up 49 similar gates (197*pf*/4*pf*) if we neglect the capacitance of interconnection.

In the case of FPGA the concepts of fan-in and fan-out are simplified and measured in number of connections. Remember that most of the logic is implemented in look-up tables (LUTs). Then the fan-in is the number of inputs a computing block has, like a two input AND gate implemented in a LUT has a fan-in of two, and a three input NAND gate a fan-in of three. The concept of fan-out is used to express the number of gates that each gate has connected at the output. In some cases there appears the concept of maximum fan-out as the maximum number of connections that we can control by a gate.

5.1.1.3 Drive Strength or Drive Capabilities

This means the amount of power a gate or circuit can output (i.e. same concept of fan-out). The larger drive strength a circuit has, the more power it can deliver and thus the more capable it is to drive a higher number of other gates. In the particular case of ASIC libraries, a typical gate has different drive strengths (x1, x2, x4, etc.) to be able to drive more gates. This allows the designer or the synthesizer to choose the right one at each time. In FPGAs the internal drive strength is not controlled by programmers. Nevertheless, the output strength of the output buffers can be controlled.

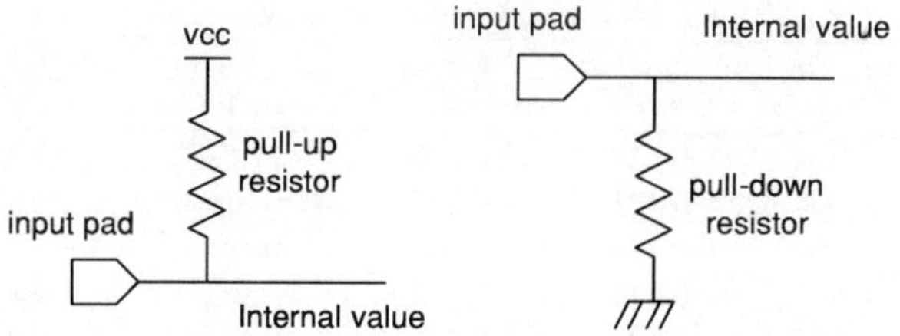

Fig. 5.2 Pull-up and pull-down resistors connected to input pads

(a)

In	G	out
0	1	0
1	1	1
-	0	Z

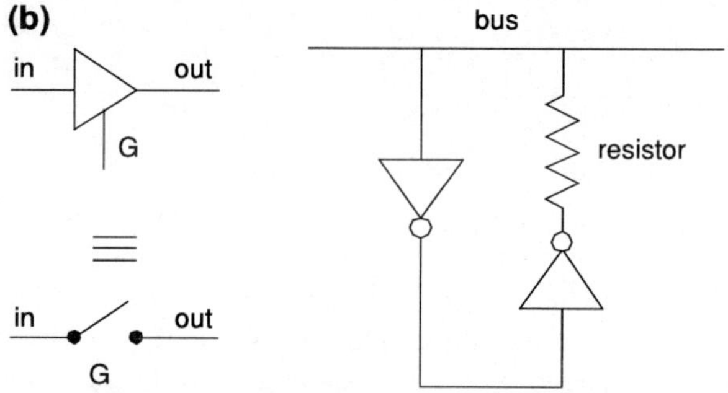

Fig. 5.3 Tri-state buffer and bus keeper

5.1.1.4 Pull-up and Pull-down Resistors

The purpose of these resistors is to force an input (or output) to a defined state. The pull-up and pull-down resistors are mainly used to avoid a circuit input from being floating if the corresponding external device is disconnected. Figure 5.2 shows a pull-up and a pull-down resistor. The pull-up resistor "weakly" pulls the internal voltage towards Vcc (logical '1') when the other components are inactive, and the pull-down resistor "weakly" pulls the internal voltage towards GND (logical '0').

Pull-up/down resistors may be discrete devices mounted on the same circuit board as the logic devices, but many microcontrollers and FGPA have internal, programmable pull-up/pull-down resistors, so that less external components are needed.

5.1.1.5 Tri-States Buffers and Bus-Keeper

Tri-state (also named as three-state or 3-state) logic allows an output port to assume a high impedance state (Z, or Hi-Z) in addition to the 0 and 1 logic levels. The tri-state buffers are used to connect multiple devices to a common bus (or line).

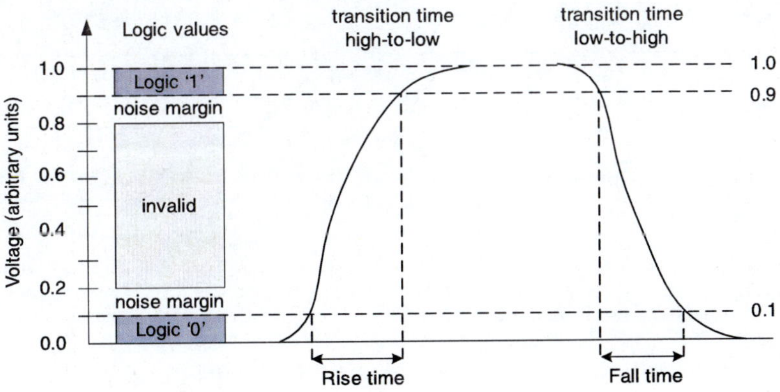

Fig. 5.4 Logic values and transition times (rise and fall time)

As is suggested in Fig. 5.3a, a tri-state buffer can be thought of as a switch. If *G* is on, the switch is closed, transmitting either 0 or 1 logic levels. If G is off, the switch is open (high impedance).

A related concept is the bus-keeper (or bus-holder). The bus-keeper is a weak latch circuit which holds the last value on a tri-state bus. The circuit is basically a delay element with the output connected back to the input through a comparatively high impedance. This is usually achieved with two inverters connected back to back. The resistor drives the bus weakly; therefore, other circuits can override the value of the bus when they are not in tri-state mode (Fig. 5.3b).

Many microcontroller and FGPA have internal, programmable tri-states buffers and the bus-keeper logic at outputs pins, so that minimal external components are needed to interface to other components.

5.1.2 Propagation Delay—Transition Time

It is the time that an electronic circuit needs to switch between two different stable states. In a logic circuit performing a change of state, it identifies the *rise-time* and the *fall-time* of the output voltage. These times are related to the times to charge and discharge capacitances. Figure 5.4 shows typical charge and discharge ramps of a CMOS gate.

5.1.2.1 Rise Time (Transition Time Low-to-High)

It refers to the time required for a signal to change from a specified low value to a specified high value. Typically, these values are 10% and 90% of the step height.

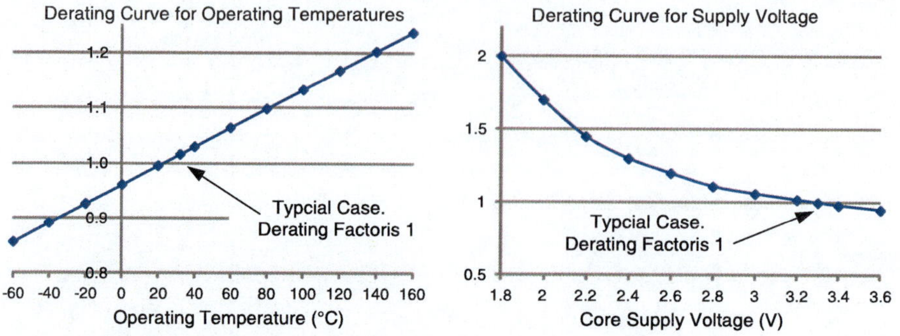

Fig. 5.5 Typical derating curves for temperature and supply voltage

5.1.2.2 Fall Time (Transition Time High-to-Low)

It is the time required for the amplitude of a pulse to decrease (fall) from a specified value, typically 90% of the peak value, to another specified value, usually 10% of the minimum value.

5.1.2.3 Slew Rate

This concept is used in linear amplifiers but also in digital circuits. In the second case, it is an approximation of the time necessary to change a logic value, without distinguishing between high-to-low and low-to-high transitions. In the FPGA arena, the output pins can be configured as slow slew rate and fast slew rate. The second one is faster, but consumes more power and is prone to transmit internal glitches to the outputs.

5.1.2.4 Propagation Delay, Intrinsic and Extrinsic Delays

The propagation delay of a gate or digital component depends on two factors, the intrinsic and the extrinsic delays. The intrinsic delay is the delay internal to the gate (also known as gate delay or internal delay). In other words, it is the time taken by the gate to produce an output after the change of an input (in ns, ps, etc.).

On the other hand, the extrinsic delay (also called, differential, load-dependent or fan-out delay) depends on the load attached to the gate. This value is expressed in ns/pf. Table 5.1 shows some intrinsic and extrinsic delays for 0.25 μm gates. The intrinsic delay in fact depends on which input is changing, but for simplicity this is not taken into account in the table. Observe that the high to low (HL) and low to high (LH) values are different. As an example, consider an inverter (INV) whose output is connected to 10 inverters. Assume that the capacitance of every

Table 5.2 A derating table for temperature and supply voltage

Vcc	Temperature (°C)								
	120	100	80	60	40	32	20	0	−20
2.6	1.39	1.36	1.32	1.27	1.24	1.20	1.19	1.15	1.12
2.8	1.29	1.25	1.22	1.18	1.14	1.11	1.10	1.07	1.03
3.0	1.23	1.20	1.17	1.12	1.09	1.06	1.05	1.02	0.99
3.2	1.18	1.15	1.12	1.08	1.05	1.02	1.01	0.98	0.95
3.3	1.16	1.13	1.10	1.06	1.03	1.00	0.99	0.96	0.93
3.4	1.14	1.11	1.08	1.04	1.01	0.98	0.97	0.94	0.91

connection is equal to 0.1 pf. What is the time required for the INV gate to propagate the transitions 0 to 1 and the 1 to 0?

$$\begin{aligned} T_{hl} &= t\_int_{hl} + t\_ext_{hl} * \text{ capacity} \\ &= 0.079\text{ns} + 2.710\text{ns/pf} * (10 * 0.003\text{pf} + 0.1\text{pf}) = 0.4313\text{ns} \end{aligned}$$

$$\begin{aligned} T_{lh} &= t\_int_{lh} + t\_ext_{lh} * \text{capacity} \\ &= 0.151\text{ns} + 4.891\text{ns/pf} * (10 * 0.003\text{pf} + 0.1\text{pf}) = 0.7868\text{ns} \end{aligned}$$

In FPGA technologies these concepts are hidden, given only a propagation delay for internal nodes (LUTs, multiplexers, nets, etc.), and do not distinguish between high to low and low to high transitions.

5.1.2.5 Timing Derating Factors

The propagation delays depend also on the temperature and the supply voltage. Variations in those parameters affect the circuit delay. The circuit manufacturers even give tables, graphics or embed the information in the static timing analyzer to adapt the timing to the operating conditions.

The derating factors are coefficients that the timing data are multiplied by, in order to estimate the delays corresponding to the operating conditions. Figure 5.5 shows typical derating curves for different operating temperature and supply voltage (the example is based on a 180 nm process). Table 5.2 is another form to show the same information.

5.1.3 Glitches in Digital Circuits

An electronics glitch is an undesired transition that occurs before the signal settles to its proposed final value. In other words, a glitch is an electrical pulse of short duration, usually unwanted, and typically produced by an imbalance in internal interconnection delays. A simple example is depicted in Fig. 5.6 where different

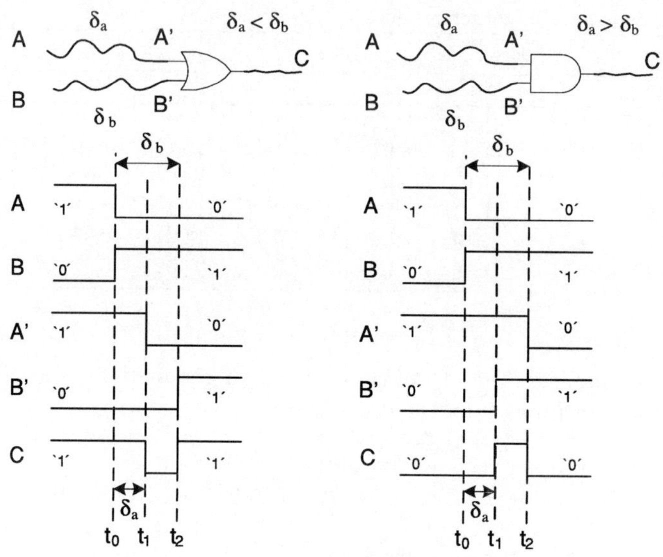

Fig. 5.6 Glitches of a simple gate due to unbalanced interconnection delay

delays in interconnections produce unnecessary changes at the output. For simplicity, in the preceding figure the gates delays are assumed to be equal to 0.

The glitch effect grows with the logic depth. As a simple example, consider two levels of XOR gates (Fig. 5.7). The four inputs change at the same time (t_0) but the net delays produce different times of arrival to the gates. In this case, changing from 1010 to 0101 should not produce any output change. Nevertheless, in the example, four transitions in output G are generated.

5.1.3.1 Runt Pulse and Spikes

A related concept to glitches is the runt pulse. It is a pulse whose amplitude is smaller than the minimum level specified for correct operation. It is a narrow pulse that, due to rise and fall times of the signal, does not reach a valid high or low level value. Typically, it has no influence on the operations, but it increases the power consumption (see Sect. 5.4).

Some authors define the concept of "spike" as being a short pulse similar to a glitch, but caused by ringing (an unwanted oscillation of a signal) or crosstalk (undesired capacitive or inductive coupling).

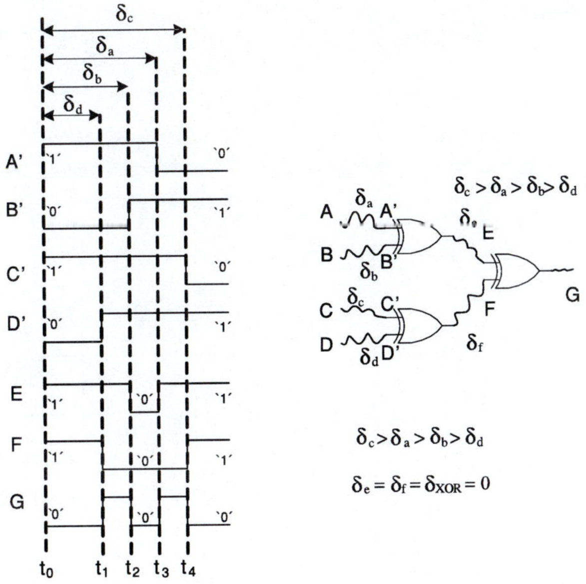

Fig. 5.7 The avalanche effect of glitches as the logic depth grows

5.2 Synchronous Design Issues

Digital circuits are mainly "synchronous". The Register Transfer Level (RTL) model recognizes register and combinational logic connected between them. The typical form, adopted to synchronize these components, is the True Single Phase Clock (TSPC). In TSPC a single clock signal is supposed to be distributed at different register levels. All the registers (flip-flops) are capturing on the rising (or falling) edge. Figure 5.8 illustrates the classical situation. A clock edge (rising in Fig. 5.8) captures the information into the first level of registers (ff_i). The contents of registers ff_i (B) are propagated into the combinational logic and generate new results. The output of the combinational circuitry (C) should be stable before the next clock arrives to the next register stage (ff_j). The following clock edge captures the content of C in ff_j and makes D available.

In an ideal synchronous circuit, every change in the logical levels of its registers (flip-flops) is simultaneous. These transitions follow the level change of the clock (*clk*) signal (positive or negative clock edge). In normal function, the input to each storage element has reached its final value before the next edge clock occurs, so the behavior of the whole circuit is exactly predictable.

Fig. 5.8 Typical single phase synchronization

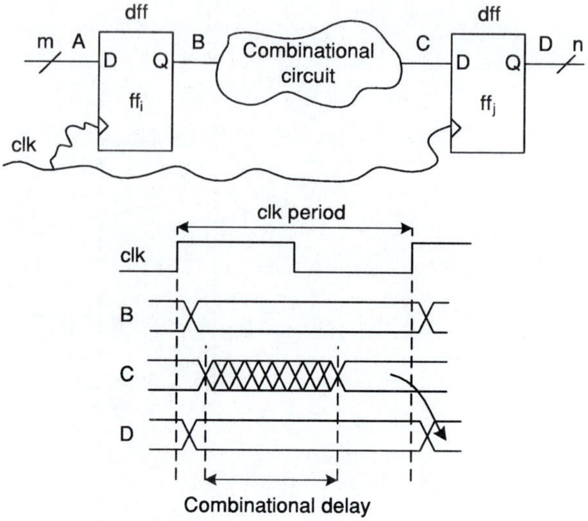

5.2.1 Edge Sensitive and Level Sensitive Registers

The edge sensitive and level sensitive registers (flip-flop and latch) are electronic circuits that have two stable states and can be used to store binary information. The circuits are made to change state by signals applied to one or more control inputs and will have one or two outputs. The latches are level sensitive circuits that pass their input D to their output Q when the clock is high (or low) (transparent mode), and the input sampled on the falling edge of the clock is held stable when the clock is low (or high)-hold mode. The flip-flops (also called edge-triggered latches) are edge sensitive circuits that sample the inputs on a clock transition (positive edge-triggered: 0 → 1 or negative edge-triggered: 1 → 0). They are built using latches (e.g., master–slave flip-flops). Historically, there have been different classes of flip-flops (FF) depending on the way they work (SR –"set-reset", D –"data", T – "toggle", and JK). Today's FPGA have registers that can be configured as latches or flip-flop, but only as a D-FF.

5.2.2 Temporal Parameters of Flip-Flops

The synchronous flip-flop requires that the data to be sampled is stable some time before and after (setup and hold times) the clock edge Fig. 5.9. If the input data changes in the setup-hold windows, metastability could occur (next section). To summarize: the *setup time* (t_{su}) is the minimum amount of time the data signal should be held steady *before* the clock event; and the *hold time* (t_h) is the minimum amount of time the data signal should be held steady after the clock event, so that

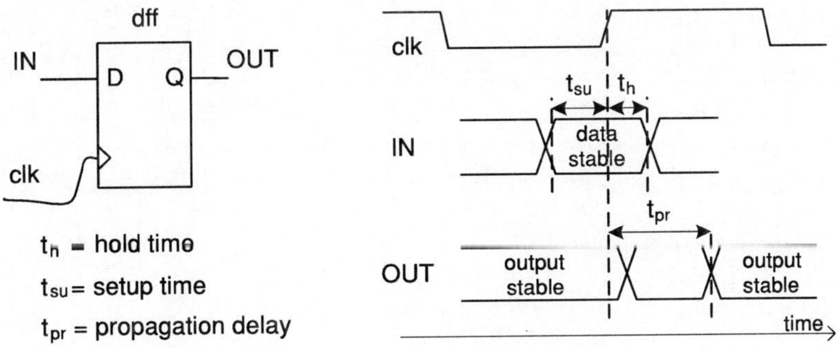

Fig. 5.9 Setup time, hold time and propagation delay of a register

the data are reliably sampled. These times are specified for any device or technology, and are typically between a few tens of picoseconds and a few nanoseconds for modern devices.

The data stored in the flip-flop is visible at its output as a *propagation delay* (t_{pr}) after the clock edge. This timing value is also known as *clock-to-output delay*.

Another related concept is the *minimum clock pulse* of a flip-flop. It is the minimum width of the clock pulse necessary to control the register.

5.2.3 Metastability

Whenever there is a setup or a hold time violation, the flip-flop could enter in a metastable state (a quasi-stable state). In this state the flip-flop output is unpredictable and it is considered as a failure of the logic design. At the end of a metastable state, when the output could reveal an "in between" value, the flip-flop settles down to either '1' or '0'. This whole process is known as metastability. Figure 5.10 illustrates this situation. The duration of the metastable state is a random variable that depends on the technology of the flip-flop. The circuit's vendors provide information about the metastability in their devices. As an example, the main vendors of FPGA provide this kind of information [3–5].

Comments 5.1

1. Not all the setup-hold window violations imply a metastable state. It is a probabilistic event.
2. Not all the metastability states cause design failures. In fact, if the data output signal resolves to a valid state before the next register captures the data, then the metastable signal does not impact negatively in the system operation.

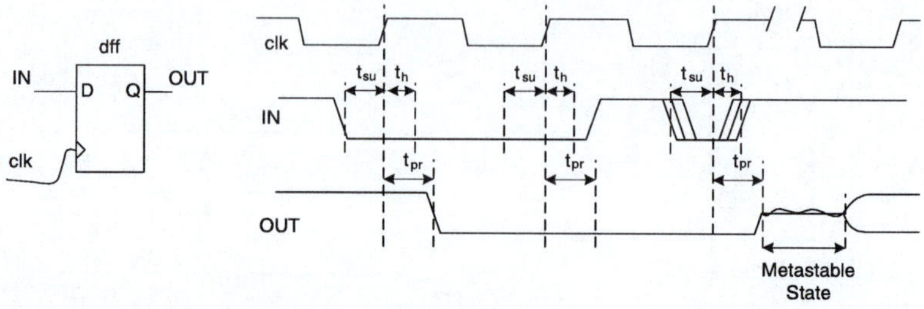

Fig. 5.10 Metastability capturing data in a flip-flop

5.2.3.1 Main Causes of Metastability

As previously mentioned, a setup or hold time violation, could produce metastability, so we have to see when signals violate this timing requirement:

- When the input signal is an asynchronous signal.
- When interfacing two domains operating at two different clock frequencies.
- When the clock skew is too high. (Sect. 5.3.1)
- When the clock slew rate is too high (rise and fall times are longer than the tolerable values).
- When interfacing two domains operating at the same frequency but with different phase.
- When the combinational delays are such that flip-flop data input change within the critical window. (setup-hold window)

Observe that that only the first two cases are real metastability problems and we will discuss possible solutions in the next sections. The other cases are related with the clock frequency and simple increases of the clock period could solve the problem. Even a static timing analysis detects these kinds of problems.

5.2.3.2 Mean Time Between Failures (MTBF) in Metastability

The Mean Time Between Failures (MTBF) is a general concept used in reliability. This concept, applied to the metastability, gives the average time interval between two successive failures due to this phenomenon. The expression that computes MTBF is:

$$MTBF = \frac{e^{T_{rec} \cdot K_2}}{K_1 \cdot f_{clk} \cdot f_{data}} \tag{5.1}$$

Where:

fclk is the frequency of the clock receiving the asynchronous signal.

fdata is the toggling frequency of the asynchronous input data signal.

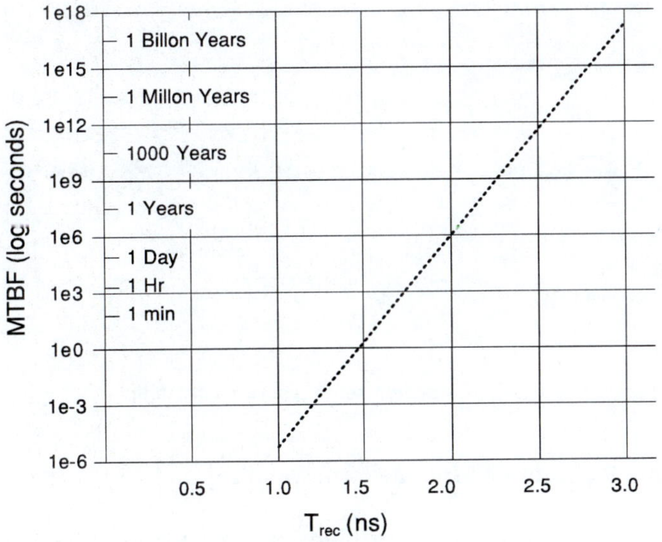

Fig. 5.11 MTBF for 300 MHz clock and 50 MHZ data in a 130 nm technology

Trec is the available metastability settling time (recovery time), or the timing until the potentially metastable signal goes to a known value '0' or '1'.

*K*1 (in *ns*) and *K*2 (in 1/*ns*) are constants that depend on the device process and on the operating conditions.

If we can tolerate more time to recover from the metastability (*Trec*) the MTBF is reduced exponentially. Faster clock frequencies (*fclk*) and faster-toggling (*fdata*) data worsen (reduce) the MTBF since the probability to have a setup-hold window violation augments. Figure 5.11 shows a typical MTBF graphic. These graphics are average data for 300 MHz clock and 50 MHz data in a 130 nm technology.

Comments 5.2

1. Observe the sensitivity to the recovery time. Following Fig. 5.11, if you are able to "wait" 2 *ns* to recover from metastability, the MTBF is less than two weeks. But if you can wait for 2.5 *ns* the MTBF is more than three thousand years.
2. Suppose for the previous data (*fclk* = 300 MHz, *fdata* = 50 MHz, T_{rec} = 2.5 *ns*) that give a MTBF of around 3200 years. But, if we have a system with 256 input bits the MTBF is reduced to 12.5 years. If we additionally produce 100,000 systems we have MTBF of 1 h!
3. Measuring the time between metastability events using real designs under real operating conditions is impractical, because it is in the order of years. FPGA vendors determine the constant parameters in the MTBF equation by characterizing the FPGA for metastability using a detector circuit designed to have a short, measurable MTBF [3, 5].

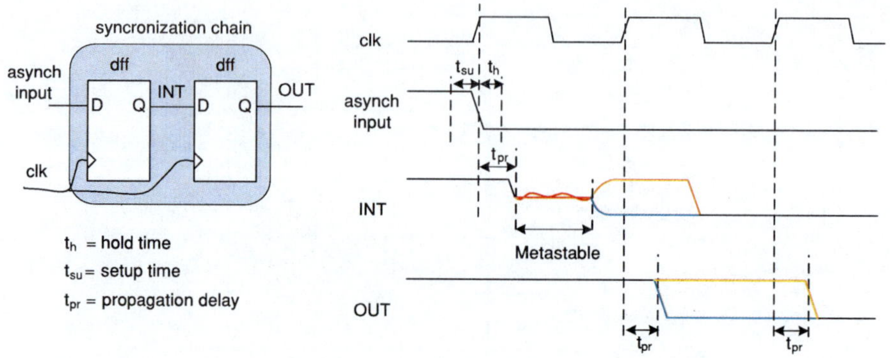

Fig. 5.12 Using a two flip-flops synchronizer to reduce metastability

5.2.3.3 How to Avoid or Mitigate Metastability

In reality, one cannot avoid metastability, without the use of tricky self-timed circuits. So a more appropriate question might be "How to mitigate the effect of metastability?"

The simplest solution is by making sure the clock period is long enough to allow the resolution of metastable states and for the delay of whatever logic may be in the path to the next flip-flop. This approach, while simple, is unpractical given the performance requirements of modern designs.

The classical solution is the use of a sequence of registers (a synchronization registers chain or synchronizer) in the destination clock domain (Fig. 5.12). The idea is to cascade one or more successive synchronizing flip-flops to the flip-flop that is interfacing with the asynchronous input.

This approach cannot guarantee that metastability cannot pass through the synchronizer; they simply reduce the probability to practical levels. In order to evaluate the effect, consider the three cases of Fig. 5.13, assuming that $t_{comb} + t_{su} \cong t_{clk}$ (combinational delay plus setup time similar to clock period). In the first case, without registering the asynchronous input, T_{rec} is near zero, raising the MTBF to an unreliable system. In the second case, with a simple flip-flop capturing the asynchronous input, $T_{rec} \cong t_{clk} - (t_{comb} + t_{su})$. In the final case, using two FFs to interface the asynchronous input, the recovery time $T_{rec} \cong t_{clk} - t_{su}$. For the data of Fig. 5.11 (i.e. $t_{clk} = 3.33$ ns) and assuming a tsu = 350 ps, the MTBF of the synchronization chain is more than 1 billon years.

5.3 Clock Distribution Network

The clock distribution network (or clock tree) distributes the clock signal(s) from a common input (an input pin) to the entire synchronous element (registers). Since this function is vital in an synchronous system, special attention is given to the

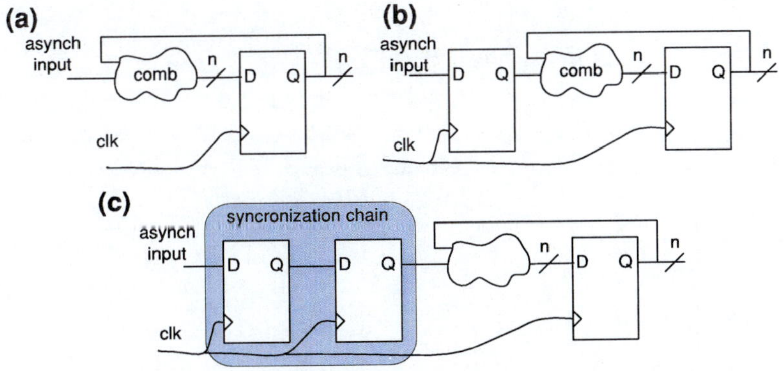

Fig. 5.13 Synchronizing asynchronous inputs in a design due to the metastability problem. **a** No registers. **b** One synchronization register. **c** Two register in synchronization chain

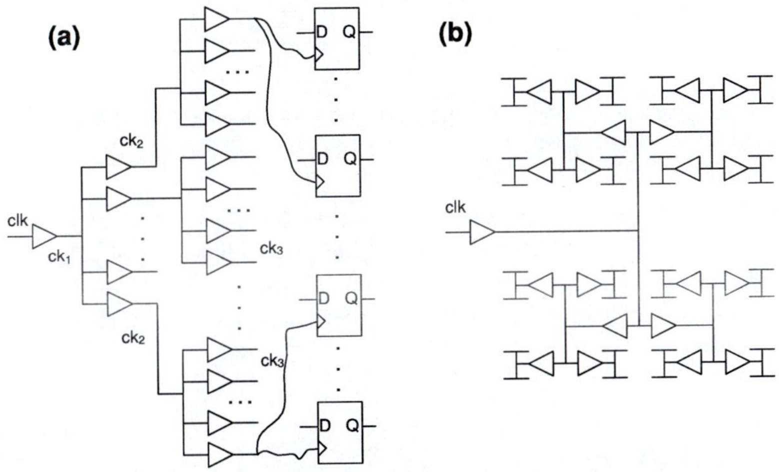

Fig. 5.14 Typical clock tree distribution. **a** Buffers tree. **b** Ideal H-routing distribution

design of this component. In ASIC design a special clock synthesizer is used (a step in physical synthesis) during the design. In FPGA several dedicated pre-fabricated clock distribution networks are present.

Clock signals are typically loaded with the greatest amount of interconnections and operate at the highest speeds of any signal in the system. Today's clock signals are easily distributed to tens of thousands of points. Due to the driving capability of the buffers, the clock distribution is designed as trees (Fig. 5.14a).

As a technological example, suppose use of the data of Table 5.1, and use the buffer BUF to distribute a clock signal. If we neglect the interconnection load (an unreal case), a simple BUF can drive 141 flip-flops (DFF) (0.425 *pf*/0.003 *pf*) and also 106 similar buffers (0.425 *pf*/0.004 *pf*). If we want to distribute the signal to

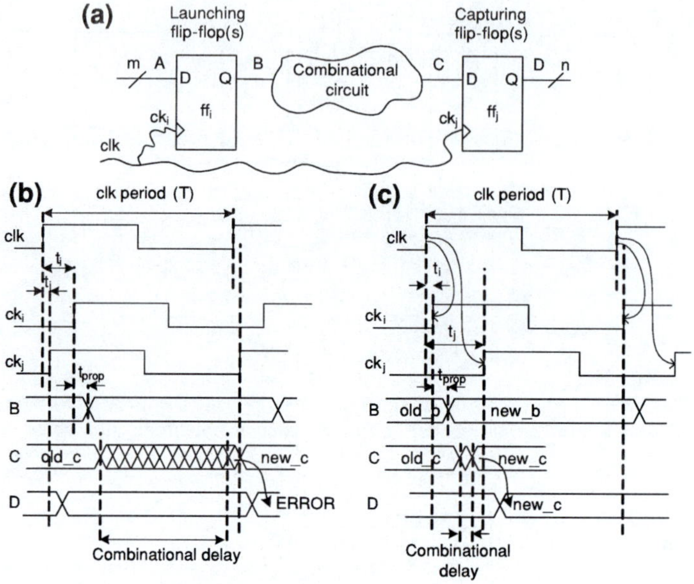

Fig. 5.15 Setup and hold violations due to clock skew. **a** Example circuit. **b** Setup violation or long-path fault. *B*. Hold violation or race-through

64,000 FF we need at least three levels of those buffers. Real clock trees use special buffers with more driving capability and with other special characteristics, and have additional problems related with the interconnections. Figure 5.14 shows a classic clock tree that present multiple levels of buffers, and the typical H-routing used to reduce the skew.

5.3.1 Clock Skew

The variation in the arrival times of a clock transition at two different locations on a chip is commonly known as the clock skew (also timing skew). Being the cause of the unbalanced delays in the clock distribution network, clock edges could arrive at clock pins ck_i and ck_j in Fig. 5.15 at different times. This spatial variation can be caused by many different things, such as wire-interconnect length, temperature variations, material imperfections, capacitance coupling, etc. The skew between two registers stages i and j could be defined as

$$skew_{i,j} = delay(ck_i) - delay(ck_i) = t_i - t_j \qquad (5.2)$$

That leads to two types of clock skew: positive skew ($skew_{i,j} > 0$) and negative skew ($skew_{i,j} < 0$). Positive skew occurs when the transmitting register (launching flip-flop, ff_i) receives the clock later than the receiving register (capturing flip-flop, ff_j). Negative skew is the opposite; the sending register gets the clock earlier than the receiving register. Two types of time violation (synchronization failures) can be caused by clock skew: setup and hold violations. They are described in what follows.

5.3.1.1 Setup Violation due to Clock Skew

If the destination flip-flop receives the clock edge earlier than the source flip-flop (positive skew), the data signal has that much less time to reach the destination flip-flop before the next clock edge. If it fails to reach the destination timeously, a *setup violation* occurs, so-called because the new data was not stable before the Tsu (setup time) of the next clock edge. This error is drawn in Fig. 5.15b and is also called *zero-clocking* (because nothing is captured properly) or also *long-path fault*.

In this case, at time t_i^{max}(assuming several FF at level i, the worst case), the clock edge arrives at flip-flops ff_i. In the worst case, to propagate through the flip-flops and the combinational logic it takes$t_{prop}^{max} + t_{comb}^{max}$. The signal must have settled down for a duration of t_{su}^{max}before the next clock edge arrives at the next flip-flop stage (ff_j) at time $t_j^{min} + T$in order to be captured properly. That transform in the following general setup time constrain:

$$t_j^{min} + T > t_i^{max} + t_{prop}^{max} + t_{comb}^{max} \tag{5.3}$$

Observe that this inequation could be always satisfied if the clock period (T) could be incremented. In other words, a positive clock skew, as shown in Fig 5.15b, places a lower bound on the allowable clock period (or an upper bound on the operating frequency) as follows:

$$T > t_{prop}^{max} + t_{comb}^{max} + t_{su}^{max} + skew_{i,j}^{max} \tag{5.4}$$

While a positive $skew_{i,j}$ decreases the maximum achievable clock frequency, a negative clock skew ($skew_{i,j} < 0$) actually increases the effective clock period. Effectively, we may have a combinational block between two adjacent flip-flops that has a propagation delay longer than the given clock period.

5.3.1.2 Hold Violation due to Clock Skew

The hold violation is caused when the clock travels more slowly than the path from one register to another (negative skew), allowing data to penetrate two registers in the same clock tick, or maybe destroying the integrity of the latched data. This is called a **hold violation** because the previous data is not held long enough at the destination flip-flop to be properly clocked through. This situation is also known as

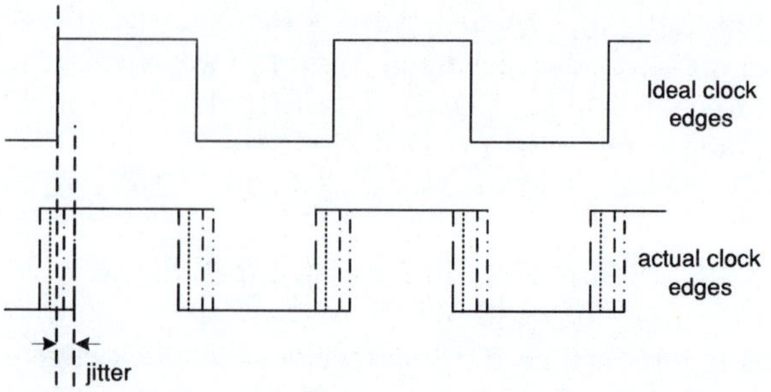

Fig. 5.16 Clock jitter example

double-clocking (because two FFs could capture with the same clock tick) or *race-through* and is common in shift registers.

To describe this situation, consider the example in Fig. 5.15c. At time t_i^{min}, the clock edge triggers flip-flops ff_i, and the signals propagates through the flip-flop and the combinational logic t_{prop}^{min} + t_{comb}^{min} (we use the shortest propagation delay, because we are considering the possibility of data racing). The input signal at ff_j has to remain stable for t_{hold}^{max} after the clock edge of the same clock cycle arrives (t_j^{max}). We are using max value since we are considering the worst case. To summarize the constraint to avoid race condition (hold violation) is:

$$t_j^{max} + t_{hold}^{max} < t_i^{min} + t_{prop}^{min} + t_{comb}^{min} \tag{5.5}$$

This can be rewritten as a constraint for the skew:

$$skew_{i,j} > t_{hold}^{max} - t_{prop}^{min} - t_{comb}^{min} \tag{5.6}$$

Observe that a hold violation is more serious than a setup violation because it cannot be fixed by increasing the clock period.

In the previous analysis we consider several launching and capturing flip-flops. If the entire FFs have the same characteristics we do not need to consider maximum or minimum setup and hold time in equations 5.3, 5.4, 5.5 and 5.6.

5.3.2 Clock Jitter

The clock edges at a flip-flop may sometimes arrive earlier and sometimes later with respect to an ideal reference clock, depending on the operating condition of the circuit. Such temporal variation in the clock period is referred to as clock jitter.

The concept of absolute jitter refers to the worst case deviation (absolute value) of the arrival time with respect to an ideal reference clock edge (Fig. 5.16).

The clock period may be shortened or lengthened by the clock jitter. For timing purposes the worst case is considered, then jitter impacts negatively in the maximum frequency of operation.

There are several sources of clock jitter. The analog component of the clock generation circuitry and the clock buffer tree in the distribution network are both significant contributors to clock jitter. There are also environmental variations that cause jitter, such as power supply noise, temperature gradients, coupling of adjacent signals, etc.

5.3.3 Clock Gating

Since the clock distribution network could take a significant fraction of the power consumed by a chip and, moreover, a large amount of power could be wasted inside computations blocks even when their output are not used (unnecessary switching activity), a typical power saving technique applied is the clock gating, which selectively switches off part of the clock tree.

Clock gating works by taking the enable conditions attached to registers, and uses them to gate the clocks (Figs. 5.17a and b). The clock gating logic can be added into a design mainly in two ways. Coding into the RTL code as enable conditions and using *automatic clock gating* by synthesis tools, or manually inserting library specific modules that implement this functionality.

Using clock gating in FPGA deserves some especial considerations. The FPGA have several low skew and low jitter dedicated clock trees; the use of this network is vital to distribute the clock and achieve high clock frequencies. In order to disable safely the clock tree, special clock buffers should be used that allow switching off the clock without glitches. Observe the potential problem to use general logic to implement gated clock in the timing diagram of Fig. 5.17c.

Today's FPGA have several clock trees and also clock buffers to disable part of it safely at different granularities (more or less logic elements). Moreover, the synthesis tools allow applying this technique automatically.

5.3.4 Clock Managers

A clock manager is a general name to describe a component that can manipulate some characteristics of the input clock. The most typical actions performed by a clock manager are:

- Delay Locked Loop (DLL), synchronizes the input clock signal with the internal clock (clock deskew, Fig. 5.18).
- Frequency Synthesis (FS): Multiply and divide an incoming clock.
- Phase Shifter (PS): a phase shift (skew) with respect to the rising edge of the input clock may be configured.
- Recondition clock signal: Reduce jitter, duty cycle correction, etc.

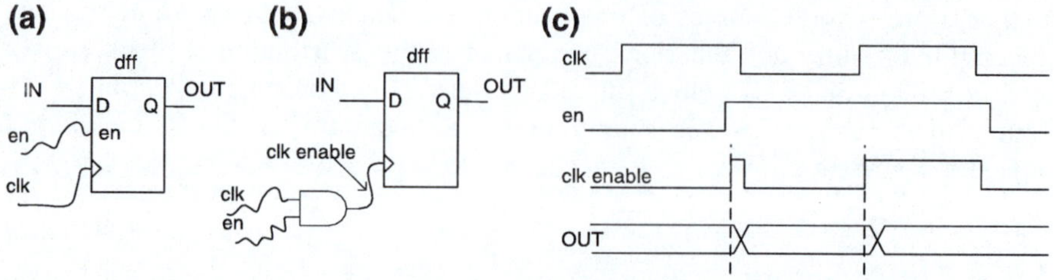

Fig. 5.17 Clock gating. **a** Register with enable. **b** Register with clock gating. **c** Clock gating risk

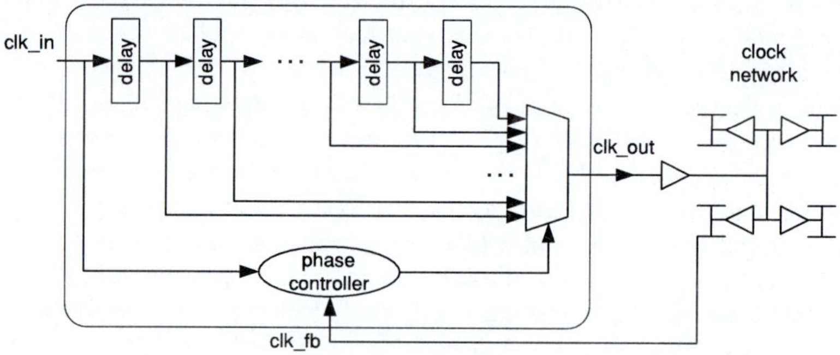

Fig. 5.18 A typical delay locked loop used to synchronize the internal clock

Today's FPGA have several of those components, for example the Xilinx Digital Clock Managers (DCM) and the Altera Phase-Locked Loops (PLL). One of the main actions performed by those components is to act as a Delay Locked Loop (DLL).

In order to illustrate the necessity of a DLL, consider the clock tree described in Sect. 5.3.1, Fig. 5.14, based on the cells of Table 5.1, that is, 64,000 FF that uses three levels of buffers. The interconnection load is neglected (an unreal case). Assume, additionally, that we decide to connect 40 buffers to the first buffer, 40 buffers at the output of each buffer of the second level, and finally 40 FFs at the output of the last 1600 buffers. We will calculate the transition time of a rising edge at input clk using the concepts of Sect. 5.1.2.4. ($T_{lh} = t\_int_{lh} + t\_ext_{lh} \cdot$ capacity). In this scenario the signal ck_1 will see the rising edge 0.44 ns later (0.056 ns + 2.399 ns/pf · (40 · 0.004pf + interconnection)). The ck_2 rising edge will be 0.44 ns after the ck_1, and finally the ck_3 will be propagated 0.344 ns after the ck_1. That is the clk signal will be at the FF 1.22 ns later (remember that for simplicity we do not consider the interconnection load). If we want to work with a clock frequency of 400 MHz (2.5 ns period) the clock will arrive half a cycle later.

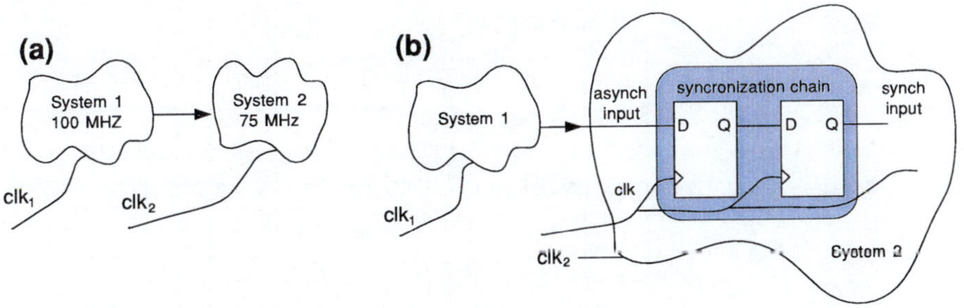

Fig. 5.19 Interfacing different clock domains using synchronization chain

5.3.4.1 Delay-Locked Loop (DLL)

A delay-locked loop (DLL) is conceptually similar to a phase-locked loop (PLL) but using a different operating principle. The main objective is to maintain the same phase of the input clock inside a device.

The main component of a DLL is a delay chain composed of many delay components. The input of the chain, and thus of the DLL, is connected to the input clock (clk_in). A multiplexer is connected to each stage of the delay chain; the selector of this multiplexer is a phase controller that compares the clock feedback (clk_fb) from the clock network and the input clock.

This circuit, after several clock cycles, ensures that the input clock rising edge is in phase with the clock feedback rising edge (offset = 360°).

5.3.5 Interfacing Different Clock Domains

Several times, a digital design needs to interface two different clock domains. This interfacing is difficult in the sense that design becomes asynchronous at the interface boundary, which could fail in setup and hold violations with the consequent metastability (see Sect. 5.2.3). Hence, particular design and interfacing techniques are necessary.

Two systems become asynchronous to each other when they operate at two different frequencies or when they operate at the same frequency, but using different and independent clock sources (Fig. 5.19a). Observe that if we use the same clock source and a multiple of the same clock (for example, divided or multiplied by 2) or a phase shifting of the clock (180°) they do not necessarily become asynchronous.

Synchronization failure is lethal and difficult to debug, so it is important to take care about this issue. If we have two systems, asynchronous to each other, and we need to transfer data between them, several methods can be considered:

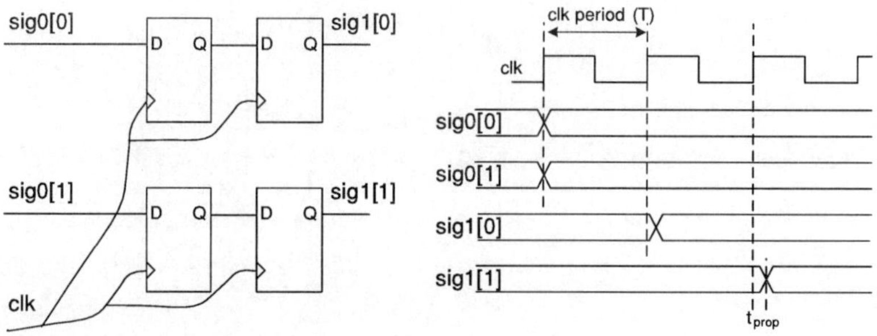

Fig. 5.20 Synchronization problems with buses

- Using synchronizer.
- Handshake signaling method.
- Asynchronous FIFO.
- Open loop communication.

5.3.5.1 Using Synchronizer

It is the simplest method, useful only for very low speed communication. As described in Sect. 5.2.3, a simple method to mitigate metastability is the use a cascade of flip-flops. Fig. 5.19b shows the simple solution using a synchronization chain (synchronizer).

This method could lead to errors when it is used to synchronize a bus. As a simple example consider Fig. 5.20. If the input bus changes at the capturing edge of the clock, different paths and different flip-flops could react differently, making a misalignment of the data in the bus. The answer to this problem is the use of a handshaking protocol.

Comments 5.3

A related synchronization pitfall could occur if a single bit is synchronized in different places. The same problem described for buses in Fig. 5.20 could appear.

5.3.5.2 Handshake Signaling

In this method system 1 sends data to system 2 based on the handshake signals *req* (request) and *ack* (acknowledge). A simple handshaking protocol (known as 4-phase) will work as follow:

- Sender outputs data and asserts *req*.
- Receiver captures data and asserts *ack*.
- Sender, after *ack* is detected, deasserts *req*.
- Receiver sees *req* deasserted, deasserts *ack* when ready to continue.

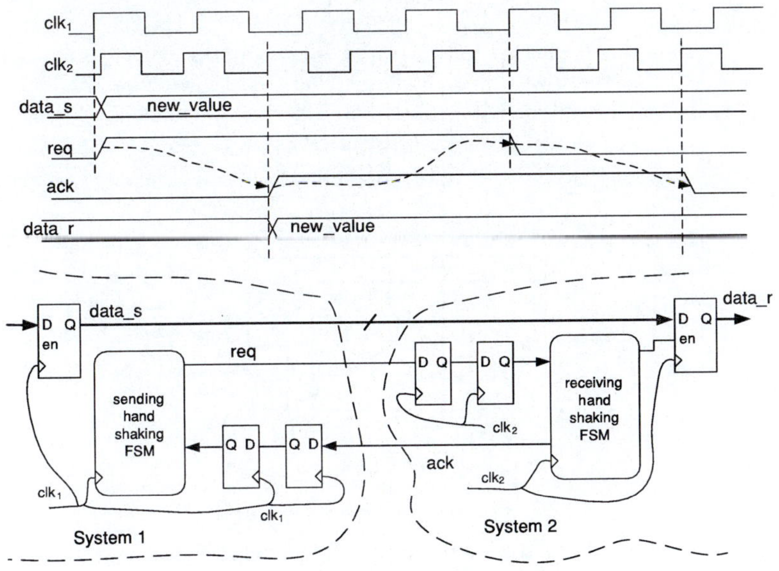

Fig. 5.21 Handshaking protocol using a two stage synchronizer

This method is straightforward, but the metastability problems could be present. In fact, when system 2 samples system 1's *req* and system 1 samples system 2's *ack* line, they use their internal clock, so setup and hold time violations could arise. To avoid this, we can use double or triple stage synchronizers, which increase the MTBF and thus reduce the metastability to a good extent. Figure 5.21 shows a handshaking protocol using a two stage synchronizer in *req* and *ack* lines.

As shown in the example, the use of double or triple stage synchronizing reduces significantly the transfer rate, due to the fact that a lot of clock cycles are wasted for handshaking.

However, a simpler handshaking is possible (2-phase or edge based). In this, the sender outputs data and changes the state of *req*; it will not change the state of *req* again until after the state of *ack* changes. The receiver latches data; once the receiver is ready for more, it changes the state of *ack*. This method (2-phase) requires one bit of state to be kept on each side of the transaction to know the *ack* state. Additionally, a reliable reset is necessary to start the synchronization.

Handshaking works great, but reduces the bandwidth at the clock crossing interface because many cycles of handshaking are wasted. The high bandwidth solution that maintains reliable communication is the use of asyncronous FIFOs.

5.3.5.3 Asynchronous FIFO

An asynchronous FIFO (First In First Out) has two interfaces, one for writing the data into the FIFO and the other for reading the data out (Fig. 5.22a). Ideal dual

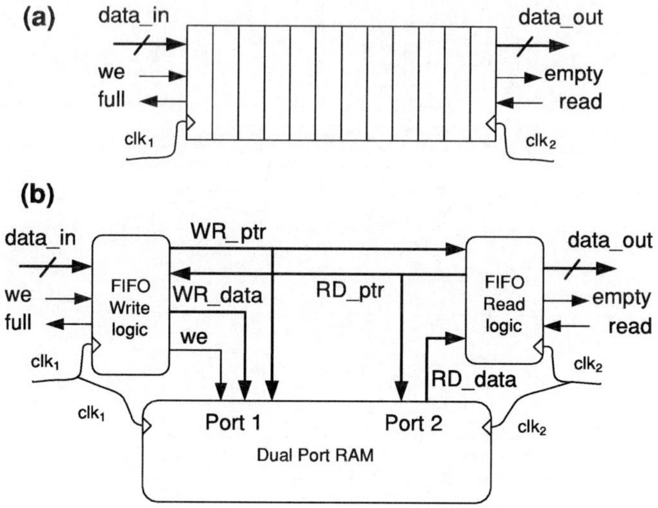

Fig. 5.22 Asynchronous FIFO. **a** Abstract FIFO design. **b** Detailed view

port FIFOs write with one clock, and read with another. The FIFO storage provides buffering to help rate match between different frequencies. Flow control is needed in case the FIFO gets totally full or totally empty. These signals are generated with respect to the corresponding clock. The *full* signal is used by system 1 (when the FIFO is full, we do not want system 1 to write data because this data will be lost or will overwrite an existing data), so it will be driven by the write clock. Similarly, the *empty* signal will be driven by the read clock.

FIFOs of any significant size are implemented using an on-chip dual port RAM (it has two independent ports). The FIFO is managed as a circular buffer using pointers. A write pointer to determine the write address and a read pointer to determine the read address are used (Fig. 5.22b). To generate *full*/*empty* conditions, the write logic needs to see the read pointer and the read logic needs to see the write pointer. That leads to more synchronization problems (in certain cases, they can produce metastability) that are solved using synchronizers and Gray encoding.

Comments 5.4

1. Asynchronous FIFOs are used at places where the performance matters, that is, when one does not want to waste clock cycles in handshake signals.
2. Most of today's FPGAs vendors offer blocks of on-chip RAMs that can also be configured as asynchronous FIFOs.
3. A FIFO is the hardware implementation of a *data stream* used in some computation models.

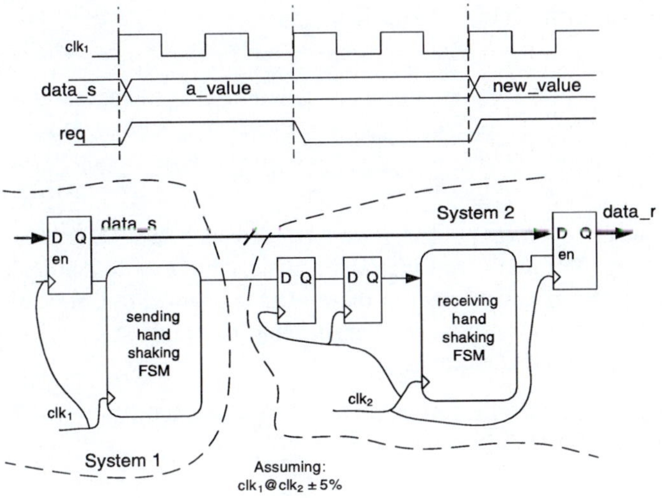

Fig. 5.23 Open loop communication (mesosynchronous)

5.3.5.4 Open Loop Communication

When two systems of bounded frequency need to communicate, open loop synchronization circuits can be used (also known as *mesosynchronous designs*). In this approach no *ack* signal is used)

The benefits of mesosynchronous designs are in less synchronization based circuitry and are of a lower latency in contrast to a 4-phase handshaking. The main drawback is that this method only works at certain frequency ratios (clk1 $\cong$ clk2 $\pm$ 5%). Figure 5.23 shows this synchronization method assuming that the sender holds two cycles of the *req* signal.

5.4 Power Consumption

Power dissipation is one of the main concerns in today's digital design. For portable devices the battery life is essential, the amount of current and energy available in a battery is nearly constant and the power dissipation of a circuit or system defines the battery life. On the other hand, the heat generated is proportional to the power dissipated by the chip or by the system; an excessive heat dissipation may increase the operating temperature and thus degrades the speed (see Sect. 5.1.2.5, derating factor) and causes circuitry malfunctions. The necessary cooling systems (fans, heatsinks, etc.) when excessive power is used, increase the total system cost. Additionally, the life of the circuitry is typically shortened when working at higher temperatures (electromigration and others).

5.4.1 Sources of Power Consumption

CMOS gates are considered very power efficient because they dissipate nearly zero power when idle. As the CMOS technology moved below sub-micron size this assumption became relative since the static power is more important. The power dissipation in CMOS circuits occurs because of two components:

- Static power dissipation. Is the power consumed when the output or input are not changing. The main contributions are the subthreshold and leakage current.
- Dynamic power dissipation. It is the power consumed during state transitions. The two main components are charging and discharging of load capacitances and the short circuit dissipation.

Thus the total power dissipation is the addition of the static power dissipation plus the dynamic power dissipation.

5.4.1.1 Static Power Consumption

In the past, the subthreshold conduction of transistors has been very small, but as transistors have been scaled down, leakages from all sources have increased. Historically, CMOS designs operated at supply voltages much larger than their threshold voltages (V_{dd} 5 V, and V_{th} around 700 mV) and the static power consumption was negligible. As transistor size is reduced, i.e. below 90 nm, the static current could be as high as 40% of the total power dissipation. The static power consumption reduction is one of the main concerns in today's technological development of new CMOS processes.

5.4.1.2 Dynamic Power Consumption

CMOS circuits dissipate power mainly by charging the different load capacitances (gates, wire capacitance, etc.) whenever they are switching (Fig. 5.24). In one complete cycle of CMOS logic, current flows from power supply V_{dd} to the load capacitance (0 to 1 transition) to charge it and then flows from the charged load capacitance to ground during discharge (1 to 0 transition). Therefore in one complete charge/discharge cycle, a total load $Q = C \cdot V_{dd}$ is thus transferred from V_{dd} to ground. If we multiply by the switching frequency to get the current, and multiply again to supply voltage we obtain the power dissipated for a gate as: $P = f \cdot C \cdot V_{dd}^2$ Some authors recognize the transition power (the power per transition) as $P = 1/2 \cdot f \cdot C \cdot V_{dd}^2$ (half power charging, half the power discharging). Since most gates do not operate/switch at the clock frequency, they are often accompanied by a factor α, called the activity factor (also switching activity). Now, the dynamic power dissipation may be re-written as $P = \alpha \cdot f \cdot C \cdot V_{dd}^2$.

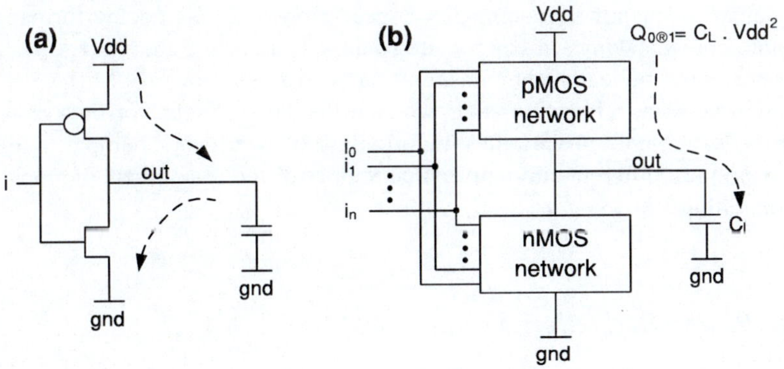

Fig. 5.24 Dynamic power consumption. **a** An inverter. **b** A general CMOS gate charging a capacitance

A clock in a system has an activity factor $\alpha = 1$, since it rises and falls every cycle. Most data has an activity factor lower than 0.5, i.e. switches less than one time per clock cycle. But real systems could have internal nodes with activity factors grater then one due to the glitches.

If correct load capacitance is calculated on each node together with its activity factor, the dynamic power dissipation at the total system could be calculated as:

$$P = \sum_i f \cdot (\alpha_i \cdot c_i) \cdot V_{dd}^2 \tag{5.7}$$

Another component in the dynamic component is the short circuit power dissipation. During transition, due to the rise/fall time both pMOS and nMOS transistors will be on for a small period of time in which current will find a path directly from V_{dd} to gnd, hence creating a short circuit current. In power estimation tools this current also can be modeled as an extra capacitance at the output of the gates.

5.4.1.3 Power and Energy

Power consumption is expressed in *Watts* and determines the design of the power supply, voltage regulators and the dimensions of the interconnections and eventually short time cooling. Moreover, the energy consumed is expressed in *Joules* and indicates the potency consumed over time, as shown in Equation 5.8.

$$\text{Energy} = \text{power} * \text{delay (joules} = \text{watts} * \text{seconds)} \tag{5.8}$$

The energy is associated with the battery life. Thus, less energy indicates less power to perform a calculation at the same frequency. Energy is thus independent of the clock frequency. Reducing clock speed alone will degrade performance, but will not achieve savings in battery life (unless you change the voltage).

That is why, typically, the consumption is expressed in mW/Mhz when comparing circuits and algorithms that produce the same amount of data results per clock cycle, and normally nJoules (nanoJoules) are used when comparing the total consumption of different alternatives that require different numbers of clock cycles for computing.

5.4.2 Power Reduction Techniques

The power reduction could be tackled at different abstraction levels. At higher abstraction levels, bigger possibilities to reduce the power exist.

The static power consumption is reduced mainly at a technological level and is beyond the scope of this book. Excellent surveys and trends are in [6] and [7].

To a designer of digital systems, Equation 5.7 gives information about the main sources of reduction. The quadratic influence of the power supply is an interesting source of power reduction. The lower supply voltage the device employs, the lower power the systems will use. Today's systems are supplied at different voltages for cores and pads in order to reduce, when possible, the power budget.

On the other hand, the operation frequency (f), the activity (α), and the capacitance (c) are linear sources of power reduction.

At circuit level, the reduction of glitches, data reordering to reduce activity, or the fan-out reduction are typical techniques. Parts of these ideas are present in the automatic synthesis for low power provided by EDA tool vendors.

At algorithm level, the complexity and the regularity of algorithms, the data representation, and other techniques, offer a big scope for optimizations. The concept of energy and power is useful in this case (Sect. 5.4.1.3); a faster algorithm consuming the same power will consume less energy.

At system level, the partition of the system allows to apply dynamic power management, that is, applying sleep modes (shut down) to part of or to the complete system. These power management techniques could be applied manually by the designer or partially automated by the synthesis tools.

The power consumption is an important issue for today's FPGA vendors. Compared to ASICs, FPGAs are not power-efficient components because they use a larger amount of transistors to provide programmability on the chip (typically an order of one magnitude more power). That leads to some specific FPGAs called "low power FPGA families", optimized for power dissipation at the cost of performance. FPGA vendors use different transistors, trading off speed versus power in their devices, and also the availability to shut off the unused components. They also offer power estimation tools and some optimization tools at synthesis level.

5.4.3 Power Measurement and Estimation

The power measurement is based on the measure of the supply voltage and the instant current (P = I * V). Some systems have supply sources that measure the current and allow measurement of the power. There also are other methods to measure the total energy consumed, but they are out of scope of the present discussion.

For power estimation, there are specific tools either for ASIC and FPGA. Some tools use estimated activities for different blocks; others are based on post place and route simulation to obtain the "exact" activity. The power estimation tool uses this activity information in Equation 5.7 to calculate the dynamic power consumption (more details are given in Chap. 6). The main drawback is the possibility to simulate a real situation to obtain the "real" activity of a circuit.

5.5 Exercises

1. Redraw Fig. 5.5 (glitches for unbalanced paths) considering: (a) NAND gate, (b) NOR gate, (c) XOR gate.
2. Redraw Fig. 5.6 (cascade effect of glitches paths) considering the transition (a) from 0000 to 1111; (b) from 0100 to 1101; (c) from 0101 to 1101;
3. Supposing one has a simple ripple carry adder (Fig. 7.1) of four bits. Analyze the glitch propagation when changing to add 0000 + 0000 to 1111 + 1111.
4. What is the rise time and fall time of an AND2 gate that connects at his output four XOR2 gates? Assume a total interconnection load of 12 pf (use data of Table 5.1).
5. How many DFF can drive a CKBUF assuming the unreal case of no interconnection capacitance? Which is the propagation delay of the rising edge?
6. Assuming the derating factors of Table 5.2, what is the delay of exercise 4 and 5 for 80°C and a supply voltage of 3.0 V.
7. Determine the MTBF for $K_1 = 0.1$ ns, $K_2 = 2$ ns^{-1}; with clock frequency of 100 MHz and data arrival at 1 MHz, for recovery time of 1, 5, 10 and 20 ns.
8. What is the MTBF expected for an asynchronous input that uses two synchronization flip-flops working at 100 MHz and using the data of the previous problem? The FFs have a setup time of one ns.
9. What is the delay of the falling edge with the data used in Sect. 5.3.4. i.e. 64,000 FF that uses three levels of BUF, neglecting the interconnection load?
10. Calculate the level of clock buffers (CKBUF) necessary to control 128,000 registers (DFF) of Table 5.1. Suppose additionally that any interconnection has a load of 0.006 pf.
11. For the previous clock tree. Determine the propagation delay from the input clock signal to the clk input of a FF.

12. Draw a timing diagram of a two stage handshaking protocol. Assume that the sending clock is faster than the receiving clock.
13. For the circuit of the figure, suppose that the FF has a propagation delay between 0.9 and 1.2 ns, a setup time between 0.4 and 0.5 ns and a hold time between 0.2 and 0.3 ns.

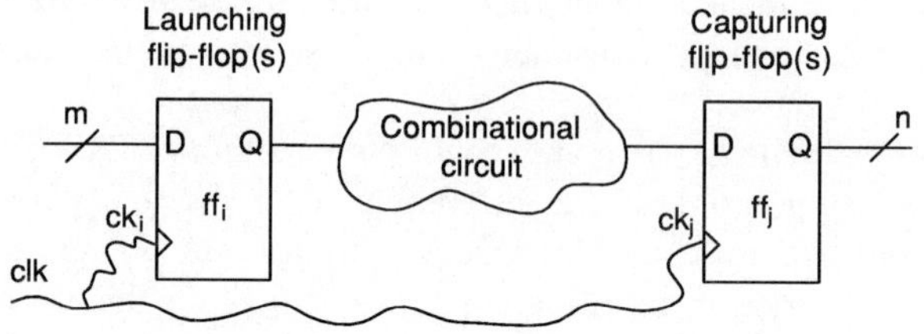

The clock arrives to the different FF of level i with a delay between 2.1 ns and 3.3 ns, and to level j with delays between 2.5 ns and 3.9 ns. What is the maximum combinational delay acceptable to work at 100 MHz?
14. Using the data of the previous exercise, what is the minimum combinational delay necessary to ensure a correct functionality?
15. A system 'A' works with a supply voltage of 1.2 V and needs 1.3 mA during 10 s to perform a computation. A second system 'B' powered at 1.0 V consumes an average of 1.2 mA and needs 40 s to perform the same task. Which consumes less power and energy?
16. In the shift register of the figure, assuming that all the flip-flops have a propagation delay of 0.9 ns, a setup time of 0.3 ns and a hold time of 0.2 ns, what is the maximum skew tolerated if the interconnection has a delay ($\delta 1$ and $\delta 2$) of 0.1 ns?

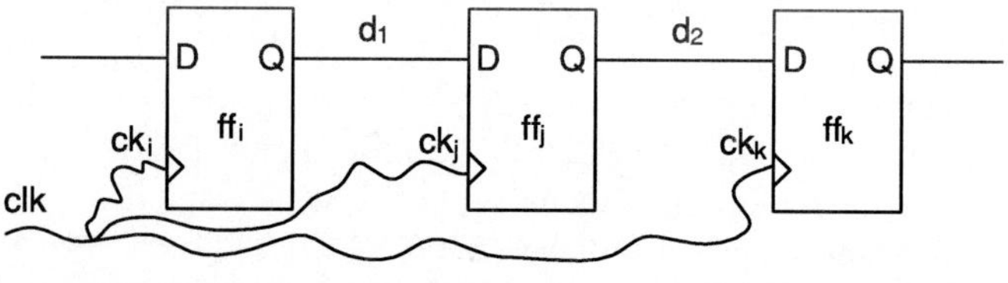

17. For the previous problem. What is the maximum frequency of operation?
18. The following FF (with same temporal parameters as in exercise 16) is used to divide the clock frequency. What is the minimum delay δ of the inverter and interconnection necessary for it to work properly?

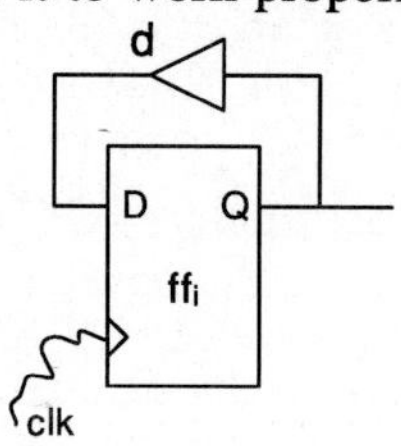

References

1. Rabaey JM, Chandrakasan A, Nikolic B (2003) Digital integrated circuits, 2nd edn. Prentice-Hall, Englewood Cliffs
2. Wakerly JF (2005) Digital design principles and practices, 4th edn. Prentice-Hall, Englewood Cliffs. ISBN 0-13-186389-4
3. Xilinx Corp (2005) XAPP094 (v3.0) Metastable Recovery in Virtex-II Pro FPGAs, XAPP094 (v3.0). http://www.xilinx.com/support/documentation/application_notes/xapp094.pdf
4. Actel Corp (2007) Metastability characterization report for Actel antifuse FPGAs; http://www.actel.com/documents/Antifuse_MetaReport_AN.pdf
5. Altera corp (2009) White paper: Understanding metastability in FPGAs. http://www.altera.com/literature/wp/wp-01082-quartus-ii-metastability.pdf
6. Pedram M (1996) Tutorial and survey paper—Power minimization in IC design: principles and applications. ACM Trans Design Autom Electron Syst 1(1):3–56
7. Rabaey JM (1996) Low power design methodologies. Kluwer Academic Publishers, Dordrecht

Chapter 6
EDA Tools

The Electronic Design Automation (EDA), is a group of software tools for designing electronic systems such as integrated circuit (ASICs), printed circuit boards (PCBs), or reprogrammable hardware as FPGA, etc. The general ideas of EDA tools and the particular for FPGA designs will be discussed in this section. Typically these tools work in a design flow that hardware and system designers use to design and analyze entire system behavior. This chapter explains the main concepts related to the EDA tools and presents an example using Xilinx ISE and Altera Quartus tools.

6.1 Design Flow in FPGA EDA Tools

The design flows are the combination of EDA tools to carry out a circuit or system design. Current digital flows are very modular and are the consequence of the evolution of the standalone tools for synthesis, placement, and routing.

Naturally, the tools are evolving, driven by Moore's law from standalone tools to integrated construction and analysis flows for design closure.

The FPGA design flow is a kind of simplification of the ASIC design flow and we will concentrate on it. The FPGA vendors groups the EDA tools in "design suites" such as Libero from Actel [5, 7] or Quartus II from Altera [2]. Figure 6.1 shows a typical design flow; in what follows we will describe the different main stages.

The design flow can be used as command-line executables or scripting, or by using the GUI (*graphical user interface*). Figures 6.5 and 6.14 shows the GUI aspects for Xilinx ISE and Altera Quartus II, respectively. The graphical interface is the preferred design entry for newcomer designers and for small projects.

J.-P. Deschamps et al., *Guide to FPGA Implementation of Arithmetic Functions*, Lecture Notes in Electrical Engineering 149, DOI: 10.1007/978-94-007-2987-2_6,

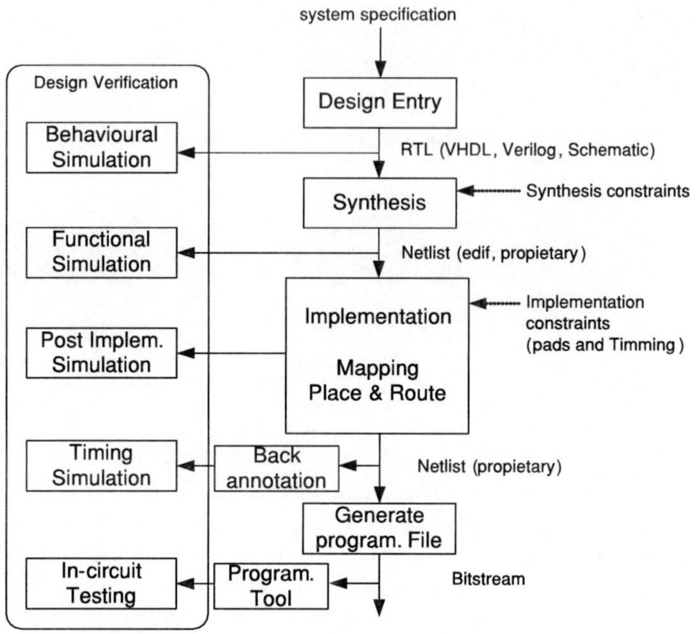

Fig. 6.1 A typical design flow for FPGA design

6.1.1 Design Entry

The design entry is the way you describe the hardware that you want to implement into an FPGA (or in an electronic device) based on a system specification. There are different methods:

- Using a Hardware Description Language (HDL).
- Using schematics.
- Using Intellectual Property (IP) blocks.
- Using Electronic System Level (ESL) languages.

Today's EDA tools allow the mixing of different design entries in a hierarchical structure. It is common to see a schematic top level, with several predesigned IPs, some specific components developed in VHDL and or Verilog and subsystems designed in an ESL language.

6.1.1.1 HDL Design Entry

There are now two industry standard hardware description languages (HDL), Verilog and VHDL. Some vendors used to have their own proprietary HDL languages but these are displaced by the use of the standard languages.

HDL is the de facto design entry in most digital designs. In ASIC designs Verilog is much more used but, FPGA designer's use either VHDL and/or Verilog. All FPGA tools support both languages and even projects mixing the use of booth languages.

The Verilog hardware description language has been used far longer than VHDL and has been used extensively since it was launched by Gateway Design Automation in 1983. Cadence bought Gateway and opened Verilog to the public domain in 1990. It became IEEE standard 1364 in 1995. It was updated in 2001 (IEEE 1364-2001) and had a minor update in 2005 (IEEE 1364-2001).

On the other hand, VHDL was originally developed at the request of the U.S Department of Defense in order to document the behavior of existing and future ASICs. VHDL stand for VHSIC-HDL (Very high speed integrated circuit Hardware Description Language) became IEEE standard 1076 in 1987. It was updated in 1993 (IEEE standard 1076-1993), and the last update in 2008 (IEEE 1076-2008, published in January 2009).

Xilinx ISE and Altera Quartus have text editors with syntax highlighting and language templates for VHDL and Verilog to help in the edition.

6.1.1.2 Schematic Design Entry

With schematic capture or schematic entry you draw on your computer using a schematic capture tool. The main advantage in the use of schematics is that it documents the design in an easily readable format. However big designs rapidly become difficult to maintain and the file formats are incompatible between vendors. A HDL code is easier to be parameterized and regular structures are easily replicated. ASIC and FPGA users rarely use schematics.

6.1.1.3 Intellectual Property (IP) Blocks

In order to make simpler the design of complex systems, there exist libraries of predefined complex functions and circuits that have been tested and optimized to speed up the design process. These predefined circuits are commonly called IP cores (or IP blocks) and are available from FPGA vendors and third-party IP suppliers. The IP Cores can be distributed as a compiled netlist or as an HDL source code. Moreover, the FPGA vendors have tools to generate most typical IP cores (Xilinx CORE generator in Xilinx, Altera megafunctions). The simplest IP cores are typically free, but the complex one rarely is without charge, and typically released under proprietary licenses.

In the FPGA arena a related concept is the idea of "hard IP core" and "soft IP core". In today's FPGA several heterogeneous block are built in the FGPA such as multipliers, blocks of memories, clock managers, transceivers, memory controllers, etc. These components are called "hard IP cores" to differentiate from the ones implemented using general purpose logic (soft IP blocks).

6.1.1.4 Electronic System Level (ESL) Languages

As FPGA applications have grown in complexity and FPGA chips have become more complex, there is an increasing need for more efficient less time-consuming development methodologies. In order to address this requirement, many chip and tool providers have developed "high-level" development tools. In this environment, "high-level" means that the input to the tool is a target independent language as C, C++or Java instead an HDL language.

To refer to these tools and methodologies, several names are used such as High-level synthesis (HLS), High Level Languages (HLL), C synthesis, C to hardware, electronic system level (ESL) synthesis, algorithmic synthesis, or behavioral synthesis and others.

The high-level synthesis seems to be the future design entry for numerous fields of applications. There are several successful products from the big EDA vendors and from small companies fighting in this field. Most of those products generate, as a result, an HDL (Verilog or VHDL) description of the circuit in order to use it in a traditional design flow.

Related to this "high-level" concept and in the DSP (Digital Signal Processing) field, the design productivity of DSP system is increased using MatLab/Simulink design entry. The big EDA companies have tools for DSP design based on MatLab/Simulink, even booth leaders in the FPGA field: Altera (DSP builder) and Xilinx (System Generator).

6.1.2 Synthesis

The synthesis (or logic synthesis) is the process by which an abstract form (an HDL description) of the circuit behavior is transformed into a design implementation in terms of logic gates and interconnections. The output is typically a netlist and various reports. In this context, a "netlist" describes the connectivity of the electronic design, using instances, nets and, perhaps, some attributes.

There are several proprietary netlist formats but, most synthesizers can generate EDIF (Electronic Design Interchange Format) that is a vendor-neutral format to store electronic netlist.

FPGA vendors have their own synthesizers (Xilinx XST, Altera Quartus II Integrated Synthesis) but, main EDA vendors have syntheses for FPGA (Precision by Mentor Graphics and Synplify by Synplicity) that can be integrated in the FPGA EDA tools.

6.1.2.1 Synthesis Optimizations

The synthesis process performs several target independent optimizations (logic simplification, state assignment, etc.) but, also the synthesis tools take into account the target technologies and make target dependent optimizations.

The optimizations available depend on the synthesizer but, most typical optimizations are present in all of them. Typical optimizations are for the area reduction, the speed optimization, the low power consumption, and the target frequency of the whole design.

Nevertheless, much more details can be controlled, such as:

- Hierarchy Preservation: control if the synthesizer flattens the design to get better results by optimizing entities and module boundaries or maintaining the hierarchy during Synthesis.
- Add I/O buffers: enables or disables the automatic input/output buffer insertion: this option is useful to synthesize a part of a design to be instantiated, later on.
- FSM Encoding: selects the Finite State Machine (FSM) coding technique. Automatic selection, one-hot, gray, Johnson, user defined, etc.
- Use of embedded components: use embedded memory or multipliers blocks or use general purpose LUTs to implement these functionalities.
- Maximum fan-out: limits the fan-out (maximum number of connections) of nets or signals.
- Register duplication: allows or limit the register duplication to reduce fan-out and to improve timing.
- Retiming or Register Balancing: automatically moves registers across combinatorial gates or LUTs to improve timing while maintaining the original behavior.

The complete description of synthesis optimization can be found at the synthesis tool documentation (for example for FPGA [3, 4, 6, 8]).

The synthesis behavior and optimization can be controlled using synthesis constraints. The constraints could be introduced using the integrated environment, a constraint file or embedding in the HDL code.

6.1.2.2 Synthesis Constraints

The synthesis optimizations are controlled globally (for the complete design) or locally for each part of the HDL code. The global optimization can be introduced in the integrated environment (then translated as switches in command line or to an additional file), or in a specific constrain file. The local optimization can be specified in a constraint file or embedded in the HDL code.

The synthesis constraint file is, typically, a plain text file having different syntax, depending on the tool. Xilinx use the xcf (Xilinx Constraint File) [8], Altera for timing use SDC files (TimeQuest Timing Constrains) and QSF (Quartus Settings File) for I/O and others [2, 3], and simplify uses SDC (Synopsys Design Compiler) constraint files [6].

The main advantage of using constraint files (generated by a graphical interface or entered in a text editor) is that it makes your source code more portable and separates the functionality described in the HDL of the synthesis detail.

The HDLs (VHDL and Verilog) allow you to embed constraints in the design code. This method is recommended for constraints that define the desired result of the HDL (encoding of FSM, type of memories or multipliers) or specific optimization (duplicate registers, registers balancing, maximum fan-out, etc.). In VHDL this is done using “attributes” defined in the declaration. The following lines of code define the maximum fan-out of signal “a_signal” to 20 connections in XST [8].

```
attribute max_fanout: string;
attribute max_fanout of a_signal: signal is "20";
```

6.1.2.3 Synthesis Reports

The synthesis reports show the results of the netlist generation synthesis process. The synthesizer gives, typically, a text report, where you can see a summary of your synthesis options, and a summary and analysis of the netlist generation. Some tools generate an HTML, XML or other proprietary formats which are easier to navigate into the information.

This report is important in order to see if the hardware generated by the synthesizer agrees with the described HDL. The main parts in a synthesis report are:

- Synthesis Options: a summary of selected optimizations used. Important to check if differences in results appear.
- HDL Compilation and Analysis: syntax errors and hierarchy are analyzed and information reported.
- HDL Synthesis: is the main part, where the inferred hardware is reported. The tool informs what is generated in each part of the code.
- Optimization reports: the advanced optimization and low level optimization are also informed.
- Final resource and timing: a summary of the device resource utilization and a preliminary timing analysis with the worst path is informed. Take into account that this information is previous to the implementation (placement and routing) and the interconnection delay is only estimated.

Additionally, most synthesis tools can generate graphical information (a schematic) of the resulted synthesis. Typically, you can see graphically the RTL view or a technological view (using the target low level components).

6.1.3 Implementation (Mapping, Placement and Routing)

The implementation step is a vendor specific tool that mainly places and routes the design in the target device. In Altera Quartus II this tool is known as a “fitter” [2, 3], meanwhile, in Xilinx ISE [7, 8] is composed of three processes (translate, map, and place & route).

The inputs to the implementation are the netlist(s) generated in synthesis and the design implementation constraints. The output is a proprietary placed and routed netlist (ncd file in Xilinx, an internal database representation in Altera) and several reports summarizing the results. Typically the design implementation uses timing and area constraints; a general description is in Sect. 6.2.

6.1.3.1 Implementation Reports

The implementation tools generate different information about that process. Typically, a text report includes detailed information about the used resources, the clock distribution (including skew), the final timing obtained with respect to the constraints and other information.

Nevertheless, more information can be obtained of the implementation, such as internal delay of the interconnections, a description of the used pads, a graphical view of the placed and routed design in the target device, a simulatable HDL file including timing information, and a power consumption estimation.

6.1.4 Programming File Generation and Programming

The programming file generation creates a bitstream file (.bit in Xilinx,.rbf in Altera) that can be downloaded to the device or can be recorded in an external EPROM. This tool in Altera Quartus II is call “assembler”, meanwhile, in Xilinx ISE “Bitgen”.

The generated bitstream can be converted in a standard format for EPROM programming or directly downloaded to an FPGA device using a JTAG connection (iMPACT in Xilinx, Programmer in Altera).

6.2 Implementation Constraints

The complexity of today’s FPGA designs and the demand for higher performance makes necessary the use of complex timing and placement constraints to meet the performance requirements. Implementation constraints are instructions given to the FPGA implementation tools to direct the mapping, placement, routing, timing or other guidelines for the implementation tools.

Implementation constraints are placed in constraint file(s) but, may exist in the HDL code, or in a synthesis constraints file and propagated for implementation. Xilinx uses the User Constraint File (.ucf), meanwhile Altera uses the Quartus II Settings File (.qsf) or, in the case of timing constraints, the Synopsys Design Constraints file (.sdc). The constraint files are plain text but, there are several graphical tools that help in the edition of such constraints avoiding the necessity to know the exact syntax.

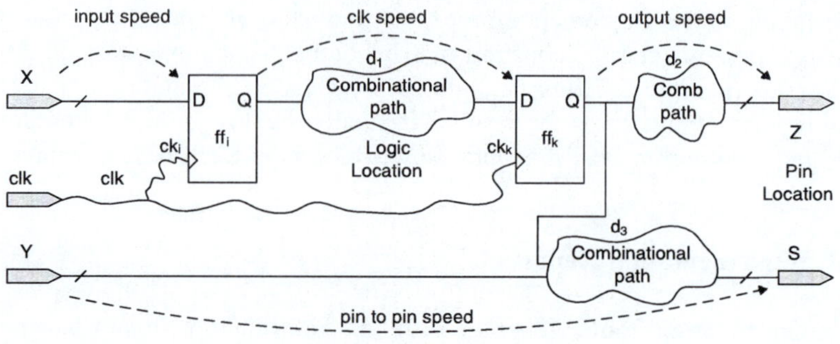

Fig. 6.2 What is possible to be constrained

Figure 6.2 shows the basic constraints of a digital synchronous design:

- Internal clock speed for one or several clocks.
- I/O speed.
- Pin to Pin timing.
- Pin Locations and Logic Locations (floor-planning).

The first three are timing constraints, meanwhile, the last are area constraints.

6.2.1 Timing Constrains

As suggested previously (Fig. 6.2) the timing constrain includes the clock definition, input and output timing requirements and the combinatorial path requirements. Creating global constraints for a design is the easiest way to provide coverage of the constrainable connections in a design, and to guide the tools to meet timing requirements for all paths. For example given a constraint of frequency of 100 MHz, we are constraining each combinational path in the design.

Nevertheless sometimes the designer needs to relax some global constraint and inform the implementation tools that some path can take more time. The typical cases are:

- False Paths: if there is a paths that is static in nature, or is not of much significance for timing.
- Multi-cycle paths: paths between registers (combinational path) that intentionally take more than one clock cycle to become stable.
- Fast or Slow path: combinational path that can work at slower speed than the global constrain.

In these cases you can use directives to eliminate the paths from timing consideration in implementation.

6.2.2 Placement and Other Constrains

The placement constraints instruct the implementation tools, where to locate the logic element into the FPGA (I/O pads, Flip-flops, ROMs, RAMs, LUTs, etc.). Since every component in the design carries a unique name, you can use these name to assign to a region in the FPGA where put this components. A representative placement constraint, always used in a design, is the position of the input/output pins. Nevertheless, there are others commonly used constraints:

- Relative placement constraints: allow to place logic blocks relative to each other to increase speed and use die resources efficiently.
- Routing constraints: mechanism of locking the routing in order to maintain timing.
- Input/output characteristics: specifies the i/o standard, the use of internal pull-ups or pull-downs, slew rate, asynchronous delay, etc.
- Dedicated block configuration: how the multipliers are configured, digital clock managers, memory blocks, SERDES, etc.
- Mapping Directives: allow eliminating the simplification of internal nets for observability, or how the logic is grouped.
- Maximun Skew: allow to control the maximum skew in a line (Sect. 5.3.1).
- Derating Factors: specifies different supply voltage and temperature. Used to derate the timing factors (Sect. 5.1.2.5).

As previously mentioned, implementation constraints (timing and placement) are placed in separated constraint file(s) but, may be directly written in the HDL code.

6.3 System Verification

Figure 6.3 shows the typical system verification flow for FPGA. The logic simulation is the most used technique in early stage of development. Nevertheless, the reprogrammable nature of FPGA gives other alternatives as in-circuit simulation, testing and debugging.

6.3.1 Simulation

Logic simulation is the primary tool used for verifying the logical correctness of a hardware design. In many cases, logic simulation is the first activity performed in the process of design. There are different simulation points were you can simulate your design, the three more relevant are:

- RTL-level (behavioral) simulation. No timing information is used.
- Post-Synthesis simulation. In order to verify synthesis result.

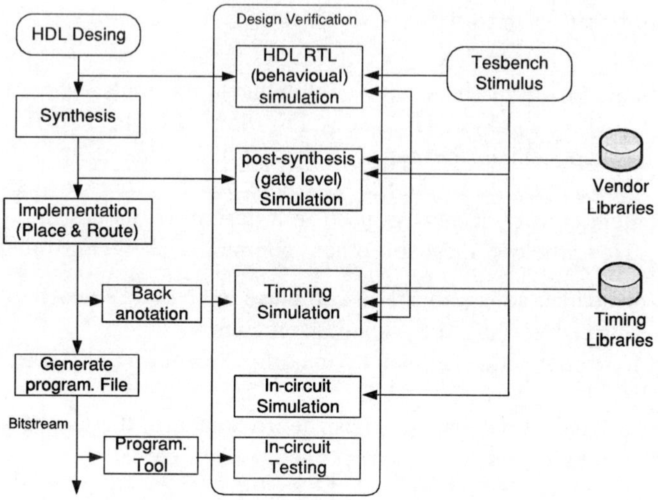

Fig. 6.3 The design verification flow in FPGAs

- Post implementation (post place & route) simulation. Also known as timing simulation because it includes blocks and nets delays.

The behavioral (RTL-level) simulation enables you to verify or simulate a description at the system level. This first pass simulation is typically performed to verify code syntax, and to confirm that the code is functioning as intended. At this step, no timing information is available, and simulation is performed in unit-delay mode to avoid the possibility of a race condition. The RTL simulation is not architecture-specific, but can contain instantiations of architecture-specific components, but in this case additional libraries are necessary to simulate.

Post-synthesis simulation, allows you to verify that your design has been synthesized correctly, and you can be aware of any differences due to the lower level of abstraction. Most synthesis tools can write out a post-synthesis HDL netlist in order to use a simulator. If the synthesizer uses architecture-specific components, additional libraries should be provided.

The timing simulation (post implementation or post Place & Route full timing) is performed after the circuit is implemented. The general functionality of the design was defined at the beginning but, timing information cannot be accurately calculated until the design has been placed and routed.

After the implementation tools, a timing simulation netlist can be created and this process is known as back annotation. The result of a back annotation is an HDL file describing the implemented design in terms of low level components and additional SDF (Standard Delay Format) file with the internal delays that allows you to see how your design behaves in the actual circuit.

Xilinx ISE has his own simulator (ISE simulator, ISim) but, can operate with Mentor Modelsim or Questa and operate with external third-party simulation tools

(Synopsys VCS-MX, Cadence NCSim, Aldec Active-HDL). Altera Quartus II uses Mentor Modelsim and can use other third party simulators.

6.3.2 Formal Verification

The formal verification is the act of proving the correctness of a system with respect to a certain formal specification, using formal mathematical methods. Since hardware complexity growth continues to follow Moore's Law, the verification complexity is even more challenging and is impossible to simulate all possible states in a design. In order to implement the formal verification, Hardware Verification Language (HVL) can be used. A HVL is a programming language used to verify the designs of electronic circuits written in a hardware description language (HDL). System-Verilog, OpenVera, and SystemC are the most commonly used HVLs. The formal verification is widely used by the big companies in the ASIC world but, is relatively new in the FPGA arena. The adoption of formal verification in FPGA flow is still poor but, increasingly important.

6.3.3 In-Circuit Co-Simulation

The idea is to simulate the system at hardware speed but, maintaining a part of the flexibility of a traditional simulation. The simulation executes the RTL code serially while a hardware implementation executes it fully in parallel. This leads to differences not only in execution time but, also in debugging. In simulation, the user can stop simulation to inspect the design state (signals and memory contents), interact with the design, and resume simulation. Downloading the design to an FPGA, the visibility and observability is greatly reduced.

EDA vendors offer products to simulate the complete or a part of the design in circuits but, controlling externally the execution. In this line, Xilinx added their simulator (Isim), the so called "Hardware co-simulation", as a complementary flow to the software-based HDL simulation. This feature allows the simulation of a design or a sub-module of the design to be offloaded to hardware (a Xilinx FPGA regular board). It can accelerate the simulation of a complex design and verify that the design actually works in hardware.

6.3.4 In-Circuit Testing and Debugging

In order to test the functionality, a simple configuration of the FPGA allows testing of the circuit. In the typical in-circuit debugging, the user employs an external logic analyzer for visibility but, can see only a limited number of signals which they determined ahead of time.

The reprogramability of FPGA opens new ways to debug the designs. It is possible to add an "internal logic analyzer" within the programmed logic. Xilinx ChipScope Pro Analyzer and Altera SignalTap II Logic Analyzer are tools that allow performing in-circuit verification, also known as on-chip debugging. They use the internal RAM to store values of internal signals and communicate to the external world using the JTAG connection.

Another intermediate solution includes inserting "probes" of internal nets anywhere in the design and connecting of the selected signals to unused pins (using Xilinx FPGA Editor, Altera Signal Probe, Actel Designer Probe Insertion).

6.3.5 Design for Test

A recurring topic in digital design is the design for test (DFT). The DFT (or also Design for Testability) is the general name for design techniques that add certain testability features to a hardware design.

In the ASIC world, tests are applied at several stages in the manufacturing flow. The tests usually are accomplished by test programs that execute in Automatic Test Equipment (ATE). For test program generation, several automatic algorithms are used as the "Stuck-at" fault model and other algorithmic methods.

One of the main issues in a test is to gain control (controllability) and observe (observability) internal nodes in order to check functionality and this leads to the concept of scan chain.

In scan chain, registers (flip-flops or latches) in the design are connected in a shift register chain in order to set the vales or read the values when a test mode is selected. Figure 6.4 shows the addition of a multiplexer to a simple register to support the scan chain mode. Observe that only using four signals (clk, Sin, Sout, test) and an option reset it is possible to gain access to any register inside a device.

The FPGA are pretested ASIC circuits that have their own circuitry to test the manufacturing procedure. Nevertheless, some techniques of DFT are useful for FPGA design debugging and testing. The main useful characteristic is the access to the internal state of the FPGA using the internal scan chain through the JTAG port. JTAG (Joint Test Action Group) is the name for standard test access port and boundary-scan architecture. It was initially devised for testing printed circuit boards using boundary but, today it is also widely used for integrated circuit debug ports.

The FPGA devices allow the accessing of the internal state (read-back) of any internal register, even the configuration ones using the JTAG port. The possibility to read and write the internal state of the device allows several standard and ad-hoc techniques for testing purposes. The access to the entire registers content allows, for example, the reconfiguration of the device, the partial reconfiguration, read internal states for in-circuit debugging (Sect. 6.3.4), communicate internal values to implement an internal logic analyzer, and several other techniques.

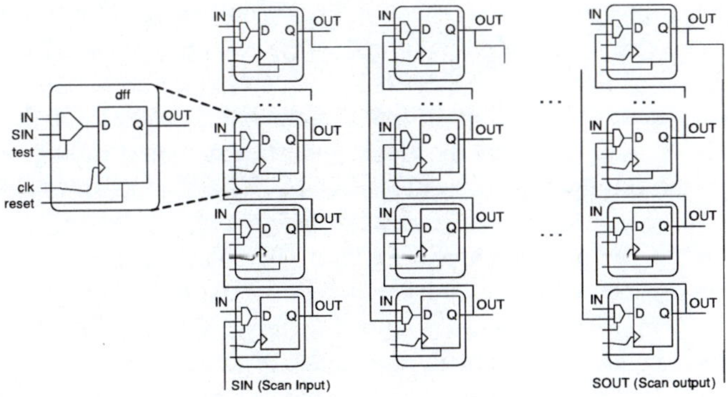

Fig. 6.4 Scan chain register and organization of a scan chain

6.4 Timing Analysis

Static timing analysis (STA or simply timing analysis) is the method of computing the expected timing of a digital circuit without requiring simulation. The word *static* refers to the fact that this timing analysis is carried out in an input independent manner. The objective is to find the worst case delay of the circuit over all possible input combinations.

The computational task is to review the interconnection graph that represents the final netlist and determine the worst case delay. The methods used are specific optimization of typical graph algorithm such as a depth-first search. Modern static timing analyzers bear similarities with project management models such as PERT (Program Evaluation and Review Technique) or CPM (Critical Path Method).

There are some common terms used in timing analysis:

- **Critical path**: is the path with the maximum delay between an input and an output (or two synchronous points).
- **Arrival time**: is the time elapsed for a signal to arrive at a certain point.
- **Required time**: is the latest time at which a signal can arrive without making the clock cycle longer than desired (or a malfunction).
- **Slack**: is the difference between the required time and the arrival time. A negative slack implies that a path is too slow. A positive slack at a node implies that the arrival time at that node may be increased without affecting the overall delay of the circuit.

In a synchronous digital system, data is supposed to move on each tick of the clock signal (Sect. 5.2). In this context, only two kinds of timing errors are possible (see Sect. 5.2.2):

- **Hold time violation**: when an input signal changes too fast, after the clock's active transition (race conditions).
- **Setup time violation**: when a signal arrives too late to a synchronous element and misses the corresponding clock's edge (long path fault).

Since the timing analysis is capable of verifying every path, it can detect the problems in consequence of the clock skew (Sect. 5.3) or due to glitches (Sect. 5.1.3).

The timing analysis can be performed interactively, asking for different paths but, typically is used to report the slack upon the specified timing requirements expressed in the timing constraint (Sect. 6.2.1).

The FPGA tools, after implementing the design, make a default timing analysis to determine the design system performance. The analysis is based on the basic types of timing paths: Clock period; input pads to first level of registers; last level of registers to output pads, and pad to pad (in asynchronous paths).

Each of these paths goes through a sequence of routing and logic. In Xilinx ISE the tool calls "Timing Analyzer" and in Altera Quartus II Altera "TimeQuest". More advanced options can be analyzed upon using the specific tool.

6.5 Power Consumption Estimation

Power dissipated in a design can be divided into static and dynamic power (Sect. 5.3). In FPGA designs, due to the reprogramability nature, the total power is also divided into three components for each power supply:

$$\text{Total Power} = \text{Device Static} + \text{Design Static} + \text{Dynamic Power}$$

Where the components are:

- Device Static: depends on manufacturing, process properties, applied voltage, and temperature.
- Design Static: blocks in an FPGA (I/O termination, transceivers, block RAM, etc.) are disabled by default and enabled depending on the design requirements. When these blocks are enabled they consume power, regardless of user design activity.
- Dynamic Power: depends on the capacitance and activity of the resources used, and also scales with the applied voltage level.

FPGA vendors offer early power estimators spreadsheet (Altera PowerPlay Early Power Estimator, Xilinx XPower Estimator) typically used the pre-design and pre-implementation phases of a project. These spreadsheets are used for architecture evaluation; device and power supply selection and helps to choose the thermal management components which may be required for the application.

For a more accurate estimation, the power analyzer tools (Xilinx XPower, Altera PowerPlay) perform power estimation, post implementation. They are more precise tools since they can read from the implemented design netlist the exact logic and routing resources used. In order to obtain good and reliable results the activity of each node should be provided. If you do not supply the activity, the software can predict the activity of each node but, the accuracy is degraded.

The power analyzer takes the design netlist, the activity of the circuit, the supply voltage and the ambient temperature and reports the consumed current (power) and the junction temperature. Nevertheless, the junction temperature itself depends on the ambient temperature, voltage level, and total current supplied. But the total current supplied includes the static current as a component that depends on temperature and voltage, so a clear circular dependency exists. The tools use a series of iterations to converge on an approximation of the static power for given operating conditions.

A significant test bench that models the real operation of the circuit provides the necessary activity of each node. The simulation result of the activity is saved in a file (SAIF or VCD file) that is later used in conjunction with the capacitance information by the power analyzer.

The Value Change Dump (VCD) file is an ASCII file containing the value change details for each step of the simulation and can be generated by most simulators. The computation time of this file can be very long, and the resulting file size is typically huge. On the other hand, the SAIF (Switching Activity Interchange format) file contains toggle counts (number of changes) on the signals of the design and is supported also for most simulators. The SAIF file is smaller than the VCD file, and recommended for power analysis.

6.5.1 Reducing the Power Consumption

Using the results of a power analyzer, having a complete system-level understanding and the accurate power model will permit the designer to make the decisions necessary to reduce the power budget (Sect. 5.3), including:

- Selecting the best device.
- Reducing the device operating voltage.
- Optimizing the clock frequencies.
- Reducing long routes in the design.
- Optimizing encodings.

With regards to the EDA automatic improvements for low power, as mentioned previously, in synthesis it is possible to have, as a target, the power reduction. In implementation, it is possible to specify optimal routing to reduce power consumption. In this case, it allows to specify an activity file (VCD or SAIF) to guide place & route when it optimizes for power reduction.

6.6 Example of EDA Tool Usage

In order to review the concepts of EDA tools we will deploy a simple example using Xilinx ISE 13.1 and Altera Quartus 11.0. The example is the combinational

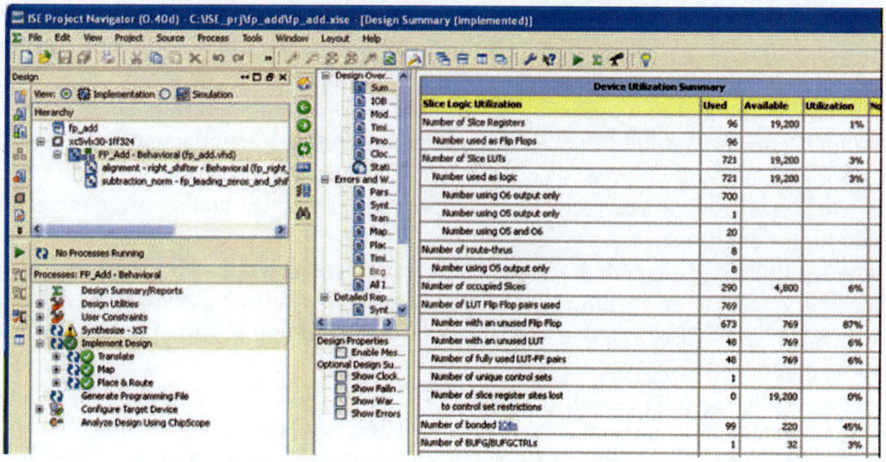

Fig. 6.5 A simple project in Xilinx ISE

floating point adder of Chap. 12 that we will implement in both vendor tools. The simple example uses three VHDL files: *FP_add.vhd*, *FP_right_shifter*, and *FP_leading_zeros_and_shift.vhd*.

If you want to begin with no previous experience, it is highly recommendable to start with the tutorial included in the tools. In the case of Quartus, follow the "Quartus II Introduction for VHDL/verilog Users" accessible form help menu. If you will start using ISE we recommend using the "ISE In-Depth Tutorial", accessible using help → Xilinx on the web → Tutorials.

6.6.1 Simple Example Using Xilinx ISE

We create a new project (file → new project…) with name fp_add. We select a Virtex 5, XC5VLX30 ff324 -1. For synthesis XST, simulation ISIM and preferred language VHDL. Then we add to the project the three VHDL source files (project → add source) (Fig. 6.5).

6.6.1.1 Design Entry and Behavioral Simulation

The design entry in this simple project is not used, since the VHDL code is provided. Nevertheless to create a new source file you can use "project → new source", and then choose the design entry type of source code. In the case of using VHDL or Verilog module you have a graphical wizard to describe the input and output of the circuit.

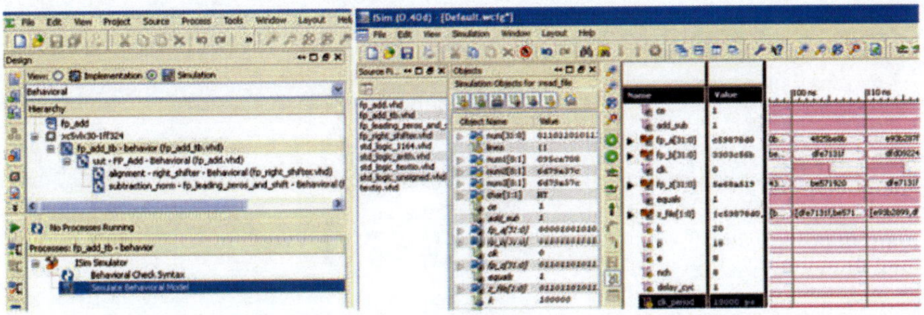

Fig. 6.6 Behavioral simulation in Xilinx ISE using ISIM simulator

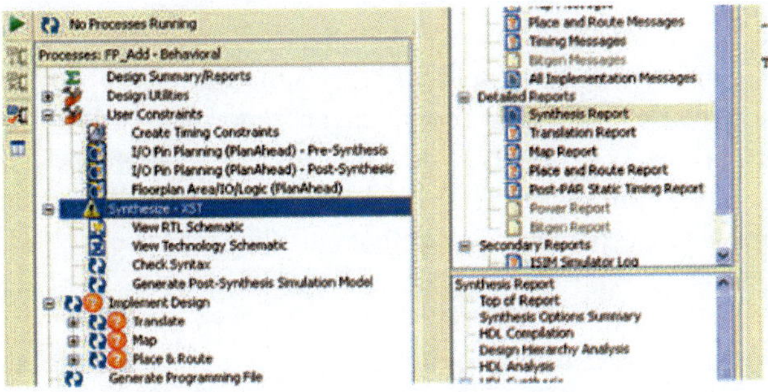

Fig. 6.7 Synthesis and synthesis report

In order to make a behavioral simulation we add to the project a test bench (fp_add_tb.vhd) selecting "project → add source". The simple test bench reads the stimuli text file (contains two encoded operands to add and the corresponding result) and verifies the results (Fig. 6.6).

In order to launch the simulation, select the test bench (fp_add_tb.vhd), and double click "Simulate Behavioral Model" in the processes window. Before that, ensure that you have selected the "simulation" view radio button, in order to be able to see the simulation process.

6.6.1.2 Synthesis and Synthesis Report

To run the synthesis, double click the "synthesis—XST" process in processes view (Fig. 6.7). The global options of the synthesis are available using the right button at the option "process option". The more specific constrain can be either embedded in the HDL code or in the Xilinx constraint file (.xcf). The xcf is a plain text file that needs to be linked in the process option. The details are available at [8].

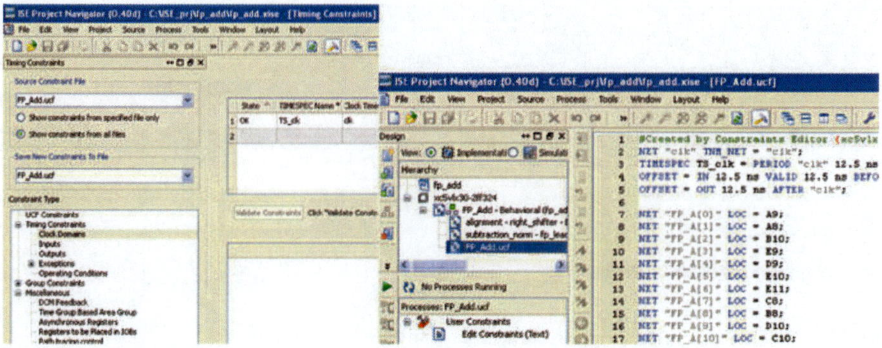

Fig. 6.8 Assigning timing constraints, and the generated ucf text file

After the synthesis you can review the graphical representation of the synthesized circuit (View RTL schematic) or review the synthesis report (.syr file). This report gives relevant information. The "HDL synthesis" part of report describes the inferred component as a function of the HDL and the "final report" summarizes the resource utilization and performs a preliminary timing report.

6.6.1.3 Implementation: Constraints and Reports

In order to implement a design, we need to generate the implementation constraints. We will firstly assign the position of the pads. We can edit manually the ucf (user constraint file) text file or double click the "I/O pin planning (plan ahead)" and use the graphical interface.

Then we can assign timing constraints using the "create timing constraint" option. We will assign a period constraints ("clock domain" in constraint type view) of 80 MHz (12.5 ns) and the same restriction for inputs and outputs ("inputs" and "outputs" in constraint type view, respectively). You can check the generated restriction editing the ucf file (Fig. 6.8).

We can assign global options to the implementation using right click over the implementation icon in processes window and selecting "process properties…"

We implement the design double clicking in "implementation". The three step in the Xilinx implementation flow will be executed (translate, map, place & route).

In ISE the design implementation, comprises the following steps:

- Translate: merges the incoming netlists from synthesis and the imple-mentation constraints into a proprietary Xilinx native generic database (NGD) file.
- Map: fits the design into the available resources on the target device. The output is a native circuit description (NCD) file.
- Place and Route: takes a mapped NCD file, places and routes the design, and produces another NCD file.

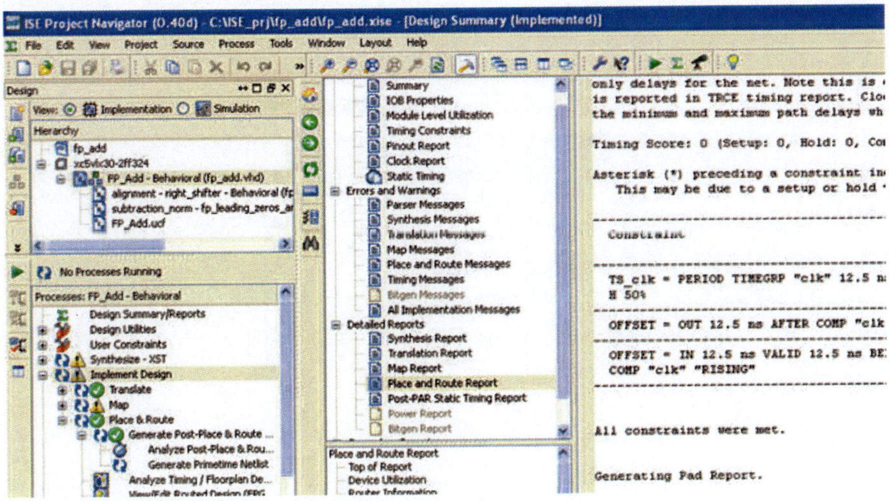

Fig. 6.9 Implemented design. Access to the reports

The result of the implementation is a proprietary placed and routed netlist (ncd file) and several reports. There are three main reports generated: the map report (.mrp), the place and route report (.par) and the "Post-PAR static timing report" (.twr) (Fig. 6.9).

The map report shows detailed information about the used resources (LUTs, slices, IOBs, etc.), the place and route report gives clock network report including the skew and informs which constraints were met and which were not. Finally, Post-PAR static timing report gives the worst case delay with respect to the specified constraints.

You can obtain additional graphical information about the place and route results using FPGA editor and PlanAhead. Use the FPGA Editor to view the actual design layout of the FPGA (in the Processes pane, expand "Place & Route", and double-click "View/Edit Routed Design (FPGA Editor)"). The PlanAhead software can be used to perform post place & route design analysis. You can observe, graphically, the timing path onto the layout, and also perform floorplaning of the design. In order to open PlanAhead in the Processes pane, expand Place & Route, and double-click "Analyze Timing/Floorplan Design (PlanAhead)".

6.6.1.4 Post Place and Route Simulation

The post place and route (or timing) simulation is accessed by selecting "simulation" in the view panel and selecting "post-route" from the drop-down list. Then select the test bench and double click "Simulate Post-Place & Route Model". This will execute the back annotation process (netgen). Then the simulation is performed. Observe that the resulting simulation gives some errors. Why? The answer is simple, the test bench generates a clock of 100 MHz (10 ns period)

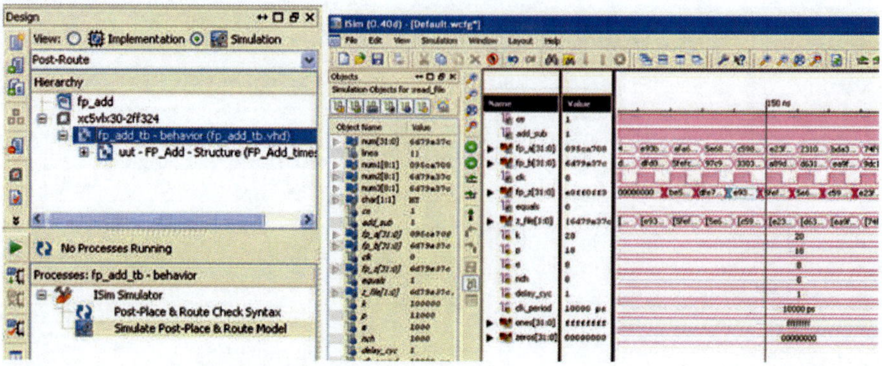

Fig. 6.10 Post place and route simulation

and the implementation that we perform is slower. Then, some input pattern combination violates setup time and gives, consequently, errors. You can modify the test bench in order to operate at a lower frequency (Fig. 6.10).

6.6.1.5 Running a Static Timing Analysis

We can review, more carefully, the timing aspects, by opening the "analyze post-Place & Route Static Timing" (expand Implement Design → Place & Route → Generate Post-Place & Route Static Timing to access this process). By default, this runs an analysis of the worst case delays with respect to the specified constraints giving the three worst paths (Fig. 6.11).

In order to control more aspects of the timing analysis you can run an analysis (Timing → Run Analysis), you can control the type of analysis (Analyze against), control the amount of path reported, control the timing derating factors (Sect. 5.1.2.5), filter nets and paths in the analysis, etc.

6.6.1.6 Generating Programming File and Programming the FPGA

After implementing the design and performed the corresponding timing analysis, you need to create configuration data. A configuration bitstream (.bit) is created for downloading to a target device or for formatting into a PROM programming file (Fig. 6.12).

6.6.1.7 Using Command Line Implementation

All the previous step could be executed in command line and group it in order to create scripts. The ISE software allows you extract the command line arguments

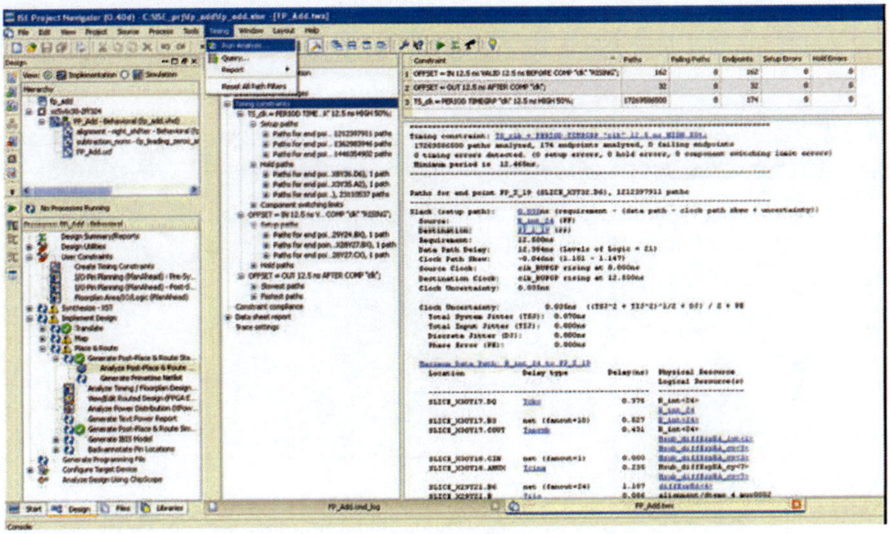

Fig. 6.11 Post place and route static timing analysis

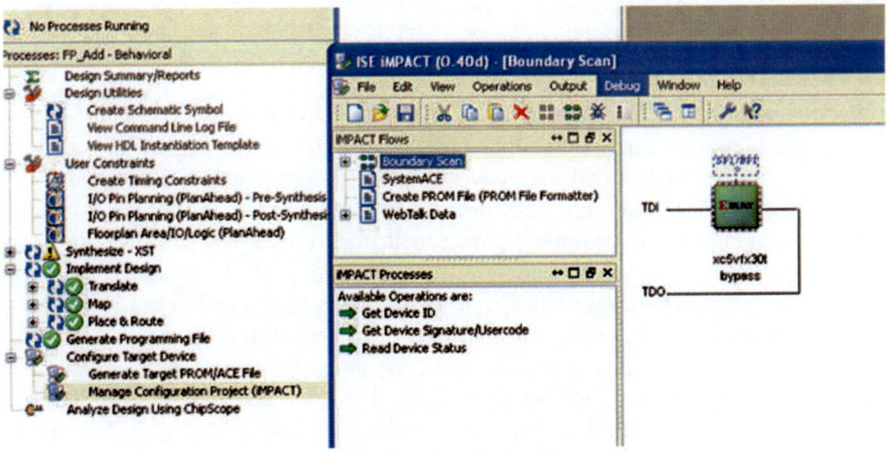

Fig. 6.12 Generating programming file and programming the FPGA

for the various steps of the implementation process (Design Utilities → View Command Line Log File). This allows you to verify the options being used or to create a command batch file to replicate the design flow.

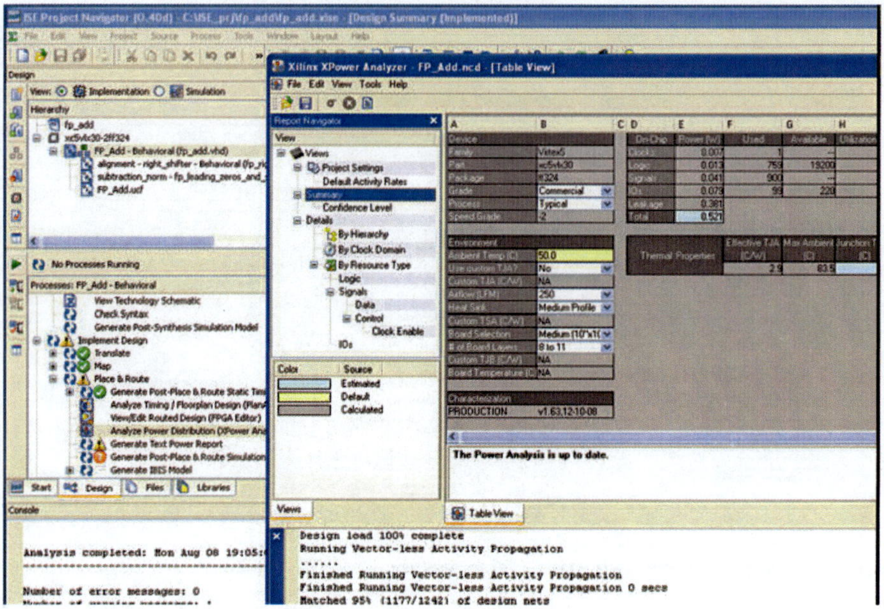

Fig. 6.13 Xpower analyzer result

6.6.1.8 Estimating the Power Consumption

After place and route of the design it is possible to obtain power consumption estimation. The accuracy of the estimation relies upon the activity given to the xpower tool. The activity file is generated with a post place & route simulation (Sect. 6.6.1.4) but instructing the ISim simulator to generate the SAIF (Switching Activity Interchange Format) file. In order to do that, right click "simulate post place and route model", click on "process properties…" then select "Generate SAIF File for Power Optimization/Estimation", it is possible to assign different file names.

In order to check the power estimation results, based on the activity computed, you can even generate a simple text report by double clicking on "Generate Text Power Report", or by opening the interactive tool Xpower double clicking "Analyze Power Distribution (Xpower Analyzer)" (Fig. 6.13). Observe that by browsing into the details you have access to the activity of each node, the fan-out of nets and the corresponding power.

6.6.2 Simple Example Using Altera Quartus II

Designing for Altera devices is very similar, in concept and practice, to designing for Xilinx devices. In most cases, the same RTL code can be compiled by Altera's

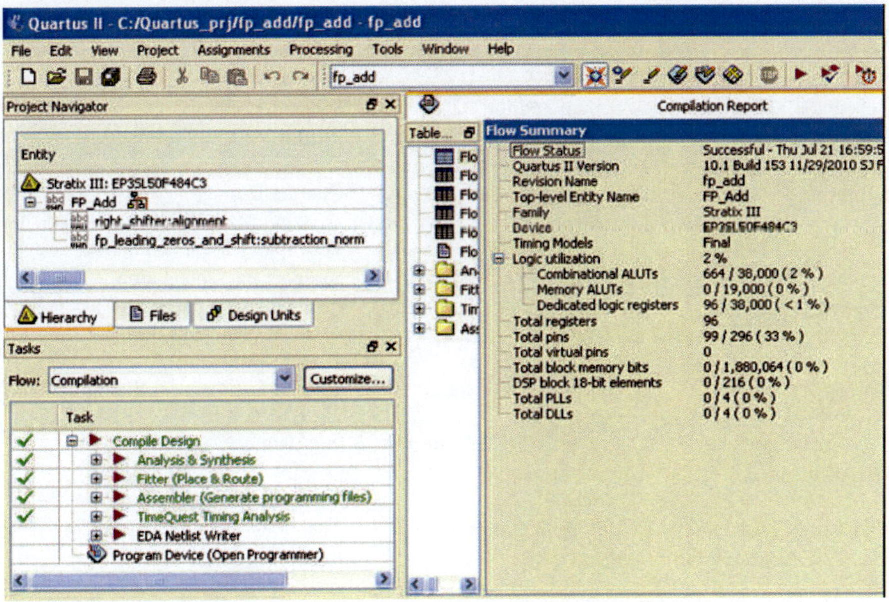

Fig. 6.14 Quartus II project

Quartus II as was explained for Xilinx ISE (Sect. 6.6.1). Altera has an application note "Altera Design Flow for Xilinx Users" [1], where this process is carefully explained.

We can start creating a new project (file → New Project Wizard) with name fp_add. We add to the project the three VHDL source files (step 2 of 5).We select a Stratix III device, EP3SL50F484C3. Leave the default synthesis and simulation options (Fig. 6.14).

For the rest of the steps in the implementation and simulation flow, Table 6.1 summarizes the GUI (graphical user interface) names for similar task in Xilinx ISE and Altera Quartus II.

6.7 Exercises

1. Implement the floating adder of Chap. 12 in Altera Quartus in a Stratix III device. What are the area results? Add implementation constraints in order to add FP numbers at 100 MHz. Constraints are met?
2. Implement the floating point multiplier of Chap. 12 in Xilinx ISE in a Virtex 5 device. What are the area results? It is possible to multiply at 100 MHz? Remember to add implementation constraints. Multiplying at 20 MHZ what is the expected power consumption? Use the XPower tool and the provided test bench.

Table 6.1 GUI names for similar tasks in Xilinx ISE and Altera Quartus II

GUI feature	Xilinx ISE	Altera quartus II
HDL design entry	HDL editor	HDL editor
Schematic entry	Schematic editor	Schematic editor
IP entry	CoreGen and architecture wizard	Megawizard plug-in manager
Synthesis	Xilinx synthesis technology (XST)	Quartus II integrated synthesis (QIS)
	Third-party EDA synthesis	Third-party EDA synthesis
Synthesis constraints	XCF (xilinx contraint file)	Same as implementation
Implementation constraint	UCF (user constraint file)	QSF (quartus II settings file) and SDC (synopsys design constraints file)
Timing constraint wizard	Create timing constraints	Quartus II timequest timing analyzer SDC editor
Pin constrain wizard	PlanAhead (PinPlanning)	Pin planner
Implementation	Translate, map, place and route	Quartus II integrated synthesis (QIS), fitter
Static timing analysis	Xilinx timing analyzer	Timequest timing analyzer
Generate programming file	BitGen	Assembler
Power estimator	XPower estimator	Powerplay early power estimator
Power analysis	XPower analyzer	Powerplay power analyzer
Simulation	ISE simulator (ISim)	Modelsim–Altera starter edition
	Third-party simulation tools	Third-party simulation tools
Co-simulation	ISim co-simulation	–
In-chip verification	Chipscope pro	SignalTap II logic analyzer
View and editing placement	PlanAhead, FPGA editor	Chip planner
Configure device	iMPACT	Programmer

3. Implement the pipelined adder of Chap. 3. Analyze the area-time-power trade off for different logic depths in the circuit. Use, for the experiment, a Virtex 5 device and compare the result with respect to the new generation Virtex 7. What happened in Altera, in comparing stratix III devices with respect to Stratix V?
4. Implement the adders of Chap. 7 (use the provided VHDL models). Add to the model input and output registers in order to estimate the maximum frequency of operation.
5. Compare the results of implementation of radix 2^k adders of Sect. 7.3 in Xilinx devices using *muxcy* and a behavioural description of the multiplexer (use the provided VHDL models of Chap. 7).
6. Implement the restoring, non-resorting, radix-2 SRT and radix 2^k SRT divider of Chap. 9 in Xilinx and Altera devices (use the provided VHDL models of Chap. 9).
7. Compare the results of square root methods of Chap. 10 in Altera and Xilinx design flow.

References

1. Altera corp (2009) AN 307: Altera design flow for Xilinx users. http://www.altera.com/literature/an/an307.pdf
2. Altera corp (2011a) Altera quartus II software environment. http://www.altera.com/
3. Altera corp (2011b) Design and synthesis. In: Quartus II integrated synthesis, quartus II handbook version 11.0, vol 1. http://www.altera.com/
4. Mentor Graphics (2011) Mentor precision synthesis reference manual 2011a. http://www.mentor.com/
5. Microsemi SoC Products Group (2010) Libero integrated design environment (IDE) v9.1. http://www.actel.com/
6. Synopsys (2011) Synopsys FPGA synthesis user guide (Synplify, Synplify Pro, or Synplify Premier). http://www.synopsys.com/
7. Xilinx inc (2011a) Xilinx ISE (Integrated Integrated software environment) design suite. http://www.xilinx.com/
8. Xilinx inc (2011b) XST (Xilinx Synthesis Technology) user guide, UG687 (v 13.1). http://www.xilinx.com/

Chapter 7
Adders

Addition is a primitive operation for most arithmetic functions, so that FPGA vendors have dedicated a particular attention to the design of optimized adders. As a consequence, in many cases the synthesis tools are able to generate fast and cost-effective adders from simple VHDL expressions. Only in the case of relatively long operands can it be worthwhile to consider more complex structures such as carry-skip, carry-select and logarithmic adders.

Another important topic is the design of multi-operand adders. In this case, the key concept is that of carry-save adder or, more generally, of parallel counter.

Obviously, the general design methods presented in Chap. 3 (pipelining, digit-serial processing, self-timing) can be applied in order to optimize the proposed circuits. Numerous examples of practical FPGA implementations are reported in Sect. 7.9.

7.1 Addition of Natural Numbers

Consider two radix-B numbers

$$x = x_{n-1} \cdot B^{n-1} + x_{n-2} \cdot B^{n-2} + \cdots + x_1 \cdot B + x_0$$

and

$$y = y_{n-1} \cdot B^{n-1} + y_{n-2} \cdot B^{n-2} + \cdots + y_1 \cdot B + y_0,$$

where all digits x_i and y_i belong to $\{0, 1,\ldots, B-1\}$, and an input carry c_0 belonging to $\{0, 1\}$. An n-digit adder generates a radix-B number

$$z = z_{n-1} \cdot B^{n-1} + z_{n-2} \cdot B^{n-2} + \cdots + z_1 \cdot B + z_0,$$

J.-P. Deschamps et al., *Guide to FPGA Implementation of Arithmetic Functions*, Lecture Notes in Electrical Engineering 149, DOI: 10.1007/978-94-007-2987-2_7,

Fig. 7.1 *n*-digit adder

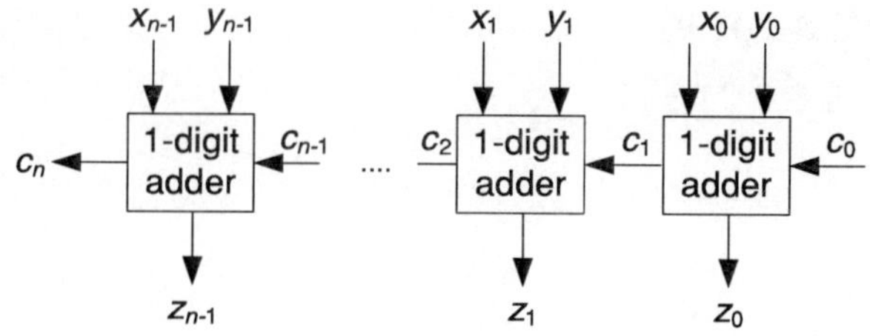

and an output carry c_n, such that

$$x + y + c_0 = c_n \cdot B^n + z.$$

Observe that $x + y + c_0 \leq 2(B^n - 1) + 1 = 2B^n - 1$, so that c_n belongs to {0, 1}.

The common way to implement an *n*-digit adder consists of connecting in series *n* 1-digit adders (Fig. 7.1). For each of them

$$x_i + y_i + c_i = c_{i+1} \cdot B + z_i,$$

where c_i and c_{i+1} belong to {0, 1}. In other words

$$z_i = (x_i + y_i + c_i) \bmod B, \; c_{i+1} = \lfloor (x_i + y_i + c_i)/B \rfloor.$$

The critical path is

$$(x_0, y_0, c_0) \rightarrow c_1 \rightarrow c_2 \rightarrow \cdots \rightarrow c_{n-1} \rightarrow (z_{n-1}, c_n),$$

so that the total computation time is approximately equal to $n \cdot T_{carry}$ where T_{carry} is the computation time of c_{i+1} in function of x_i, y_i and c_i.

In order to reduce T_{carry}, it is convenient to compute two binary functions *p* (*propagate*) and *g* (*generate*) of x_i and y_i:

$$p(x_i, y_i) = 1 \text{ if } x_i + y_i = B - 1, p(x_i, y_i) = 0 \text{ otherwise;}$$

$g(x_i, y_i) = 1$ if $x_i + y_i \geq B$, $g(x_i, y_i) = 0$ if $x_i + y_i \leq B - 2$, otherwise, any value. So, c_{i+1} can be expressed under the following way:

$$c_{i+1} = p(x_i, \; y_i) \cdot c_i + \overline{p(x_i, \; y_i)} \cdot g(x_i, \; y_i)$$

The last relation expresses that if $x_i + y_i = B-1$, then c_{i+1} is equal to c_i; if $x_i + y_i \geq B$, then $c_{i+1} = 1$; if $x_i + y_i \leq B-2$, then $c_{i+1} = 0$. The corresponding implementation is shown in Fig. 7.2. It is made up of two 2-operand combinational circuits that compute $p(x_i, y_i)$ and $g(x_i, y_i)$, and a multiplexer. In an *n*-digit adder (Fig. 7.1), all functions $p(x_i, y_i)$ and $g(x_i, y_i)$ are computed in parallel, so that the value of T_{carry} is practically equal to the multiplexer delay T_{mux}.

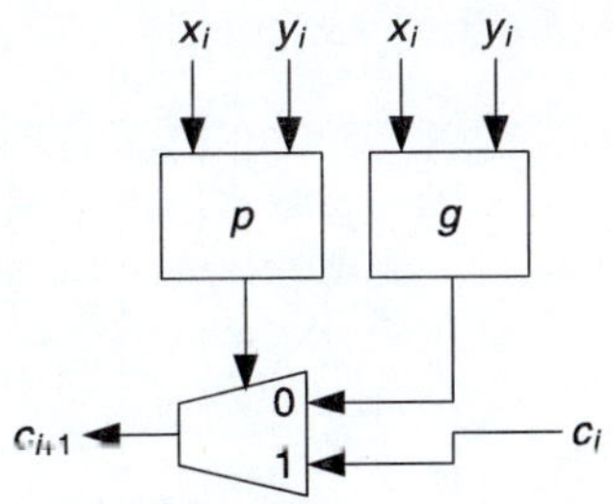

Fig. 7.2 Carry computation

7.2 Binary Adder

If $B = 2$, then $p(x_i, y_i) = x_i$ XOR y_i, and $g(x_i, y_i)$ can be chosen equal to x_i (or y_i). A complete n-bit adder is shown in Fig. 7.3. Its computation time is equal to

$$T_{adder}(n) = T_{xor} + (n - 1) \cdot T_{mux} + \max\{T_{mux}, T_{xor}\}, \tag{7.1}$$

and the delay from the input carry to the output carry is equal to

$$T_{carry-to-carry}(n) = n \cdot T_{mux}. \tag{7.2}$$

Comment 7.1
Most FPGA's include the basic components to implement the structure of Fig. 7.3, and the synthesis tools automatically generate this optimized adder from a simple VHDL expression such as

```
z <= x + y + c0;
```

7.3 Radix-2^k Adder

If $B = 2^k$, then $p(x_i, y_i) = 1$ if $x_i + y_i = 2^k - 1$, that is, if the k less significant bits of $s_i = x_i + y_i$ are equal to 1, and $g(x_i, y_i) = 1$ if $x_i + y_i \geq 2^k$, that is, if the most significant bit of s_i is equal to 1. The iterative cell of a radix-2^k adder is shown in Fig. 7.4. The critical path of the part of the circuit that computes $g(x_i, y_i)$ and $p(x_i, y_i)$ has been shaded. Its computation time is equal to $T_{adder}(k) + T_{and}$. An m-digit radix-2^k adder is equivalent to an n-bit adder with $n = m \cdot k$. The total computation time is

$$T_{adder}(n) = T_{adder}(k) + T_{and} + (m - 1) \cdot T_{mux} + T_{half-adder}(k), \tag{7.3}$$

and the delay from the input carry to the output carry to

$$T_{carry-to-carry}(n) = m \cdot T_{mux}. \tag{7.4}$$

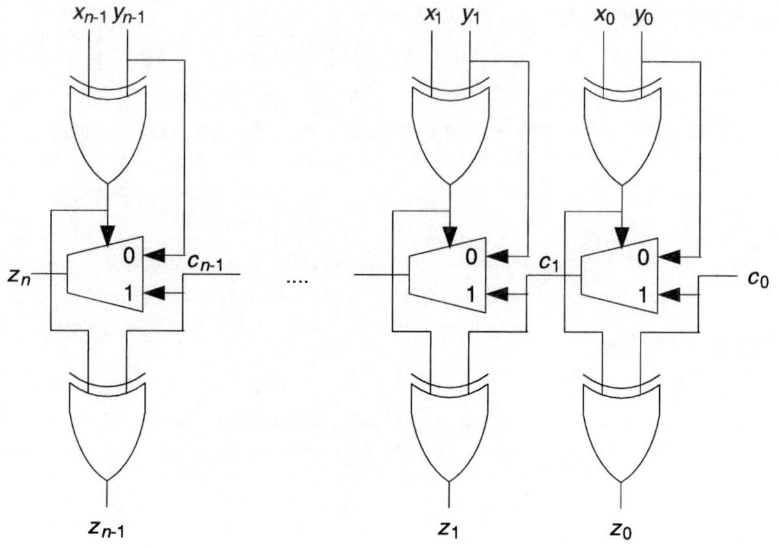

Fig. 7.3 *n*-bit adder

The following VHDL model describes the basic cell of Fig. 7.4.

```
s <= '0'&x + y;
t(1) <= s(0);
and_gates: FOR i in 1 TO k-1 GENERATE
  t(i+1) <= t(i) AND s(i);
END GENERATE;

p <= t(k);
WITH p SELECT c_out <= c_in WHEN '1', s(k) WHEN OTHERS;
z <= s(k-1 DOWNTO 0) + c_in;
```

An alternative way of computing $p = s_0 \cdot s_1 \cdot \ldots \cdot s_{k-1}$ is

```
t <= '0'&s(k-1 DOWNTO 0) + '1';
p <= t(k);
```

The corresponding circuit is a *k*-bit half adder that computes $t = (s \bmod 2^k) + 1$. The most significant bit t_k of t is equal to 1 if, and only if, all the bits of $(s \bmod 2^k)$ are equal to 1. As mentioned above (Comment 7.1) most FPGA's include the basic components to implement the structure of Fig. 7.3. In the particular case where $x = 0$, $y = s$ and $c_0 = 1$, the circuit of Fig. 7.4 is obtained. The apparently unnecessary XOR gates are included because there is generally no direct connection between the adder inputs and the multiplexer control inputs. Actually, the

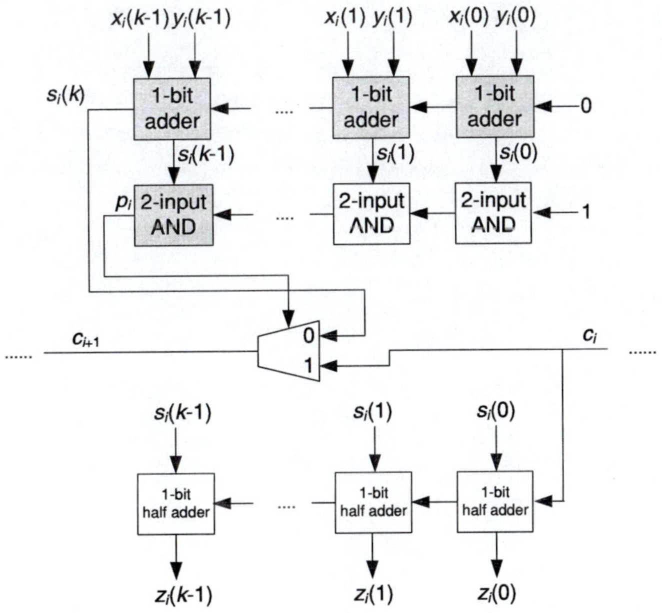

Fig. 7.4 Radix 2^k adder

Fig. 7.5 FPGA implementation of a k-input AND

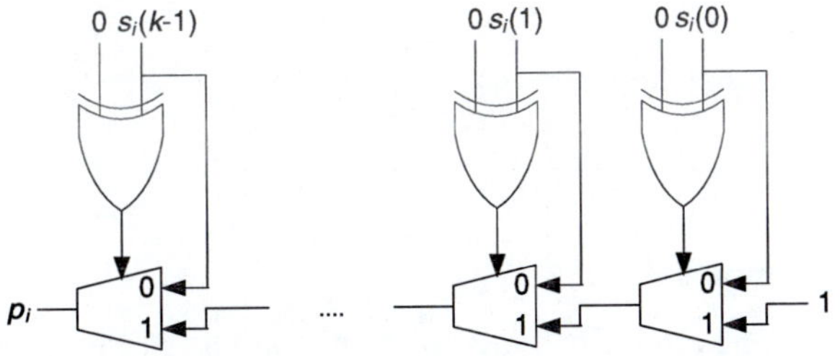

XOR gates are LUTs whose outputs are permanently connected to the carry-logic multiplexers.

A complete generic model *base_2 k_adder.vhd* is available at the Authors' web page and examples of FPGA implementations are given in Sect. 7.9.

According to (7.3), the non-constant terms of $T_{adder}(n)$ are:

- $m{\cdot}T_{mux}$,
- $k{\cdot}T_{mux}$ included in $T_{adder}(k)$ according to (7.1),
- $k{\cdot}T_{mux}$ included in $T_{half\text{-}adder}(k)$ according to (7.1).

Thus, the sum of the non-constant terms of $T_{adder}(n)$ is equal to $(2k + m){\cdot}T_{mux}$. The value of $2k + m$, with $m{\cdot}k = n$, is minimum when $2k \cong m$, that is, when $k \cong (n/2)^{1/2}$. With this value of k, the sum of the non-constant terms of $T_{adder}(n)$ is equal to $(8n)^{½}{\cdot}T_{mux}$. Thus, the computation time is $O(n)^{½}$ instead of $O(n)$.

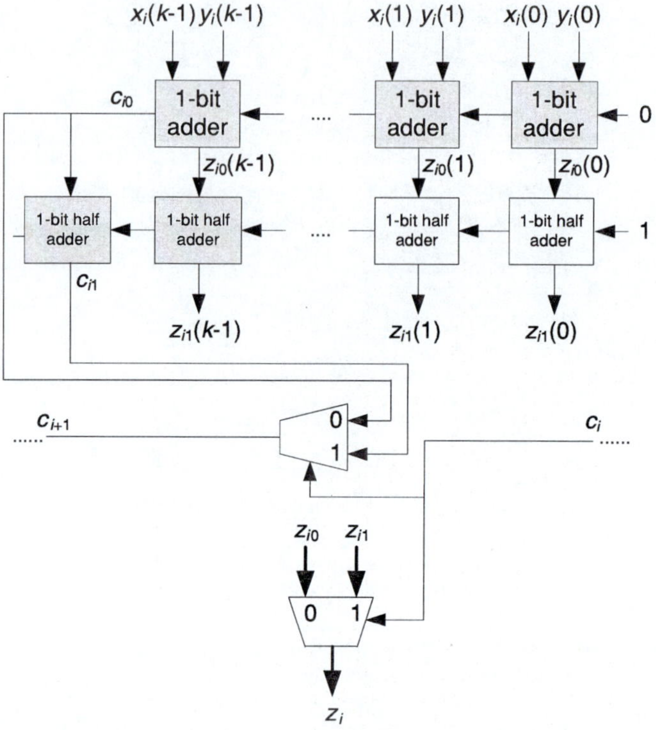

Fig. 7.6 Carry select adder

Comments 7.2

1. The circuit of Fig. 7.4 is an example of *carry-skip adder*. For every group of k bits, both the carry-propagate and carry-generate functions are computed. If the carry-propagate function is equal to 1, the input carry is directly propagated to the carry output of the k-bit group, thus skipping k bits.
2. A mixed-radix numeration system could be used. Assume that $n = k_1 + k_2 + \ldots + k_m$; then a radix

$$(2^{k_1}, 2^{k_2}, \cdots, 2^{k_m})$$

representation can be considered. The corresponding adder consists of m blocks, similar to that of Fig. 7.3, whose sizes are k_1, k_2,…, and k_m, respectively. Nevertheless, within an FPGA it is generally better to use adders that fit within a single column. Assuming that the chosen device has r carry-logic cells per column, a good option could be a fixed-radix adder with $k \leq r$. In order to minimize the computation time, k must be approximately equal to $(n/2)^{1/2}$, so that n must be smaller than $2r^2$, which is a very large number.

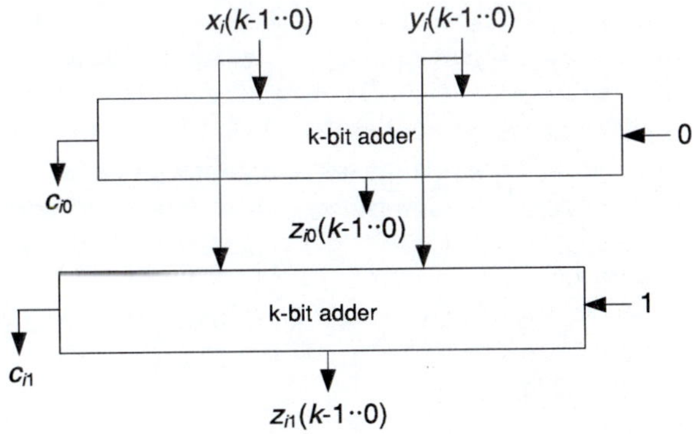

Fig. 7.7 Carry-select adder (second version)

7.4 Carry Select Adders

Another way of reducing the computation time of a radix-2^k adder consists in computing, at each step, the next carry and the output digit for both values of the input carry. The corresponding circuit is shown in Fig. 7.6.

The critical path of the part of the circuits that computes the two possible values of the next carry and of the output digit has been shaded. Its computation time is equal to $T_{adder}(k) + T_{adder}(2)$. The total computation time is ($n = m{\cdot}k$)

$$T_{adder}(n) = T_{adder}(k) + T_{half\_adder}(2) + (m - 1) \cdot T_{mux} + T_{mux}, \tag{7.5}$$

and the delay from the input carry to the output carry to

$$T_{carry-to-carry}(m \cdot k) = m \cdot T_{mux}. \tag{7.6}$$

The following VHDL model describes the basic cell of Fig. 7.6.

```
t0 <= '0'&x + y;
t1 <= '0'&x + y + '1';
c0 <= t0(k);
c1 <= t1(k);
z0 <= t0(k-1 DOWNTO 0);
z1 <= t1(k-1 DOWNTO 0);
WITH c_in SELECT c_out <= c0 WHEN '0', c1 WHEN OTHERS;
WITH c_in SELECT z <= z0 WHEN '0', z1 WHEN OTHERS;
```

A complete generic model *carry_select_adder.vhd* is available at the Authors' web page and examples of FPGA implementations are given in Sect. 7.9.

The non-constant term of $T_{adder}(n)$ is equal to $(k + m){\cdot}T_{mux}$. The minimum value is obtained when $k \cong m$, that is $k \cong (n)^{1/2}$. With this value of k, the non-constant term of $T_{adder}(n)$ is equal to $(4n)^{1/2}{\cdot}T_{mux}$. Thus, the computation time is $O(n)^{1/2}$ instead of $O(n)$.

Comment 7.3
As before (Comments 7.2) a mixed-radix numeration system could be considered.

As a matter of fact, the FPGA implementation of a half-adder is generally not more cost-effective than the implementation of a full adder. So, the circuit of Fig. 7.6 could be slightly modified: instead of computing c_{i0} and c_{i1} with a full adder and a half adder, two independent full adders of any type can be used (Fig. 7.7).

The following VHDL model describes the modified cell:

```
t0 <= '0'&x + y;
t1 <= t0 + '1';
c0 <= t0(k);
c1 <= t1(k);
z0 <= t0(k-1 DOWNTO 0);
z1 <= t1(k-1 DOWNTO 0);
WITH c_in SELECT c_out <= c0 WHEN '0', c1 WHEN OTHERS;
WITH c_in SELECT z <= z0 WHEN '0', z1 WHEN OTHERS;
```

The computation time of the modified circuit is

$$T_{adder}(n) = T_{adder}(k) + (m - 1) \cdot T_{mux} + T_{mux} = T_{adder}(k) + m \cdot T_{mux}. \quad (7.7)$$

A complete generic model *carry_select_adder2.vhd* is available at the Authors' web page and examples of FPGA implementations are given in Sect. 7.9.

7.5 Logarithmic Adders

Several types of adders whose computation time are proportional to the logarithm of n have been proposed. For example: carry-lookahead adders ([1], Chap. 6), Ling adders [2], Brent-Kung prefixed adders [3], Ladner-Fischer prefixed adders [4]. Nevertheless, their FPGA implementations are generally not as fast as what could be theoretically expected. There are two reasons for that. On the one hand, the special purpose carry-logic included in most FPGAs is very fast, so that ripple-carry adders are fast. Their computation time is approximately equal to $a + b{\cdot}n$, where a and b are very small constants: a is the delay of a LUT and b is the delay of a multiplexer belonging to the carry logic. On the other hand, the structure of most logarithmic adders is not so regular as the structure of ripple-carry adders, so that they include long connections which in turn introduce long additional delays. The practical result is that, except for very great values of n, the adders described in Sects. 7.2–7.4 are faster and more cost-effective.

Obviously, any optimization method that allows the dividing up of an n-bit adder into smaller k-bit and m-bit adders, with $k{\cdot}m = n$, in such a way that

$$T_{adder}(n) \cong T_{adder}(k) + T_{adder}(m),$$

can be recursively used in order to generate a logarithmic adder. As an example, consider again a carry-select adder. According to (7.7)

$$T_{adder}(n) = T_{adder}(k) + m.T_{mux}.$$

Assume that $k = r \cdot s$. Then each k bit adder (Fig. 7.7) can in turn be decomposed in such a way that

$$T_{adder}(k) = T_{adder}(r) + s.T_{mux},$$

so that the computation time of the corresponding 2-level carry-select adder is

$$T_{adder}(n) = T_{adder}(r) + (s + m) \cdot T_{mux},$$

where $n = r \cdot s \cdot m$. Finally, if $n = n_1 \cdot n_2 \cdot \ldots \cdot n_t$, then a ($t$-1)-level carry-select adder, whose computation time is equal to

$$T_{adder}(n_1 \cdot n_2 \cdot \ldots \cdot n_t) = T_{adder}(n_1) + (n_2 + \ldots + n_t) \cdot T_{mux} = O(n_1 + n_2 + \ldots + n_t),$$

can be generated.

Example 7.1

The following VHDL model describes an n-bit 2-level carry-select adder with $n = n_1 \cdot n_2 \cdot n_3$. First, define the basic cell *carry_select_step3*, in which two 1-level carry-select adders, with $k = n_1$ and $m = n_2$, are used:

```
first_adder: carry_select_adder2
GENERIC MAP(k => n1, m => n2)
PORT MAP(x => x, y => y, c_in => '0', z => z0, c_out => c0);
second_adder: carry_select_adder2
GENERIC MAP(k => n1, m => n2)
PORT MAP(x => x, y => y, c_in => '1', z => z1, c_out => c1);
WITH c_in SELECT c_out <= c0 WHEN '0', c1 WHEN OTHERS;
WITH c_in SELECT z <= z0 WHEN '0', z1 WHEN OTHERS;
```

The complete circuit is made up of n_3 basic cells:

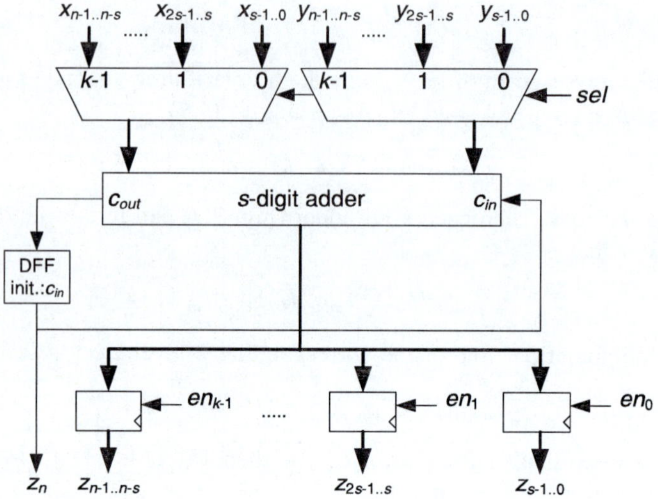

Fig. 7.8 Long-operand adder

```
carries(0) <= c_in;
iteration: FOR i IN 0 TO n3-1 GENERATE
  main_component:carry_select_step3
  GENERIC MAP(n1 => n1, n2 => n2)
  PORT MAP(x => x(n1*n2*i+n1*n2-1 DOWNTO n1*n2*i),
  y => y(n1*n2*i+n1*n2-1 DOWNTO n1*n2*i), c_in => carries(i),
  z => z(n1*n2*i+n1*n2-1 DOWNTO n1*n2*i),
  c_out => carries(i+1));
END GENERATE;
c_out <= carries(n3);
```

A complete generic model *carry_select_adder3.vhd* is available at the Authors' web page and examples of FPGA implementations are given in Sect. 7.9.

7.6 Long-Operand Adder

In the case of long-operand additions, the n-digit operands can be broken down into s-digit groups and the addition computed according to the following algorithm in which *natural_addition* is a procedure that computes

$$z_i = (x_i + y_i + c_i) \bmod B^s \text{ and } c_{i+1} = \lfloor (x_i + y_i + c_i) / B^s \rfloor,$$

where x_i, y_i and z_i are s-digit numbers, and c_i and c_{i+1} are bits.

Algorithm 7.1: Long-operand addition

```
c0 := cin;
for i in 0 .. n/s - 1 loop
  natural_addition(ci, x(i·s+s-1..i·s), y(i·s+s-1..i·s), ci+1,
    z(i·s+s-1..i·s));
end loop;
zn := cn/s;
```

The complete circuit (Fig. 7.8, with $k = n/s$) is made up of an s digit adder, connection resources (k-to-1 multiplexers) giving access to the s-digit groups, a D-flip-flop which stores the carries (c_i in Algorithm 7.1), an output register storing z, and a control unit whose main component is a k-state counter.

The following VHDL model describes the circuit of Fig. 7.8 ($B = 2$).

```
adder_in1 <= x(sel*s+s-1 DOWNTO sel*s);
adder_in2 <= y(sel*s+s-1 DOWNTO sel*s);
sum <= '0'&adder_in1 + adder_in2 + q;
adder_out <= sum(s-1 DOWNTO 0);
next_q <= sum(s);
registers: FOR i IN 0 TO k-1 GENERATE
  PROCESS(clk)
  BEGIN
    IF clk'EVENT AND clk = '1' THEN
      IF (update = '1') AND (sel = i) THEN
        z(i*s+s-1 DOWNTO i*s) <= adder_out;
      END IF;
    END IF;
  END PROCESS;
END GENERATE;
flip_flop: PROCESS(clk)
BEGIN
  IF clk'EVENT AND clk = '1' THEN
    IF load = '1' THEN q <= c_in;
    ELSIF update = '1' THEN q <= next_q; END IF;
  END IF;
END PROCESS;
```

A complete generic model *long_operand_adder.vhd* is available at the Authors' web page.

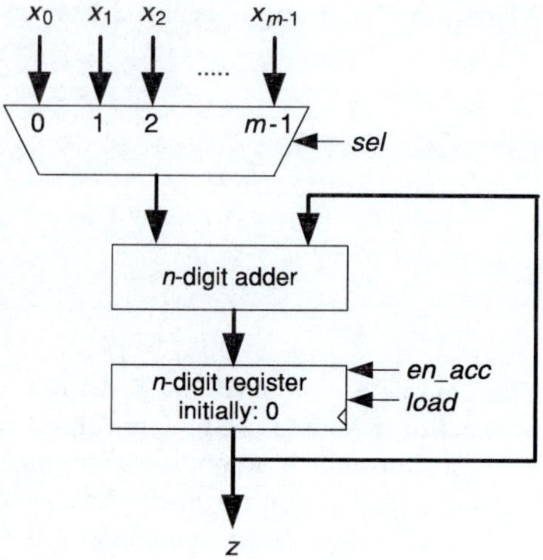

Fig. 7.9 Multioperand addition

7.7 Multioperand Adders

Consider m natural numbers x_0, x_1,..., x_{m-1}. A multioperand adder computes

$$z = x_0 + x_1 + \cdots + x_{m-1}. \tag{7.8}$$

7.7.1 Sequential Multioperand Adders

In order to compute (7.8), the following (obvious) algorithm can be used.

Algorithm 7.2: Basic multioperand addition

```
accumulator := 0;
for j in 0 .. m-1 loop
  accumulator := accumulator + xj;
end loop;
z := accumulator;
```

The corresponding sequential circuit (Fig. 7.9) is made up of an n-digit adder, an n-digit register, an m-to-1 n-digit multiplexer, and a control unit whose main component is an m-state counter.

The following VHDL model describes the circuit of Fig. 7.9 ($B = 2$). The $n{\cdot}m$-bit vector x is the concatenation of x_0, x_1,..., x_{m-1}.

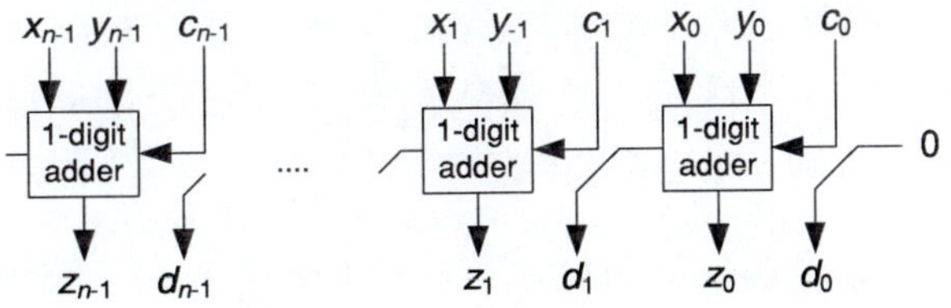

Fig. 7.10 Carry-save adder

```
adder_in1 <= x(sel*n+n-1 DOWNTO sel*n);
adder_out <= adder_in1 + accumulator;
z <= accumulator;
accumulator_register: PROCESS(clk)
BEGIN
  IF clk'EVENT AND clk = '1' THEN
    IF load = '1' THEN accumulator <= (OTHERS => '0');
    ELSIF en_acc = '1' THEN accumulator <= adder_out; END IF;
  END IF;
END PROCESS;
```

A complete generic model *multioperand_adder.vhd* is available at the Authors' web page.

The computation time of the preceding m-operand n-digit sequential adder is approximately equal to

$$T_{sequential}(m, n) \cong m \cdot T_{adder}(n). \tag{7.9}$$

In order to reduce the computation time, a *carry-save adder* can be used. The basic component is shown in Fig. 7.10: it consists of n 1-digit adders working in parallel. Given two n-digit numbers x and y, and an n-bit number c, it expresses the sum $(x + y + c) \bmod B^n$ under the form $z + d$, where z is an n-digit number and d an n-bit number. In other words, the carries are stored within the output binary vector d instead of being propagated (*stored-carry encoding*). As all cells work in parallel the computation time is independent of n.Let *CSA* be the function implemented by the circuit of Fig. 7.10, that is

$$CSA(x,\ y,\ c) = (z,\ d),$$

where

$$z_i = (x_i + y_i + c_i) \bmod B, d_i = \lfloor (x_i + y_i + c_i)\,/B \rfloor, \forall i \in \{0,\ 1, \ldots, n-1\}.$$

Assume that at every step of Algorithm 7.2 the value of *accumulator* is represented under the form $u + v$, where u is an n-digit number and v an n-bit number. Then, at step j, the following operation must be executed:

$$(u, v) := CSA(u, x_j, v).$$

The following formal algorithm computes z.

Fig. 7.11 Carry save adder

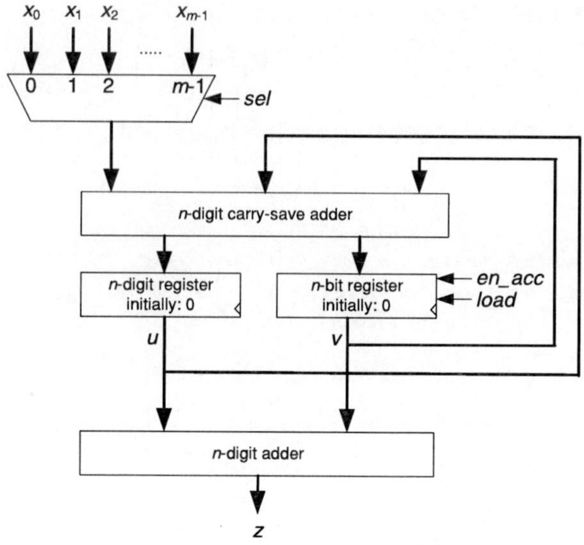

Algorithm 7.3: Multioperand addition with stored-carry encoding

```
u0 := 0; v0 := 0;
for j in 0 .. m-1 loop
  (uj+1, vj+1) := CSA(uj, xj, vj);
end loop;
z := um + vm;
```

The sequential circuit corresponding to Algorithm 7.3 (Fig. 7.11) is made up of an n-digit carry-save adder (Fig. 7.10), an n-digit register, an n-bit register, an m-to-1 n-digit multiplexer, a conventional n-digit adder implementing the last step of Algorithm 7.3, and a control unit whose main component is an m-state counter.The following VHDL model describes the circuit of Fig. 7.10 ($B = 2$). As before, x is the concatenation of $x_0, x_1, \ldots, x_{m-1}$.

```
adder_in1 <= x(sel*n+n-1 DOWNTO sel*n);
next_v(0) <= '0';
csa: FOR i IN 0 TO n-2 GENERATE
  next_u(i) <= adder_in1(i) XOR u(i) XOR v(i);
  next_v(i+1) <=
    (adder_in1(i) AND u(i)) OR (adder_in1(i) AND v(i))
    OR (u(i) AND v(i));
END GENERATE;
```

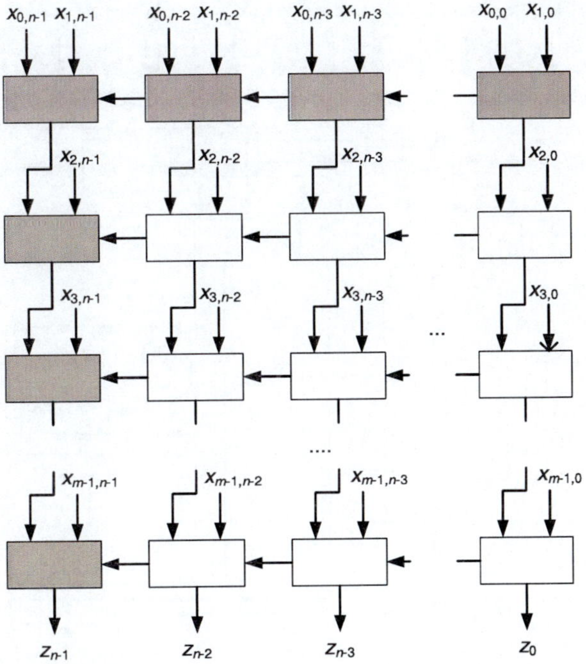

Fig. 7.12 Multioperand addition array

```
next_u(n-1) <= adder_in1(n-1) XOR u(n-1) XOR v(n-1);
z <= u + v;
accumulator_register: PROCESS(clk)
BEGIN
  IF clk'EVENT AND clk = '1' THEN
    IF load = '1' THEN
      u <= (OTHERS => '0'); v <= (OTHERS => '0');
    ELSIF en_acc = '1' THEN u <= next_u; v <= next_v; END IF;
  END IF;
END PROCESS;
```

A complete generic model *CSA_multioperand_adder.vhd* is available at the Authors' web page.

Taking into account that the computation time of the circuit of Fig. 7.10 is independent of the number n of digits, the computation time of the circuit of Fig. 7.10 is approximately equal to

$$T_{sequential\_csa}(m, n) \cong m \cdot T_{adder}(1) + T_{adder}(n). \qquad (7.10)$$

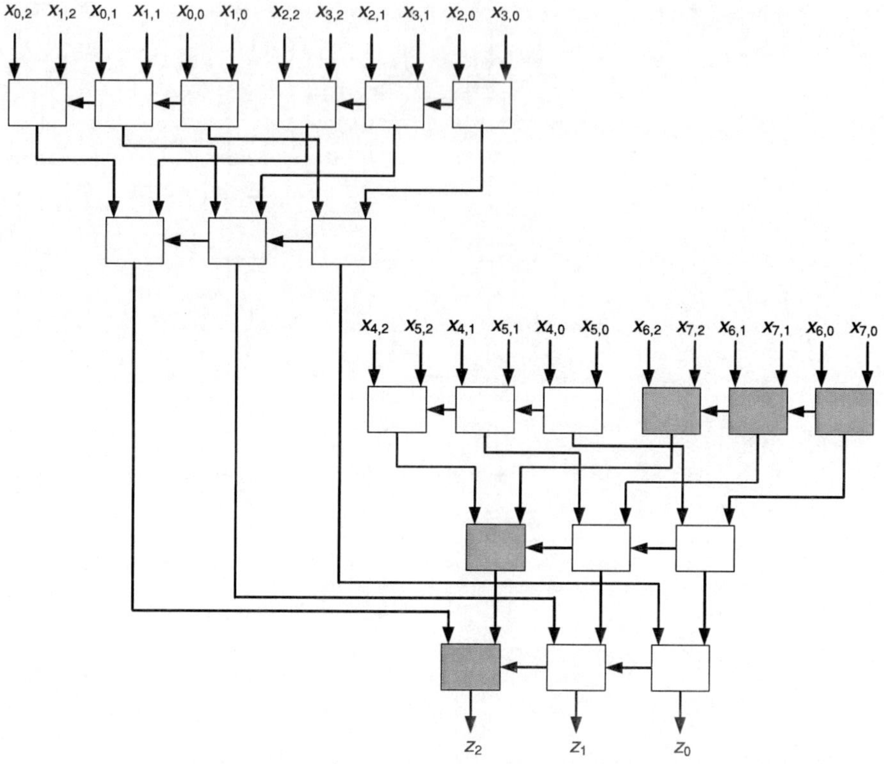

Fig. 7.13 Multioperand addition tree

7.7.2 Combinational Multioperand Adders

The combinational circuit that corresponds to Algorithm 7.2 is an iterative circuit made up of m-1 2-operand n-digit adders. If every adder is a simple ripple-carry adder, then the complete circuit is a 2-dimensional array made up of $(m\text{-}1)\cdot n$ one-digit adders, as shown in Fig. 7.12 in which one of the critical paths has been shaded. The corresponding computation time is equal to

$$T_{combinational}(m,\ n) = (m + n - 2) \cdot T_{adder}(1). \tag{7.11}$$

The following VHDL model describes the circuit of Fig. 7.12 ($B = 2$). As before, x is the concatenation of x_0, x_1,..., $x_{m\text{-}1}$.

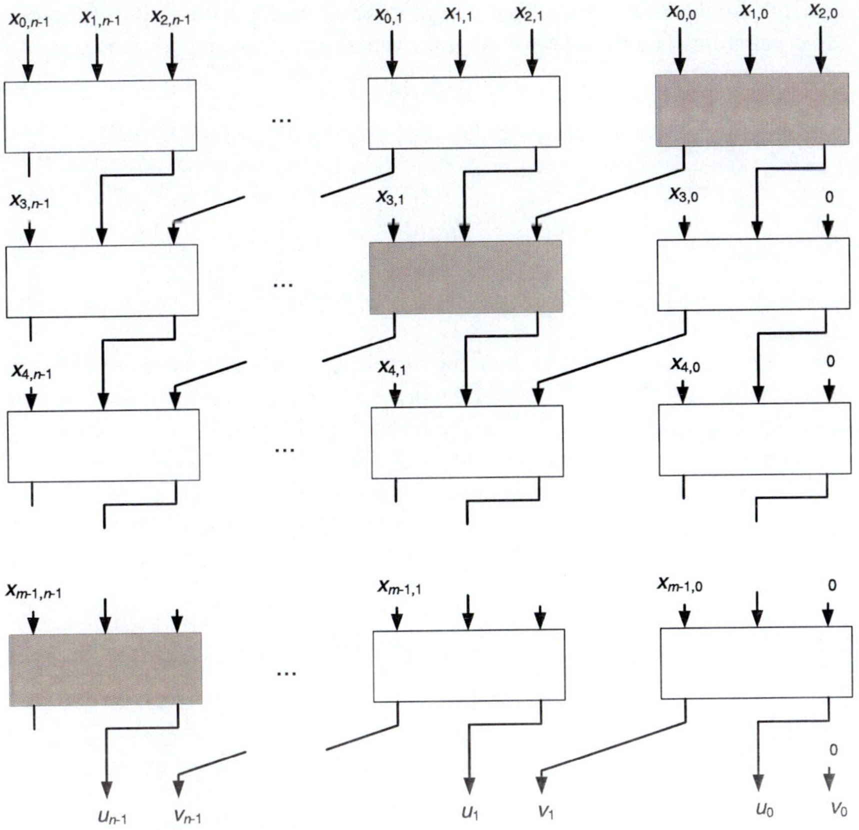

Fig. 7.14 Combinational carry-save adder

```
y(2*n-1 DOWNTO n) <= x(2*n-1 DOWNTO n) + x(n-1 DOWNTO 0);
iteration: FOR i in 2 TO m-1 GENERATE
  y(i*n+n-1 DOWNTO i*n) <=
    y(i*n-1 DOWNTO i*n-n) + x(i*n+n-1 DOWNTO i*n);
END GENERATE;
z <= y(m*n-1 DOWNTO m*n-n);
```

A complete generic model *comb_multioperand_adder.vhd* is available at the Authors' web page.

A most time-effective solution is a binary tree of 2-operand n-digit adders instead of an iterative circuit. An example, with $n = 3$ and $m = 8$, is shown in Fig. 7.13:

$$x_0 = x_{0,2}x_{0,1}x_{0,0}, x_1 = x_{1,2}x_{1,1}x_{1,0}, \ldots, x_7 = x_{7,2}x_{7,1}x_{7,0}.$$

The depth of the tree is equal to $\lceil log_2 m \rceil$ and its computation time (one of the critical paths has been shaded) is approximately equal to

$$T_{adder-tree}(m, n) \cong (n + \log_2 m - 1) \cdot T_{adder}(1). \tag{7.12}$$

The following VHDL model describes the circuit of Fig. 7.13 ($B = 2$).

```
y0 <= x0 + x1; y1 <= x2 + x3; y2 <= x4 + x5; y3 <= x6 + x7;
y4 <= y0 + y1; y5 <= y2 + y3;
z <= y4 + y5;
```

A complete generic model *eight_operand_adder.vhd* is available at the Authors' web page.

Another way to reduce the computation time, with an iterative architecture similar to that of Fig. 7.12, is to use the *carry-save* principle. An m-operand carry-save array (Algorithm 7.3) is shown in Fig. 7.14 (if $B > 2$, x_2 must be an n-bit number or an initial file that computes $x_0 + x_1 + 0$ must be added). The result is the sum of two n-digit numbers u and v. In order to get the actual result, an additional 2-operand n-digit adder is necessary for computing $u + v$ (last instruction of Algorithm 7.3). The corresponding computation time is equal to

$$T_{combinational\_csa}(m, n) = (m - 2) \cdot T_{adder}(1) + T_{adder}(n). \tag{7.13}$$

The following VHDL model describes a 2-operand carry-save adder, also called 3-to-2 *counter* (Sect. 7.7.3). It corresponds to a file of the circuit of Fig. 7.14.

```
v(0) <= '0';
iteration: FOR i IN 0 TO n-2 GENERATE
  u(i) <= a(i) XOR b(i) XOR c(i);
  v(i+1) <=
    (a(i) AND b(i)) OR (a(i) AND c(i)) OR (b(i) AND c(i));
END GENERATE;
u(n-1) <= a(n-1) XOR b(n-1) XOR c(n-1);
```

The complete circuit is made up of m-2 3-to-2 counters:

```
first_row: three_to_two_counter GENERIC MAP(n => n)
PORT MAP(a => x(n-1 DOWNTO 0), b => x(n+n-1 DOWNTO n),
  c => x(2*n+n-1 DOWNTO 2*n), u => u(n-1 DOWNTO 0),
  v => v(n-1 DOWNTO 0));
iteration: FOR i IN 1 TO m-3 GENERATE
  following_rows: three_to_two_counter GENERIC MAP(n => n)
  PORT MAP(a => x((i+2)*n+n-1 DOWNTO (i+2)*n),
    b => u((i-1)*n+n-1 DOWNTO (i-1)*n),
    c => v((i-1)*n+n-1 DOWNTO (i-1)*n),
    u => u(i*n+n-1 DOWNTO i*n), v => v(i*n+n-1 DOWNTO i*n));
END GENERATE;
z <= u((m-3)*n+n-1 DOWNTO (m-3)*n) +
  v((m-3)*n+n-1 DOWNTO (m-3)*n);
```

A complete generic model *comb_CSA_mutioperand_adder.vhd* is available at the Authors' web page and examples of FPGA implementations are given in Sect. 7.9.

Comment 7.4
In all of the previously described multioperand adders, the operands, as well as the result, were assumed to be n-digit numbers. If all of the operands belong to the same range, and the result is known to be an n-digit number, whatever the value of the operands, then the operands can be represented with $(n\text{–}k)$ digits where $k \cong log_B m$, and the previously described circuits can be pruned.

7.7.3 Parallel Counters

Given two n-digit numbers x and y, and an n-bit number c, the carry-save adder of Fig. 7.10 allows the expression of the sum $(x + y + c)$ mod B^n under the form $z + d$, where z is an n-digit numbers and d an n-bit number. In other words, it reduces the sum of three digits x, y and c to the sum of two digits z and d. For that reason, it is also called a 3-to-2 *counter*.

This 3-to-2 counter can be used as a computation resource for reducing the sum of m digits $x_0, x_1, \ldots, x_{m-1}$ to the sum of two digits u and v as shown in Fig. 7.14. Thus, the circuit of Fig. 7.14 could be considered as an m-to-2 counter.

This type of construction can be generalized. As an example, consider an adder that computes the sum of 6 bits $x_0, x_1, \ldots, x_5$. The result, smaller than or equal to 6, is a 3-bit number. Thus, this 6-operand 1-bit adder computes

$$x_0 + x_1 + \cdots + x_5 = 4z_2 + 2z_1 + z_0 \tag{7.14}$$

and can be implemented by three 6-input Look Up Tables (LUT6) working in parallel:

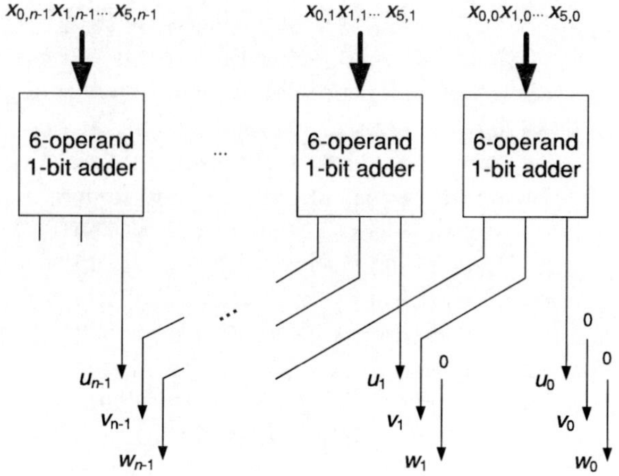

Fig. 7.15 6-to-3 counter

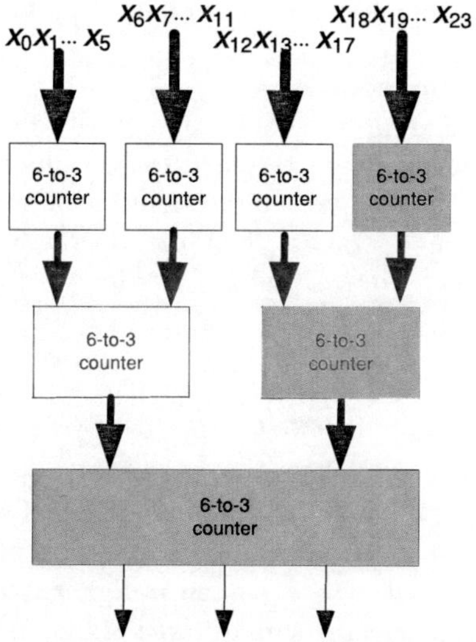

Fig. 7.16 24-to-3 counter

```
first_lut: lut6
GENERIC MAP(truth_vector => x"000101170117177f")
PORT MAP(a => x, b => z(2));
second_lut: lut6
GENERIC MAP(truth_vector => x"177e7ee87ee8e881")
PORT MAP(a => x, b => z(1));
third_lut: lut6
GENERIC MAP(truth_vector => x"6996966996696996")
PORT MAP(a => x, b => z(0));
```

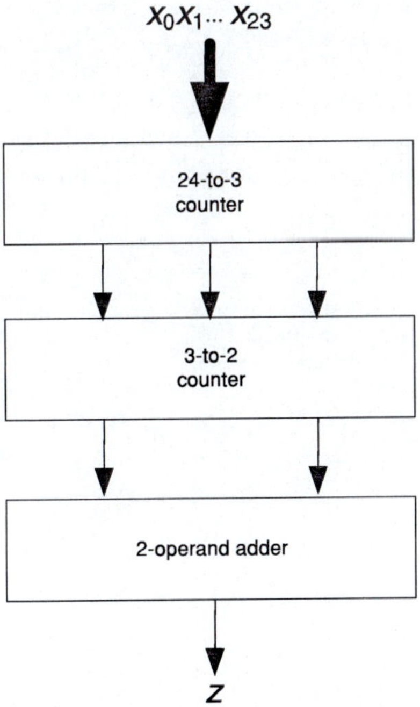

Fig. 7.17 24-operand adder

Then, by connecting in parallel *n* circuits of this type, a binary 6-to-3 counter is obtained (Fig. 7.15):

```
iteration1: FOR i IN 0 TO n-1 GENERATE
  x(i) <= x0(i)&x1(i)&x2(i)&x3(i)&x4(i)&x5(i);
END GENERATE;
iteration2: FOR i IN 0 TO n-1 GENERATE
  cells: six_to_three_counter_cell
  PORT MAP(x => x(i), z => z(i));
  u(i) <= z(i)(0);
END GENERATE;
v(0) <= '0';
iteration3: FOR i IN 1 TO n-1 GENERATE
  v(i) <= z(i-1)(1);
END GENERATE;

w(0) <= '0'; w(1) <= '0';
iteration4: FOR i IN 2 TO n-1 GENERATE
  w(i) <= z(i-2)(2);
END GENERATE;
```

The counter of Fig. 7.15 can in turn be used as a building block for generating

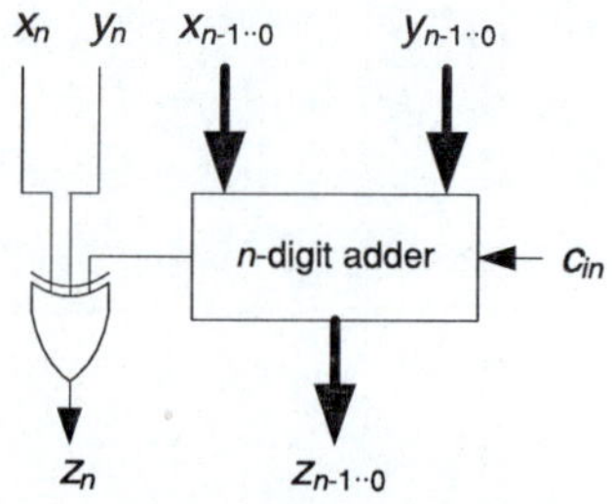

Fig. 7.18 Radix-*B B*'s complement adder

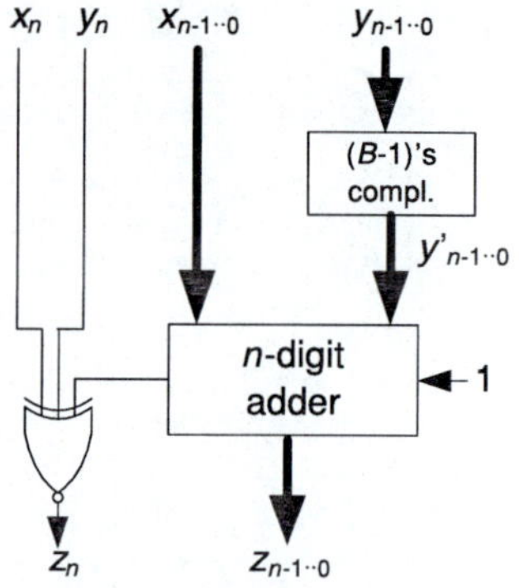

Fig. 7.19 Radix-*B B*'s complement subractor

more complex counters. As an example, the circuit of Fig. 7.16 is a 24-to-3 counter.

The computation time of the circuit of Fig. 7.16 is equal to $3T_{LUT6}$. More generally, a tree made up of 2^k-1 6-to-3 counters generates a $6{\cdot}2^{k-1}$-to-3 counter, with a computation time equal to $k{\cdot}T_{LUT6}$. In the case of Fig. 7.16, $k = 3$ and $6{\cdot}2^{k-1} = 24$.

Finally, with an additional 3-to-2 counter and an *n*-bit adder a 24-operand adder is obtained (Fig. 7.17). Complete VHDL models *six_to_three_counter.vhd* and *twenty_four_operand_adder.vhd* are available at the Authors' web page and examples of FPGA implementations are given in Sect. 7.9.

To summarize, an *m*-operand adder, with $m = 6{\cdot}2^{k-1}$, can be synthesized with 2^k-1 6-to-3 counters plus a 3-to-2 counter and an *n*-bit adder. Its computation time is

$$T(m, n) \cong k \cdot T_{LUT6} + T_{FA} + T_{adder}(n),$$

where $k = log_2 m + 1 - log_2 6 < log_2 m$.

Comment 7.5

More complex types of counters have been proposed (see, for example, Chap. 8 of [1], Chap. 3 of [5], Chap. 11 of [6]). Nevertheless, they do not necessarily give high performance FPGA implementations. As a matter of fact, in many cases the best FPGA implementations are based on relatively simple algorithms, to which correspond regular circuits that allow taking advantage of the special purpose carry

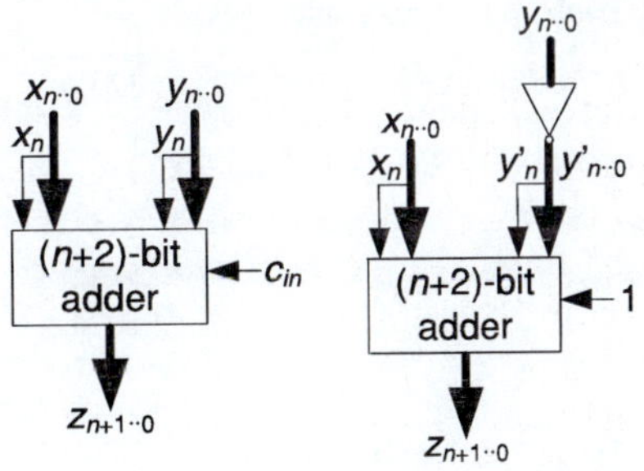

Fig. 7.20 2′s complement adder and subractor

logic circuitry, and permit the use of efficient design techniques such as pipelining and digit-serial processing.

7.8 Subtractors and Adder–Subtractors

Given two radix-B naturals x and y, the difference $z = x - y$ could be negative. So, the subtraction operation must be considered over the set of integers. A convenient way to represent integers is *B's complement*: the vector

$$x_n x_{n-1} x_{n-2} \ldots x_1 x_0, \text{ with } x_n \in \{0, 1\} \text{ and } x_i \in \{0, 1, \ldots, B-1\} \forall i < n,$$

represents

$$x = -x_n \cdot B^n + x_{n-1} \cdot B^{n-1} + x_{n-2} \cdot B^{n-2} + \cdots + x_1 \cdot B + x_0.$$

Thus, x_n is a *sign bit*: if $x_n = 0$, x is a non-negative integer (a natural), and if $x_n = 1$, x is a negative integer. The range of represented integers is

$$-B^n \le x < B^n.$$

Let $x_n\, x_{n-1}\, x_{n-2} \ldots x_1\, x_0$ and $y_n\, y_{n-1}\, y_{n-2} \ldots y_1\, y_0$ be the B's complement representations of x and y. If the sum $z = x + y + c_{in}$, being c_{in} an initial carry, belongs to the interval $-B^n \le z < B^n$, then z is represented by the vector $z_n\, z_{n-1}\, z_{n-2} \ldots z_1\, z_0$ generated by the mixed-radix adder of Fig. 7.18 (all radix-B digits but the most significant binary digits x_n, y_n and z_n).

If the difference $z = x - y$ belongs to the interval $-B^n \le z < B^n$, then z is represented by the vector $z_n\, z_{n-1}\, z_{n-2} \ldots z_1\, z_0$, generated by the circuit of Fig. 7.19 in which y_i' is the (B-1)'s complement of y_i, $\forall i < n$.

The sum $z = x + y$ or the difference $z = x - y$ could lie outside the interval $-B^n \le z < B^n$ (an *overflow* situation). In order to avoid overflows, both x and y should be represented with an additional digit. In the case of B's complement representations, *digit extension* is performed as follows:

$$x_n x_{n-1} x_{n-2} \ldots x_1 x_0 \rightarrow x_n w_n x_{n-1} x_{n-2} \ldots x_1 x_0, \text{ with } w_n = x_n \cdot (B-1).$$

For example, if $B = 10$ and $x = 249$, then x is represented by

Table 7.1 Binary adders

n	LUTs	Delay
32	32	2.25
64	64	2.98
128	128	4.44
256	256	7.35
512	512	13.1
1024	1024	24.82

Table 7.2 Radix-2^k n-bit adders

n	k	LUTs	Delay
32	4	88	2.92
64	4	176	3.11
64	5	143	3.05
64	8	152	3.64
64	16	140	4.95
128	8	304	3.85
128	12	286	4.73
128	16	280	5.04
256	16	560	5.22
256	12	572	4.98
256	13	551	4.99
512	16	1120	5.59
1024	16	2240	6.31
1024	22	2303	6.15
1024	23	2295	6.13
1024	32	2304	6.41

$$0249,\ 00249,\ 000249,\ \text{etc.}$$

If $B = 10$ and $x = -249$, then x is represented by

$$1751,\ 19751,\ 199751,\ \text{etc.}$$

Observe that if $B = 2$, then the *bit extension* operation amounts to repeating the most significant bit. In Fig. 7.20 a 2′s complement adder and a 2′s complement subtractor are shown. In both cases the comparison of bits z_{n+1} and z_n allows the detection of overflows: if $z_{n+1} \neq z_n$ then the result does not belong to the interval $-B^n \leq z < B^n$.

The following VHDL models describe the circuits of Fig. 7.20.

```
z <= (x(n)&x) + (y(n)&y) + c_in;
```

```
z <= (x(n)&x) - (y(n)&y);
```

Generic models *two_s_comp_adder.vhd* and *two_s_comp_subtractor.vhd* are available at the Authors' web page.

Table 7.3 *n*-bit carry-select adders

n	k	LUTs	Delay
32	6	84	4.83
32	8	72	3.99
32	4	60	3.64
64	8	144	6.06
64	16	199	4.17
64	4	120	4.03
128	12	417	5.37
128	16	399	4.87
256	16	799	5.69
256	32	783	5.26
256	13	819	5.64
512	16	1599	6.10
512	32	1567	6.09
512	23	1561	6.16
1024	16	3199	6.69
1024	64	3103	6.74
1024	32	3135	6.52

Table 7.4 *n*-bit carry-select adders (version 2)

n	k	Delay
32	8	3.99
256	16	5.69
512	32	6.09
1024	32	6.52

Table 7.5 *n*-bit adders with $n = n_1 \cdot n_2 \cdot n_3$

n	n_1	n_2	n_3	LUTs	Delay
256	16	4	4	1452	6.32
256	4	16	4	684	6.81
512	8	8	8	2120	10.20
512	4	16	8	1364	7.40
512	16	4	8	2904	7.33
1024	16	4	16	5808	10.33
1024	16	16	4	6242	7.79

7.9 FPGA Implementations

Several adders have been implemented within a Virtex 5-2 device. All along this section, the times are expressed in *ns* and the costs in numbers of Look Up Tables (LUTs) and flip-flops (FF's). All VHDL models as well as several test benches are available at the Authors' web page.

Table 7.6 Long-operand adders

n	s	k	FF	LUTs	Period	Total time
128	8	16	135	107	3.21	51.36
128	16	8	134	97	3.14	25.12
128	32	4	133	132	3.18	12.72
128	64	2	132	137	3.45	6.90
256	16	16	263	187	3.40	54.40
256	32	8	262	177	3.51	28.08
256	64	4	261	234	3.87	15.48
512	16	32	520	381	3.92	125.44
512	32	16	519	347	3.78	60.48
512	64	8	518	337	4.26	34.08
1024	16	64	1033	757	4.20	268.80
1024	32	32	1034	717	4.32	138.24
1024	64	16	1031	667	4.55	72.80
2048	32	64	2063	1427	4.45	284.80
2048	64	32	2056	1389	5.04	161.28

Table 7.7 Sequential multioperand adders

n	m	FF	LUTs	Period	Total time
8	4	12	23	2.25	9.00
8	8	13	32	2.37	18.96
16	16	22	90	2.71	43.36
16	8	21	56	2.57	20.56
32	32	39	363	3.72	119.04
32	16	38	170	3.09	49.44
32	64	40	684	3.89	248.96
64	64	72	1356	4.62	295.68
64	32	71	715	4.48	143.36
64	16	70	330	4.41	70.56

7.9.1 Binary Adder

The circuit is shown in Fig. 7.3. The synthesis results for several numbers n of bits are given in Table 7.1.

7.9.2 Radix 2^k Adders

The circuit is shown in Fig. 7.4. The synthesis results for several numbers $n = 2^k$ of bits are given in the Table 7.2. In these implementations, the carry propagation multiplexer *muxcy* has been explicitly instantiated within the VHDL description.

Table 7.8 Sequential carry-save adders

n	m	FF's	LUTs	Period	Total time
8	4	19	37	1.81	7.24
8	8	20	46	1.81	14.48
16	16	37	120	1.87	29.92
16	8	36	86	1.84	14.72
32	32	70	425	2.57	82.24
32	16	69	232	1.88	30.08
32	64	71	746	2.68	171.52
64	64	135	1482	2.69	172.16
64	32	134	841	2.61	83.52
64	16	133	456	1.9	30.40

Table 7.9 Multioperand addition array

n	m	LUTs	Delay
8	4	21	2.82
8	8	47	5.82
16	16	219	11.98
16	8	103	6.00
32	32	947	24.32
32	8	215	6.36
32	16	459	12.35
32	64	1923	47.11
64	64	3939	49.98
64	32	1939	25.04
64	16	939	13.07

7.9.3 Carry Select Adder

The circuit is shown in Fig. 7.6. The synthesis results for several numbers $n = m \cdot k$ of bits are given in Table 7.3.

The alternative circuit of Fig. 7.7 has also been implemented for several values of n. The results are given in Table 7.4.

7.9.4 Logarithmic Adders

The synthesis results for several numbers $n = n_1 \cdot n_2 \cdot n_3$ of bits are given in Table 7.5.

7.9.5 Long Operand Adder

The circuit is shown in Fig. 7.8. The synthesis results for several numbers $n = k \cdot s$ of bits are given in Table 7.6. Both the clock period T_{clk} and the total delay ($k \cdot T_{clk}$) are given.

Table 7.10 Combinational carry-save adder

n	m	LUTs	Delay
8	4	22	2.93
8	8	68	5.49
16	16	314	10.26
16	8	135	5.59
32	32	1388	20.03
32	8	279	5.95
32	16	649	10.65
32	64	2868	37.75
64	64	5844	39.09
64	32	2828	20.31
64	16	1321	11.35

Table 7.11 8-operand addition trees

n	LUTs	Delay
8	50	3.78
16	106	3.97
32	218	4.33
64	442	5.06

Table 7.12 24-operand adders based on 6-to-3 counters

n	LUTs	Delay
8	157	4.59
16	341	4.77
24	525	4.95
32	709	5.13
64	1445	5.86

7.9.6 Sequential Multioperand Adders

The circuit is shown in Fig. 7.9. The synthesis results for several numbers n of bits and m of operands are given in Table 7.7. Both the clock period T_{clk} and the total delay ($m{\cdot}T_{clk}$) are given.

The carry-save adder of Fig. 7.10 has also been implemented. The results are given in Table 7.8.

7.9.7 Combinational Multioperand Adders

The circuit of Fig. 7.12 has been implemented. The synthesis results for several numbers n of bits and m of operands are given in Table 7.9.

The carry-save adder of Fig. 7.14 has also been implemented. The results are given in Table 7.10.

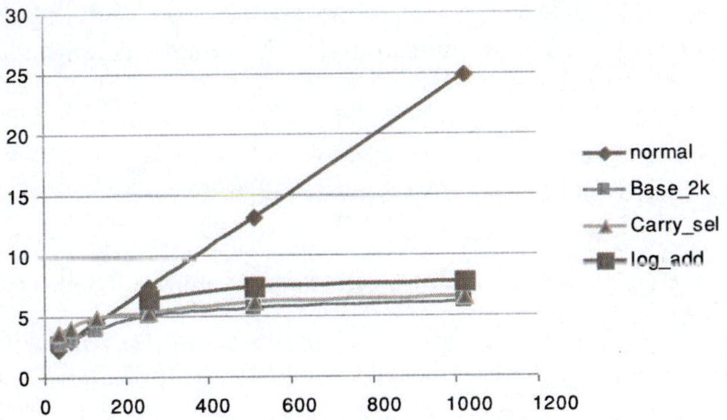

Fig. 7.21 Delay in function of the number of bits for several 2-operand adders

As an example of multioperand adddition trees (Fig. 7.13), several 8-bit adders have been implemented, with the results given in Table 7.11.

Examples of implementation results for 24-operand adders based on 6-to-3 counters (Fig. 7.17) are given in Table 7.12.

7.9.8 Comparison

A comparison between four types of 2-operand adders, namely binary (normal), radix-2^k, carry-select and logarithmic adders, has been done: Fig. 7.21 gives the corresponding adder delays (*ns*) in function of the number *n* of bits.

7.10 Exercises

1. Generate a generic model of a 2′s complement adder–subtractor with overflow detection.
2. An integer x can be represented under the form $(-1)^s \cdot m$ where s is the sign of x and m its magnitude (absolute value). Design an n-bit sign-magnitude adder–subtractor.
3. Design several n-bit counters, for example

   7-to-3,
   31-to-3,
   5-to-2,
   26-to-2.

4. Design a self-timed 64-bit adder with end of computation detection (*done* signal).

5. Generate several generic models of an *incrementer/decrementer*, that is, a circuit that computes $x \pm 1 \bmod m$ under the control of an *upb/down* binary signal.

References

1. Parhami B (2000) Computer arithmetic: algorithms and hardware design. Oxford University Press, New York
2. Ling H (1981) High-speed binary adder. IBM J Res Dev 25(3):156–166
3. Brent R, Kung HT (1982) A regular layout for parallel adders. IEEE Trans Comput C-31:260–264
4. Ladner RE, Fischer MJ (1980) Parallel prefix computation. J ACM 27:831–838
5. Ercegovac MD, Lang T (2004) Digital arithmetic. Morgan Kaufmann, San Francisco
6. Deschamps JP, Bioul G, Sutter G (2006) Synthesis of arithmetic circuits. Wiley, New York

Chapter 8
Multipliers

Multiplication is a basic arithmetic operation whose execution is based on 1-digit by 1-digit multipliers and multi-operand adders. Most FPGA families include the basic components for implementing fast and cost-effective multipliers. Furthermore, they also include optimized fixed-size multipliers which, in turn, can be used for implementing larger-size multipliers.

The basic multiplication algorithm is described in Sect. 8.1. Several combinational implementations are proposed in Sect. 8.2. They correspond to different types of multi-operand adders: iterative ripple-carry adders, carry-save adders, multi-operand adders based on counters, radix-2^k and mixed-radix adders. Sequential implementations are proposed in Sect. 8.3. They used the shift and add method implemented with either a ripple-carry adder or a carry-save adder. If integer operands are considered, several options are proposed in Sect. 8.4. A first method consists of multiplying B's complement integers as they are naturals; the drawback of this conceptually simple method is that the operands must be represented, and multiplied, with as many digits as the final result. Better options are a modification of the shift and add algorithm, multiplication of naturals followed by a post-correction, and the Booth algorithms. The last section describes a LUT-based method for implementing a constant multiplier, that is to say, circuits that compute $c \cdot y + u$, where c is a constant.

8.1 Basic Algorithm

Consider two radix-B numbers

$$x = x_{n-1} \cdot B^{n-1} + x_{n-2} \cdot B^{n-2} + \ldots + x_1 \cdot B + x_0 \text{ and}$$
$$y = y_{m-1} \cdot B^{m-1} + y_{m-2} \cdot B^{m-2} + \ldots + y_1 \cdot B + y_0,$$

J.-P. Deschamps et al., *Guide to FPGA Implementation of Arithmetic Functions*, Lecture Notes in Electrical Engineering 149, DOI: 10.1007/978-94-007-2987-2_8,

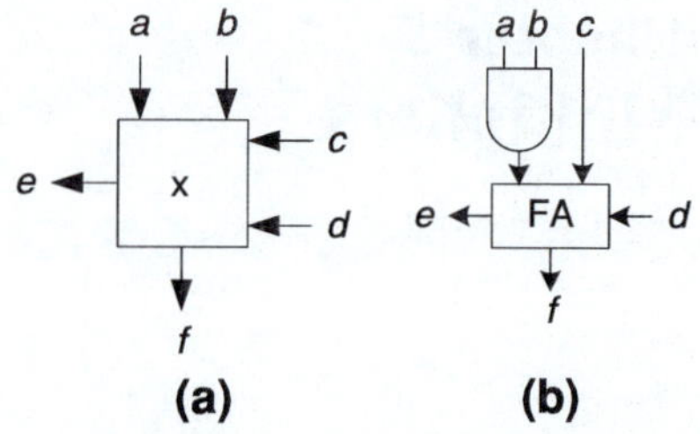

Fig. 8.1 1-digit by 1-digit multiplier. **a** Symbol, **b** internal structure ($B = 2$)

where x_i and y_i belong to {0, 1,..., B −1}. An n-digit by m-digit multiplier generates a radix-B number

$$z = z_{n+m-1} \cdot B^{n+m-1} + z_{n+m-2} \cdot B^{n+m-2} + \ldots + z_1 \cdot B + z_0$$

such that

$$z = x \cdot y.$$

A somewhat more general definition considers the addition of two additional numbers

$$u = u_{n-1} \cdot B^{n-1} + u_{n-2} \cdot B^{n-2} + \ldots + u_1 \cdot B + u_0 \text{ and}$$
$$v = v_{m-1} \cdot B^{m-1} + v_{m-2} \cdot B^{m-2} + \ldots + v_1 \cdot B + v_0,$$

so that

$$z = x \cdot y + u + v. \quad (8.1)$$

Observe that the maximum value of z is

$$(B^n - 1)(B^m - 1) + (B^n - 1) + (B^m - 1) = B^{n+m} - 1.$$

In order to compute (8.1), first define a 1-digit by 1-digit multiplier: given four B-ary digits a, b, c and d, it generates two B-ary digits e and f such that

$$a \cdot b + c + d = e \cdot B + f \quad (8.2)$$

(Fig. 8.1a).

If $B = 2$, it amounts to a 2-input AND gate and a 1-digit adder (Fig. 8.1b).

An n-digit by 1-digit multiplier made up of n 1-digit by 1-digit multipliers is shown in Fig. 8.2. It computes as

$$z = x \cdot b + u + d \quad (8.3)$$

where x and u are n-digit numbers, b and d are 1-digit numbers, and z is an $(n + 1)$-digit number. Observe that the maximum value of z is

$$(B^n - 1)(B - 1) + (B^n - 1) + (B - 1) = B^{n+1} - 1.$$

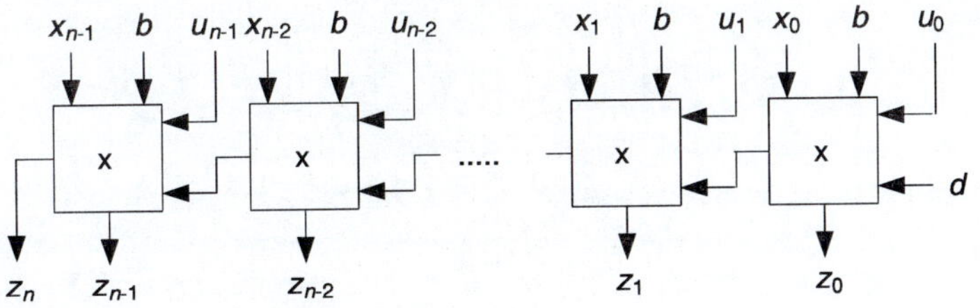

Fig. 8.2 *n*-digit by 1-digit multiplier

Using the iterative circuit of Fig. 8.2 as a computation resource, the computation of (8.1) amounts to computing the m n-digit by 1-digit products

$$\begin{aligned} z^{(0)} &= x \cdot y_0 + u + v_0, \\ z^{(1)} &= (x \cdot y_1 + v_1)B, \\ z^{(2)} &= (x \cdot y_2 + v_2)B^2, \\ &\ldots \\ z^{(m-1)} &= (x \cdot y_{m-1} + v_{m-1})B^{m-1}, \end{aligned} \tag{8.4}$$

and to adding them, that is

$$z = z^{(0)} + z^{(1)} + z^{(2)} + \ldots + z^{(m-1)} = x \cdot y + u + v. \tag{8.5}$$

For that, one of the multioperand adders of Sect. 7.7 can be used. As an example, if Algorithm 7.2 is used, then z is computed as follows.

Algorithm 8.1: Multiplication, right to left algorithm

```
accumulator := u;
for j in 0 .. m-1 loop
  accumulator := accumulator + (x·yj + vj)·B^j;
end loop;
z := accumulator;
```

8.2 Combinational Multipliers

8.2.1 Ripple-Carry Parallel Multiplier

The combinational circuit of Fig. 8.3 implements Algorithm 8.1 (with $n = 4$ and $m = 3$). One of its critical paths has been shaded. Its computation time is equal to

$$T_{multiplier}(n, m) = (n + 2m - 2) \cdot T_{multiplier}(1, 1). \tag{8.6}$$

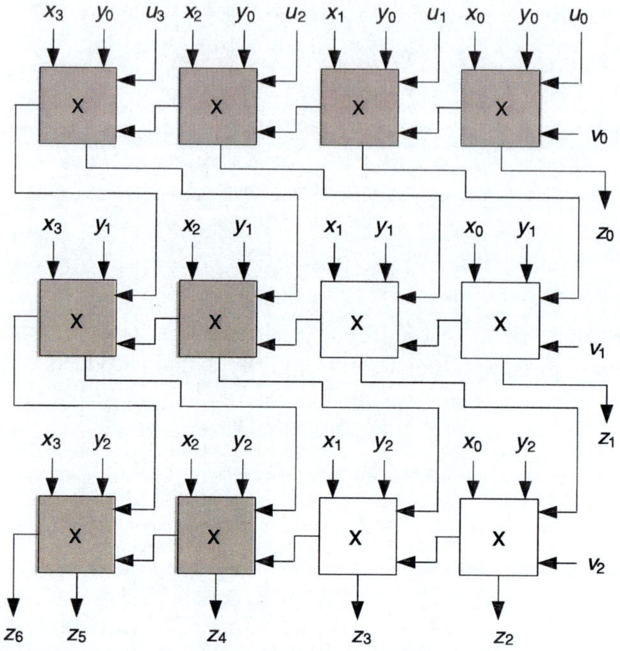

Fig. 8.3 Combinational multiplier

The following VHDL model describes the circuit of Fig. 8.3 ($B = 2$).

```
main_iteration: FOR i IN 0 TO m-1 GENERATE
  internal_iteration: FOR j IN 0 TO n-1 GENERATE
    f(i)(j) <= (x(j) AND y(i)) XOR c(i)(j) XOR d(i)(j);
    e(i)(j) <= (x(j) AND y(i) AND c(i)(j))
      OR (x(j) AND y(i) AND d(i)(j))
      OR (c(i)(j) AND d(i)(j));
  END GENERATE;
END GENERATE;
connections1: FOR j IN 0 TO n-1 GENERATE c(0)(j) <= u(j);
END GENERATE;
connections2: FOR i IN 1 TO m-1 GENERATE
  connections3: FOR j IN 0 TO n-2 GENERATE
      c(i)(j) <= f(i-1)(j+1);
  END GENERATE;
  c(i)(n-1) <= e(i-1)(n-1);
END GENERATE;
connections4: FOR i IN 0 TO m-1 GENERATE
  d(i)(0) <= v(i);
```

```
    connections5: FOR j IN 1 TO n-1 GENERATE
        d(i)(j) <= e(i)(j-1);
    END GENERATE;
  END GENERATE;
  outputs: FOR j IN 0 TO m-1 GENERATE z(j) <= f(j)(0);
  END GENERATE;
  z(m+n-2 DOWNTO m) <= f(m-1)(n-1 DOWNTO 1);
  z(m+n-1) <= e(m-1)(n-1);
```

A complete generic model *parallel_multiplier.vhd* is available at the Authors' web page.

8.2.2 Carry-Save Parallel Multiplier

A straightforward modification of the multiplier of Fig. 8.3, similar to the carry-save principle, is shown in Fig. 8.4. The circuit is made up of an n-by-m array of 1-by-1 multipliers, whose computation time is equal to $n \cdot T(1,1)$, plus an m-digit output adder. Its critical path has been shaded. Its computation time is equal to

$$T_{multiplier}(n,m) = n \cdot T_{multiplier}(1,1) + m \cdot T_{adder}(1) \leq (n+m) \cdot T_{multiplier}(1,1). \tag{8.7}$$

The following VHDL model describes the circuit of Fig. 8.4 ($B = 2$).

```
main_iteration: FOR i IN 0 TO m-1 GENERATE
  internal_iteration: FOR j IN 0 TO n-1 GENERATE
    f(i)(j) <= (x(j) AND y(i)) XOR c(i)(j) XOR d(i)(j);
    e(i)(j) <= (x(j) AND y(i) AND c(i)(j))
      OR (x(j) AND y(i) AND d(i)(j))
      OR (c(i)(j) AND d(i)(j));
  END GENERATE;
END GENERATE;
connections1: FOR j IN 0 TO n-1 GENERATE c(0)(j) <= u(j);
END GENERATE;
connections2: FOR i IN 1 TO m-1 GENERATE
  connections3: FOR j IN 0 TO n-2 GENERATE
      c(i)(j) <= f(i-1)(j+1);
  END GENERATE;
  c(i)(n-1) <= e(i-1)(n-1);
END GENERATE;
```

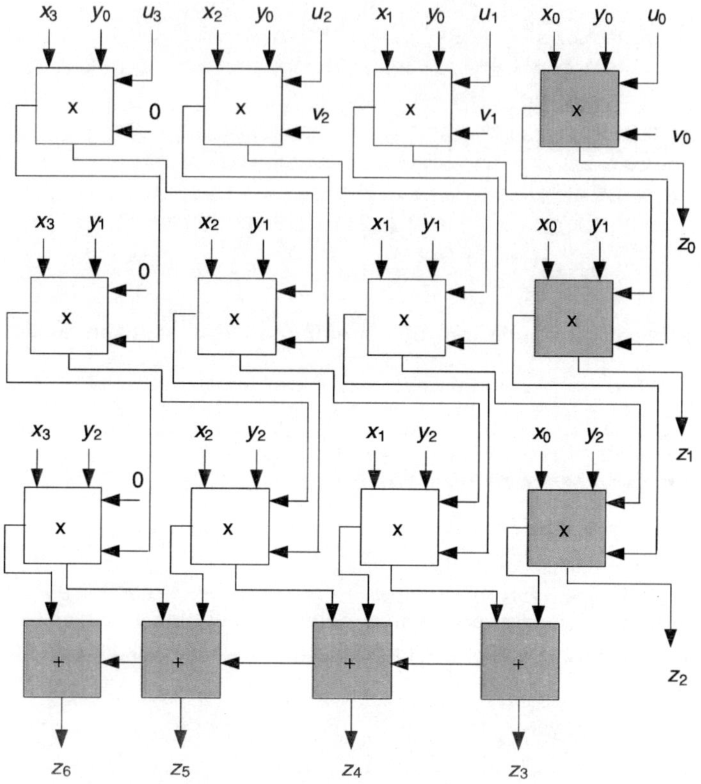

Fig. 8.4 Carry-save combinational multiplier

```
connections4: FOR i IN 0 TO m-1 GENERATE
  d(i)(0) <= v(i);
  connections5: FOR j IN 1 TO n-1 GENERATE
      d(i)(j) <= e(i)(j-1);
  END GENERATE;
END GENERATE;
outputs: FOR j IN 0 TO m-1 GENERATE z(j) <= f(j)(0);
END GENERATE;
first_operand<= f(m-1)(n-1 DOWNTO 1);
second_operand <= e(m-1);
z(n+m-1 DOWNTO m) <= first_operand + second_operand;
```

A complete generic model *parallel_csa_multiplier.vhd* is available at the Authors' web page.

8.2.3 Multipliers Based on Multioperand Adders

A straightforward implementation of Eqs. (8.4) and (8.5) can also be considered (Fig. 8.5). For that, any type of multioperand adder can be used.

Example 8.1
Consider an n-bit by 7-bit multiplier. The 7-operand adder can be divided up into a 7-to-3 counter, a 3-to-2 counter and a ripple-carry adder. The complete structure is shown in Fig. 8.6 and is described by the following VHDL model:

```
yy0 <= (OTHERS => y(0));
...
yy6 <= (OTHERS => y(6));
w0 <= (x AND yy0) + ('0'&u) + v(0);
w1 <= (x AND yy1) + zero + v(1);
...
w6 <= (x AND yy6) + zero + v(6);
z0 <= "000000"&w0;
z1 <= "00000"&w1&'0';
...
z6 <= w6&"000000";
first_component: seven_to_three1 GENERIC MAP(n => n+7)
PORT MAP(x1 => z0, x2 => z1, x3 => z2, x4 => z3, x5 => z4,
x6 => z5, x7 => z6, y1 => x1, y2 => x2, y3 => x3);
second_component: csa GENERIC MAP(n => n+7)
PORT MAP(x1 => x1, x2 => x2, x3 => x3, y1 => y1, y2 => y2);
z <= y1 + y2;
```

A complete generic model *N_by_7_multiplier.vhd* is available at the Authors' web page.

Numerous multipliers, based on trees of counters, have been proposed and reported, among others the Wallace and Dadda multipliers (Wallace [4]; Dadda [3]). Nevertheless, as already mentioned before (Comment 7.5), in many cases the best FPGA implementations are based on relatively simple algorithms, to which correspond regular circuits that allow taking advantage of the special purpose carry logic circuitry. To follow, an example of efficient FPGA implementation is described.

Consider the set of equations (8.4). If two successive steps are merged within an only step (loop unrolling), the new set of equations is:

$$
\begin{aligned}
z^{(1,0)} &= (x \cdot y_1 + v_1)B + x \cdot y_0 + u + v_0, \\
z^{(3,2)} &= [(x \cdot y_3 + v_3)B + (x \cdot y_2 + v_2)]B^2, \\
&\cdots \\
z^{(m-1,m-2)} &= [(x \cdot y_{m-1} + v_{m-1})B + (x \cdot y_{m-2} + v_{m-2})]B^{m-2},
\end{aligned}
\tag{8.8}
$$

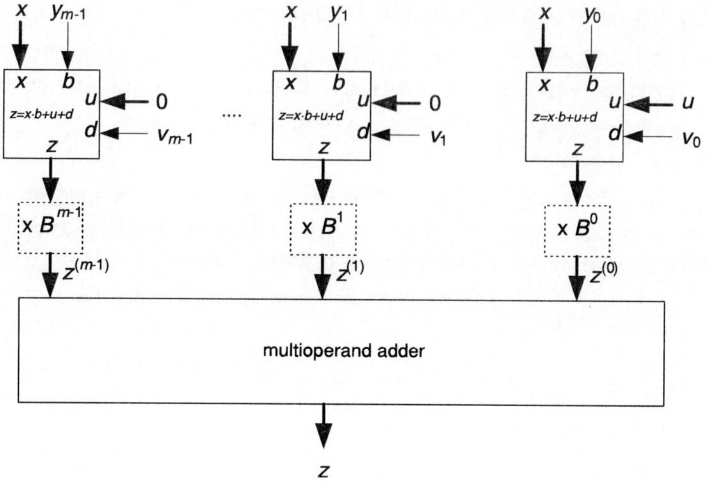

Fig. 8.5 Multiplier with a multioperand adder

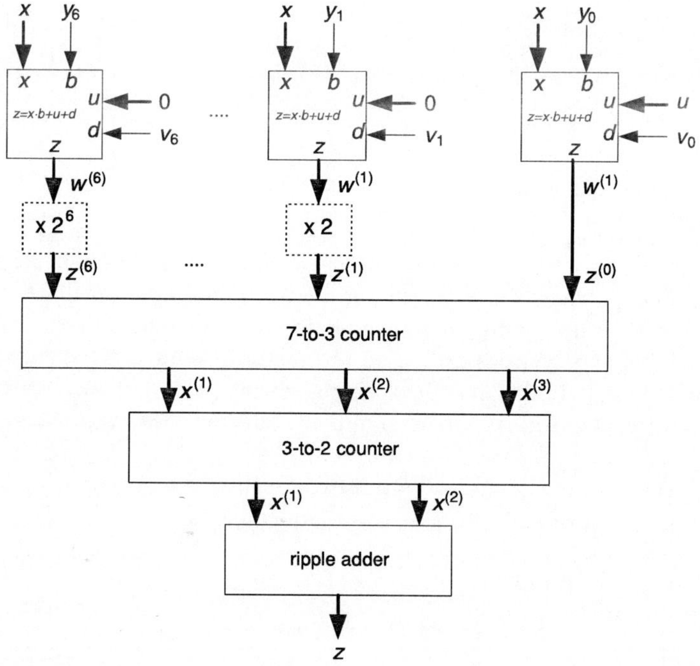

Fig. 8.6 An *n*-bit by 7-bit multiplier

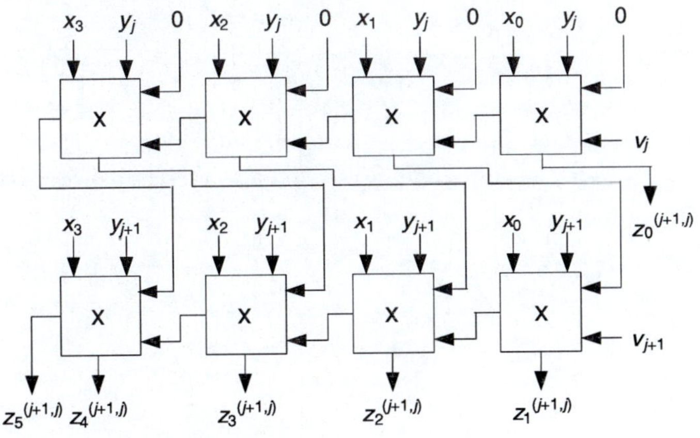

Fig. 8.7 4-digit by 2-digit multiplier

and the product is equal to

$$z = z^{(1,0)} + z^{(3,2)} + \ldots + z^{(m-1,m-2)}.$$

Assuming that $u = 0$, the basic operation to implement (8.8) is

$$z^{(i+1,i)} = (x \cdot y_{j+1} + v_{j+1})B + (x \cdot y_j + v_j)$$

to which corresponds the circuit of Fig. 8.7 (with $n = 4$).

The circuit of Fig. 8.7 can be decomposed into $n + 1$ vertical slices of the type shown in Fig. 8.8a (with obvious simplifications regarding the first and last slices). Finally, if $B = 2$ and $v_j = 0$, the carries of the first line are equal to 0, so that the circuit of Fig. 8.8a can be implemented as shown in Fig. 8.8b.

Comment 8.1

Most FPGA's include the basic components for implementing the structure of Fig. 8.8b, and the synthesis tools have the capability to generate optimized multipliers from a simple VHDL expression, such as

```
z <= x * y;
```

Furthermore, many FPGA's also include fixed-size multiplier blocks.

8.2.4 *Radix-2^k and Mixed-Radix Parallel Multipliers*

The basic multiplication algorithm (Sect. 8.1) and the corresponding ripple-carry and carry-save multipliers (Sects. 8.2.1 and 8.2.2) have been defined for any radix-B. In particular, radix-2^k multipliers can be defined. This allows the synthesis of $n \cdot k$-bit by $m \cdot k$-bit multipliers using k-bit by k-bit multipliers as building blocks.

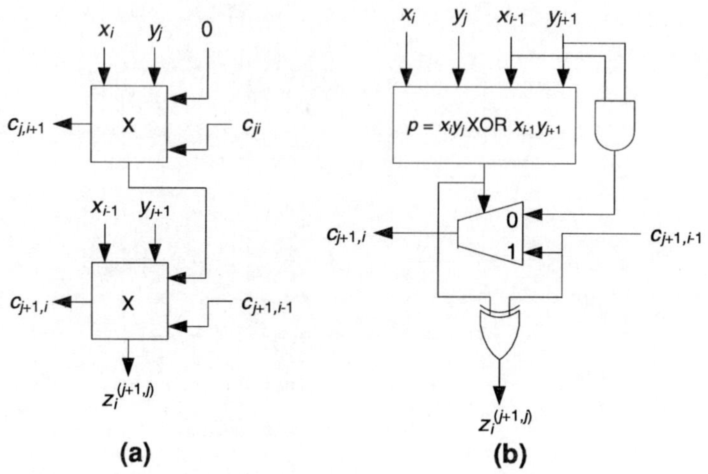

Fig. 8.8 Iterative cell of a parallel multiplier

The following VHDL model defines a radix-2^k ripple-carry parallel multiplier. The main iteration consists of $m \cdot n$ instantiations of any type of k-bit by k-bit combinational multiplier that computes $z = a \cdot b + c + d$ and represents z under the form $z_H \cdot 2^k + z_L$, where z_H and z_L are k-bit numbers:

```
main_iteration: FOR i IN 0 TO m-1 GENERATE
  internal_iteration: FOR j IN 0 TO n-1 GENERATE
    a_multiplier: k_by_k_parallel_multiplier
    GENERIC MAP(k => k)
    PORT MAP(a => x(j*k+k-1 DOWNTO j*k),
    b => y(i*k+k-1 DOWNTO i*k), c => c(i,j),
    d => d(i,j), zH => e(i,j), zL => f(i,j));
  END GENERATE;
END GENERATE;

--connections similar to those of the binary ripple-carry
--multiplier of Section 8.2.1

outputs1: FOR j IN 0 TO m-1 GENERATE
  z(j*k+k-1 DOWNTO j*k ) <= f(j,0);
END GENERATE;
outputs2: FOR j IN m TO m+n-2 GENERATE
  z(j*k+k-1 DOWNTO j*k ) <= f(m-1,j-m+1);
END GENERATE;
z(m*k+n*k-1 DOWNTO m*k+n*k-k) <= e(m-1,n-1);
```

A complete generic model *base_2k_parallel_multiplier.vhd* is available at the Authors' web page.

The stored-carry encoding can also be applied. Once again, the main iteration consists of $m \cdot n$ instantiations of any type of k-bit by k-bit multiplier, and the connections are similar to those of the carry-save multiplier of Sect. 8.2.2. A complete generic model *base_2k_csa_multiplier.vhd* is available at the Authors' web page.

A straightforward generalization of relations (8.2) to (8.5) allows defining mixed-radix combinational multipliers. First consider the circuit of Fig. 8.1a, assuming that

$$a, c \in \{0, 1, \ldots, B_1 - 1\}, \text{ and } b, d \in \{0, 1, \ldots, B_2 - 1\}.$$

Then

$$z = a \cdot b + c + d \leq (B_1 - 1) \cdot (B_2 - 1) + (B_1 - 1) + (B_2 - 1) = B_1 \cdot B_2 - 1,$$

so that z can be expressed under the form

$$z = e \cdot B_1 + f, \text{ with } e \in \{0, 1, \ldots, B_2 - 1\}, f \in \{0, 1, \ldots, B_1 - 1\}.$$

Then, consider the circuit of Fig. 8.2, assuming that x and u are n-digit radix-B_1 numbers, and b and d are 1-digit radix-B_2 numbers. Thus,

$$x \cdot b + u + d = z_n \cdot B_1^n + z_{n-1} \cdot B_1^{n-1} + \ldots + z_1 \cdot B_1 + z_0, \tag{8.9}$$

with

$$z_n \in \{0, 1, \ldots, B_2 - 1\} \text{ and } z_i \in \{0, 1, \ldots, B_1 - 1\}, \forall i \text{ in } \{0, 1, \ldots, n - 1\}.$$

Finally, given two n-digit radix-B_1 numbers x and u, and two m-digit radix-B_2 numbers y and v, compute

$$\begin{aligned} z^{(0)} &= x \cdot y_0 + u + v_0, \\ z^{(1)} &= (x \cdot y_1 + v_1)B_2, \\ z^{(2)} &= (x \cdot y_2 + v_2)B_2^2, \\ &\ldots \\ z^{(m-1)} &= (x \cdot y_{m-1} + v_{m-1})B_2^{m-1}. \end{aligned} \tag{8.10}$$

Then

$$z = z^{(0)} + z^{(1)} + z^{(2)} + \ldots + z^{(m-1)} = x \cdot y + u + v. \tag{8.11}$$

Consider the case where

$$B_1 = 2^{k_1}, \; B_2 = 2^{k_2}.$$

An easy way to define a VHDL model of the corresponding multiplier consists in first modelling a circuit that implements (8.9). The main iteration consists of n instantiations of any type of k_1-bit by k_2-bit combinational multiplier that computes

$$a \cdot b + c + d = z_H \cdot 2^{k_1} + z_L,$$

where z_H is a k_2-bit number and z_L a k_1-bit number:

```
first_cell: k1_by_k2_parallel_multiplier
GENERIC MAP(k1 => k1, k2 => k2)
PORT MAP(a => x(k1-1 DOWNTO 0), b => b,
c => u(k1-1 DOWNTO 0), d => d, zL => z(k1-1 DOWNTO 0),
zH => e(0));
iteration: FOR i IN 1 TO n-1 GENERATE
  other_cells: k1_by_k2_parallel_multiplier
  GENERIC MAP(k1 => k1, k2 => k2)
  PORT MAP(a => x(i*k1+k1-1 DOWNTO i*k1), b => b,
  c => u(i*k1+k1-1 DOWNTO i*k1), d => e(i-1),
  zL => z(i*k1+k1-1 DOWNTO i*k1), zH => e(i));
END GENERATE;
z(n*k1+k2-1 DOWNTO n*k1) <= e(n-1);
```

Then, it remains to instantiate m rows:

```
first_row: MR_multiplier_row
GENERIC MAP(n => n, k1 => k1, k2 => k2)
PORT MAP(x => x, u => u, b => y(k2-1 DOWNTO 0),
  d => v(k2-1 DOWNTO 0), z => zzz(0));
  z(k2-1 DOWNTO 0) <= zzz(0)(k2-1 DOWNTO 0);
next_rows: FOR i IN 1 TO m-1 GENERATE
  another_row: MR_multiplier_row
  GENERIC MAP(n => n, k1 => k1, k2 => k2)
  PORT MAP(x => x, u => zzz(i-1)(n*k1+k2-1 DOWNTO k2),
    b => y(i*k2+k2-1 DOWNTO i*k2),
    d => v(i*k2+k2-1 DOWNTO i*k2), z => zzz(i));
    z(i*k2+k2-1 DOWNTO i*k2) <= zzz(i)(k2-1 DOWNTO 0);
END GENERATE;
z(n*k1+m*k2 -1 DOWNTO m*k2) <= zzz(m-1)(n*k1+k2-1 DOWNTO k2);
```

A complete generic model *MR_parallel_multiplier.vhd* is available at the Authors' web page.

The circuit defined by the preceding VHDL model is a bidirectional array similar to that of Fig. 8.3, but with more complex connections. As an example, with $k_1 = 4$ and $k_2 = 2$, the connections corresponding to cell (j, i) are shown in

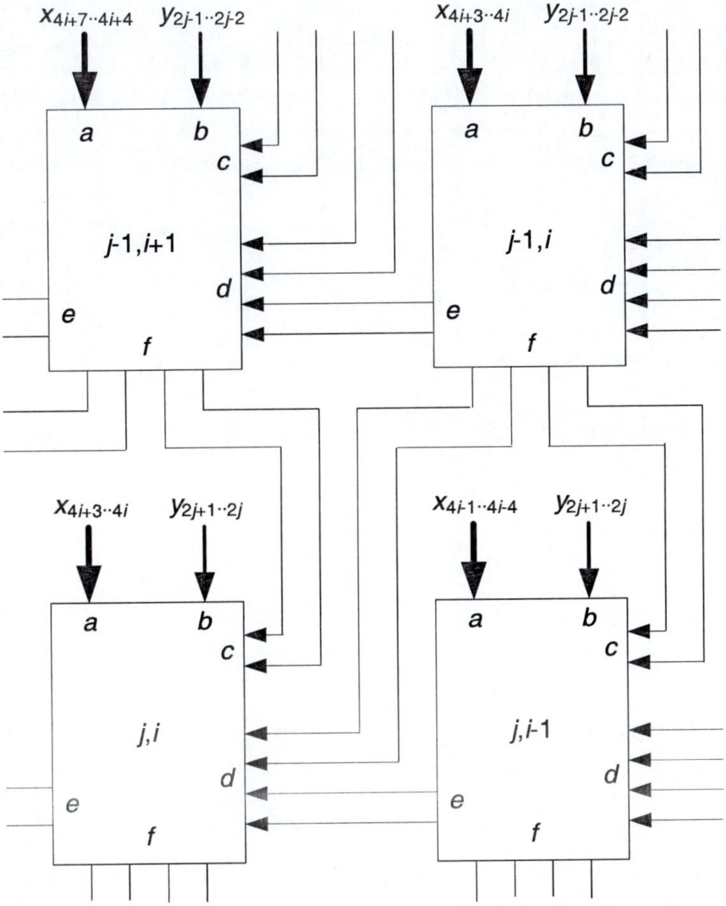

Fig. 8.9 Part of a $4n$-bit by $2n$-bit multiplier using 4-bit by 2-bit multiplication blocks

Fig. 8.9. As before, a stored-carry encoding circuit could also be designed, but with an even more complex connection pattern. It is left as an exercise.

8.3 Sequential Multipliers

8.3.1 Shift and Add Multiplier

In order to synthesize sequential multipliers, the basic algorithm of Sect. 8.1 can be modified. For that, Eq. (8.4) are substituted by the following:

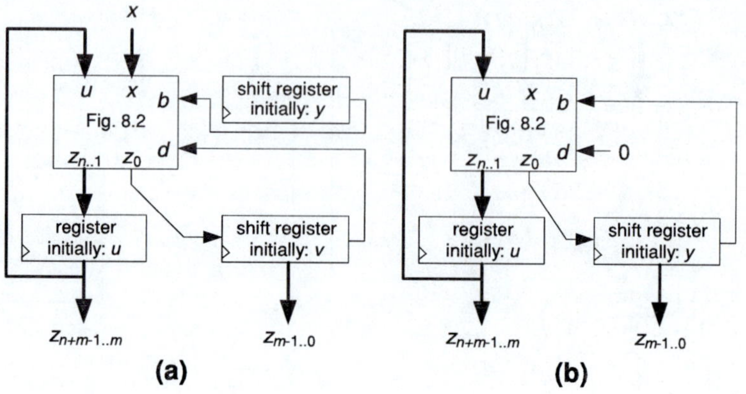

Fig. 8.10 Shift and add multipliers

$$
\begin{aligned}
z^{(0)} &= (u + x \cdot y_0 + v_0)/B, \\
z^{(1)} &= \left(z^{(0)} + x \cdot y_1 + v_1\right)/B, \\
z^{(2)} &= \left(z^{1)} + x \cdot y_2 + v_2\right)/B, \\
&\ldots \\
z^{(m-1)} &= \left(z^{(m-2)} + x \cdot y_{m-1} + v_{m-1}\right)/B.
\end{aligned}
\tag{8.12}
$$

Multiply the first equation by B, the second by B^2, and so on, and add the so obtained equations. The result is

$$
\begin{aligned}
z^{(m-1)}B^m &= u + x \cdot y_0 + v_0 + (x \cdot y_1 + v_1)B + \ldots + (x \cdot y_{m-1} + v_{m-1})B^{m-1} \\
&= xy + u + v.
\end{aligned}
$$

Algorithm 8.2: Shift and add multiplication

```
accumulator := u;
for j in 0 .. m-1 loop
  accumulator := (accumulator + x·yj + vj)/B;
end loop;
z := accumulator·B^m;
```

A data path for executing Algorithm 8.2 is shown in Fig. 8.10a. The following VHDL model describes the circuit of Fig. 8.10a ($B = 2$).

```
carries(0) <= acc_0(0);
main_iteration: FOR i IN 0 TO n-1 GENERATE
  product(i) <=
    (x(i) AND int_y(0)) XOR acc_1(i) XOR carries(i);
  carries(i+1) <=
    (x(i) AND int_y(0) AND acc_1(i)) OR
    (x(i) AND int_y(0) AND carries(i)) OR
    (acc_1(i) AND carries(i));
END GENERATE;
product(n) <= carries(n);
z <= acc_1 & acc_0;
parallel_register: PROCESS(clk)
BEGIN
  IF clk'EVENT and clk = '1' THEN
    IF load = '1' THEN acc_1 <= u;
    ELSIF update = '1' THEN acc_1 <= product(n DOWNTO 1);
    END IF;
  END IF;
END PROCESS;
shift_register1: PROCESS(clk)
BEGIN
  IF clk'EVENT and clk = '1' THEN
    IF load = '1' THEN acc_0 <= v;
    ELSIF update = '1' THEN
      acc_0 <= product(0)&acc_0(m-1 DOWNTO 1);
    END IF;
  END IF;
END PROCESS;
shift_register2: PROCESS(clk)
BEGIN
  IF clk'EVENT and clk = '1' THEN
    IF load = '1' THEN int_y <= y;
    ELSIF update = '1' THEN
      int_y <= '0'& int_y(m-1 DOWNTO 1);
    END IF;
  END IF;
END PROCESS;
```

The complete circuit also includes an m-state counter and a control unit. A complete generic model *shift_and_add_multiplier.vhd* is available at the Authors' web page.

If $v = 0$, the same shift register can be used for storing both y and the least significant bits of z. The modified circuit is shown in Fig. 8.10b. A complete generic model *shift_and_add_multiplier2* is also available.

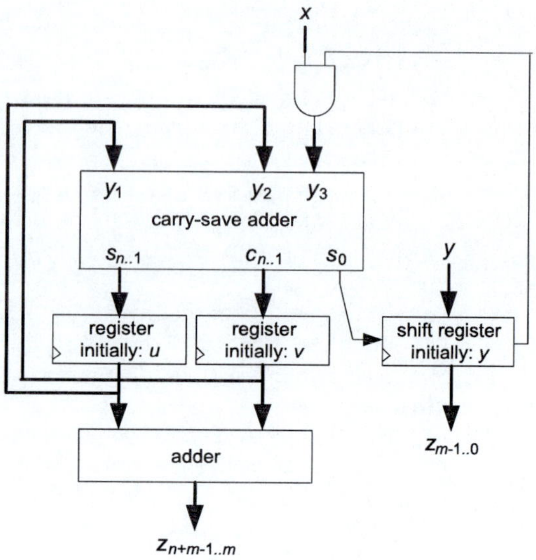

Fig. 8.11 Sequential carry-save multiplier

The computation time of the circuits of Fig. 8.10 is approximately equal to

$$T_{multiplier}(n, m) = m \cdot T_{multiplier}(n, 1) = m \cdot n \cdot T_{multiplier}(1, 1). \tag{8.13}$$

8.3.2 Shift and Add Multiplier with CSA

The shift and add algorithm can also be executed with stored-carry encoding. After m steps the result is obtained under the form

$$s_{n-1}\, B^{n+m-1} + (c_{n-2} + s_{n-2})B^{n+m-2} + \ldots + (c_0 + s_0)B^m + z_{m-1}\, B^{m-1} + \ldots + z_1 B + z_0,$$

and an additional n-digit adder computes

$$s_{n-1}\, B^{n-1} + (c_{n-2} + s_{n-2})B^{n-2} + \ldots + (c_0 + s_0) = z_{m+n-1}\, B^{n-1} + \ldots + z_{m+1}\, B + z_m.$$

The corresponding data path is shown in Fig. 8.11. The carry-save adder computes

$$y_1 + y_2 + y_3 = s + c,$$

where y_1, y_2 and y_3 are n-bit numbers, and s and c are $(n + 1)$-bit numbers. At the end of step i, the less significant bit of s is z_i, and the n most significant bits of s and c are transmitted to the next step:

```
xy <= '0'&(x AND y_i);
main_component: csa GENERIC MAP(n => n+1)
PORT MAP(y1 => xy, y2 => s , y3 => c, s => next_s,
  c => next_c);
register_s: PROCESS(clk) ...
register_c: PROCESS(clk) ...
shift_register: PROCESS(clk)
BEGIN
  IF clk'EVENT and clk = '1' THEN
    IF load = '1' THEN int_y <= y;
    ELSIF update = '1' THEN
      int_y <= next_s(0)& int_y(m-1 DOWNTO 1);
    END IF;
  END IF;
END PROCESS;
y_i <= (OTHERS => int_y(0));
z(m-1 DOWNTO 0) <= int_y(m-1 DOWNTO 0);
z(m+n-1 DOWNTO m) <= s(n-1 DOWNTO 0) + c(n-1 DOWNTO 0);
```

The complete circuit also includes an m-state counter and a control unit. A complete generic model *sequential_CSA_multiplier.vhd* is available at the Authors' web page. The minimum clock period is equal to the delay of a 1-bit by 1-bit multiplier. Thus, the total computation time is equal to

$$T_{multiplier}(n,m) = m \cdot T_{multiplier}(1,1) + T_{adder}(n) \leq (n+m) \cdot T_{multiplier}(1,1). \tag{8.14}$$

Comment 8.2

In *sequential_CSA_multiplier.vhd* the *done* flag is raised as soon as the final values of the adder inputs are available. A more correct control unit should raise the flag k cycles later, being $k \cdot T_{clk}$ an upper bound of the n-bit adder delay. The value of k could be defined as a generic parameter (Exercise 8.3).

8.4 Integers

Given four B's complement integers

$$x = x_n x_{n-1} x_{n-2} \ldots x_0,\ \ y = y_m y_{m-1} y_{m-2} \ldots y_0,\ \ u = u_n u_{n-1}\ u_{n-2} \ldots u_0,$$
$$v = v_m v_{m-1} v_{m-2} \ldots v_0,$$

belonging to the ranges

$$-B^n \leq x < B^n, -B^m \leq y < B^m, -B^n \leq u < B^n, -B^m \leq v < B^m,$$

then $z = x \cdot y + u + v$ belongs to the interval

$$-B^{n+m+1} \leq z < B^{n+m+1}.$$

Thus, z is a B's complement number of the form

$$z = z_{n+m+1}\, z_{n+m}\, z_{n+m-1} \ldots z_1 z_0.$$

8.4.1 Mod $2B^{n+m}$ Multiplication

The integer represented by a vector $x_n\, x_{n-1}\, x_{n-2} \ldots x_1\, x_0$ is

$$x = -x_n B^n + x_{n-1}\, B^{n-1} + x_{n-2}\, B^{n-2} + \ldots + x_1 B + x_0,$$

while the natural *natural*(x) represented by this same vector is

$$natural\,(x) = x_n B^n + x_{n-1}\, B^{n-1} + x_{n-2}\, B^{n-2} + \ldots + x_1 B + x_0.$$

As $x_n \in \{0, 1\}$, either *natural*(x) = x or *natural*(x) = $x + 2B^n$. So,

$$natural(x) = x \bmod 2B^n.$$

The following method can be used to compute $z = x \cdot y + u + v$. First, represent the operands x, y, u and v with the same number of digits ($n + m + 2$) as the result z (digit extension, Sect. 7.8). Then, compute $z = x \cdot y + u + v$ as if x, y, u and v were naturals:

$$z = natural(x) \cdot natural(y) + natural(u) + natural(v) = natural(x \cdot y + u + v).$$

Finally, reduce z modulo $2B^{n+m+1}$. Assume that before the mod $2B^{n+m+1}$ reduction

$$z = \ldots + z_{n+m+1}\, B^{n+m+1} + z_{n+m}\, B^{n+m} + z_{n+m-1}\, B^{n+m-1} + \ldots + z_1 B + z_0;$$

then

$$z \bmod 2B^{n+m+1} = (\ldots + z_{n+m+1} \bmod 2) B^{n+m+1} + z_{n+m}\, B^{n+m} + z_{n+m+1}\, B^{n+m-1} + \ldots + z_1 B + z_0.$$

In particular, if B is even,

$$z \bmod 2B^{n+m+1} = (z_{n+m+1}\;\; 2) B^{n+m+1} + z_{n+m}\, B^{n+m} + z_{n+m-1}\, B^{n+m-1} + \ldots + z_1 B + z_0.$$

Example 8.2

Assume that $B = 10$, $n = 4$, $m = 3$, $x = 7918$, $y = -541$, $u = -7017$, $v = 742$, and compute $z = 7918 \cdot (-541) + (-7017) + 742$. In 10's complement: $x = 07918$, $y = 1459$, $u = 12983$, $v = 0742$.

1. Express all operands with 9 digits: $x = 000007918$, $y = 199999459$, $u = 199992983$, $v = 000000742$.

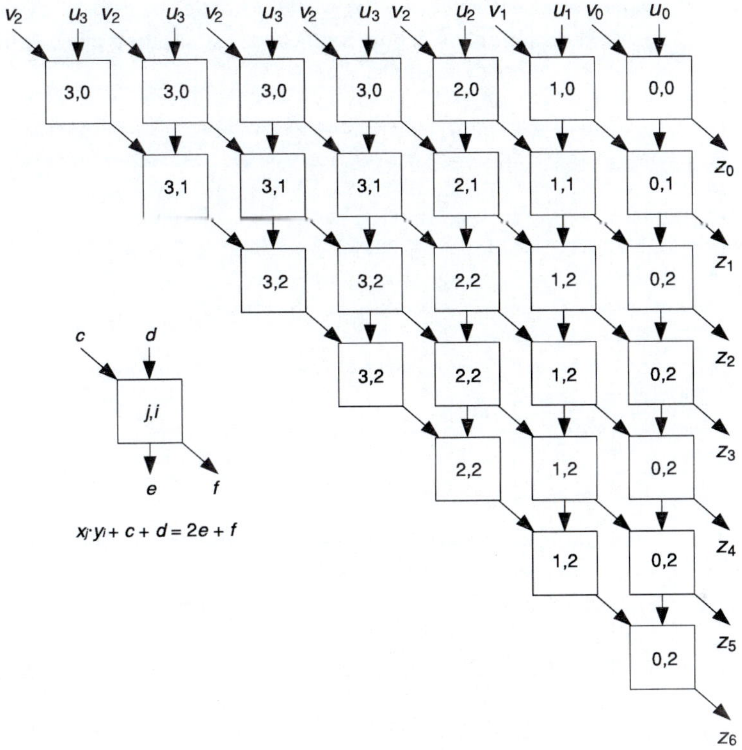

Fig. 8.12 Carry-save multiplier for integers ($n = 3$, $m = 2$)

2. Compute $x \cdot y + u + v$: 000007918·199999459 + 199992983 + 000000742 = 1583795710087.
3. Reduce 1583795710087 modulo $2 \cdot 10^8$: (1583795710087) mod $2 \cdot 10^8$ = (7 mod 2)·10^8 + 95710087 = 195710087.

The result 195710087 is the 10's complement representation of −4289913.

Thus, any multiplier for natural numbers can be used. As an example, an $(n + m + 2)$-digit by $(n + m + 2)$-digit carry-save multiplier could be used (Fig. 8.4). As the result is reduced modulo $2B^{n+m+1}$, only the rightmost part of the circuit is used (if B is even), so that there is no output adder, and the most significant digit is reduced mod 2. An example with $n = 3$ and $m = 2$ is shown in Fig. 8.12. The corresponding computation time is equal to

$$(n + m + 2) \cdot T_{multiplier}(1, 1). \tag{8.15}$$

This delay is practically the same as that of a carry-save combinational multiplier (8.7). Nevertheless, the number of 1-digit by 1-digit multiplication cells is equal to $1 + 2 + 3 + \ldots + (n + m + 2) = (n + m + 2)(n + m + 3)/2$ instead of $n \cdot m$.

A very simple way to generate a VHDL model consists of defining $(n + m + 2)$-bit representations of all operands and instantiating an $(n + m + 2)$-bit by $(n + m + 2)$-bit carry-save multiplier:

```
long_x(n+m+1 DOWNTO n+1) <= (OTHERS => x(n));
long_x(n DOWNTO 0) <= x;
long_u(n+m+1 DOWNTO n+1) <= (OTHERS => u(n));
long_u(n DOWNTO 0) <= u;
long_y(n+m+1 DOWNTO m+1) <= (OTHERS => y(m));
long_y(m DOWNTO 0) <= y;
long_v(n+m+1 DOWNTO m+1) <= (OTHERS => v(m));
long_v(m DOWNTO 0) <= v;
main_component: parallel_csa_multiplier
GENERIC MAP(n => n+m+2, m => n+m+2)
PORT MAP(x => long_x, u => long_u, y => long_y, v => long_v,
  z => long_z);
z <= long_z(n+m+1 DOWNTO 0);
```

Only $n + m + 2$ output bits of the carry-save multiplier are connected to output ports, and the synthesis program will prune the circuit accordingly.

A complete generic model *integer_CSA_multiplier.vhd* is available at the Authors' web page.

To conclude, this approach is conceptually attractive because any type of multiplier for natural numbers can be used. Nevertheless, the cost of the corresponding circuits is very high.

8.4.2 Modified Shift and Add Algorithm

Consider again four B's complement integers

$$x = x_n\, x_{n-1}\, x_{n-2} \ldots x_0, y = y_m y_{m-1}\, y_{m-2} \ldots y_0,\ u = u_n\, u_{n-1}\, u_{n-2} \ldots u_0,$$
$$v = v_m\, v_{m-1}\, v_{m-2} \ldots v_0.$$

A set of equations similar to (8.12) can be defined:

$$\begin{aligned}
z^{(0)} &= (u + x \cdot y_0 + v_0)/B,\\
z^{(1)} &= \left(z^{(0)} + x \cdot y_1 + v_1\right)/B,\\
z^{(2)} &= \left(z^{1)} + x \cdot y_2 + v_2\right)/B,\\
&\ldots\\
z^{(m-1)} &= \left(z^{(m-2)} + x \cdot y_{m-1} + v_{m-1}\right)/B,\\
z^{(m)} &= \left(z^{(m-1)} - x \cdot y_m - v_m\right)/B.
\end{aligned} \tag{8.16}$$

Multiply the first equation by B, the second by B^2, and so on, and add the $m + 1$ so obtained equations. The result is

$$\begin{aligned} z^{(m)}B^{m+1} &= u + x \cdot y_0 + v_0 + (x \cdot y_1 + v_1)B + \ldots + (x \cdot y_{m-1} + v_{m-1})B^{m-1} \\ &\quad - (x \cdot y_m + v_m)B^m \\ &= xy + u + v. \end{aligned}$$

Algorithm 8.3: Modified shift and add multiplication

```
accumulator := u;
for j in 0 .. m-1 loop
  accumulator := (accumulator + x·yj + vj)/B;
end loop;
accumulator := (accumulator - x·ym - vm)/B;
z := accumulator·B^(m+1);
```

In what follows it is assumed that $v_m = 0$, that is to say $v \geq 0$; so, in order to implement Algorithm 8.3, the two following computation primitives must be defined:

$$z = u + x \cdot b + d \tag{8.17}$$

and

$$z = u - x \cdot b, \tag{8.18}$$

where

$$-B^n \leq x < B^n, -B^n \leq u < B^n, 0 \leq b < B, 0 \leq d < B.$$

Thus, in the first case,

$$-B^{n+1} \leq z < B^{n+1},$$

and in the second case

$$-B^{n+1} + (B-1) \leq z < B^{n+1},$$

so that in both cases z is an $(n + 2)$-digit B's complement integer and *natural*$(z) = z \bmod 2B^{n+1}$.

The first primitive (8.17) is implemented by the circuit of Fig. 8.13 and the second (8.18) by the circuit of Fig. 8.14. In both, circuit z_{n+1} is computed modulo 2.

As an example, the combinational circuit of Fig. 8.15 implements Algorithm 8.3 (with $n = m = 2$). Its cost and computation time are practically the same as in the case of a ripple-carry multiplier for natural numbers. It can be described by the following VHDL model.

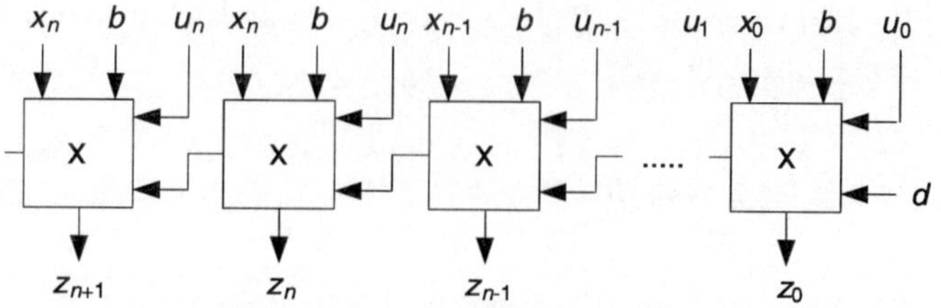

Fig. 8.13 First computation primitive

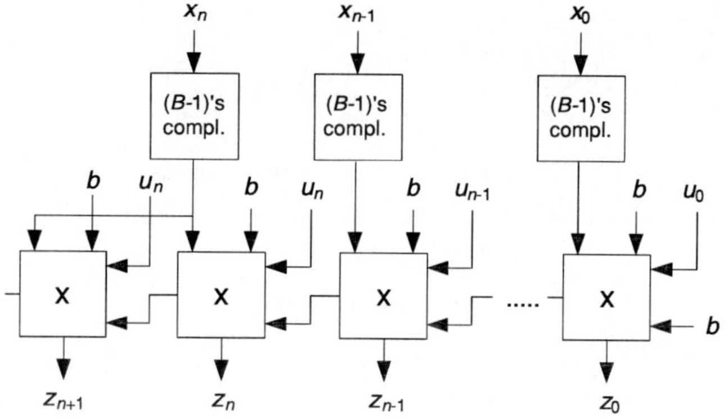

Fig. 8.14 Second computation primitive

```
main_iteration: FOR i IN 0 TO m-1 GENERATE
  internal_iteration: FOR j IN 0 TO n GENERATE
    f(i)(j) <= (x(j) AND y(i)) XOR c(i)(j) XOR d(i)(j);
    e(i)(j) <= (x(j) AND y(i) AND c(i)(j))
      OR (x(j) AND y(i) AND d(i)(j))
      OR (c(i)(j) AND d(i)(j));
  END GENERATE;
  f(i)(n+1) <= (x(n) AND y(i)) XOR c(i)(n+1) XOR d(i)(n+1);
  e(i)(n+1) <= (x(n) AND y(i) AND c(i)(n+1))
    OR (x(n) AND y(i) AND d(i)(n+1))
    OR (c(i)(n+1) AND d(i)(n+1));
END GENERATE;
last_row: FOR j IN 0 TO n GENERATE
  f(m)(j) <= (NOT(x(j)) AND y(m)) XOR c(m)(j) XOR d(m)(j);
  e(m)(j) <= (NOT(x(j)) AND y(m) AND c(m)(j)) OR
    (NOT(x(j)) AND y(m) AND d(m)(j)) OR
    (c(m)(j) AND d(m)(j));
END GENERATE;
```

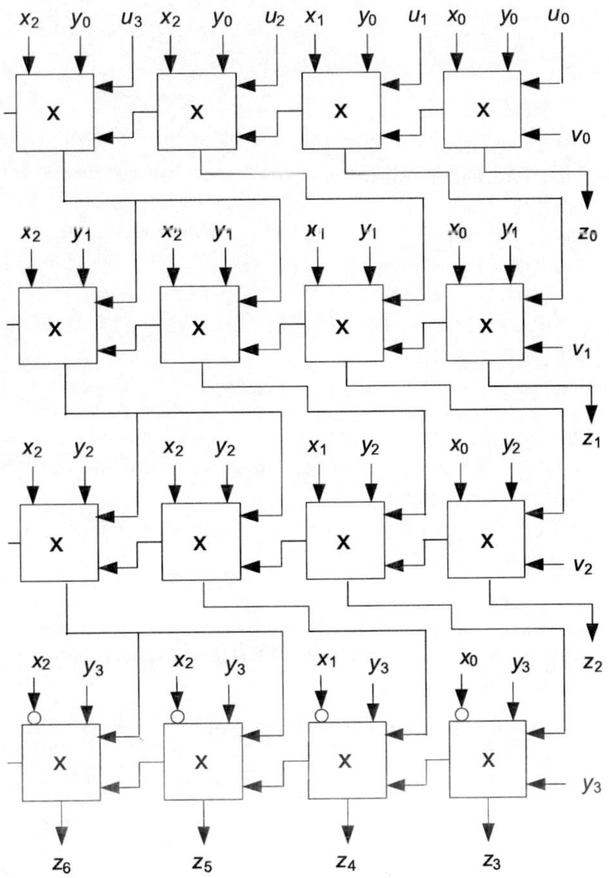

Fig. 8.15 Combinational multiplier for integers ($B = 2$, $m = n = 2$)

```
f(m)(n+1) <=
  (NOT(x(n)) AND y(m)) XOR c(m)(n+1) XOR d(m)(n+1);
e(m)(n+1) <= (NOT(x(n)) AND y(m) AND c(m)(n+1))
  OR (NOT(x(n)) AND y(m) AND d(m)(n+1))
  OR (c(m)(n+1) AND d(m)(n+1));
connections1: FOR j IN 0 TO n GENERATE c(0)(j) <= u(j);
END GENERATE;
c(0)(n+1) <= u(n);
connections2: FOR i IN 1 TO m GENERATE
  connections3: FOR j IN 0 TO n GENERATE
    c(i)(j) <= f(i-1)(j+1);
  END GENERATE;
  c(i)(n+1) <= f(i-1)(n+1);
END GENERATE;
```

```
connections4: FOR i IN 0 TO m-1 GENERATE
  d(i)(0) <= v(i);
  connections5: FOR j IN 1 TO n+1 GENERATE
    d(i)(j) <= e(i)(j-1);
  END GENERATE;
END GENERATE;
d(m)(0) <= y(m);
connections6: FOR j IN 1 TO n+1 GENERATE
  d(m)(j) <= e(m)(j-1);
END GENERATE;
outputs: FOR j IN 0 TO m GENERATE z(j) <= f(j)(0);
END GENERATE;
z(m+n+1 DOWNTO m+1) <= f(m)(n+1 DOWNTO 1);
```

A complete generic model *modified_parallel_multiplier.vhd* is available at the Authors' web page.

The design of a sequential multiplier based on Algorithm 8.3 is left as an exercise.

8.4.3 Post Correction Multiplication

Given four B's complement integers

$$x = x_n\, x_{n-1}\, x_{n-2} \ldots x_0, y = y_m\, y_{m-1}\, y_{m-2} \ldots y_0, u = u_n u_{n-1}\, u_{n-2} \ldots u_0,$$
$$v = v_m v_{m-1}\, v_{m-2} \ldots v_0,$$

then $z = x \cdot y + u + v$, belonging to the interval $-B^{n+m+1} \le z < B^{n+m+1}$, can be expressed under the form

$$z = (X_0 \cdot Y_0 + U_0 + V_0) + x_n \cdot y_m \cdot B^{n+m} - (x_n \cdot Y_0 + u_n) \cdot B^n - (y_m \cdot X_0 + v_m) \cdot B^n,$$

where X_0, Y_0, U_0 and V_0 are four naturals

$$X_0 = x_{n-1}\, x_{n-2} \ldots x_0, Y_0 = y_{m-1}\, y_{m-2} \ldots y_0,\; U_0 = u_{n-1}\, u_{n-2} \ldots u_0,$$
$$X_0 = v_{m-1} v_{m-2} \ldots v_1 v_0$$

deduced from x, y, u and v by eliminating the sign bits. Thus, the computation of z amounts to the computation of

$$Z_0 = X_0 \cdot Y_0 + U_0 + V_0,$$

that can be executed by any type of multiplier for naturals, plus a post correction that consists of several additions and left shifts.

If $B = 2$ and $u = v = 0$, then

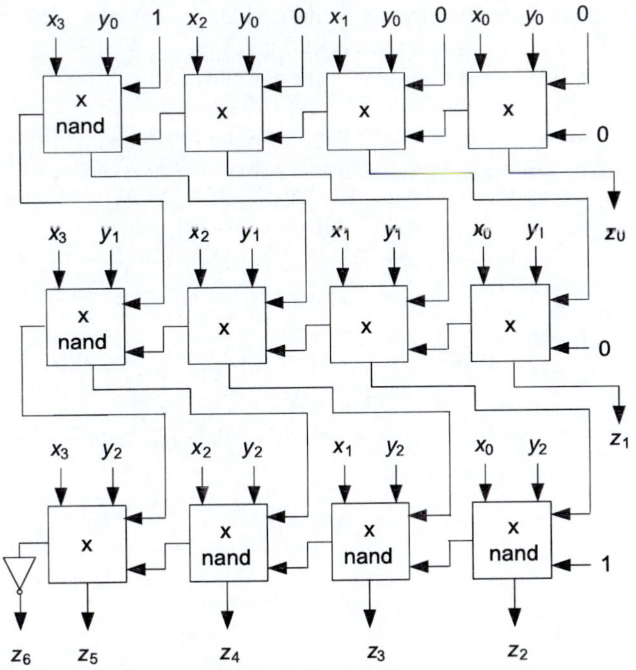

Fig. 8.16 Multiplier with post correction

$$z = x \cdot y = X_0 \cdot Y_0 + x_n \cdot y_m \cdot 2^{n+m} - x_n \cdot Y_0 \cdot 2^n - y_m \cdot X_0 \cdot 2^n.$$

The $(n + m + 2)$-bit 2's complement representations of $-x_n \cdot Y_0 \cdot 2^n$ and $-y_m \cdot X_0 \cdot 2^m$ are

$$(2^{m+1} + 2^m + \overline{(x_n \cdot y_{m-1})} \cdot 2^{m-1} + \ldots + \overline{(x_n \cdot y_0)} \cdot 2^0 + 1) \cdot 2^n \bmod 2^{n+m-2},$$

and

$$(2^{n+1} + 2^n + \overline{(y_m \cdot x_{n-1})} \cdot 2^{n-1} + \ldots + \overline{(y_m \cdot x_0)} \cdot 2^0 + 1) \cdot 2^m \bmod 2^{n+m-2},$$

so that the representation of $x_n \cdot y_m \cdot 2^{n+m} - x_n \cdot Y_0 \cdot 2^n - y_m \cdot X_0 \cdot 2^n$ is

$$(2^{n+m+1} + x_n \cdot y_m \cdot 2^{n+m} + \overline{(x_n \cdot y_{m-1})} \cdot 2^{n+m-1} + \ldots + \overline{(x_n \cdot y_0)} \cdot 2^n$$
$$+2^n + \overline{(y_m \cdot x_{n-1})} \cdot 2^{n+m-1} + \ldots + \overline{(y_m \cdot x_0)} \cdot 2^m + 2^m) \bmod 2^{n+m+2}.$$

A simple modification of the combinational multipliers of Fig. 8.3 and 8.4 allows computing $x \cdot y$, where x is an $(n + 1)$-bit 2's complement integer and y an $(m + 1)$-bit 2's complement integer. An example is shown in Fig. 8.16 ($n = 3$,

$m = 2$). The *nand* multiplication cells are similar to that of Fig. 8.1b, but for the substitution of the AND gate by a NAND gate [1].
The following VHDL model describes the circuit of Fig. 8.16.

```
main_iteration: FOR i IN 0 TO m-1 GENERATE
  internal_iteration: FOR j IN 0 TO n-1 GENERATE
    f(i)(j) <= (x(j) AND y(i)) XOR c(i)(j) XOR d(i)(j);
    e(i)(j) <= (x(j) AND y(i) AND c(i)(j)) OR
      (x(j) AND y(i) AND d(i)(j)) OR (c(i)(j) AND d(i)(j));
  END GENERATE;
END GENERATE;
first_column: FOR i IN 0 TO m-1 GENERATE
  f(i)(n) <= (x(n) NAND y(i)) XOR c(i)(n) XOR d(i)(n);
  e(i)(n) <= ((x(n) NAND y(i)) AND c(i)(n)) OR
    ((x(n) NAND y(i)) AND d(i)(n)) OR (c(i)(n) AND d(i)(n));
END GENERATE;
last_row: FOR j IN 0 TO n-1 GENERATE
  f(m)(j) <= (x(j) NAND y(m)) XOR c(m)(j) XOR d(m)(j);
  e(m)(j) <= ((x(j) NAND y(m)) AND c(m)(j)) OR
    ((x(j) NAND y(m)) AND d(m)(j)) OR (c(m)(j) AND d(m)(j));
END GENERATE;
f(m)(n) <= (x(n) AND y(m)) XOR c(m)(n) XOR d(m)(n);
e(m)(n) <= (x(n) AND y(m) AND c(m)(n)) OR
  (x(n) AND y(m) AND d(m)(n)) OR (c(m)(n) AND d(m)(n));

--connections similar to those of a multiplier for naturals
--(Fig.8.3)

outputs: FOR j IN 0 TO m GENERATE z(j) <= f(j)(0);
END GENERATE;
z(m+n DOWNTO m+1) <= f(m)(n DOWNTO 1);
z(m+n+1) <= NOT(e(m)(n));
```

A complete generic model *postcorrection_multiplier.vhd* is available at the Authors' web page.

8.4.4 Booth Multiplier

Given an $(m + 1)$-bit 2's complement integer $y = -y_m \cdot 2^m + y_{m-1} \cdot 2^{m-1} + \ldots + y_1 \cdot 2 + y_0$, define

$$y'_0 = -y_0 \text{ and } y'_j = -y_j + y_{j-1},\ \forall i \text{ in } \{1, 2, \ldots, m\},$$

so that all coefficients y_i' belong to $\{-1, 0, 1\}$. Then y can be represented under the form

$$y = y'_m \cdot 2^m + y'_{m-1} \cdot 2^{m-1} + \ldots + y'_1 \cdot 2 + y'_0,$$

the so-called Booth's encoding of y (Booth [2]. Unlike the 2's complement representation in which y_m has a specific function, all coefficients y_i' have the same function. Formally, the Booth's representation of an integer is the same as the binary representation of a natural. The basic multiplication algorithm (Algorithm 8.1), with $v = 0$, can be used.

Algorithm 8.4: Booth multiplication, z = x·y + u

```
accumulator := u;
for j in 0 .. m-1 loop
  accumulator := accumulator + x·yj'·2^j;
end loop;
z := accumulator;
```

The following VHDL model describes a combinational circuit based on Algorithm 8.4.

```
a(0) <= ((u(n)&u) - (x(n)&x)) WHEN y(0) = '1'
  ELSE u(n)&u;
z(0) <= a(0)(0);
main_iteration: FOR i IN 1 TO m GENERATE
  a(i) <= ((a(i-1)(n+1)&a(i-1)(n+1 DOWNTO 1)) - (x(n)&x))
    WHEN (y(i-1) = '0' AND y(i) = '1')
  ELSE ((a(i-1)(n+1)&a(i-1)(n+1 DOWNTO 1)) + (x(n)&x))
    WHEN (y(i-1) = '1' AND y(i) = '0')
  ELSE a(i-1)(n+1)&a(i-1)(n+1 DOWNTO 1);
  z(i) <= a(i)(0);
END GENERATE;
z(n+m+1 DOWNTO m+1) <= a(m)(n+1 DOWNTO 1);
```

A complete generic model *Booth1_multiplier.vhd* is available at the Authors' web page.

Higher radix Booth multipliers can be defined. Given an $(m + 1)$-bit 2's complement integer $y = -y_m \cdot 2^m + y_{m-1} \cdot 2^{m-1} + \ldots + y_1 \cdot 2 + y_0$, where m is odd, define

$$y'_0 = -2 \cdot y_1 + y_0, y'_i = -2 \cdot y_{2 \cdot i+1} + y_{2 \cdot i} + y_{2 \cdot i-1}, \forall i \text{ in}\{1, 2, \ldots, (m-1)/2\},$$

so that all coefficients y_i' belong to $\{-2, -1, 0, 1, 2\}$. Then y can be represented under the form

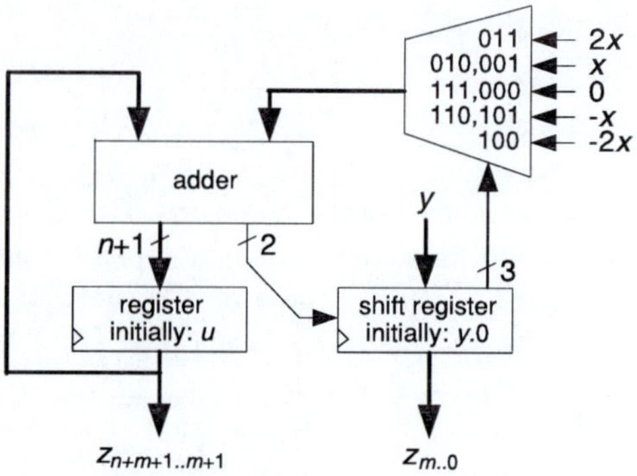

Fig. 8.17 Sequential radix-4 Booth multiplier

$$y = y'_{(m-1)/2} \cdot 4^{(m-1)/2} + y'_{(m-1)/2-1} \cdot 4^{(m-1)/2-1} + \ldots + y'_1 \cdot 4 + y'_0,$$

the so-called Booth-2 encoding of y.

Example 8.3
Consider the case where $m = 9$ and thus $(m-1)/2 = 4$. The 2's complement representation of -137 is 1101110111. The corresponding Booth-2 encoding is $-1\ 2\ -1\ 2\ -1$ and, indeed, $-4^4 + 2 \cdot 4^3 - 4^2 + 2 \cdot 4 - 1 = -137$.The basic radix-4 multiplication algorithm, with $v = 0$, can be used.

Algorithm 8.5: Radix-4 Booth multiplication, z = x · y + u

```
accumulator := u;
for j in 0 .. (m-1)/2 - 1 loop
  accumulator := accumulator + x·yj'·2j;
end loop;
z := accumulator;
```

A sequential implementation is shown in Fig. 8.17. It includes a shift register whose content is shifted two positions at each step, a parallel register and an adder whose second operand is $-2x$, $-x$, 0, x or $2x$ depending on the three least significant bits $(y_{2\cdot i+1}, y_{2\cdot i}, y_{2\cdot i-1})$ of the shift register. At each step, two output bits are generated. Hence, the total computation time is equal to $(m + 1)/2 \cdot T_{clk}$, where T_{clk} must be greater than the computation time of an $(n + 3)$-bit adder. Thus,

$$T(n, m) \cong \frac{m+1}{2} \cdot T_{adder}(n+3).$$

With respect to a radix-2 shift and add multiplier (Sect. 8.2.1), the computation time has been divided by 2.

The following VHDL model describes the circuit of Fig. 8.17.

```
long_x <= x(n)&(x(n)&x); minus_x <= NOT(long_x)+1;
two_x <= x(n)&(x&'0'); minus_two_x <= NOT(two_x)+1;
zero <= (OTHERS => '0');
yyy <= acc0(2 DOWNTO 0);
WITH yyy SELECT second_operand <= zero WHEN "000"|"111",
long_x WHEN "001"|"010", two_x WHEN "011",
  minus_two_x WHEN "100", minus_x WHEN OTHERS;
product <= acc1(n)&(acc1(n)&acc1) + second_operand;
z <= acc1 & acc0(m+1 DOWNTO 1);
parallel_register: PROCESS(clk)
BEGIN
  IF clk'EVENT and clk = '1' THEN
    IF load = '1' THEN acc1 <= u;
    ELSIF update = '1' THEN acc1 <= product(n+2 DOWNTO 2);
    END IF;
  END IF;
END PROCESS;
shift_register: PROCESS(clk)
BEGIN
  IF clk'EVENT and clk = '1' THEN
    IF load = '1' THEN acc0 <= y&'0';
    ELSIF update = '1' THEN
      acc0 <= product(1 DOWNTO 0)&acc0(m+1 DOWNTO 2);
    END IF;
  END IF;
END PROCESS;
```

The complete circuit also includes an $(m + 1)/2$-state counter and a control unit. A complete generic model *Booth2_sequential_multiplier.vhd* is available at the Authors' web page.

8.5 Constant Multipliers

Given an n-bit constant natural c and an m-bit natural y, the computation of $c \cdot y$ can be performed with any n-bit by m-bit multiplier whose first operand is connected to the constant value c. Then, the synthesis tool will eliminate useless components. In the case of FPGA implementations, an alternative method is to store the constant c within the LUTs.

Assume that the technology at hand includes k-input LUTs. The basic component is a circuit that computes $w = c \cdot b$, where b is a k-bit natural. The maximum value of w is

$$(2^n - 1)(2^k - 1) = 2^{n+k} - 2^k - 2^n + 1,$$

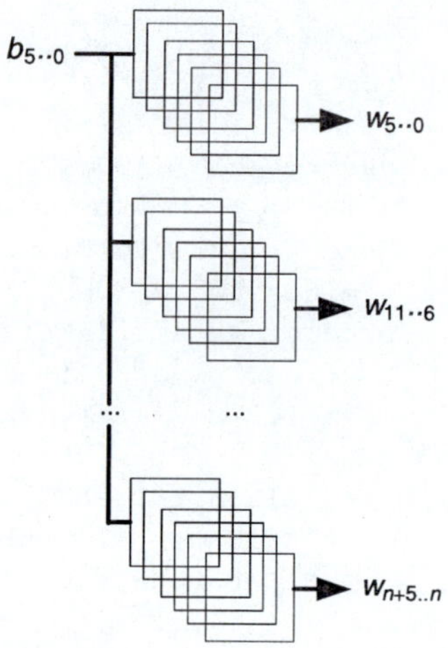

Fig. 8.18 LUT implementation of a k-bit by n-bit constant multiplier

so w is an $(n + k)$-bit number. The circuit is shown in Fig. 8.18, with $k = 6$. It is made up of $n + 6$ LUT-6, each of them being programmed in such a way that

$$w_{6j+5\ldots 6j}(b) = [c_1 \cdot b]_{6j+5\ldots 6j}.$$

Its computation time is equal to T_{LUT6}.

The following VHDL model describes the circuit of Fig. 8.18.

```
main_iteration: FOR i IN 0 TO n+5 GENERATE
  LUT_instantiation: lut6
  GENERIC MAP (truth_vector => LUT_definition(c)(i))
  PORT MAP(a => b, b => w(i));
END GENERATE;
```

The function *LUT_definition* defines the LUT contents.

```
TYPE vectors IS ARRAY (0 TO n+5)
  OF STD_LOGIC_VECTOR(0 TO 63);
FUNCTION LUT_definition(c: NATURAL) RETURN vectors IS
  VARIABLE zz: NATURAL;
  VARIABLE zzz: STD_LOGIC_VECTOR(n+5 DOWNTO 0);
  VARIABLE truth_vector: vectors;
BEGIN
  FOR i IN 0 to 63 LOOP
    zz := c*i;
    zzz := CONV_STD_LOGIC_VECTOR(zz, n+6);
      FOR j IN 0 TO n+5 LOOP
        truth_vector(j)(i) := zzz(j);
      END LOOP;
    END LOOP;
  RETURN truth_vector;
END LUT_definition;
```

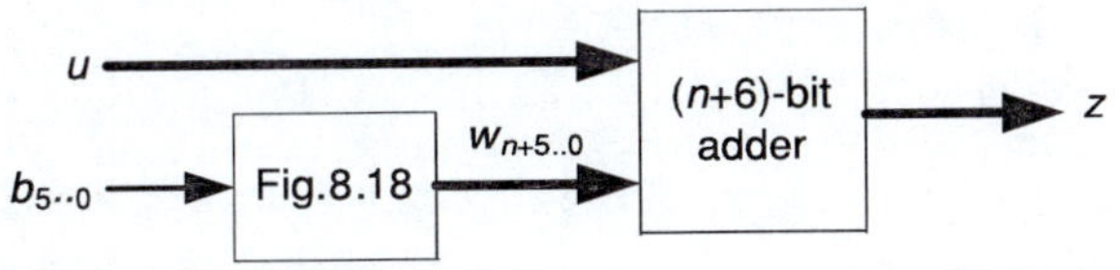

Fig. 8.19 Computation of $w = c \cdot b + u$

The circuit of Fig. 8.18 can be used as a component for generating constant multipliers. As an example, a sequential n-bit by m-bit constant multiplier is synthesized. First define a component similar to that of Fig. 8.2, with x constant. It computes $z = c \cdot b + u$, where c is an n-bit constant natural, b a k-bit natural, and u an n-bit natural. The maximum value of z is

$$(2^n - 1)(2^k - 1) + 2^n - 1 = 2^{n+k} - 2^k,$$

so it is an $(n + k)$-bit number. It consists of a k-bit by n-bit multiplier (Fig. 8.18) and an $(n + k)$-bit adder (Fig. 8.19).

Finally, the circuit of Fig. 8.19 can be used to generate a radix-2^k shift and add multiplier that computes $z = c \cdot y + u$, where c is an n-bit constant natural, y an m-bit natural, and u an n-bit natural. The maximum value of z is

$$(2^n - 1)(2^m - 1) + 2^n - 1 = 2^{n+m} - 2^m,$$

so z is an $(n + m)$-bit number. Assume that the radix-2^k representation of y is $Y_{m/k-1}\ Y_{m/k-2\ldots}\ Y_0$, where each Y_i is a k-bit number. The circuit implements the following set of equations:

$$\begin{aligned}
z^{(0)} &= (u + c \cdot Y_0)/2^k,\\
z^{(1)} &= \left(z^{(0)} + c \cdot Y_1\right)/2^k,\\
z^{(2)} &= \left(z^{1)} + c \cdot Y_2\right)/2^k,\\
&\ldots\\
z^{(m/k-1)} &= \left(z^{(m/k-2)} + c \cdot Y_{m/k-1}\right)/2^k.
\end{aligned} \tag{8.19}$$

Thus,

$$z^{(m/k-1)} \cdot \left(2^k\right)^{m/k} = u + c \cdot Y_0 + c \cdot Y_1 \cdot 2^k + \ldots + c \cdot Y_{m/k-1} \cdot \left(2^k\right)^{m/k-1},$$

that is to say

$$z^{(m/k-1)} \cdot 2^m = c \cdot y + u.$$

The circuit is shown in Fig. 8.20.

The computation time is approximately equal to

$$T \cong (m/k) \cdot (T_{LUT-k} + T_{adder}(n + k).$$

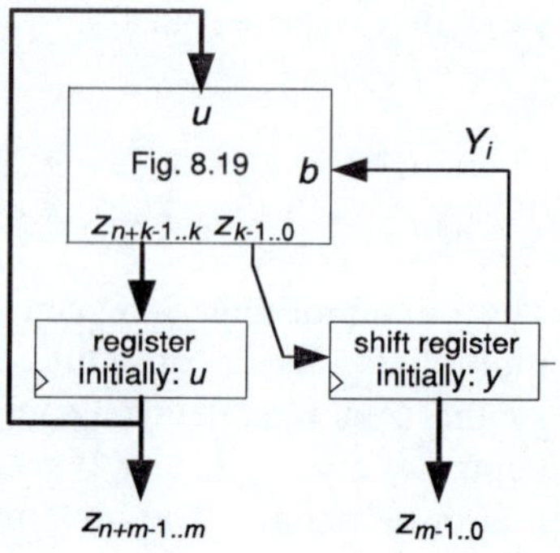

Fig. 8.20 *n*-bit by *m*-bit constant multiplier

The following VHDL model describes the circuit of Fig. 8.20 ($k = 6$).

```
main_component: digit_by_constant_multiplier
  GENERIC MAP(n => n, c => c)
  PORT MAP(b => acc_0(5 DOWNTO 0), u => acc_1, z => product);
z <= acc_1 & acc_0;
parallel_register: PROCESS(clk)
BEGIN
  IF clk'EVENT and clk = '1' THEN
    IF load = '1' THEN acc_1 <= u;
    ELSIF update = '1' THEN acc_1 <= product(n+5 DOWNTO 6);
    END IF;
  END IF;
END PROCESS;
shift_register: PROCESS(clk)
BEGIN
  IF clk'EVENT and clk = '1' THEN
    IF load = '1' THEN acc_0 <= y;
    ELSIF update = '1' THEN
      acc_0 <= product(5 DOWNTO 0)& acc_0(m-1 DOWNTO 6);
    END IF;
  END IF;
END PROCESS;
```

A complete model *sequential_constant_multiplier.vhd* is available at the Authors' web page.

The synthesis of constant multipliers for integers is left as an exercise.

Table 8.1 Combinational multiplier

m	*n*	*LUTS*	*Delay*
8	8	96	13.29
16	16	384	28.26
32	16	771	36.91
32	32	1536	57.46
64	32	3073	74.12
64	64	6181	119.33

Table 8.2 Carry-save combinational multiplier

m	*n*	*LUTS*	*Delay*
8	8	102	8.05
16	16	399	15.42
32	16	788	16.99
32	32	1580	29.50
64	32	3165	32.08
64	64	6354	60.90

8.6 FPGA Implementations

Several multipliers have been implemented within a Virtex 5-2 device. Those devices include Digital Signal Processing (DSP) slices that efficiently perform multiplications (25 bits by 18 bits), additions and accumulations. Apart from multiplier implementations based on LUTs and FFs, more efficient implementations, taking advantage of the availability of DSP slices, are also reported. As before, the times are expressed in *ns* and the costs in numbers of Look Up Tables (LUTs), flip-flops (FFs) and DSP slices. All VHDL models as well as several test benches are available at the Authors' web page.

8.6.1 Combinational Multipliers

The circuit is shown in Fig. 8.3. The synthesis results for several numbers n and m of bits are given in Table 8.1.

A faster implementation is obtained by using the carry-save method (Fig. 8.4; Table 8.2).

If multipliers based on the cell of Fig. 8.8b are considered, more efficient circuits can be generated. It is the "by default" option of the synthesizer (Table 8.3).

Finally, if DSP slices are used, better implementations are obtained (Table 8.4).

Table 8.3 Optimized combinational multiplier

n	*m*	*LUTs*	*Delay*
8	8	113	5.343
16	16	435	6.897
32	16	835	7.281
32	32	1668	7.901
64	64	6460	11.41
64	32	3236	9.535
32	64	3236	9.535

Table 8.4 Combinational multiplier with DSP slices

n	*m*	*LUTs*	*DSPs*	*Delay*
8	8	0	2	4.926
16	16	0	2	4.926
32	16	77	2	6.773
32	32	93	4	9.866
64	64	346	12	12.86
64	32	211	6	11.76
32	64	211	6	11.76

Table 8.5 Radix-2^k parallel multipliers

m	*n*	*k*	$m \cdot k$	$n \cdot k$	*LUTs*	*Delay*
2	2	8	16	16	452	10.23
4	4	4	16	16	448	17.40
2	2	16	32	32	1740	12.11
4	4	8	32	32	1808	20.29
4	2	16	64	32	3480	15.96
4	4	16	64	64	6960	22.91

8.6.2 Radix-2^k Parallel Multipliers

Several $m \cdot k$ bits by $n \cdot k$ bits multipliers (Sect. 8.2.4) have been implemented (Table 8.5).

A faster implementation is obtained by using the carry-save method (Table 8.6).

The same circuits have been implemented with DSP slices. The implementation results are given in Tables 8.7, 8.8

8.6.3 Sequential Multipliers

Several shift and add multipliers have been implemented. The implementation results are given in Tables 8.9, 8.10. Both the clock period T_{clk} and the total delay ($m \cdot T_{clk}$) are given.

Table 8.6 Carry-save radix-2^k parallel multipliers

m	*n*	*k*	$m \cdot k$	$n \cdot k$	*LUTs*	*Delay*
2	2	8	16	16	461	8.48
4	4	4	16	16	457	10.09
2	2	16	32	32	1757	10.36
4	4	8	32	32	3501	11.10
4	2	16	64	32	1821	12.32
4	4	16	64	64	6981	14.93

Table 8.7 Radix-2^k parallel multipliers with DSPs

m	n	k	$m \cdot k$	$n \cdot k$	DSPs	LUTs	Delay
2	2	8	16	16	8	0	12.58
4	4	4	16	16	32	0	27.89
2	2	16	32	32	8	0	12.58
4	4	8	32	32	32	0	27.89
4	2	16	64	32	16	0	16.70
4	4	16	64	64	32	0	27.89

Table 8.8 Carry-save radix-2^k parallel multipliers with DSPs

m	*n*	*k*	$m \cdot k$	$n \cdot k$	*DSPs*	*LUTs*	*Delay*
2	2	8	16	16	8	15	9.90
4	4	4	16	16	32	15	17.00
2	2	16	32	32	8	31	10.26
4	4	8	32	32	32	31	17.36
4	2	16	64	32	16	63	11.07
4	4	16	64	64	32	63	18.09

Table 8.9 Shift and add multipliers

n	*m*	*FFs*	*LUTs*	*Period*	*Total time*
8	8	29	43	2.87	23.0
8	16	46	61	2.87	45.9
16	8	38	72	4.19	33.5
16	16	55	90	4.19	67.0
32	16	71	112	7.50	120.0
32	32	104	161	7.50	240.0
64	32	136	203	15.55	497.6
64	64	201	306	15.55	995.2

8.6.4 Combinational Multipliers for Integers

A carry-save multiplier for integers is shown in Fig. 8.12. The synthesis results for several numbers n and m of bits are given in Table 8.11.

Table 8.10 Sequential carry-save multipliers

n	*m*	*FFs*	*LUTs*	*Period*	*Total time*
8	8	29	43	1.87	15.0
16	8	47	64	1.88	15.0
16	16	56	74	1.92	30.7
32	16	88	122	1.93	30.9
32	32	106	139	1.84	58.9
64	32	170	235	1.84	58.9
64	64	203	268	1.84	117.8

Table 8.11 Carry-save mod 2^{n+m+1} multipliers

n	*m*	*LUTs*	*Delay*
8	8	179	12.49
8	16	420	18.00
16	8	421	20.41
16	16	677	25.86
32	16	1662	42.95
32	32	2488	55.69

Table 8.12 Modified shift and add algorithm

n	m	LUTs	Delay
8	8	122	15.90
8	16	230	27.51
16	8	231	20.20
16	*16*	*435*	*31.81*
32	16	844	39.96
32	32	1635	62.91

Table 8.13 Multipliers with post correction

n	*m*	*LUTs*	*Delay*
8	8	106	14.18
8	16	209	24.60
16	8	204	18.91
16	16	407	30.60
32	16	794	39.43
32	32	1586	62.91

Another option is the modified shift and add algorithm of Sect. 8.4.2 (Fig. 8.15; Table 8.12).

In Table 8.13, examples of post correction implementations are reported.

As a last option, several Booth multipliers have been implemented (Table 8.14).

Table 8.14 Combinational Booth multipliers

n	m	*LUTs*	*Delay*
8	8	188	13.49
8	16	356	25.12
16	8	332	13.68
16	16	628	25.31
32	16	1172	25.67
32	32	2276	49.09

Table 8.15 Sequential radix-4 Booth multipliers

n	m	*FFs*	*LUTs*	*Period*	*Total time*
8	9	25	58	2.90	26.1
8	17	34	68	2.90	49.3
16	9	33	125	3.12	28.1
16	17	42	135	3.12	53.0
32	17	58	231	3.48	59.2
32	33	75	248	3.48	114.8
64	33	107	440	4.22	139.1
64	65	140	473	4.22	274.0

8.6.5 Sequential Multipliers for Integers

Several radix-4 Booth multipliers have been implemented (Fig. 8.17). Both the clock period T_{clk} and the total delay ($m \cdot T_{clk}$) are given (Table 8.15).

8.7 Exercises

1. Generate the VHDL model of a mixed-radix parallel multiplier (Sect. 8.2.4).
2. Synthesize a $2n$-bit by $2n$-bit parallel multiplier using n-bit by n-bit multipliers as building blocks.
3. Modify the VHDL model *sequential_CSA_multiplier.vhd* so that the *done* flag is raised when the final result is available (Comment 8.2).
4. Generate the VHDL model of a carry-save multiplier with post correction (Sect. 8.4.3).
5. Synthesize a sequential multiplier based on Algorithm 8.3.
6. Synthesize a parallel constant multiplier (Sect. 8.5).
7. Generate models of constant multipliers for integers.
8. Synthesize a constant multiplier that computes $z = c_1 \cdot y_1 + c_2 \cdot y_2 + \ldots + c_s \cdot y_s + u$.

References

1. Baugh CR, Wooley BA (1973) A two's complement parallel array multiplication algorithm. IEEE Trans Comput C 31:1045–1047
2. Booth AD (1951) A signed binary multiplication technique. Q J Mech Appl Mech 4:126–140
3. Dadda L (1965) Some schemes for parallel multipliers. Alta Frequenza 34:349–356
4. Wallace CS (1964) A suggestion for fast multipliers. IEEE Trans Electron Comput EC-13:14–17

Chapter 9
Dividers

Division is a basic arithmetic operation whose execution is based on 1-digit by m-digit multiplications and subtractions. Nevertheless, unlike addition and multiplication, division is generally not included as a predefined block within FPGA families. So, in many cases, the circuit designer will have to generate dividers by choosing some division algorithm and implementing it with adders and multipliers.

The basic digit-recurrence algorithm is described in Sect. 9.1. Among others, it introduces the Robertson and PD diagrams [1, 2]. Several radix-2 dividers are proposed in Sect. 9.2: non-restoring and restoring dividers, binary SRT dividers [2, 3]), and radix-2^k dividers. Section 9.3 gives some information about radix-B division. As a matter of fact, unless $B = 2^k$, the main and perhaps unique application of radix-B operations is decimal arithmetic (Chap. 11). The last section describes two convergence algorithms, namely the Newton-Raphson and Goldschmidt algorithms, which could be considered in the case of real number operations (Chap. 12).

9.1 Basic Digit-Recurrence Algorithm

Consider an integer x and a natural $y > 0$ such that $-y \leq x < y$. The quotient q and the remainder r, with an accuracy of p fractional radix-B digits, are defined by the following relation:

$$x \cdot B^p = q \cdot y + r, \text{ with } -y \leq r < y. \tag{9.1}$$

In fact, Eq. (9.1) has two solutions: one with $r \geq 0$ and another with $r < 0$.

J.-P. Deschamps et al., *Guide to FPGA Implementation of Arithmetic Functions*, Lecture Notes in Electrical Engineering 149, DOI: 10.1007/978-94-007-2987-2_9,

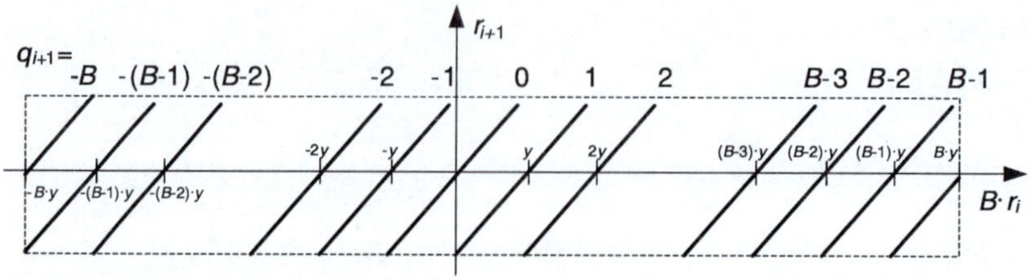

Fig. 9.1 Robertson diagram

The following set of equations generates q and r:

$$\begin{aligned} r_0 &= x, \\ B \cdot r_0 &= q_1 \cdot y + r_1, \\ B \cdot r_1 &= q_2 \cdot y + r_2, \\ &\dots \\ B \cdot r_{p-1} &= q_p \cdot y + r_p. \end{aligned} \tag{9.2}$$

From (9.2) the following relation is obtained:

$$x \cdot B^p = \left(q_1 \cdot B^{p-1} + q_2 \cdot B^{p-2} + \cdots + q_{p-1} \cdot B^1 + q_p \cdot B^0\right) \cdot y + r_p, \text{ with } -y \leq r_p < y. \tag{9.3}$$

Thus,

$$q = q_1 \cdot B^{p-1} + q_2 \cdot B^{p-2} + \cdots + q_{p-1} \cdot B^1 + q_p \cdot B^0 \text{ and } r = r_p.$$

At each step of (9.2) r_{i+1} and q_{i+1} are computed in function of y and r_i:

$$r_{i+1} = B \cdot r_i - q_{i+1} \cdot y, \text{ where } -y \leq r_{i+1} < y. \tag{9.4}$$

The *Robertson diagram* of Fig. 9.1 defines the set of possible solutions: the dotted lines define the domain $\{(B \cdot r_i, r_{i+1}) | -B \cdot y \leq B \cdot r_i < B \cdot y \text{ and } -y \leq r_{i+1} < y\}$, and the diagonals correspond to the equations $r_{i+1} = B \cdot r_i - k \cdot y$ with $k \in \{-B, -(B-1), \ldots, -1, 0, 1, \ldots, B-1, B\}$. If $k \cdot y \leq B \cdot r_i < (k+1) \cdot y$, there are two possible solutions for q_{i+1}, namely k and $k+1$. To the first one corresponds a non-negative value of r_{i+1}, and to the second one a negative value.

If the values $q_{i+1} = -B$ and $q_{i+1} = B$ are discarded, the solutions of (9.4) are the following:

- if $B \cdot r_i \geq (B-1) \cdot y$ then $q_{i+1} = B - 1$,
- if $k \cdot y \leq B \cdot r_i < (k+1) \cdot y$ then $q_{i+1} = k$ or $k+1, \forall k$ in $\{-(B-1), \ldots, -1, 0, 1, \ldots, B-2\}$,
- if $B \cdot r_i < -(B-1) \cdot y$ then $q_{i+1} = -(B-1)$,

so all q_i's are radix-B digits or negative radix-B digits. The quotient q is obtained under the form

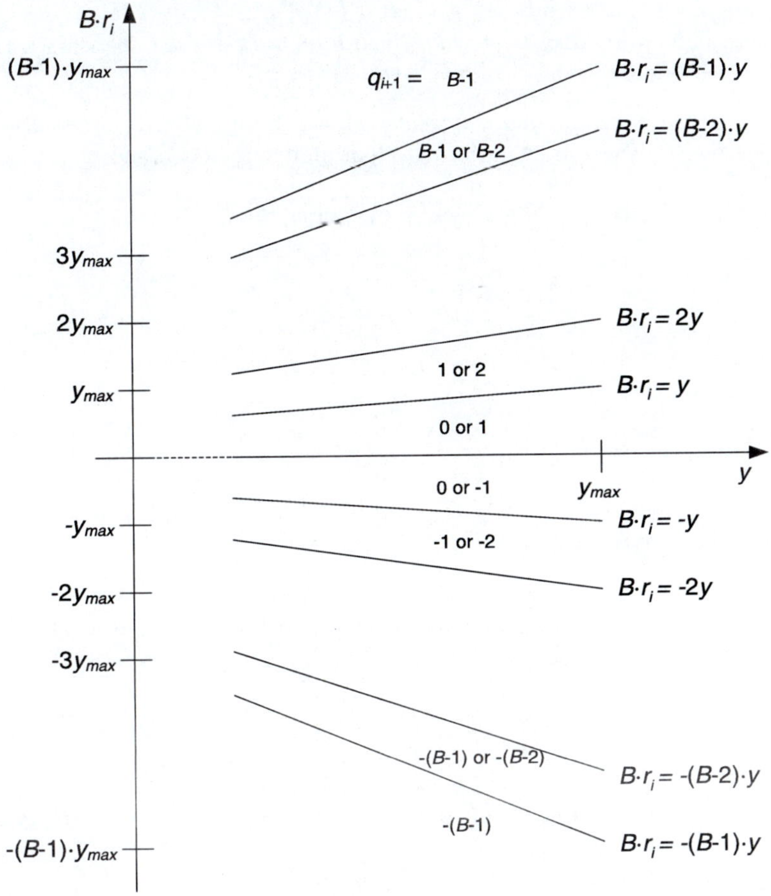

Fig. 9.2 P-D diagram

$$q = q_1 \cdot B^{p-1} + q_2 \cdot B^{p-2} + \cdots + q_{p-1} \cdot B^1 + q_p \cdot B^0, \text{ with } q_i \text{ in } \{-(B-1), \ldots, -1, 0, 1, \ldots, B-1\},$$

that is to say a signed-digit radix-B number. According to (9.3)

$$x/y = 0 \cdot q_1 q_2 \ldots q_p + r_p \cdot B^{-p}/y, \text{ with } -B^{-p} \le r_p \cdot B^{-p}/y < B^{-p}. \tag{9.5}$$

In other words, $0.q_1 q_2 \ldots q_p$ is an approximation of x/y with an absolute error smaller than or equal to B^{-p}.

Another useful diagram is the Partial remainder-Divisor diagram (*P-D diagram*) of Fig. 9.2. It defines zones of the $(B \cdot r_i, y)$ plane and the corresponding solutions for q_{i+1}:

- beyond the line defined by $B \cdot r_i = (B-1) \cdot y, q_{i+1} = B-1$,
- between the lines defined by $B \cdot r_i = k \cdot y$ and $B \cdot r_i = (k+1) \cdot y, q_{i+1} = k$ or $k+1$,
- below the line defined by $B \cdot r_i = -(B-1) \cdot y, q_{i+1} = -(B-1)$.

Assume that a function *quotient_selection* has been defined, compatible with Figs. 9.1 or 9.2. The following formal algorithm computes q and r.

Algorithm 9.1: Division, digit-recurrence algorithm

```
r0 := x; q := 0;
for i in 1 .. p loop
  qi := quotient_selection(ri-1, y);
  ri := B·ri-1 - qi·y ; q := q + qi·B^-i;
end loop ;
r := rp;
```

9.2 Radix-2 Division

The Robertson and P-D diagram corresponding to $B = 2$ are shown in Fig. 9.3.

9.2.1 Non-Restoring Divider

According to the diagrams of Fig. 9.3, a possible choice of the *quotient_selection* function of Algorithm 1 is $q_{i+1} = -1$ if $r_i < 0$ and $q_{i+1} = 1$ if $r_i \geq 0$. The corresponding algorithm is

Algorithm 9.2: Division, non-restoring algorithm, version 1

```
r0 := x; q := 0;
for i in 1 .. p loop
  if ri-1 < 0  then ri := 2·ri-1 + y ; q := q - 2^-i;
  else ri := 2·ri-1 - y ; q := q + 2^-i; end if;
end loop;
r := rp;
```

In this case, a slight modification allows avoiding the conversion of a signed-digit number to an unsigned-digit one. Let $q = q_1 \cdot 2^{p-1} + q_2 \cdot 2^{p-2} + \cdots + q_{p-1} \cdot 2^1 + q_p \cdot 2^0$, with q_i in $\{-1, 1\}$, be the quotient in signed-digit form computed with the previously defined *quotient_selection* function. Define

$$q'_i = (1 + q_{i+1})/2, \forall i \text{ in } \{0, 1, \ldots, p-1\}.$$

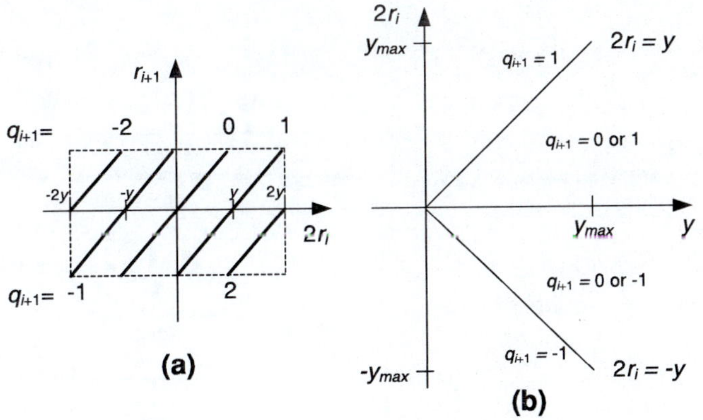

Fig. 9.3 Robertson diagram (**a**) and P-D diagram (**b**), when $B = 2$

Then $q'_i \in \{0, 1\}$ and $q = -(1 - q'_0) \cdot 2^p + q'_1 \cdot 2^{p-1} + q'_2 \cdot 2^{p-2} + \cdots + q'_{p-1} \cdot 2^1 + 2^0$. In other words, the vector

$$(1 - q'_0) q'_1\, q'_2 \cdots q'_{p-1}\, 1$$

is the 2's complement representation of q. The corresponding modified algorithm is the following.

Algorithm 9.3: Division, non-restoring algorithm, version 2

```
r0 := x;
for i in 1 .. p loop
  if ri-1 < 0  then ri := 2·ri-1 + y ; qi-1 := 0;
  else ri := 2·ri-1 - y ; qi-1 := 1; end if;
end loop;
q0 := 1 - q0; qp := 1; r := rp;
```

The corresponding circuit is shown in Fig. 9.4.

The following VHDL model describes the circuit of Fig. 9.4. The dividend $x = x_n\, x_{n-1} \cdots x_0$ and the remainder $r = r_n\, r_{n-1} \cdots r_0$ are $(n + 1)$-bit 2's complement integers, the divisor $y = y_{n-1}\, y_{n-2} \cdots y_0$ is an n-bit natural, the quotient $q = q_0\, q_1 \cdots q_p$ is a $(p + 1)$-bit 2's complement integer, and condition

$$-y \le x < y \tag{9.6}$$

must hold. Then $2^p \cdot x = q \cdot y + r$, with $-y \le r < y$, and

$$x = (q_0 \cdot q_1\, q_2 \cdots q_p) \cdot y + (r/y) \cdot 2^{-p}, \text{ with } -2^{-p} \le (r/y) \cdot 2^{-p} < 2^{-p}.$$

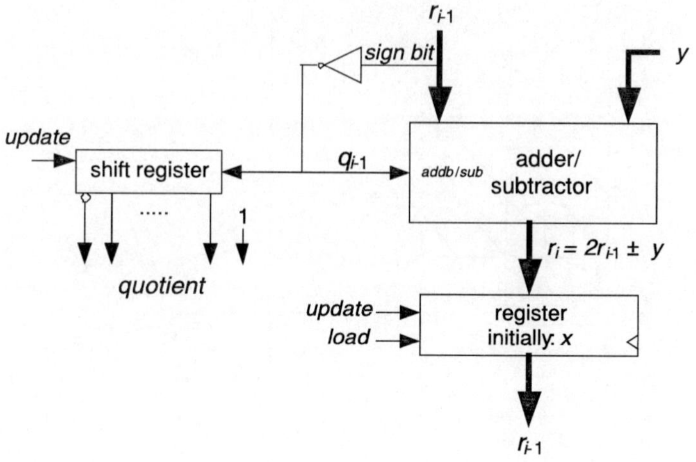

Fig. 9.4 Non-restoring divider

```
long_y <= '0'&y;
two_r <= r(n-1 DOWNTO 0)&'0';
WITH r(n) SELECT next_r <= two_r + long_y WHEN '1',
  two_r - long_y WHEN OTHERS;
remainder_register: PROCESS(clk)
BEGIN
  IF clk'EVENT AND clk = '1' THEN
    IF load = '1' THEN r <= x;
    ELSIF update = '1' THEN r <= next_r;
    END IF;
  END IF;
END PROCESS;
remainder <= r;
quotient_register: PROCESS(clk)
BEGIN
  IF clk'EVENT AND clk = '1' THEN
    IF load = '1' THEN q <= (others => '0');
    ELSIF update = '1' THEN q <= q(1 TO p-1)&NOT(r(n));
    END IF;
  END IF;
END PROCESS;
quotient <= NOT(q(0)) & q(1 TO p-1)&'1';
```

The complete circuit also includes a p-state counter and a control unit. A complete generic model *non_restoring.vhd* is available at the Authors' web page.

The computation time is approximately equal to

$$T_{divider}(n, p) = p \cdot T_{adder}(n) = p \cdot n \cdot T_{adder}(1). \tag{9.7}$$

Comment 9.1
The least significant bit is always equal to 1 while the sign of the remainder could be different from that of the dividend. For some applications a final correction step could be necessary.

9.2.2 Restoring Divider

If x is a natural, so that $r_0 \geq 0$, the *quotient_selection* function can be chosen as follows (Fig. 9.3):

- if $2 \cdot r_i < y$ then $q_{i+1} = 0$ and $r_{i+1} = 2 \cdot r_i$,
- if $2 \cdot r_i \geq y$ then $q_{i+1} = 1$ and $r_{i+1} = 2 \cdot r_i - y$

So, all along the algorithm execution the remainders r_i are non-negative.

Algorithm 9.4: Division, restoring algorithm

```
r0 := x;
for i in 1 .. p loop
  dif := 2*ri-1 - y;
  if dif < 0 then qi := 0; ri := 2*ri-1;
  else qi := 1; ri := dif; end if;
end loop ;
r := rp;
```

The corresponding circuit is shown in Fig. 9.5.

The following VHDL model describes the circuit of Fig. 9.5. The dividend $x = x_{n-1} \ldots x_0$, the divisor $y = y_{n-1}\, y_{n-2} \ldots y_0$ and the remainder $r = r_{n-1} \ldots r_0$ are n-bit naturals, the quotient $q = q_1\, q_2 \ldots q_p$ is a p-bit natural, and the condition $x < y$ must hold. Then $2^p \cdot x = q \cdot y + r$, with $r < y$, so

$$x = (0 \cdot q_1\, q_2 \ldots q_p) \cdot y + (r/y) \cdot 2^{-p}, \text{ with } (r/y) \cdot 2^{-p} < 2^{-p}.$$

```
long_y <= '0'&y;
two_r <= r&'0';
dif <= two_r - long_y;
WITH dif(n) SELECT next_r <= dif(n-1 DOWNTO 0) WHEN '0',
  two_r(n-1 DOWNTO 0) WHEN OTHERS;
remainder_register: PROCESS(clk)
```

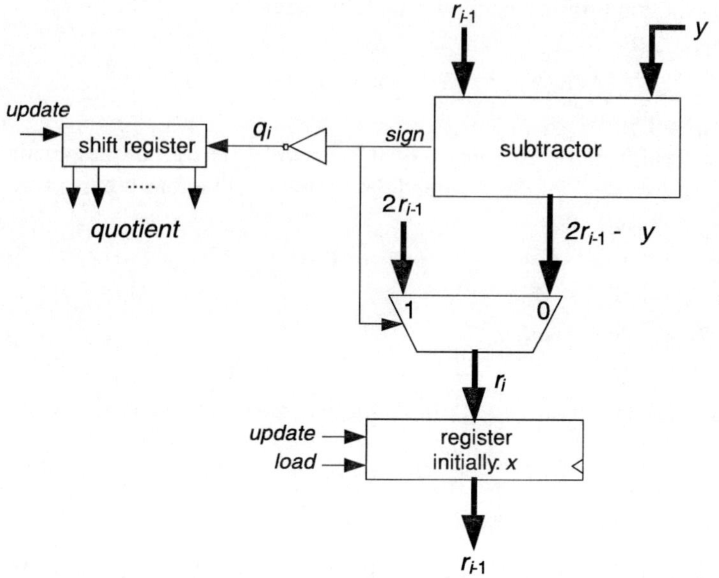

Fig. 9.5 Restoring divider

```
BEGIN
  IF clk'EVENT AND clk = '1' THEN
    IF load = '1' THEN r <= x;
    ELSIF update = '1' THEN r <= next_r;
    END IF;
  END IF;
END PROCESS;
remainder <= r;
quotient_register: PROCESS(clk)
BEGIN
  IF clk'EVENT AND clk = '1' THEN
    IF load = '1' THEN q <= (others => '0');
    ELSIF update = '1' THEN q <= q(2 TO p)&NOT(dif(n));
    END IF;
  END IF;
END PROCESS;
quotient <= q;
```

The complete circuit also includes a p-state counter and a control unit. A complete generic model *restoring.vhd* is available at the Authors' web page.

Fig. 9.6 P-D diagram, $B = 2, y \geq y_{min}$

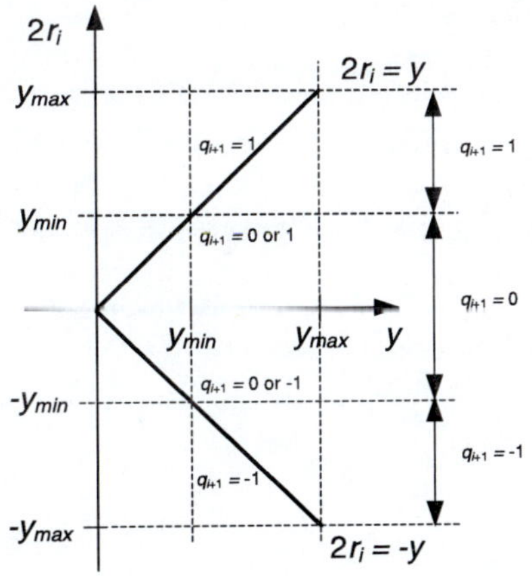

The computation time is approximately the same as that of a non-restoring divider (9.7).

9.2.3 Binary SRT Divider

Assume that a lower bound y_{min} of y is known. Then, according to the P-D diagram of Fig. 9.6, the *quotient_selection* function can be defined as follows:

- if $2 \cdot r_i \geq y_{\min}$ then $q_{i+1} = 1$,
- if $-y_{\min} \leq 2 \cdot r_i < y_{\min}$ then $q_{i+1} = 0$,
- if $2 \cdot r_i < -y_{\min}$ then $q_{i+1} = -1$.

Algorithm 9.5: Division, binary SRT algorithm, version 1

```
r0 := x; q := 0;
for i in 1 .. p loop
  if 2·ri-1 ≥ ymin then ri := 2·ri-1 - y; q := q + 2^-i;
  elsif 2·ri-1 < -ymin then ri := 2·ri-1 + y; q := q - 2^-i;
  else ri := 2·ri-1; end if;
end loop;
r := rp;
```

Assume that x is an $(n + 1)$-bit 2's complement integer and y a normalized n-bit natural, that is $2^{n-1} \leq y < 2^n$, so $y_{\min} = 2^{n-1}$. The condition $2 \cdot r \geq y_{min}$ is equivalent to $r/2^{n-2} \geq 1$, and the condition $2 \cdot r < -y_{min}$ to $r/2^{n-2} < -1$. If $r = r_n\, r_{n-1}\, r_{n-2}\, r_{n-3} \ldots r_0$, then $2 \cdot r \geq y_{min}$ if $r_n\, r_{n-1}\, r_{n-2} \geq 1$, and $2 \cdot r < -y_{min}$ if $r_n\, r_{n-1}\, r_{n-2} < -1$.

In order to convert the signed-bit representation of q to a 2's complement representation, the following *on-the-fly conversion* algorithm can be used. Assume that $q = q_1 \cdot 2^{-1} + q_2 \cdot 2^{-2} + \cdots + q_p \cdot 2^{-p}$, with q_i in $\{-1, 0, 1\}$, and define

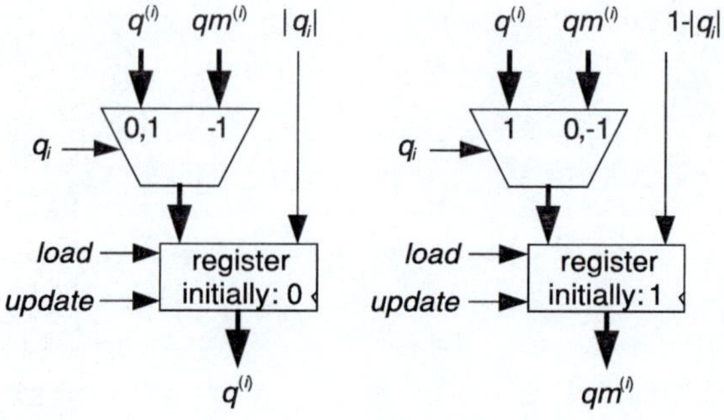

Fig. 9.7 On-the-fly conversion

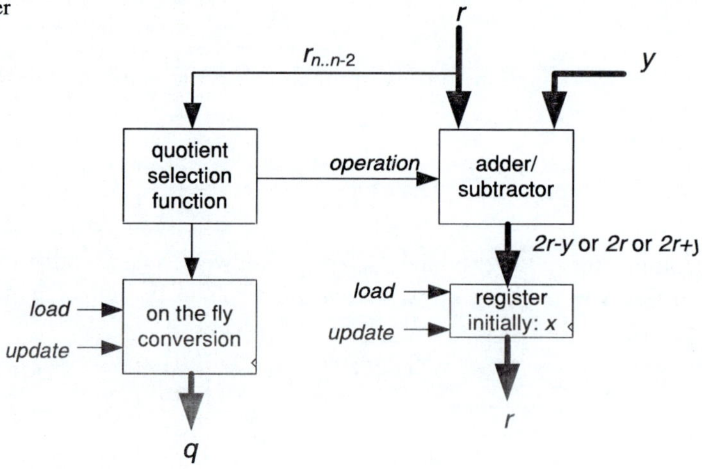

Fig. 9.8 SRT divider

$$q^{(0)} = 0, q^{(i)} = q_1 \cdot 2^{-1} + q_2 \cdot 2^{-2} + \cdots + q_i \cdot 2^{-i}, qm^{(0)} = 1,\ qm^{(i)} = q^{(i)} - 2^{-i},$$

$\forall i$ in $\{1, 2, \ldots, p\}$. Then

- if $q_i = 1 : q^{(i)} = q^{(i-1)} + 2^{-i}, qm^{(i)} = q^{(i-1)}$;
- if $q_i = 0 : q^{(i)} = q^{(i-1)}, qm^{(i)} = q^{(i-1)} - 2^{-i} = q^{(i-1)} - 2^{-(i-1)} + 2^{-i} = qm^{(i-1)} + 2^{-i}$;
- if $q_i = -1 : q^{(i)} = q^{(i-1)} - 2^{-i} = qm^{(i-1)} + 2^{-i}, qm^{(i)} = qm^{(i-1)}$.

The corresponding circuit is shown in Fig. 9.7.

The complete circuit is shown in Fig. 9.8.

The following VHDL model describes the circuit of Fig. 9.8. The dividend $x = x_n x_{n-1} \ldots x_0$ and the remainder $r = r_n r_{n-1} \ldots r_0$ are $(n + 1)$-bit 2's complement integers, the divisor $y = 1y_{n-2} \ldots y_0$ is an n-bit normalized natural, the quotient $q = q_0 q_1 \ldots q_p$ is a $(p + 1)$-bit 2's complement integer, and condition (9.6) must hold.

```
long_y <= '0'&y;
two_r <= r(n-1 DOWNTO 0)&'0';
plus1 <= NOT(r(n)) AND (r(n-1) OR r(n-2));
minus1 <= r(n) AND (NOT(r(n-1)) OR NOT(r(n-2)));
operation <= plus1 & minus1;
WITH operation SELECT next_r <= two_r - long_y WHEN "10",
  two_r + long_y WHEN "01", two_r WHEN OTHERS;
remainder_register: PROCESS(clk)
BEGIN
  IF clk'EVENT AND clk = '1' THEN
    IF load = '1' THEN r <= x;
    ELSIF update = '1' THEN r <= next_r;
    END IF;
  END IF;
END PROCESS;
remainder <= r;
q_register: PROCESS(clk)
BEGIN
  IF clk'EVENT AND clk = '1' THEN
    IF load = '1' THEN q <= (OTHERS => '0');
    ELSIF update = '1' THEN
      IF plus1 = '1' THEN
        q(0 TO p-1) <= q(1 TO p); q(p) <= '1';
      ELSIF minus1 = '1' THEN
        q(0 TO p-1) <= qm(1 TO p); q(p) <= '1';
      ELSE q(0 TO p-1) <= q(1 TO p); q(p) <= '0';
      END IF;
    END IF;
  END IF;
END PROCESS;
qm_register: PROCESS(clk)
BEGIN
  IF clk'EVENT AND clk = '1' THEN
    IF load = '1' THEN
      qm(0 TO p-1) <= (OTHERS => '0'); qm(p) <= '1';
    ELSIF update = '1' THEN
      IF plus1 = '1' THEN
        qm(0 TO p-1) <= q(1 TO p); qm(p) <= '0';
      ELSIF minus1 = '1' THEN
        qm(0 TO p-1) <= qm(1 TO p); qm(p) <= '0';
      ELSE qm(0 TO p-1) <= qm(1 TO p); qm(p) <= '1';
```

```
      END IF;
    END IF;
  END IF;
END PROCESS;
quotient <= q;
```

The complete circuit also includes a p-state counter and a control unit. A complete generic model *srt_divider.vhd* is available at the Authors' web page.

The computation time is approximately the same as that of a non-restoring divider (9.7).

9.2.4 Binary SRT Divider with Carry-Save Adder

Assume again that a lower bound y_{min} of y is known and that an integer estimation of $v_i = 2 \cdot r_i/y_{min}$ satisfying

$$2 \cdot r_i/y_{min} - 2 < v_i \le 2 \cdot r_i/y_{min} \tag{9.8}$$

is computed at each step. The *quotient_selection* function can be defined as follows:

- if $v_i \le -2$ then $2 \cdot r_i/y_{min} < 0$; select $q_{i+1} = -1$;
- if $v_i = -1$ then $-1 \le 2 \cdot r_i/y_{min} < 1$; select $q_{i+1} = 0$;
- if $v_i \ge 0$ then $2 \cdot r_i/y_{min} \ge 0$; select $q_{i+1} = 1$.

In the following algorithm the function *estimated_value* generates an integer v_i belonging to (9.8).

Algorithm 9.6: Division, binary SRT algorithm, version 2

```
r0 := x; q := 0;
for i in 1 .. p loop
  vi-1 := estimated_value(ri-1);
  if vi-1 ≤ -2 then ri := 2·ri-1 + y; q := q - 2^-i;
  elsif vi-1 ≥ 0 then ri := 2·ri-1 - y; q := q + 2^-i;
  else ri := 2·ri-1; end if;
end loop;
r := rp;
```

In what follows x is an $(n + 1)$-bit 2's complement integer and y a normalized n-bit natural, that is $2^{n-1} \le y < 2^n$, so that $y_{min} = 2^{n-1}$. Assume that all remainders r_i are represented in stored-carry form, that is $r_i = c_i + s_i$. Define

$$v_i = \lfloor c_i/2^{n-2} \rfloor + \lfloor s_i/2^{n-2} \rfloor.$$

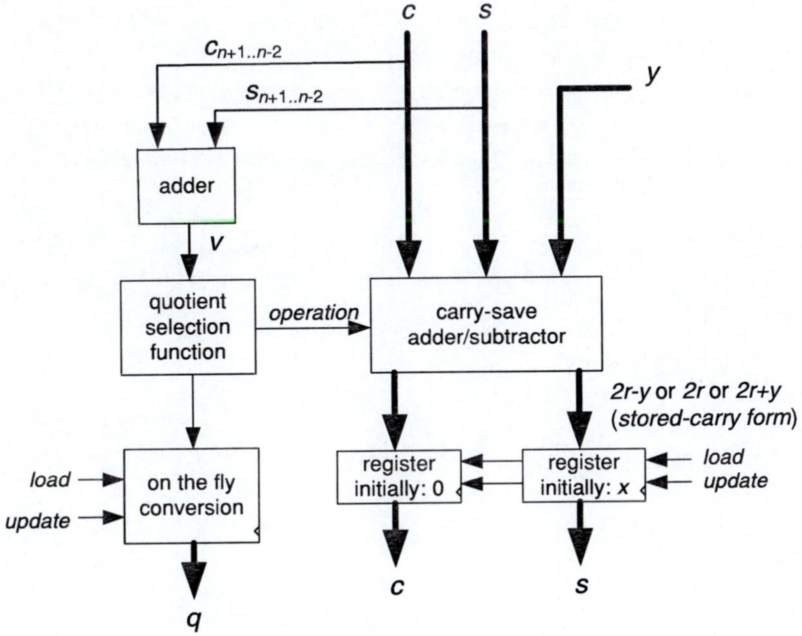

Fig. 9.9 SRT divider with carry-save adder

Thus $v_i \leq (c_i/2^{n-2}) + (s_i/2^{n-2}) = r_i/2^{n-2}$ and $v_i > (c_i/2^{n-2}) - 1 + (s_i/2^{n-2}) - 1 = (r_i/2^{n-2}) - 2$, so that (9.8), with $y_{min} = 2^{n-1}$, holds. Lower and upper bounds of v_i are computed as follows:

$$v_i \leq r_i/2^{n-2} < 2^n/2^{n-2} = 4 \text{ and } v_i > (r_i/2^{n-2}) - 2 \geq (-2^n/2^{n-2}) - 2 = -6.$$

Thus, v_i belongs to the range $-5 \leq v_i \leq 3$ and is a 4-bit 2's complement integer. In order to compute v_i, both c_i and s_i must be represented with $4 + (n - 2) = n + 2$ bits.

The complete circuit is shown in Fig. 9.9.

The following VHDL model describes the circuit of Fig. 9.9. The dividend $x = x_n\ x_{n-1} \ldots x_0$ and the remainder $r = r_n\ r_{n-1} \ldots r_0$ are $(n + 1)$-bit 2's complement integers, the divisor $y = 1\ y_{n-2} \ldots y_0$ is an n-bit normalized natural, the quotient $q = q_0\ q_1 \ldots q_p$ is a $(p + 1)$-bit 2's complement integer, and condition (9.6) must hold.

```
long_y <= "00"&y;
minus_y <= zero - long_y;
two_c <= c(n DOWNTO 0)&'0';
two_s <= s(n DOWNTO 0)&'0';
v <= c(n+1 DOWNTO n-2) + s(n+1 DOWNTO n-2);
plus1 <= NOT(v(3));
```

```
minus1 <= v(3) AND (NOT(v(2)) OR NOT(v(1)) OR NOT(v(0)));
operation <= plus1 & minus1;
WITH operation SELECT third_operand <= minus_y WHEN "10",
  long_y WHEN "01", zero WHEN OTHERS;
next_c(0) <= '0';
carry_save_adder: FOR i IN 0 TO n GENERATE
  next_s(i) <= two_c(i) XOR two_s(i) XOR third_operand(i);
  next_c(i+1)  <=  (two_c(i)  AND  two_s(i))  OR  (two_c(i)  AND
third_operand(i)) OR (two_s(i) AND third_operand(i));
END GENERATE;
next_s(n+1)    <=    two_c(n+1)    XOR    two_s(n+1)    XOR
third_operand(n+1);
s_register: PROCESS(clk)
BEGIN
  IF clk'EVENT AND clk = '1' THEN
    IF load = '1' THEN s <= x(n) & x;
    ELSIF update = '1' THEN s <= next_s;
    END IF;
  END IF;
END PROCESS;
c_register: PROCESS(clk)
BEGIN
  IF clk'EVENT AND clk = '1' THEN
    IF load = '1' THEN c <= (OTHERS => '0');
    ELSIF update = '1' THEN c <= next_c;
    END IF;
  END IF;
END PROCESS;
remainder <= c(n DOWNTO 0) + s(n DOWNTO 0);
q_register: PROCESS(clk) ... END PROCESS;
qm_register: PROCESS(clk) ... END PROCESS;
quotient <= q;
```

The complete circuit also includes a p-state counter and a control unit. A complete generic model *srt_csa_divider.vhd* is available at the Authors' web page.

The computation time is approximately equal to

$$T_{divider}(n,p) = p \cdot T_{adder}(1) + T_{adder}(n) = (n+p) \cdot T_{adder}(1). \qquad (9.9)$$

9.2.5 Radix-2^k SRT Dividers

Another way to accelerate the quotient calculation consists of generating k quotient bits at each step. For that, the digit-recurrence algorithm can be executed in base 2^k. As an example, an SRT divider with $k = 2$ is defined.

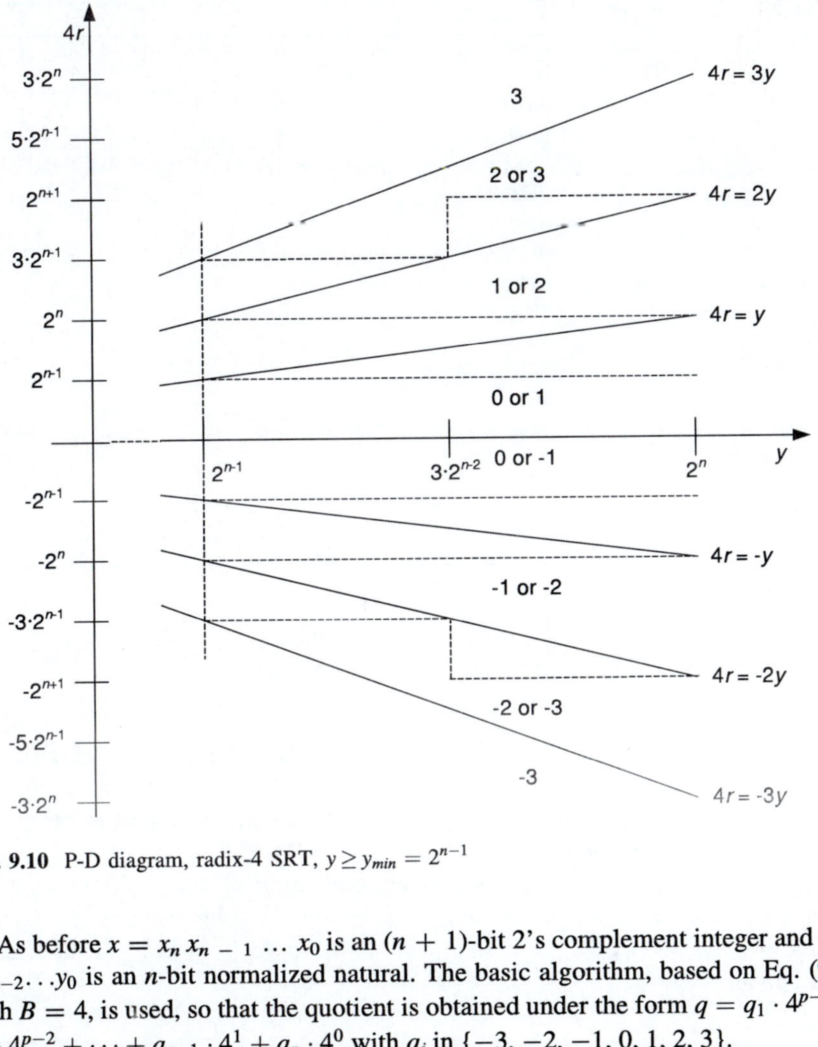

Fig. 9.10 P-D diagram, radix-4 SRT, $y \geq y_{min} = 2^{n-1}$

As before $x = x_n\, x_{n-1} \ldots x_0$ is an $(n+1)$-bit 2's complement integer and $y = 1y_{n-2} \ldots y_0$ is an n-bit normalized natural. The basic algorithm, based on Eq. (9.2) with $B = 4$, is used, so that the quotient is obtained under the form $q = q_1 \cdot 4^{p-1} + q_2 \cdot 4^{p-2} + \cdots + q_{p-1} \cdot 4^1 + q_p \cdot 4^0$ with q_i in $\{-3, -2, -1, 0, 1, 2, 3\}$.

According to the P-D diagram of Fig. 9.10, the *quotient_selection* function can be defined as follows:

- if $4 \cdot r_i \geq 2^{n+1}$ then $q_{i+1} = 3$,
- if $3 \cdot 2^{n-1} \leq 4 \cdot r_i < 2^{n+1}$ and $y < 3 \cdot 2^{n-2}$ then $q_{i+1} = 3$,
- if $3 \cdot 2^{n-1} \leq 4 \cdot r_i < 2^{n+1}$ and $y \geq 3 \cdot 2^{n-2}$ then $q_{i+1} = 2$,
- if $2^n \leq 4 \cdot r_i < 3 \cdot 2^{n-1}$ then $q_{i+1} = 2$,
- if $2^{n-1} \leq 4 \cdot r_i < 2^n$ then $q_{i+1} = 1$,
- if $-2^{n-1} \leq 4 \cdot r_i < 2^{n-1}$ then $q_{i+1} = 0$,
- if $-2^n \leq 4 \cdot r_i < -3 \cdot 2^{n-1}$ then $q_{i+1} = -1$,
- if $-3 \cdot 2^{n-1} \leq 4 \cdot r_i < -2^n$ then $q_{i+1} = -2$,

Table 9.1 Digit selection

$r_n\, r_{n-1}\, r_{n-2}\, r_{n-3}$	$r/2^{n-2}$	$q_{-(i+1)}$
0000	0	0
0001	1	1
0010	2	2
0011	3	2 if $y_{n-2} = 1$, 3 if $y_{n-2} = 0$
0100	4	3
0101	5	3
0110	6	3
0111	7	3
1000	−8	−3
1001	−7	−3
1010	−6	−3
1011	−5	−3
1100	−4	−3 if $y_{n-2} = 0$, −2 if $y_{n-2} = 1$
1101	−3	−2
1110	−2	−1
1111	−1	0

- if $-2^{n+1} \leq 4 \cdot r_i < -3 \cdot 2^{n-1}$ and $y \geq 3 \cdot 2^{n-2}$ then $q_{i+1} = -2$,
- if $-2^{n+1} \leq 4 \cdot r_i < -3 \cdot 2^{n-1}$ and $y < 3 \cdot 2^{n-2}$ then $q_{i+1} = -3$,
- if $4 \cdot r_i < -2^{n+1}$ then $q_{i+1} = -3$.

Given a remainder $r = r_n\, r_{n-1}\, r_{n-2}\, r_{n-3} \ldots r_1\, r_0$, then the next digit value is a function of $4 \cdot r/2^{n-1} = r/2^{n-3}$ and of $y/2^{n-2}$, that is $(r_n\, r_{n-1}\, r_{n-2}\, r_{n-3})$ and $(y_{n-1}\, y_{n-2}) = (1 y_{n-2})$. Table 9.1 defines the value of the next digit in function of $r_n, r_{n-1,1}, r_{n-1,0}, r_{n-2,1}$ and y_{n-2}.

In order to convert the signed-digit representation of q to a 2's complement representation, the following *on-the-fly conversion* algorithm can be used. Assume that $q = q_1 \cdot 4^{-1} + q_2 \cdot 4^{-2} + \cdots + q_p \cdot 4^{-p}$, with q_i in {−3, −2, −1, 0, 1, 2, 3}, and define

$$q^{(0)} = 0, q^{(i)} = q_1 \cdot 4^{-1} + q_2 \cdot 4^{-2} + \cdots + q_i \cdot 4^{-i}, qm^{(0)} = 3, qm^{(i)} = q^{(i)} - 4^{-i},$$

$\forall i$ in $\{1, 2, \ldots, p\}$. Then

- if $q_i > 0 : q^{(i)} = q^{(i-1)} + q_i \cdot 4^{-i}, qm^{(i)} = q^{(i-1)} + (q_i - 1) \cdot 4^{-i}$;
- if $q_i = 0 : q^{(i)} = q^{(i-1)}, qm^{(i)} = q^{(i-1)} - 4^{-i} = q^{(i-1)} - 4^{-(i-1)} + 3 \cdot 4^{-i} = qm^{(i-1)} + 3 \cdot 4^{-i}$;
- if $q_i < 0 : q^{(i)} = q^{(i-1)} + q_i \cdot 4^{-i} = q^{(i-1)} + (4 + q_i) \cdot 4^{-i} - 4^{-(i-1)} = qm^{(i-1)} + (4 + q_i) \cdot 4^{-i}$,
- $qm^{(i)} = q^{(i)} - 4^{-i} = qm^{(i-1)} + (3 + q_i) \cdot 4^{-i}$.

The corresponding circuit is shown in Fig. 9.11.

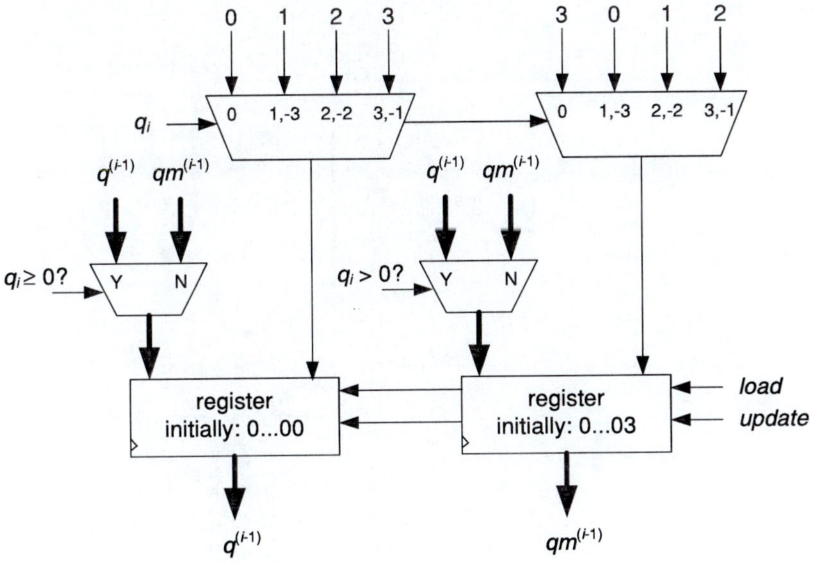

Fig. 9.11 Radix-4 on-the-fly conversion

The complete circuit is shown in Fig. 9.12. The dividend $x = x_n\, x_{n-1} x_{n-2} \cdots x_1 x_0$ and the remainder $r = r_n\, r_{n-1}\, r_{n-2} \cdots r_1\, r_0$ are $(n + 1)$-bit 2's complement integers, the divisor $y = 1\, y_{n-2} \cdots y_1\, y_0$ is an n-bit natural, the quotient $q = q_0\, q_1\, q_2 \cdots q_{2p-1}\, q_{2p}$ is a $(2p + 1)$-bit 2's complement integer, and condition (9.6) must hold.

The following VHDL model describes the circuit of Fig. 9.12.

```
zero_y <= (OTHERS => '0'); one_y <= '0'&y;
two_y <= one_y + one_y; three_y <= one_y + two_y;
four_r <= r(n-2 DOWNTO 0)&"00";
two_or_three <= "01"&NOT(y(n-2));
minus_two_or_three <= '1'&y(n-2)&NOT(y(n-2));
short_r <= r(n DOWNTO n-3);
WITH short_r SELECT signed_digit <=
  "000" WHEN "0000", "001" WHEN "0001", "010" WHEN "0010",
  two_or_three WHEN "0011", "011" WHEN "0100",
  "011" WHEN "0101", "011" WHEN "0110",
  "011" WHEN "0111", "101" WHEN "1000",
  "101" WHEN "1001", "101" WHEN "1010",
  "101" WHEN "1011", minus_two_or_three WHEN "1100",
  "110" WHEN "1101", "111" WHEN "1110", "000" WHEN OTHERS;
```

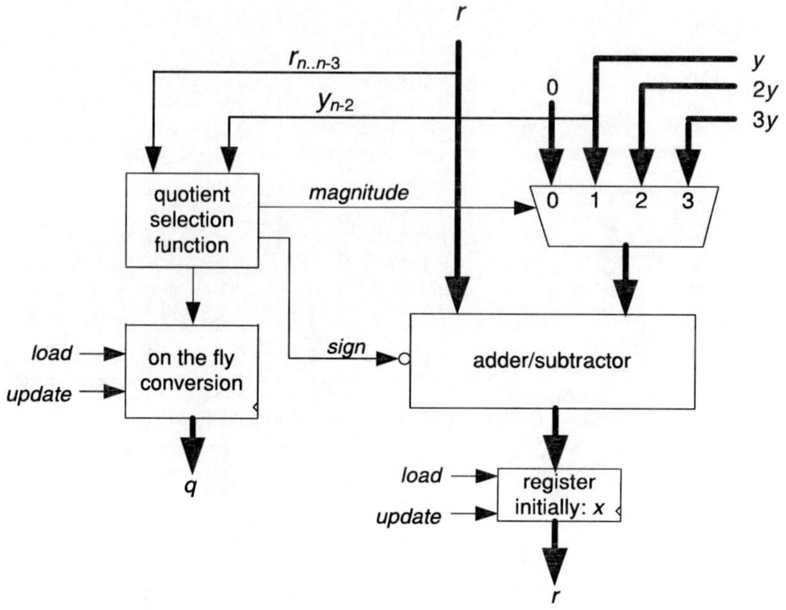

Fig. 9.12 Radix-4 SRT divider

```
WITH signed_digit SELECT second_operand <=
   zero_y WHEN "000", one_y WHEN "001", two_y WHEN "010",
   three_y WHEN "011", three_y WHEN "101", two_y WHEN "110",
   one_y WHEN OTHERS;
 WITH signed_digit(2) SELECT next_r <=
   four_r - second_operand WHEN '0',
   four_r + second_operand WHEN OTHERS;
 remainder_register: PROCESS(clk) ... END PROCESS;
 remainder <= r;
 q_register: PROCESS(clk)
 BEGIN
   IF clk'EVENT AND clk = '1' THEN
     IF load = '1' THEN q <= (OTHERS => '0');
     ELSIF update = '1' THEN
      CASE signed_digit IS
        WHEN "000" => q(0 TO 2*p-2) <= q(2 TO 2*p);
          q(2*p-1 TO 2*p) <= "00";
        WHEN "001" => q(0 TO 2*p-2) <= q(2 TO 2*p);
          q(2*p-1 TO 2*p) <= "01";
        WHEN "010" => q(0 TO 2*p-2) <= q(2 TO 2*p);
          q(2*p-1 TO 2*p) <= "10";
        WHEN "011" => q(0 TO 2*p-2) <= q(2 TO 2*p);
```

```
        q(2*p-1 TO 2*p) <= "11";
      WHEN "101" => q(0 TO 2*p-2) <= qm(2 TO 2*p);
        q(2*p-1 TO 2*p) <= "01";
      WHEN "110" => q(0 TO 2*p-2) <= qm(2 TO 2*p);
        q(2*p-1 TO 2*p) <= "10";
      WHEN OTHERS => q(0 TO 2*p-2) <= qm(2 TO 2*p);
        q(2*p-1 TO 2*p) <= "11";
    END CASE;
   END IF;
  END IF;
END PROCESS;
qm_register: PROCESS(clk)
BEGIN
  IF clk'EVENT AND clk = '1' THEN
    IF load = '1' THEN qm(0 TO 2*p-2) <= (OTHERS => '0');
      qm(2*p-1 TO 2*p) <= "11";
    ELSIF update = '1' THEN
     CASE signed_digit IS
      WHEN "000" => qm(0 TO 2*p-2) <= qm(2 TO 2*p);
        qm(2*p-1 TO 2*p) <= "11";
      ...
        END CASE;
       END IF;
    END IF;
  END PROCESS;
  quotient <= q;
```

The complete circuit also includes a *p*-state counter and a control unit. A complete generic model *radix_four_divider.vhd* is available at the Authors' web page.

9.3 Radix-*B* Dividers

When B is greater than 2, the *quotient_selection* function of Algorithm 9.1 is not easy to implement. According to the Robertson diagram of Fig. 9.1, the definition of q_{i+1} is based on the knowledge of the interval to which belongs r_i: if $k \cdot y \leq B \cdot r_i < (k+1) \cdot y$, then q_{i+1} can be chosen equal to k or $k+1$. A conceptually simple method would consist of computing $B \cdot r_i - k \cdot y$ for all k in $\{B-1, \ldots, -1, 0, 1, \ldots, B-1\}$ and choosing q_{i+1} in function of the signs of the results. Better methods would use approximate values of r_i and y, taking advantage of the fact that for each interval there are two possible values of q_{i+1}. In other words, larger intervals can be considered: if $k \cdot y \leq B \cdot r_i < (k+2) \cdot y$, then $q_{i+1} = k+1$. Algorithms based on truncated values of r_i and y have been proposed in the case where $B = 10$ [4, 5], and the method proposed in the second reference can be generalized to the case of any radix B [6]. Nevertheless, the corresponding FPGA implementations are generally very costly in terms of used cells.

Another method for executing radix-B division consists of converting the radix-B operands to binary operands, executing the division with any type of binary divider, and converting the binary results to radix-B results. Radix-B to binary and binary to radix-B converters are described in Chap. 10. Decimal dividers ($B = 10$) based on this method are reported in Chap. 12.

An alternative option, similar to the previous one, is to use a classical binary algorithm (Sect. 9.2) and to execute all the operations in base B. All along this section it is assumed that B is even.

Consider the following digit-recurrence algorithm (Algorithm 9.1 with $B = 2$):

Algorithm 9.7: Division, binary digit-recurrence algorithm

```
r0 := x; q := 0;
for i in 1 .. p loop
  qi := quotient_selection(ri-1, y);
  ri := 2·ri-1 - qi·y ;
  q := q + qi·2^-i;
end loop ;
r := rp;
```

It generates q and r_p such that

$$x/y = q \cdot 2^{-p} + (r_p/y)2^{-p}, \text{ with } -2^{-p} \leq (r_p/y)2^{-p} < 2^{-p}. \tag{9.10}$$

Algorithm 9.7 can be executed whatever the radix used for representing the numbers. If B's complement radix-B representation is used, then the following operations must be available: radix-B doubling, adding, subtracting and halving.

Algorithm 9.8: Division, binary digit-recurrence algorithm, version 2

```
r0 := x; q := 0; ulp := 2^-1;
for i in 1 .. p loop
  two_r := doubling(ri-1); sum := adding(two_r, y);
  dif := subtracting(two_r, y);
  if quotient_selection(ri-1, y) = 1 then
    ri := dif; q := adding(q, ulp);
  elsif quotient_selection(ri-1, y) = -1 then
    ri := sum; q := subtracting(q, ulp);
  else
    ri := two_r;
  end if;
  ulp := halving(ulp);
  end loop ;
  r := rp;
```

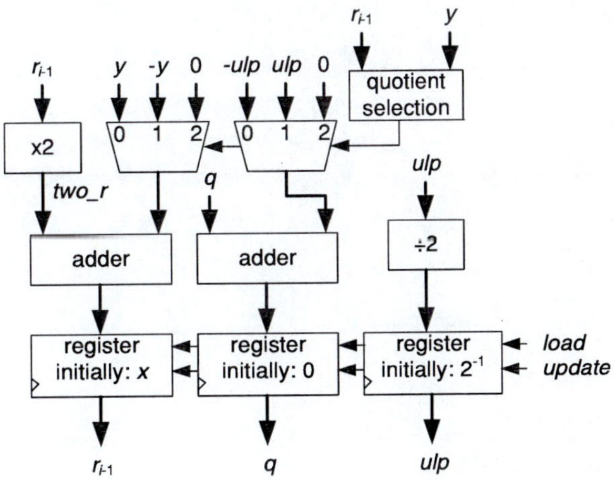

Fig. 9.13 Digit-recurrence divider

The corresponding data path is shown in Fig. 9.13.

Radix-B adders and subtractors can be generated using the general principles presented in Sects. 7.1 and 7.8. The p and g blocks of Fig. 7.2 compute binary functions of radix-B variables, namely $p = 1$ if $x_i + y_i = B-1$, $p = 0$ otherwise, $g = 1$ if $x_i + y_i \geq B$, $g = 0$ if $x_i + y_i \leq B-2$, otherwise, any value. The carry chain is a binary circuit, whatever the radix. An output block computes $(x_i + y_i + c_i) \bmod B$. Furthermore, in the case of a subtractor, additional $(B-1)$'s complement cells, are necessary. Decimal adders and subtractors ($B = 10$) are described and implemented in Chap. 12. In order to synthesize the circuit of Fig. 9.13, it remains to generate circuits which compute $2x$ and $x/2$.

Consider a radix-B natural $x = x_{n-1}B^{n-1} + x_{n-2}B^{n-2} + \cdots + x_0B^0$. First define a digit-doubling cell: given a radix-B digit x_i, the cell computes the radix-B representation of $2x_i$. As $x_i \leq B-1$, then $2x_i \leq 2B - 2 = B + (B-2)$, so that $2x_i$ can be expressed under the form

$$2x_i = d_iB + u_i, \text{ where } d_i\epsilon\{0,1\} \text{ and } u_i \leq B - 2(\text{if } B \text{ is even, so is } u_i).$$

Then, $2x$ can be expressed under the form

$$2x = d_{n-1}B^n + (u_{n-1} + d_{n-2})B^{n-1} + (u_{n-2} + d_{n-3})B^{n-2} + \cdots + (u_1 + d_0)B^1 + u_0B^0.$$

As all coefficients d_i belong to $\{0, 1\}$ and all coefficients u_i are even, then all sums $u_i + d_{i-1}$ amount to a simple concatenation, that is $u_i + d_{i-1} = (u_j/2)\,\&d_{i-1}$, and no carry is generated. The corresponding circuit is shown in Fig. 9.14.

Assume that radix-B digits are represented as k-bit naturals and that k-input LUT's are available. Then, every cell of Fig. 9.14 can be implemented with k LUT-k. As an example, if $B = 10$, the following model defines the doubling cell of Fig. 9.14.

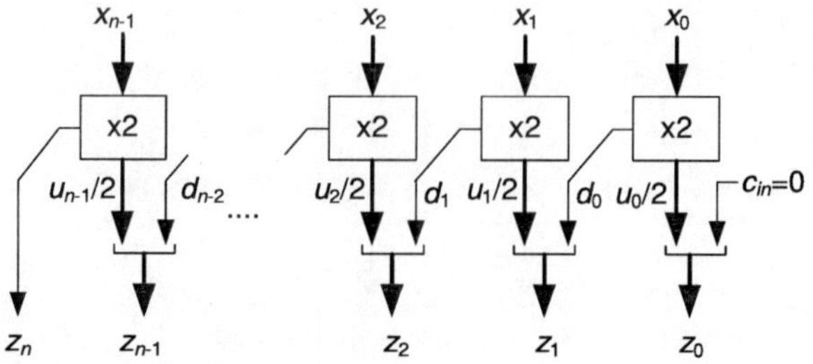

Fig. 9.14 Doubling circuit

```
lut_d: lut4 GENERIC MAP(truth_vector => "0000011111000000")
  PORT MAP(a => a, b => d);
lut_u3: lut4 GENERIC MAP(truth_vector => "0000100001000000")
  PORT MAP(a => a, b => u(3));
lut_u2: lut4 GENERIC MAP(truth_vector => "0011000110000000")
  PORT MAP(a => a, b => u(2));
lut_u1: lut4 GENERIC MAP(truth_vector => "0101001010000000")
  PORT MAP(a => a, b => u(1));
u(0) <= '0';
```

The complete circuit is defined by instantiating n doubling cells. Its computation time is equal to $T_{LUT\text{-}k}$. A complete model *doubling_circuit2.vhd* is available at the Authors' web page ($B = 10$).

As B is even, halving is equivalent to multiplying by $B/2$ and dividing by B. Let $a_{k-1}\, a_{k-2}\ldots a_0$ be the binary representation of a radix-B digit a, that is $a = a_{k-1}2^{k-1} + a_{k-2}2^{k-2} + \cdots + a_0 2^0$. Then

$$a(B/2) = dB + u, \text{ where } d = a_{k-1}2^{k-2} + a_{k-2}2^{k-3} + \cdots + a_1 2^0 \text{ and } u = a_0(B/2).$$

Observe that $d \le a/2 < B/2$ and $u \le B/2$.

Consider now a radix-B natural $x = x_{n-1} \cdot B^{n-1} + x_{n-2} \cdot B^{n-2} + \cdots + x_0 \cdot B^0$. The product $(B/2) \cdot x_i$ can be represented under the form $d_i B + u_i$ where $d_i < B/2$ and $u_i \le B/2$, so that

$$(B/2) \cdot x = d_{n-1}B^n + (u_{n-1} + d_{n-2})B^{n-1} + (u_{n-2} + d_{n-3})B^{n-2} + \cdots + (u_1 + d_0)B^1 + u_0 B^0.$$

All sums $u_i + d_{i-1}$ are smaller than B, so that no carry is generated. Finally

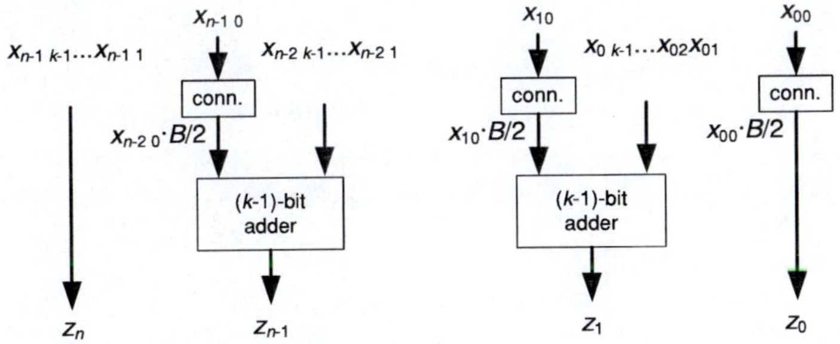

Fig. 9.15 Multiplication by $B/2$ ("conn." only includes connections)

$$x/2 = d_{n-1}B^{n-1} + (u_{n-1} + d_{n-2})B^{n-2} + (u_{n-2} + d_{n-3})B^{n-3} + \cdots + (u_1 + d_0)B^0 + u_0B^{-1}.$$

The circuit of Fig. 9.15 computes $z = (B/2) \cdot x$.

If k-input LUT's are available, then every cell of Fig. 9.15 can be implemented with k LUT-k. As an example, if $B = 10$, the following model defines the multiply-by-five cell of Fig. 9.15. It computes $b = 5a_0 + (4a_3 + 2a_2 + a_1)$.

```
lut_b3: lut4 GENERIC MAP(truth_vector => "0000000101000000")
  PORT MAP(a => a, b => b(3));
lut_b2: lut4 GENERIC MAP(truth_vector => "0101010010000000")
  PORT MAP(a => a, b => b(2));
lut_b1: lut4 GENERIC MAP(truth_vector => "0001111000000000")
  PORT MAP(a => a, b => b(1));
lut_b0: lut4 GENERIC MAP(truth_vector => "0110011001000000")
  PORT MAP(a => a, b => b(0));
```

The complete circuit is defined by instantiating $n - 1$ multiply-by-five cells. Its computation time is equal to $T_{LUT\text{-}k}$. A complete model *multiply_by_five.vhd* is available at the Authors' web page ($B = 10$).

Example 9.1

A decimal divider ($B = 10$) based on the circuit of Fig. 9.13 can be synthesized. The quotient selection is done according to the non-restoring algorithm method, that is to say $r_{i+1} = 2 \cdot r_i + y$ if r_i is negative, and $r_{i+1} = 2 \cdot r_i - y$ if r_i is non-negative. The number p of steps is defined as follows: according to 9.10, the maximum value of the error $|x/y - q|$ is smaller than or equal to 2^{-p}. If q is rounded in such a way that the final result $q_{rounded}$ has m fractional digits, the additional rounding error is smaller than 10^{-m}, and the final error $|x/y - q_{rounded}|$ is smaller than $2^{-p} + 10^{-m}$. By choosing p so that $2^{-p} \cong 10^{-m}$, that is $p \cong 3.3\,m$, then $|x/y - q_{rounded}| < 2 \cdot 10^{-m}$. In the following VHDL models, behavioral descriptions of the decimal adders have been used. Real implementations are proposed in Chap. 12.

```
a_doubling_circuit: doubling_circuit2 GENERIC MAP(n => n)
  PORT MAP(x => r(4*n-1 DOWNTO 0), z => two_r);
complement_y: FOR i IN 0 TO n-1 GENERATE
  complement_of_y(4*i+3 DOWNTO 4*i) <=
    "1001" - y(4*i+3 DOWNTO 4*i);
END GENERATE;
complement_of_y(4*n) <= '1';
WITH r(4*n) SELECT second_operand <=
  ('0'&y) WHEN '1', complement_of_y WHEN OTHERS;
carry1(0) <= NOT(r(4*n));

--behavioral description of a decimal adder:
a_decimal_adder: FOR i IN 0 TO n-1 GENERATE
  carry1(i+1) <= '1' WHEN ('0'&two_r(4*i+3 DOWNTO 4*i)
  + second_operand(4*i+3 DOWNTO 4*i)
  + carry1(i)) >= "01010" ELSE '0';
  next_r(4*i+3 DOWNTO 4*i) <=
    two_r(4*i+3 DOWNTO 4*i)
    + second_operand(4*i+3 DOWNTO 4*i)
    + carry1(i) WHEN carry1(i+1) = '0'
    ELSE two_r(4*i+3 DOWNTO 4*i)
    + second_operand(4*i+3 DOWNTO 4*i)
    + carry1(i) + "0110";
END GENERATE;
next_r(4*n) <=
  two_r(4*n) XOR second_operand(4*n) XOR carry1(n);

complement_ulp: FOR i IN 0 TO p-1 GENERATE
  complement_of_ulp(4*i+3 DOWNTO 4*i) <= "1001" - ulp(4*i+3
DOWNTO 4*i);
END GENERATE;
complement_of_ulp(4*p) <= '1';
WITH r(4*n) SELECT signed_ulp <=
  ('0'&ulp) WHEN '0', complement_of_ulp WHEN OTHERS;
carry2(0) <= r(4*n);

--next_q = long_q + signed_ulp + carry2(0)
--behavioral description of another decimal adder: ...

a_multiply_by_five_circuit: multiply_by_five
  GENERIC MAP(n => p)
  PORT MAP(x => ulp, z => next_ulp_by_ten);
next_ulp <= next_ulp_by_ten(4*p+3 DOWNTO 4);
```

```
register_r: PROCESS(clk) ...
register_q: PROCESS(clk) ...
q <= long_q(4*p DOWNTO 4*(p-m));
register_ulp: PROCESS(clk) ...
```

A complete generic model *decimal_divider.vhd* is available at the Authors' web page.

9.4 Convergence Algorithms

Instead of using digit-recurrence algorithms, an alternative option is the use of convergence algorithms. This type of algorithm is mainly attractive in the case of real number operations (Chap. 12).

Given a real number x belonging to the interval $1 \le x < 2$, the inverse y of x belongs to the interval $1/2 < y \le 1$. The following *Newton-Raphson algorithm* generates successive approximations $y_0, y_1, y_2, \ldots$ of $y = 1/x$.

Algorithm 9.9: Newton-Raphson inversion algorithm

```
y0 := 0.75;
for i in 1 .. m loop
   yi := yi-1·(2-x·yi-1);
end loop;
```

If all operations were executed with full precision, then the maximum error after i steps would be given by the following relation:

$$y - y_i < 1/2^{2^i}. \tag{9.11}$$

Another option is the *Goldschmidt algorithm*. Given two real numbers x and y belonging to the intervals $1 \le x < 2$ and $1 \le y < 2$, it generates two sequences of real numbers a_0, a_1, a_2, ... and b_0, b_1, b_2, ... , and the second one constitutes a sequence of successive approximations of x/y.

Algorithm 9.10: Goldschmidt's algorithm

```
a0 := y/2; b0 := x/2;
for i in 1 .. m loop
   c := 2-ai-1; ai := ai-1·c; bi := bi-1·c;
end loop;
z := bm;
```

If all operations were executed with full precision, then the maximum error after i steps would be defined by the following relation:

$$\frac{x}{y} - b_i < 1/2^{(2^i-1)}. \tag{9.12}$$

The following VHDL model describes the corresponding circuit; n is the number of fractional bits of x and y, and p is the number of fractional bits of the internal variables a, b and c.

```
not_a <= NOT(a);
c(p-1 DOWNTO 0) <= not_a + 1; c(p) <= '1';
ac <= a*c; bc <= b*c;
next_a <= ac(2*p-1 DOWNTO p); next_b <= bc(2*p DOWNTO p);
register_a: PROCESS(clk)
BEGIN
  IF clk'EVENT and clk = '1' THEN
    IF load = '1' THEN
      a(p-1 DOWNTO p-n-1) <= y;
      a(p-n-2 DOWNTO 0) <= (OTHERS => '0');
    ELSIF update = '1' THEN a <= next_a; END IF;
  END IF;
END PROCESS;
register_b: PROCESS(clk)
BEGIN
  IF clk'EVENT and clk = '1' THEN
    IF load = '1' THEN
      b(p) <= '0'; b(p-1 DOWNTO p-n-1) <= x;
      b(p-n-2 DOWNTO 0) <= (OTHERS => '0');
    ELSIF update = '1' THEN b <= next_b; END IF;
  END IF;
END PROCESS;
quotient <= b;
```

A complete generic model *goldschmidt.vhd* is available at the Authors' web page. It includes an m-state counter, where m is the number of steps (a generic parameter), and a control unit.

9.5 FPGA Implementations

Several dividers have been implemented within a Virtex 5-2 device. As before, the times are expressed in *ns* and the costs in numbers of Look Up Tables (LUTs) and flip-flops (FFs). All VHDL models as well as several test benches are available at the Authors' web page.

Table 9.2 Non-restoring dividers

n	p	FFs	LUTs	Period	Total time
8	8	22	28	2.21	19.9
16	16	39	46	2.43	41.3
24	27	59	64	2.63	73.6
32	16	55	78	2.82	47.9
32	32	72	79	2.82	93.1
53	56	118	123	3.30	188.1
64	32	104	143	3.55	117.2
64	64	137	144	3.55	230.8

Table 9.3 Restoring dividers

n	p	FFs	LUTs	Period	Total time
8	8	21	26	2.46	22.1
16	16	38	44	2.68	45.6
24	27	58	62	2.90	81.2
32	16	54	76	3.08	52.4
32	32	71	77	3.08	101.6
53	56	117	121	3.56	202.9
64	32	103	141	3.82	126.1
64	64	136	142	3.82	248.3

Table 9.4 Binary SRT dividers

n	p	FFs	LUTs	Period	Total time
8	8	33	44	2.43	21.9
16	16	58	78	2.57	43.7
24	27	89	118	2.76	77.3
32	16	75	110	2.94	50.0
32	32	107	143	2.94	97.0
53	56	177	235	3.42	194.9
64	32	139	207	3.66	120.8
64	64	204	272	3.66	237.9

9.5.1 Digit-Recurrence Algorithms

Tables 9.2, 9.3, 9.4, 9.5 give implementation results for several

- non-restoring dividers,
- restoring dividers,
- binary SRT dividers,
- binary SRT dividers with carry-save adder.

Radix-4 SRT dividers have also been implemented. The quotient is a $(2p + 1)$-bit 2's complement integer (Table 9.6).

Examples of decimal dividers are given in Chap. 11.

Table 9.5 Binary SRT dividers with carry-save adders

n	p	FFs	LUTs	Period	Total time
8	8	40	86	2.81	25.3
16	16	73	143	2.86	48.6
24	27	112	215	2.86	80.1
32	16	105	239	2.86	48.6
32	32	138	272	2.86	94.4
53	56	229	448	2.87	163.6
64	32	202	464	2.87	94.7
64	64	267	529	2.87	186.6

Table 9.6 Radix-4 SRT dividers

n	p	$2p$	FFs	LUTs	Period	Total time
8	4	8	30	70	3.53	17.7
16	8	16	55	112	3.86	34.7
24	14	28	88	170	4.08	61.2
32	8	16	71	175	4.27	38.4
32	16	32	104	209	4.27	72.6
53	28	56	174	343	4.76	138.0
64	16	32	136	337	5.01	85.2
64	32	64	201	402	5.01	165.3

Table 9.7 Dividers based on the Goldschmidt's algorithm

n	p	m	FFs	LUTs	DSPs	Period	Total time
8	10	4	25	38	2	5.76	23.0
8	16	5	38	42	2	5.68	28.4
16	18	5	43	119	2	7.3	36.5
16	32	7	71	76	8	10.5	73.5
24	27	6	62	89	8	10.5	63.0
53	56	8	118	441	24	14.42	115.4

9.5.2 Convergence Algorithms

Table 9.7 gives implementation results of several dividers based on the Goldschmidt's algorithm. In this case, DSP slices have been used.

9.6 Exercises

1. Synthesize several types of combinational dividers: restoring, non-restoring, SRT, radix-2^k.
2. Generate a generic VHDL model of a Newton-Raphson inverter.

3. Synthesize several dividers using a Newton-Raphson inverter and different types of multipliers.
4. Generate generic models of combinational circuits with three n-bit inputs x, y and m, and an n-bit output $z = x \cdot y \bmod m$. For that, use any type of combinational multiplier to compute $p = x \cdot y$ and any type of combinational divider to compute $z = p \bmod m$.
5. Generate generic models of sequential circuits that implement the same function as in exercise 4.

References

1. Freiman CV (1961) Statistical analysis of certain binary division algorithms. IRE Proc 49:91–103
2. Robertson JE (1958) A new class of division methods. IRE Trans Electron Comput EC-7:218–222
3. Cocke J, Sweeney DW (1957) High speed arithmetic in a parallel device. IBM technical report, Feb 1957
4. Deschamps JP, Sutter G (2010) Decimal division: algorithms and FPGA implementations. In: 6th southern conference on programmable logic (SPL)
5. Lang T, Nannarelli A (2007) A radix-10 digit-recurrence division unit: algorithm and architecture. IEEE Trans Comput 56(6):1–13
6. Deschamps JP (2010) A radix-*B* divider. Contact: jp.deschamps@telefonica.net

Chapter 10
Other Operations

This chapter is devoted to arithmetic functions and operations other than the four basic ones. The conversion of binary numbers to radix-B ones, and conversely, is dealt with in Sects. 10.1 and 10.2. An important particular case is $B = 10$ as human interfaces generally use decimal representations while internal computations are performed with binary circuits. In Sect. 10.3, several square rooting circuits are presented, based on digit-recurrence or convergence algorithms. Logarithms and exponentials are the topics of Sects. 10.4 and 10.5. Finally, the computation of trigonometric functions, based on the CORDIC algorithm [2, 3], is described in Sect. 10.6.

10.1 Binary to Radix-B Conversion (B even)

Assume that B is even and greater than 2. Consider the binary representation of x:

$$x = x_{n-1} \cdot 2^{n-1} + x_{n-2} \cdot 2^{n-2} + \cdots + x_1 \cdot 2 + x_0. \tag{10.1}$$

Taking into account that $B > 2$, the bits x_i can be considered as radix-B digits, and a simple conversion method consists of computing (10.1) in radix-B.

Algorithm 10.1: Binary to radix-B conversion

```
z := 0;
for i in 1 .. n loop
  z := z·2 + x_{n-i};
end loop;
x := z;
```

J.-P. Deschamps et al., *Guide to FPGA Implementation of Arithmetic Functions*, Lecture Notes in Electrical Engineering 149, DOI: 10.1007/978-94-007-2987-2_10,

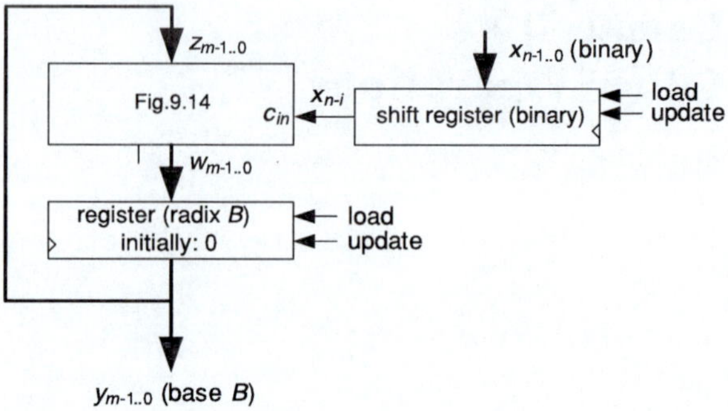

Fig. 10.1 Binary to radix-*B* converter

In order to compute $z \cdot 2 + x_{n-i}$ the circuit of Fig. 9.14, with $c_{in} = x_{n-i}$ instead of 0, can be used. A sequential binary to radix-*B* converter is shown in Fig. 10.1. It is described by the following VHDL model.

```
main_component: doubling_circuit2 GENERIC MAP(n => m)
PORT MAP(x => z, z => w, c_in => xNminusI);
register_z: PROCESS(clk) ...
y <= z;
shift_register_x: PROCESS(clk) ...
xNminusI <= int_x(n-1);
```

The complete circuit also includes an *n*-state counter and a control unit. A complete model *BinaryToDecimal2.vhd* is available at the Authors' web page ($B = 10$).

The computation time of the circuit of Fig. 10.1 is equal to $n \cdot T_{clk}$ where T_{clk} must be greater than T_{LUT-k} (Sect. 9.3). Thus

$$T_{\text{binary-radix-}B} \cong n \cdot T_{LUT-k}. \tag{10.2}$$

10.2 Radix-*B* to Binary Conversion (*B* even)

Given a natural *z*, smaller than 2^n, its binary representation is deduced from the following set of integer divisions

$$z = q_1 \cdot 2 + r_0,$$
$$q_1 = q_2 \cdot 2 + r_1,$$
$$\ldots$$
$$q_{n-1} = q_n \cdot 2 + r_{n-1},$$

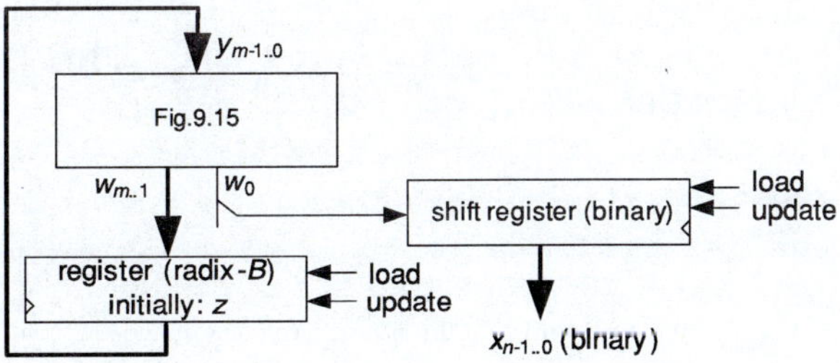

Fig. 10.2 Decimal to binary converter

so

$$z = q_n \cdot 2^n + r_{n-1} \cdot 2^{n-1} + r_{n-2} \cdot 2^{n-2} + \cdots + r_1 \cdot 2 + r_0.$$

As z is smaller than 2^n, then $q_n = 0$, and the binary representation of z is constituted by the set of remainders $r_{n-1}\, r_{n-2} \ldots r_1\, r_0$.

Algorithm 10.2: Radix-B to binary conversion

```
q0 := z;
for i in 0 .. n-1 loop
  ri := qi mod 2; qi+1 := ⌊qi/2⌋;
end loop;
x := rn-1 rn-2 ... r1 r0;
```

Observe that if $q_i = q_{i+1} \cdot 2 + r_i$, then $q_i \cdot (B/2) = q_{i+1} \cdot B + r_i \cdot (B/2)$ where $r_i \cdot (B/2) < 2 \cdot (B/2) = B$.

Algorithm 10.3: Radix-B to binary conversion, version 2

```
q0 := z;
for i in 0 .. n-1 loop
  yi := (B/2)·qi;
  qi+1 := ⌊yi/B⌋; ri := (yi mod B) mod 2;
end loop;
x := rn-1 rn-2 ... r1 r0;
```

In order to compute $(B/2) \cdot q_i$ the circuit of Fig. 9.15 can be used. A sequential radix-B to binary converter is shown in Fig. 10.2. It is described by the following VHDL model.

```
main_component: multiply_by_five GENERIC MAP(n => m)
PORT MAP(x => q, z => w);
r <= w(0);
register_y: PROCESS(clk)
BEGIN
  IF clk'EVENT AND clk = '1' THEN
    IF load = '1' THEN q <= x;
    ELSIF update = '1' THEN q <= w(4*m+3 DOWNTO 4);
    END IF;
  END IF;
END PROCESS;
shift_register_z: PROCESS(clk)
BEGIN
  IF clk'EVENT AND clk = '1' THEN
    IF update = '1' THEN z <= r&z(n-1 DOWNTO 1);
    END IF;
  END IF;
END PROCESS;
```

The complete circuit also includes an n-state counter and a control unit. A complete model *DecimalToBinary2.vhd* is available at the Authors' web page ($B = 10$).

The computation time of the circuit of Fig. 10.2 is equal to $n \cdot T_{clk}$ where T_{clk} must be greater than T_{LUT-k} (Sect. 9.3). Thus

$$T_{\text{radix}-B-\text{binary}} \cong n \cdot T_{LUT-k}. \tag{10.3}$$

10.3 Square Rooters

Consider a $2n$-bit natural $X = x_{2n-1} \cdot 2^{2n-1} + x_{2n-2} \cdot 2^{2n-2} + \cdots + x_1 \cdot 2 + x_0$, and compute $Q = \lfloor X^{1/2} \rfloor$. Thus, $Q^2 \leq X < (Q+1)^2$, and the difference $R = X - Q^2$ belongs to the range

$$0 \leq R \leq 2Q. \tag{10.4}$$

10.3.1 Restoring Algorithm

A digit recurrence algorithm consisting of n steps is defined. At each step two numbers are generated:

$$Q_i = q_{n-1} \cdot 2^{i-1} + q_{n-2} \cdot 2^{i-2} + \cdots + q_{n-i+1} \cdot 2 + q_{n-i} \text{ and } R_i = X - \left(Q_i \cdot 2^{n-i}\right)^2,$$

such that

$$0 \leq R_i < (1 + 2Q_i)2^{2(n-i)}. \qquad (10.5)$$

After n steps, $0 \leq R_n < 1 + 2Q_n$, that is (10.4) with $Q = Q_n$ and $R = R_n$.

Initially define $Q_0 = 0$ and $R_0 = X$, so condition (10.5) amounts to $0 \leq X < 2^{2n}$. Then, at step i, compute Q_i and R_i in function of Q_{i-1} and R_{i-1}:

$$Q_i = 2Q_{i-1} + q_{n-i} \text{ where } q_{n-i} \in \{0, 1\},$$
$$R_i = X - \left(Q_i \cdot 2^{n-i}\right)^2 = X - \left((2Q_{i-1} + q_{n-i})2^{n-i}\right)^2$$
$$= X - \left(Q_{i-1} \cdot 2^{n-i+1}\right)^2 - (q_{n-i} + 4Q_{i-1})q_{n-i} \cdot 2^{2(n-i)} = R_{i-1} - (q_{n-i} + 4Q_{i-1})q_{n-i}2^{2(n-i)}.$$

The value of q_{n-i} is chosen in such a way that condition (10.5) holds. Consider two cases:

- If $R_{i-1} < (1 + 4Q_{i-1})2^{2(n-i)}$, then $q_{n-i} = 0$, $Q_i = 2Q_{i-1}, R_i = R_{i-1}$.
  As $R_i = R_{i-1} < (1 + 4Q_{i-1})2^{2(n-i)} = (1 + 2Q_i)2^{2(n-i)}$ and $R_i = R_{i-1} \geq 0$, condition (10.5) holds.
- If $R_{i-1} \geq (1 + 4Q_{i-1})2^{2(n-i)}$, then $q_{n-i} = 1$, $Q_i = 2Q_{i-1} + 1$, $R_i = R_{i-1} - (1 + 4Q_{i-1})2^{2(n-i)}$, so that $R_i \geq 0$ and $R_i < (1 + 2Q_{i-1})2^{2(n-i+1)} - (1 + 4Q_{i-1}) 2^{2(n-i)} = (3 + 4Q_{i-1})2^{2(n-i)} = (1 + 2Q_i)2^{2(n-i)}$.

Algorithm 10.4: Square root, restoring algorithm

```
Q0 := 0; R0 := X;
for i in 1 to n loop
  Pi-1 := (1 + 4·Qi-1)·2^2(n-i);
  if Pi-1 ≤ Ri-1 then qn-i := 1; Ri := Ri-1 - Pi-1;
  else qn-i := 0; Ri := Ri-1;
  end if;
  Qi := 2·Qi-1 + qn-i;
end loop;
Q := Qn;
```

Q_i is an i-bit number, $R_i < (1 + 2Q_i)2^{2(n-i)} = Q_i \,\&\, 1 \& 00 \cdots 0$ is a $(2n-i+1)$-bit number, and $P_{i-1} = (1 + 4Q_{i-1})2^{2(n-i)} = Q_i \,\&\, 01 \& 00 \cdots 0$ a $(2n + 2 - i)$-bit number.

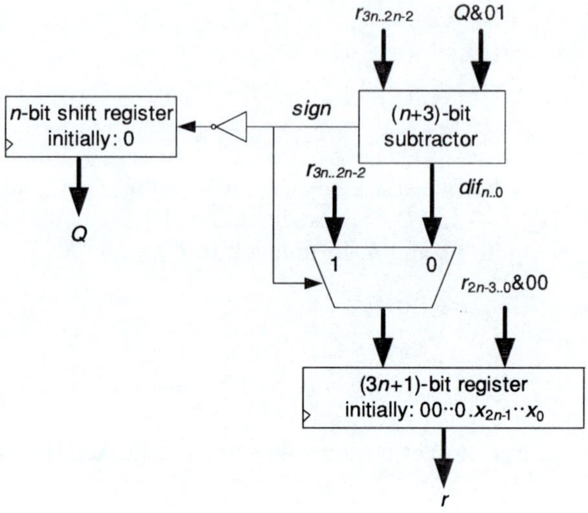

Fig. 10.3 Square root computation: data path

An equivalent algorithm is obtained if P_i and R_i are replaced by $p_{i-1} = P_{i-1}/2^{2(n-i)}$ and $r_i = R_i/2^{2(n-i)}$.

Algorithm 10.5: Square root, restoring algorithm, version 2

```
Q0 := 0; r0 := X/2^2n;
for i in 1 to n loop
  pi-1 := 1 + 4Qi-1;
  if pi-1 ≤ 4ri-1 then
    qn-i := 1; ri := 4ri-1 - pi-1;
  else qn-i := 0; ri := 4ri-1;
  end if;
  Qi := 2Qi-1 + qn-i;
end loop;
Q := Qn; R := rn;
```

As before, Q_i is an i-bit number, r_i is a $(2n-i+1)$-bit fixed-point number with $2(n-i)$ fractional bits and an $(i+1)$-bit integer part, and p_{i-1} a $(2n-i+2)$-bit fixed-point number with $2(n-i)$ fractional bits and an $(i+2)$-bit integer part.

A sequential implementation is shown in Fig. 10.3. It can be described by the following VHDL model.

```
dif <= r(3*n DOWNTO 2*n-2) - ('0'&q&"01");
WITH dif(n+2) SELECT next_r <=
  r(3*n-2 DOWNTO 0)&"00" WHEN '1',
  dif(n DOWNTO 0)&r(2*n-3 DOWNTO 0) &"00" WHEN OTHERS;
remainder_register: PROCESS(clk)
BEGIN
  IF clk'EVENT AND clk = '1' THEN
    IF load = '1' THEN r(2*n-1 DOWNTO 0) <= x;
      r(3*n DOWNTO 2*n) <= (OTHERS => '0');
    ELSIF update = '1' THEN r <= next_r;
    END IF;
  END IF;
END PROCESS;
remainder <= r(3*n DOWNTO 2*n);
quotient_register: PROCESS(clk)
BEGIN
  IF clk'EVENT AND clk = '1' THEN
    IF load = '1' THEN q <= (OTHERS => '0');
    ELSIF update = '1' THEN
      q <= q(n-2 DOWNTO 0)&NOT(dif(n+2));
    END IF;
  END IF;
END PROCESS;
root <= q;
```

The only computational resource is an (n+3)-bit subtractor so that the computation time is approximately equal to $n \cdot T_{adder}(n)$. The complete circuit includes an n-bit counter and a control unit. A generic VHDL model *SquareRoot.vhd* is available at the Authors' web page.

Another equivalent algorithm is obtained if P_i, Q_i and R_i are replaced by $p_i = P_i/2^{2n-i-1}, q_i = Q_i/2^i, r_i = R_i/2^{2n-i}$.

Algorithm 10.6: Square root, restoring algorithm, version 3

```
q0 := 0; r0 := 0.x2n-1x2n-2...x0;
for i in 1 to n loop
  pi-1 := 2qi-1 + 2^-i;
  if pi-1 ≤ 2ri-1 then qi := qi-1+ 2^-i; ri := 2ri-1 - pi-1;
  else ri := 2ri-1;
  end if;
end loop;
Q := qn2^n; R := rn2^n;
```

Algorithm 10.6 is similar to the restoring algorithm defined in Chap. 21 of [1]. Its implementation is left as an exercise.

10.3.2 Non-Restoring Algorithm

Instead of computing R_i and Q_i as in Algorithm 10.4, an alternative option is the following. Define $R_i = R_{i-1} - (1+4Q_{i-1})2^{2(n-i)}$, whatever the sign of R_i. If R_i is non-negative, then its value is the same as before. If R_i is negative then it is equal to $R_{i\,restoring} - (1+4Q_{i-1})2^{2(n-i)}$ where $R_{i\ restoring}$ is the value that would have been computed with Algorithm 10.4. Then, at the next step, Q_i and R_{i+1} are computed as follows:

- if R_i is non-negative, then $Q_i = 2Q_{i-1} + 1$ and $R_{i+1} = R_i - (1+4Q_i)\,2^{2(n-i-1)}$,
- if R_i is negative, then $Q_i = 2Q_{i-1}$ and $R_{i+1} = R_{i\,restorig} - (1+4Q_i)2^{2(n-i-1)} = R_i + (1+4Q_{i-1})2^{2(n-i)} - (1+4Q_i)2^{2(n-i-1)} = R_i + (1+2Q_i)2^{2(n-i)}$ $-(1+4Q_i)2^{2(n-i-1)} = R_i + (3+4Q_i)2^{2(n-i-1)}$.

Algorithm 10.7: Square root, non-restoring algorithm

```
Q0 := 0; R0 := X;
R1 := R0 - 2^(2(n-1));
for i in 1 to n loop
  if Ri ≥ 0 then Qi := 2·Qi-1 + 1; Ri+1 = Ri-(1 + 4·Qi)·2^(2(n-i-1));
  else Qi := 2·Qi-1; Ri+1 = Ri + (3 + 4·Qi)·2^(2(n-i-1));
  end if;
end loop;
Q := Qn;
```

Q_i is an i-bit number and R_i is an (i+2)-bit signed number.

An equivalent algorithm is obtained if P_i and R_i are replaced by $p_{i-1} = P_{i-1}/2^{2(n-i)}, r_i = R_i/2^{2(n-i)}$.

Algorithm 10.8: Square root, non-restoring algorithm, version 2

```
Q0 := 0; r0 := 0.x2n-1 x2n-2 ... x0;
r1 := 4·r0 - 1;
for i in 1 to n loop
  if ri ≥ 0 then Qi := 2·Qi-1 + 1; ri+1 = 4·ri - (1 + 4·Qi);
  else Qi := 2·Qi-1; ri+1 = 4·ri + (3 + 4·Qi);
  end if;
end loop;
Q := Qn;
```

As before, Q_i is an i-bit number and r_i a (2n+1)-bit fixed-point number $a_{2n} \cdot a_{2n-1}\,a_{2n-2}\ldots a_0$ initially equal to $0.x_{2n-1}\,x_{2n-2}\ldots x_0$.

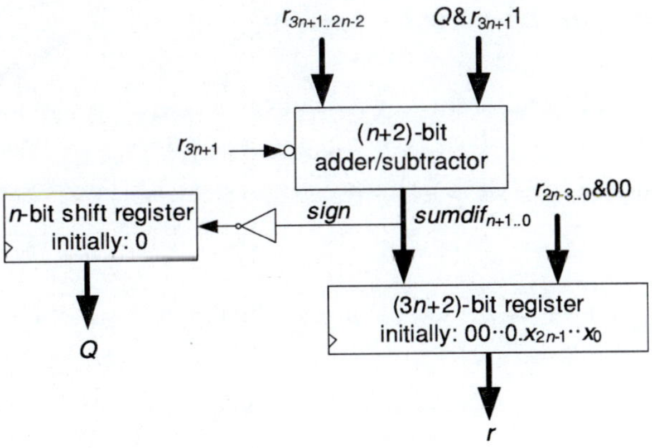

Fig. 10.4 Square root computation: non-restoring algorithm

In this case $r_n 2^n$ is the remainder only if r_n is non-negative. In fact, the remainder is equal to $(r_{n-i} \cdot 4^i) \cdot 2^n$ where r_{n-i} is the last non-negative remainder.

A sequential implementation is shown in Fig. 10.4. It can be described by the following VHDL model.

```
left_operand <= r(3*n-1 DOWNTO 2*n-2);
right_operand <= q&r(3*n+1)&'1';
WITH r(3*n+1) SELECT sumdif <=
  left_operand - right_operand WHEN '0',
  left_operand + right_operand WHEN OTHERS;
next_r <= sumdif&r(2*n-3 DOWNTO 0) &"00";
remainder_register: PROCESS(clk) ...
remainder <= r(3*n DOWNTO 2*n);
quotient_register: PROCESS(clk)
BEGIN
  IF clk'EVENT AND clk = '1' THEN
    IF load = '1' THEN q <= (OTHERS => '0');
    ELSIF update = '1' THEN
      q <= q(n-2 DOWNTO 0)&NOT(sumdif(n+1));
    END IF;
  END IF;
END PROCESS;
root <= q;
```

The only computation resource is an $(n+2)$-bit adder/subtractor so that the computation time is again approximately equal to $n \cdot T_{adder}(n)$. The complete circuit includes an n-bit counter and a control unit. A generic VHDL model *SquareRoot3.vhd* is available at the Authors' web page.

10.3.3 Fractional Numbers

Assume that X is a $2(n+p)$-bit fractional number $x_{2n-1}\, x_{2n-2} \cdots x_1\, x_0 . x_{-1}\, x_{-2} \cdots x_{-2p}$. The square root Q of X, with an accuracy of p fractional bits, is defined as follows:

$$Q = q_{n-1}2^{n-1} + q_{n-2}2^{2n-2} + \ldots + q_0 + q_{-1}2^{-1} + q_{-2}2^{-2} + \ldots + q_{-p}2^{-p},$$
$$Q^2 \leq X \text{ and } (Q + 2^{-p})^2 > X,$$

so that the remainder $R = X - Q^2$ belongs to the range $0 \leq R < 2^{1-p}Q + 2^{-2p}$, that is to say

$$0 \leq R \leq Q \cdot 2^{1-p}.$$

In order to compute Q, first substitute X by $X' = X \cdot 2^{2p}$, which is a natural, and then compute the square root $Q' = q_{n+p-1}\, q_{n+p-2} \ldots q_1\, q_0$ of X', and the corresponding remainder $R' = r_{n+p}\, r_{n+p-1} \ldots r_1\, r_0$, using for that one of the previously defined algorithms. Thus, $X' = (Q')^2 + R'$, with $0 \leq R' \leq 2Q'$, so that $X = (Q' \cdot 2^{-p})^2 + R' \cdot 2^{-2p}$, with $0 \leq R' \cdot 2^{-2p} \leq 2Q' \cdot 2^{-2p}$. Finally, define $Q = Q' \cdot 2^{-p}$ and $R = R' \cdot 2^{-2p}$. Thus

$$X = Q^2 + R, \text{ with } 0 \leq R \leq Q \cdot 2^{1-p},$$

where $Q = q_{n+p-1}\, q_{n+p-2} \ldots q_p \cdot q_{p-1} \ldots q_1\, q_0$ and $R = r_{n+p}\, r_{n+p-1} \ldots r_{2p} \cdot r_{2p-1} \ldots r_1\, r_0$ is smaller than or equal to $Q \cdot 2^{1-p}$.

Comment 10.1

The previous method can also be used for computing the square root of a natural $X = x_{2n-1}\, x_{2n-2} \ldots x_1\, x_0$ with an accuracy of p bits: represent X as an $(n+p)$-bit fractional number $x_{2n-1}\, x_{2n-2} \ldots x_1\, x_0 \cdot 00 \ldots 0$ and use the preceding method.

10.3.4 Convergence Methods (Newton–Raphson)

Instead of a digit-recurrence algorithm, an alternative option is the Newton–Raphson convergence method. The following iteration can be used for computing $X^{1/2}$

$$x_{i+1} = (1/2) \cdot (x_i + X/x_i).$$

It corresponds to the graphical construction of Fig. 10.5.

First check that $(1/2) \cdot (x + X/x)$ is a function whose minimum value, within the half plane $x > 0$, is equal to $X^{1/2}$, and is obtained when $x_i = X^{1/2}$. Thus, whatever the initial value x_0, x_i is greater than or equal to $X^{1/2}$ for all $i > 0$. Furthermore, if $x_i > X^{1/2}$ then $x_{i+1} < (1/2) \cdot (x_i + X/X^{1/2}) = (1/2) \cdot (x_i + X^{1/2}) < (1/2) \cdot (x_i + x_i) = x_i$. Thus, either $X^{1/2} < x_{i+1} < x_i$ or $X^{1/2} = x_{i+1} = x_i$. For x_0 choose a first rough approximation of $X^{1/2}$. As regards the computation of X/x_i, observe that if $x_i \geq X^{1/2}$ and $X < 2^{2n}$, then $x_i \cdot 2^n > X^{1/2} \cdot X^{1/2} = X$. So, compute $q \cong X/(x_i \cdot 2^n)$, with an

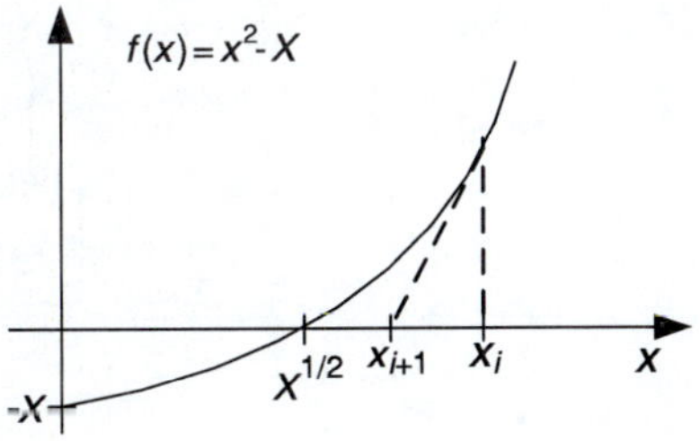

Fig. 10.5 Newton–Raphson method: computation of $X^{1/2}$

accuracy of $p+n$ fractional bits, using any division algorithm, so that $X \cdot 2^{n+p} = q \cdot x_i \cdot 2^n + r$, with $r < x_i \cdot 2^n$, and $X = Q \cdot x_i + R$, where $Q = q \cdot 2^{-p}$ and $R = (r/x_i) \cdot 2^{-(n+p)} < 2^{-p}$.

An example of implementation *SquareRootNR4.vhd* is available at the Authors' web page. The corresponding data path is shown in Fig. 10.6. The initial value x_0 must be defined in such a way that $x_0 \cdot 2^n > X$. In the preceding example, *initial_y* = x_0 is defined so that *initial_y(n+p .. n+p−4)*$\cdot 2^{-2}$ is an approximation of the square root of $X_{2n-1\,..\,2n-4}$ and that *initial_y*$\cdot 2^n$ is greater than X.

```
first_bits <= x(2*n-1 DOWNTO 2*n-4);
initial_y(n+p DOWNTO n+p-4) <=
  table_x0(CONV_INTEGER(first_bits));
```

table_x0 is a constant array defined within a user package:

```
TYPE table IS ARRAY(0 TO 15) OF STD_LOGIC_VECTOR(4 DOWNTO 0);
CONSTANT table_x0: table := (
"00001",
"00101",
"00110",
"00111",
"01001",
"01001",
"01010",
"01011",
"01100",
"01101",
"01101",
"01110",
"01110",
"01111",
"01111",
"10000");
```

The end of computation is detected when $x_{i+1} = x_i$.

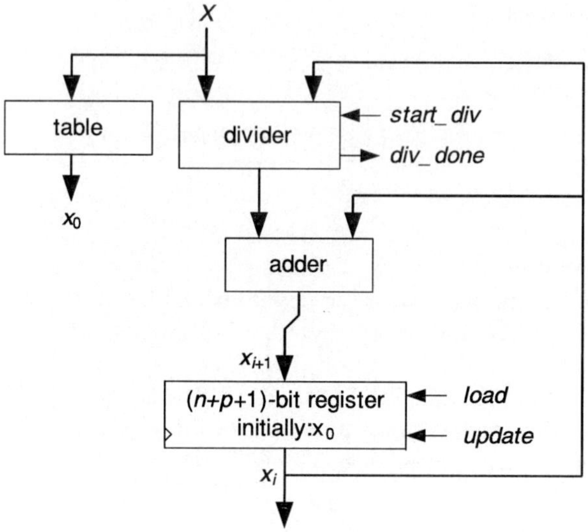

Fig. 10.6 Newton–Raphson method: data path

Comment 10.2
Every iteration step includes a division, an operation whose complexity is similar to that of a complete square root computation using a digit recurrence algorithm. Thus, this type of circuit is generally not time effective.

Another method is to first compute $X^{-1/2}$. A final multiplication computes $X^{1/2} = X^{-1/2} \cdot X$. The following iteration can be used for computing $X^{-1/2}$

$$x_{i+1} = (x_i/2) \cdot \left(3 - x_i^2 \cdot X\right),$$

where the initial value x_0 belongs to the range $0 < x_0 \leq X^{-1/2}$. The corresponding graphical construction is shown in Fig. 10.7.

The corresponding circuit does not include dividers, only multipliers and an adder. The implementation of this second convergence algorithm is left as an exercise.

10.4 Logarithm

Given an n-bit normalized fractional number $x = 1{\cdot}x_{-1}\, x_{-2} \cdots x_{-n}$, compute $y = log_2 x$ with an accuracy of p fractional bits. As x belongs to the interval $1 \leq x < 2$, its base-2 logarithm is a non-negative number smaller than 1, so $y = 0.y_{-1}\, y_{-2} \cdots y_{-p}$.

If $y = log_2 x$, then $x = 2^{0 \cdot y_{-1} y_{-2} \ldots y_{-p} \cdots}$, so that $x^2 = 2^{y_{-1} \cdot y_{-2} \cdots y_{-p} \cdots}$. Thus

- if $x^2 \geq 2 : y_{-1} = 1$ and $x^2/2 = 2^{0.y_{-2} \cdots y_{-p} \cdots}$;
- if $x^2 < 2 : y_{-1} = 0$ and $x^2 = 2^{0.y_{-2} \cdots y_{-p} \cdots}$.

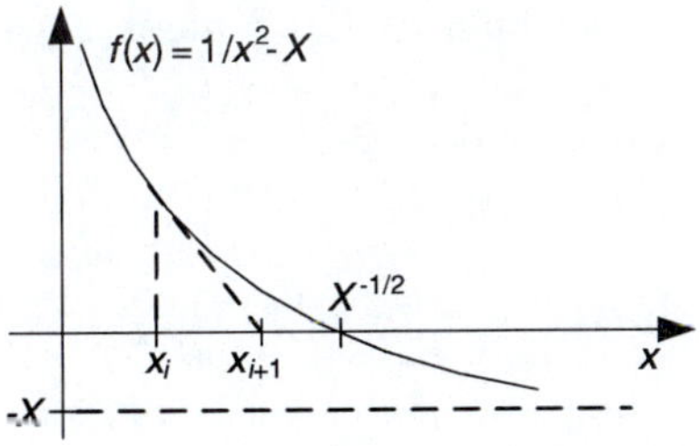

Fig. 10.7 Newton–Raphson method: computation of $X^{-1/2}$

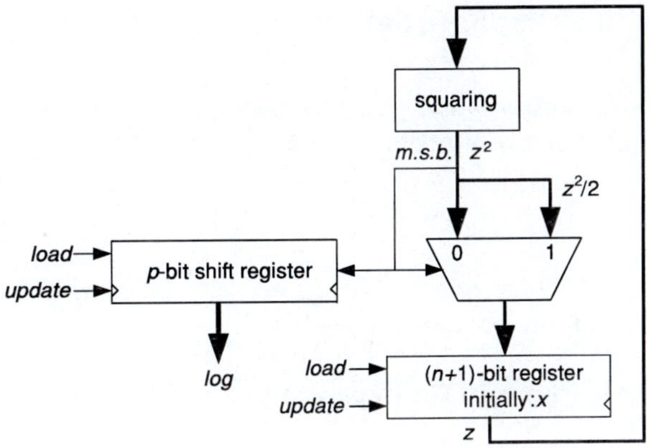

Fig. 10.8 Logarithm: data path

The following algorithm computes y:

Algorithm 10.9: Base-2 logarithm

```
z = x;
for i in 1 to p loop
  z := z²;
  if z ≥ 2 then y-i := 1; z := z/2;
  else y-i := 0;
end loop;
```

The preceding algorithm can be executed by the data path of Fig. 10.8 to which corresponds the following VHDL model.

```
square <= z*z;
WITH square(2*n+1) SELECT next_z <=
  square(2*n+1 DOWNTO n+1) WHEN '1',
  square(2*n DOWNTO n) WHEN OTHERS;
register_z: PROCESS(clk) ...
shift_register: PROCESS(clk) ...
```

A complete VHDL model *Logarithm.vhd* is available at the Authors' web page.

Comments 10.3

1. If x belongs to the interval $2^{n-1} \leq x < 2^n$, then it can be expressed under the form $x = 2^{n-1} \cdot y$ where $1 \leq y < 2$, so that $log_2 x = n - 1 + log_2 y$.
2. If the logarithm in another base, say b, must be computed, then the following relation can be used: $log_b x = log_2 x / log_2 b$.

10.5 Exponential

Given an n-bit fractional number $x = 0.x_{-1}x_{-2}\ldots x_{-n}$, compute $y = 2^x$ with an accuracy of p fractional bits. As $x = x_{-1}2^{-1} + x_{-2}2^{-2} + \ldots + x_{-n}2^{-n}$, then

$$2^x = \left(2^{2^{-1}}\right)^{x_{-1}} \left(2^{2^{-2}}\right)^{x_{-2}} \ldots \left(2^{2^{-n}}\right)^{x_{-n}}.$$

If all the constant values $a_i = 2^{2^{-i}}$ are computed in advance, then the following algorithm computes 2^x.

Algorithm 10.10: Exponential 2^x

```
z := 1;
for i in 1 to n loop
  if x-i = 1 then z := z·ai; end if;
end loop;
```

The preceding algorithm can be executed by the data path of Fig. 10.9.

The problem is accuracy. Assume that all a_i's are computed with m fractional bits so that the actual operand a_i' is equal to $a_i - \varepsilon_i$, where ε_i belongs to the range

$$0 \leq \varepsilon_i < 2^{-m}. \tag{10.6}$$

Consider the worst case, that is $y = 2^{0.11 \cdots 1}$. Then the obtained value is $y' = (a_1 - \varepsilon_1)(a_2 - \varepsilon_2)\ldots(a_n - \varepsilon_n)$. If second or higher order products $\varepsilon_i\ \varepsilon_j \cdots \varepsilon_k$ are not taken into account, then $y' \cong y - (\varepsilon_1 a_2 \cdots a_n + a_1 \varepsilon_2 \cdots a_n + a_1 \cdots a_{n-1} \varepsilon_n)$. As all products $p_1 = a_2 \cdots a_n$, $p_2 = a_1 a_3 \cdots a_n$, etc., belong to the range $1 < p_i < 2$, and ε_i to (10.6), then

$$y - y' < 2 \cdot n \cdot 2^{-m-}. \tag{10.7}$$

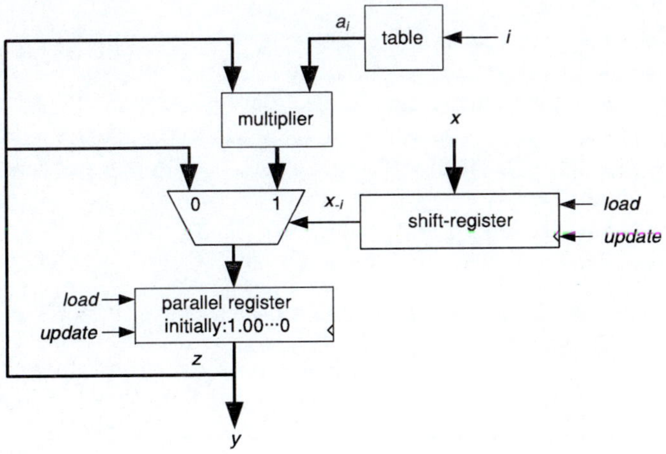

Fig. 10.9 Exponential: data path

Relation (10.7) would define the maximum error if all products were computed exactly, but it is not the case. At each step the obtained product is rounded. Thus Algorithm 10.8 successively computes

$$z_2 > a_1^{'} \cdot a_2^{'} - 2^m,$$
$$z_3 > \left(a_1^{'} \cdot a_2^{'} - 2^m\right) \cdot a_3^{'} - 2^m = a_1^{'} \cdot a_2^{'} \cdot a_3^{'} - 2^{-m}\left(1 + a_3^{'}\right),$$
$$z_4 > \left(a_1^{'} \cdot a_2^{'} \cdot a_3^{'} - 2^{-m}\left(1 + a_3^{'}\right)\right) \cdot a_4^{'} - 2^m = a_1^{'} \cdot a_2^{'} \cdot a_3^{'} \cdot a_4^{'} - 2^{-m}\left(1 + a_4^{'} + a_3^{'} \cdot a_4^{'}\right),$$

and so on. Finally

$$\begin{aligned} z_n &> y^{'} - 2^{-m}\left(1 + a_n^{'} + a_{n-1}^{'} \cdot a_n^{'} + \cdots + a_3^{'} \cdot a_4^{'} \cdots a_n^{'}\right) \\ &> y^{'} - 2^{-m}(1 + 2(n-2)) > y^{'} - 2 \cdot n \cdot 2^{-m}. \end{aligned} \tag{10.8}$$

Thus, from (10.7) and (10.8), the maximum error $y - z_n$ is smaller than $4{\cdot}n{\cdot}2^{-m}$. In order to obtain the result y with p fractional bits, the following relation must hold true: $4{\cdot}n{\cdot}2^{-m-} \leq 2^{-p}$, and thus

$$m \geq p + log_2 n + 2. \tag{10.9}$$

As an example, with $n = 8$ and $p = 12$, the internal data must be computed with $m = 17$ fractional bits.

The following VHDL model describes the circuit of Fig. 10.9.

```
a <= powers(count)(23 DOWNTO 24 - m);
product <= ('1'&a) * z;
WITH int_x(n-1) SELECT next_z <=
  product(2*m DOWNTO m) WHEN '1', z WHEN OTHERS;
register_z: PROCESS(clk) ...
y <= z(m DOWNTO m-p);
shift_register_x: PROCESS(clk) ...
```

powers is a constant array defined within a user package; it stores the fractional part of a_i with 24 bits:

```
TYPE table IS ARRAY(0 TO 7) OF STD_LOGIC_VECTOR(23 DOWNTO 0);
CONSTANT powers: table := (
 x"6a09e6",
 x"306fed",
 x"172b83",
 x"0b5586",
 x"059b0d",
 x"02c9a3",
 x"0163da",
 x"00b1af");
```

A complete VHDL model *Exponential.vhd* is available at the Authors' web page.

Instead of storing the constants a_i, an alternative solution is to store a_n, and to compute the other values on the fly:

$$a_{i-1} = 2^{2^{-i+1}} = \left(2^{2^{-i}}\right)^2 = a_i^2.$$

Algorithm 10.11: Exponential 2 ˣ, version 2

```
z := 1; a := aₙ;
for i in 0 to n-1 loop
  if xₙ₋ᵢ = 1 then z := z·a; end if;
  a := a·a;
end loop;
```

The preceding algorithm can be executed by the data path of Fig. 10.9 in which the table is substituted by the circuit of Fig. 10.10.

Once again, the problem is accuracy. In this case there is an additional problem: in order to get all coefficients a_i with an accuracy of m fractional bits, they must be computed with an accuracy of $k > m$ bits. Algorithm 10.11 successively computes

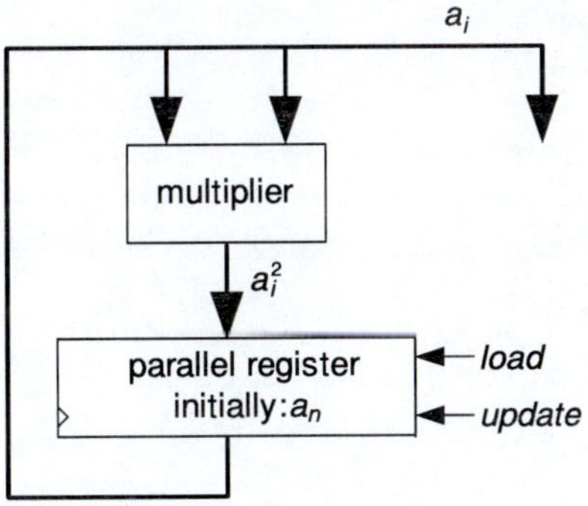

Fig. 10.10 Computation of a_i on the fly

$$a'_n > a_n - 2^{-k}, a'_{n-1} > (a_n - 2^{-k})^2 - 2^{-k} \cong a_n^2 - 2a_n 2^{-k} - 2^{-k} = a_{n-1} - 2^{-k}$$
$$(1 + 2a_n), a'_{n-2} > \left(a_{n-1} - 2^{-k}(1 + 2a_n)\right)^2 - 2^{-k} \cong a_{n-1}^2 - 2a_{n-1} 2^{-k}(1 + 2a_n)$$
$$-2^{-k} = a_{n-2} - 2^{-k}(1 + 2a_{n-1} + 4a_{n-1} a_n), a'_{n-3} > \left(a_{n-2} - 2^{-k}\right.$$
$$\left.(1 + 2a_{n-1} + 4a_{n-1}a_n)\right)^2 \quad -2^{-k} \cong a_{n-2}^2 - 2a_{n-2}2^{-k}(1 + 2a_{n-1} + 4a_{n-1}a_n) - 2^{-k}$$
$$= a_{n-3} \ -2^{-k}(1 + 2a_{n-2} + 4a_{n-2} a_{n-1} + 8a_{n-2} a_{n-1} a_n),$$

and so on. Finally

$$\begin{aligned} a'_1 &> a_1 - 2^{-k}\left(1 + 2a_2 + 4a_2a_3 + \cdots + 2^{n-2}a_2 a_3 \ldots a_n\right) \\ &> a_1 - 2^{-k}\left(1 + 4 + 8 + \cdots + 2^{n-1}\right) \\ &= a_1 - 2^{-k}(2^n - 3). \end{aligned}$$

In conclusion, $a_1 - a_1' < 2^{-k}(2^n - 3) < 2^{n-k}$. The maximum error is smaller than 2^{-m} if $n-k \leq -m$, that is $k \geq n + m$. Thus, according to (10.9)

$$k \geq n + p + log_2 n + 2.$$

As an example, with $n = 8$ and $p = 8$, the coefficients a_i (Fig. 10.10) are computed with $k = 21$ fractional bits and z (Fig. 10.9) with 13 fractional bits.

A complete VHDL model *Exponential2.vhd*, in which a_n, expressed with k fractional bits, is a generic parameter, is available at the Authors' web page.

Comment 10.3

Given an n-bit fractional number x and a number $b > 2$, the computation of $y = b^x$, with an accuracy of p fractional bits, can be performed with Algorithm 10.10 if the constants a_i are defined as follows:

$$a_i = b^{2^{-i}}.$$

So, the circuit is the same, but for the definition of the table which stores the constants a_i. In particular, it can be used for computing e^x or 10^x.

Fig. 10.11 Pseudo-rotation

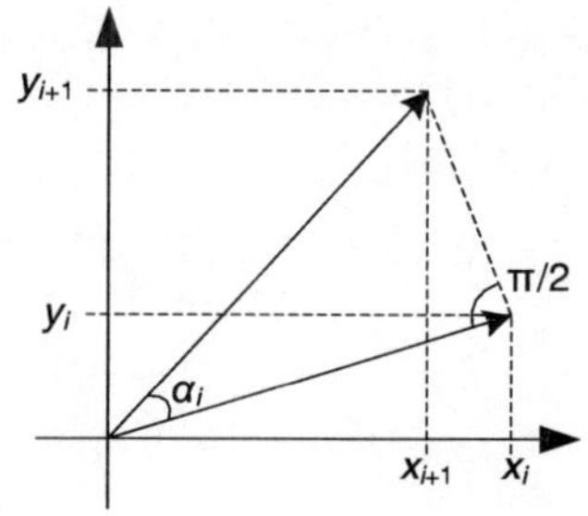

10.6 Trigonometric Functions

A digit-recurrence algorithm for computing $e^{jz} = \cos z + j{\cdot}\sin z$, with $0 \le z \le \pi/2$, similar to Algorithm 10.10 can be defined. In the modified version, the operations are performed over the complex field.

Algorithm 10.12: Exponential $e^{jz}, z = z_0 \cdot z_{-1} z_{-2} \ldots z_{-n}$

```
(eR, eI) := (1, 0);
for i in 0 to n loop
  if z-i = 1 then (eR, eI) := (eR·aRi-eI·aIi, eR·aIi+eI·aRi);
  end if;
end loop;
```

The constants a_{Ri} and a_{Ii} are equal to cos 2^{-i} and sin 2^{-i}, respectively. The synthesis of the corresponding circuit is left as an exercise.

A more efficient algorithm, which does not include multiplications, is CORDIC [2, 3]. It is a convergence method based on the graphical construction of Fig. 10.11. Given a vector (x_i, y_i), then a pseudo-rotation by α_i radians defines a rotated vector (x_{i+1}, y_{i+1}) where

$$x_{i+1} = x_i - y_i \cdot \tan \alpha_i = (x_i \cdot \cos \alpha_i - y_i \cdot \sin \alpha_i) \cdot (1 + \tan^2 \alpha_i)^{0.5},$$

$$y_{i+1} = y_i + x_i \cdot \tan \alpha_i = (y_i \cdot \cos \alpha_i + x_i \cdot \sin \alpha_i) \cdot (1 + \tan^2 \alpha_i)^{0.5}.$$

In the previous relations, $x_i{\cdot}\cos \alpha_i - y_i{\cdot}\sin \alpha_i$ and $y_i{\cdot}\cos \alpha_i + x_i{\cdot}\sin \alpha_i$ define the vector obtained after a (true) rotation by α_i radians. Therefore, if an initial vector (x_0, y_0) is rotated by successive angles $\alpha_0, \alpha_1, \cdots, \alpha_{n-1}$, then the final vector is (x_n, y_n) where

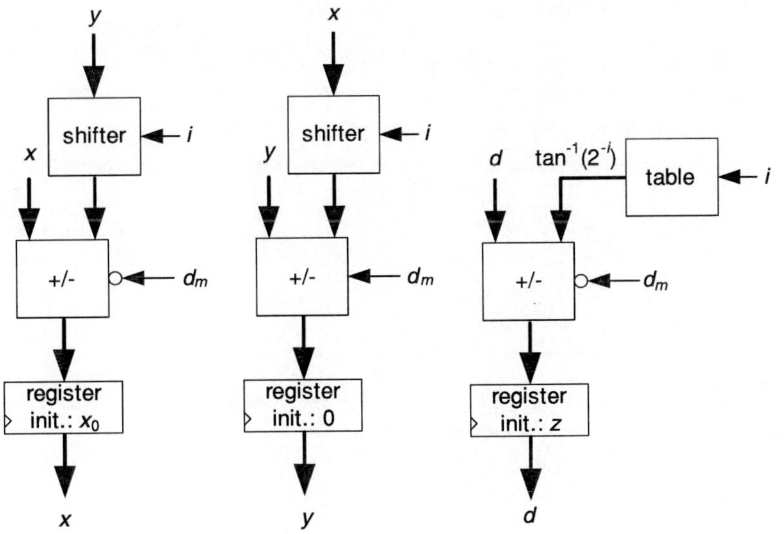

Fig. 10.12 Data path executing CORDIC

$$x_n = (x_0 \cdot \cos\alpha - y_0 \cdot \sin\alpha)(1 + \tan^2\alpha_0)^{0.5}(1 + \tan^2\alpha_1)^{0.5} \ldots (1 + \tan^2\alpha_{n-1})^{0.5},$$
$$y_n = (y_0 \cdot \cos\alpha + x_0 \cdot \sin\alpha)(1 + \tan^2\alpha_0)^{0.5}(1 + \tan^2\alpha_1)^{0.5} \ldots (1 + \tan^2\alpha_{n-1})^{0.5},$$

with $\alpha = \alpha_0 + \alpha_1 + \cdots + \alpha_{n-1}$. The CORDIC method consists in choosing angles α_i such that

- $x_i - y_i \cdot \tan \alpha_i$ and $y_i + x_i \cdot \tan \alpha_i$ are easy to calculate,
- $\alpha_0 + \alpha_1 + \alpha_2 + \cdots$ tends toward z.

As regards the first condition, α_i is chosen as follows: $\alpha_i = \tan^{-1}2^{-i}$ or $-\tan^{-1}2^{-i}$, so that $\tan \alpha_i = 2^{-i}$ or -2^{-i}. In order to satisfy the second condition, α_i is chosen in function of the difference $d_i = z - (\alpha_0 + \alpha_1 + \cdots + \alpha_{i-1})$: if $d_i < 0$, then $\alpha_i = \tan^{-1}2^{-i}$; else $\alpha_i = -\tan^{-1}2^{-i}$. The initial values are

$$x_0 = 1/k, \text{ where } k = (1+1)^{0.5}(1+2^{-2})^{0.5} \ldots (1+2^{-2(n-1)})^{0.5}, y_0 = 0, d_0 = z.$$

Thus, if $z \cong \alpha_0 + \alpha_1 + \ldots + \alpha_{n-1}$, then $x_n \cong \cos z$ and $y_n \cong \sin z$. It can be shown that the error is less than 2^{-n}. In the following algorithm, x_0 has been computed with $n = 16$ and 32 fractional bits.

Algorithm 10.13: CORDIC, $x = \cos z$, $y = \sin z$

```
x_0 := 0.607252935; y_0 := 0; d_0 := z;
for i in 0 to n-1 loop
  if d_i < 0 then
    d_(i+1) := d_i + tan^-1 2^-i; x_(i+1) := x_i + y_i·2^-i; y_(i+1) := y_i - x_i·2^-i;
  else
    d_(i+1) := d_i - tan^-1 2^-i; x_(i+1) := x_i - y_i·2^-i; y_(i+1) := y_i + x_i·2^-i;
  end if;
end loop;
x := x_n; y := y_n;
```

A circuit for executing Algorithm 10.13 is shown in Fig. 10.12. It can be described by the following VDL model.

```
a <= angles(count)(31 DOWNTO 32 - m);
WITH d(m) SELECT next_d <=
  d + ('0'&a) WHEN '1', d - ('0'&a) WHEN OTHERS;
shifter_x: shifter GENERIC MAP(m => m)
PORT MAP(a => x, shift => count, b => shifted_x);
shifter_y: shifter GENERIC MAP(m => m)
PORT MAP(a => y, shift => count, b => shifted_y);
WITH d(m) SELECT next_x <=
  x + shifted_y WHEN '1', x - shifted_y WHEN OTHERS;
WITH d(m) SELECT next_y <=
  y - shifted_x WHEN '1', y + shifted_x WHEN OTHERS;
register_d: PROCESS(clk) ...
register_x: PROCESS(clk) ...
cos <= x(m DOWNTO m-p);
register_y: PROCESS(clk) ...
sin <= y(m DOWNTO m-p);
```

angles is a constant array defined within a user package; it stores $\tan^{-1}2^{-i}$, for i up to 15, with 32 bits:

```
CONSTANT angles: table := (
 x"c90fdaa2",
 x"76b19c15",
 x"3eb6ebf2",
 x"1fd5ba9a",
 x"0ffaaddb",
 x"07ff556e",
 x"03ffeaab",
 x"01fffd55",
 x"00ffffaa",
 x"007ffff5",
 x"003ffffe",
 x"001fffff",
 x"000fffff",
 x"0007ffff",
 x"0003ffff",
 x"0001ffff");
```

shifter is a previously defined component that computes $b = a \cdot 2^{-shift}$. A complete VHDL model *cordic2.vhd* is available at the Authors' web page. It includes an *n*-state counter, which generates the index *i* of Algorithm 10.13, and a control unit.

CORDIC can be used for computing other functions. In fact, Algorithm 10.13 is based on *circular* CORDIC *rotations*, defined in such a way that the difference $d_i = z - (\alpha_0 + \alpha_1 + \ldots + \alpha_{i-1})$ tends to 0. Another CORDIC mode, called *circular vectoring*, can be used. As an example, assume that at each step the value of α_i is chosen in such a way that y_i tends toward 0: if $sign(x_i) = sign(y_i)$, then $\alpha_i = -\tan^{-1}2^{-i}$; else, $\alpha_i = \tan^{-1}2^{-i}$. Thus, if $y_n \cong 0$, then x_n is the length of the initial vector multiplied by *k*. The following algorithm computes $(x^2 + y^2)^{0.5}$.

Algorithm 10.14: CORDIC, $z = (x^2 + y^2)^{0.5}$

```
x0 := x; y0 := y;
for i in 0 to n-1 loop
  if sign(xi) = sign(yi) then
    xi+1 := xi + yi·2^-i; yi+1 := yi - xi·2^-i;
  else
    xi+1 := xi - yi·2^-i; yi+1 := yi + xi·2^-i;
  end if;
end loop;
z := 0.607252935·xn;
```

Table 10.1 Binary to decimal converters

n	m	FFs	LUTs	Period	Total time
8	3	27	29	1.73	15.6
16	5	43	45	1.91	32.5
24	8	54	56	1.91	47.8
32	10	82	82	1.83	60.4
48	15	119	119	1.83	89.7
64	20	155	155	1.83	119.0

Table 10.2 Decimal-to-binary converters

n	m	FFs	LUTs	Period	Total time
8	3	26	22	1.80	16.2
16	5	43	30	1.84	31.3
24	8	65	43	1.87	46.8
32	10	81	51	1.87	61.7
48	15	118	72	1.87	91.6
64	20	154	92	1.87	121.6

A complete VHDL model *norm_cordic.vhd* corresponding to the previous aslgorithm is available at the Authors' web page.

10.7 FPGA Implementations

Several circuits have been implemented within a Virtex 5-2 device. The times are expressed in *ns* and the costs in numbers of Look Up Tables (LUTs) and flip-flops (FFs). All VHDL models are available at the Authors' web page.

10.7.1 Converters

Table 10.1 gives implementation results of several binary-to-decimal converters. They convert n-bit numbers to m-digit numbers.

In the case of decimal-to-binary converters, the implementation results are given in Table 10.2.

Table 10.3 Square rooters: restoring algorithm

n	FFs	LUTs	Period	Total time
8	38	45	2.57	20.6
16	71	79	2.79	44.6
24	104	113	3.00	72.0
32	136	144	3.18	101.8

Table 10.4 Square rooters: non-restoring algorithm

n	FFs	LUTs	Period	Total time
8	39	39	2.61	20.9
16	72	62	2.80	44.8
24	105	88	2.98	71.5
32	137	111	3.16	101.1

Table 10.5 Square rooter: Newton–Raphson method

n	p	FFs	LUTs	Period
8	0	42	67	2.94
8	4	51	78	3.50
8	8	59	90	3.57
16	8	92	135	3.78
16	16	108	160	3.92
32	16	173	249	4.35
32	32	205	301	4.67

Table 10.6 Base-2 logarithm

n	p	FFs	LUTs	DSPs	Period	Total time
8	10	16	20	1	4.59	45.9
16	18	25	29	1	4.59	82.6
24	27	59	109	2	7.80	210.5
32	36	44	46	4	9.60	345.6

10.7.2 Square Rooters

Three types of square rooters have been considered, based on the restoring algorithm (Fig. 10.3), the non-restoring algorithm (Fig. 10.4) and the Newton–Raphson method (Fig. 10.6). The implementation results are given in Tables 10.3, 10.4.

In the case of the Newton–Raphson method, the total time is data dependent. In fact, as was already indicated above, this type of circuit is generally not time effective (Table 10.5).

Table 10.7 Exponential 2^x

n	p	m	FFs	LUTs	DSPs	Period	Total time
8	8	13	27	29	1	4.79	38.3
16	16	23	46	48	2	6.42	102.7

Table 10.8 Exponential 2^x, version 2

n	p	m	k	FFs	LUTs	DSPs	Period	Total time
8	8	13	21	49	17	3	5.64	45.1
16	16	23	39	86	71	10	10.64	170.2

Table 10.9 CORDIC: sine and cosine

n	p	m	FFs	LUTs	Period	Total time
16	8	16	57	134	3.58	57,28
32	16	32	106	299	4.21	134.72
32	24	32	106	309	4.21	134.72

Table 10.10 CORDIC: $z = (x^2 + y^2)^{0.5}$

n	p	m	FFs	LUTs	DSPs	Period	Total time
8	8	16	43	136	1	3.39	27.12
16	16	32	76	297	2	4.44	71.04
48	24	48	210	664	5	4.68	224.64

10.7.3 Logarithm and Exponential

Table 10.6 gives implementation results of the circuit of Fig. 10.8. DSP slices have been used.

The circuit of Fig. 10.9 and the alternative circuit using a multiplier instead of a table (Fig. 10.10) have been implemented. In both cases DSP slices have been used (Tables 10.7, 10.8)

10.7.4 Trigonometric Functions

Circuits corresponding to algorithms 10.13 and 10.14 have been implemented. The results are summarized in Tables 10.9, 10.10.

10.8 Exercises

1. Generate VHDL models of combinational binary-to-decimal and decimal-to-binary converters.
2. Synthesize binary-to-radix-60 and radix-60-to-binary converters using LUT-6.

3. Implement Algorithm 10.6.
4. Implement the second square rooting convergence algorithm (based on Fig. 10.7).
5. Synthesize circuits for computing ln, log_{10}, e^x and10^x.
6. Generate a circuit which computes $e^{jx} = cos\,x + j \cdot sin\,x$.

References

1. Parhami B (2000) Computer arithmetic: algorithms and hardware design. Oxford University Press, New York
2. Volder JE (1959) The CORDIC trigonometric computing technique. IRE Trans Electron Comput EC8:330–334
3. Volder JE (2000) The birth of CORDIC. J VLSI Signal Process Sys 25:101–105

Chapter 11
Decimal Operations

In a number of computer arithmetic applications, decimal systems are preferred to the binary ones. The reasons come, not only from the complexity of coding/decoding interfaces but, mostly from the lack of precision in the results of the binary systems.

For the following circuits the input and output operators are supposed to be encoded in Binary Encoded Digits (BCD).

11.1 Addition

Addition is a primitive operation for most arithmetic functions, and then it deserves special attention. The general principles for addition are in Chap. 7, and in this section we examine special consideration for efficient implementation of decimal addition targeting FPGAs.

11.1.1 Decimal Ripple-Carry Adders

Consider the base-B representations of two n-digit numbers

$$x = x_{n-1} \cdot B^{n-1} + x_{n-2} \cdot B^{n-2} + \cdots + x_0 \cdot B^0,$$
$$y = y_{n-1} \cdot B^{n-1} + y_{n-2} \cdot B^{n-2} + \cdots + y_0 \cdot B^0.$$

The following (pencil and paper) Algorithm 11.1 computes the $(n + 1)$-digit representation of the sum $z = x + y + c_{in}$ where c_{in} is the initial carry.

J.-P. Deschamps et al., *Guide to FPGA Implementation of Arithmetic Functions*, Lecture Notes in Electrical Engineering 149, DOI: 10.1007/978-94-007-2987-2_11,

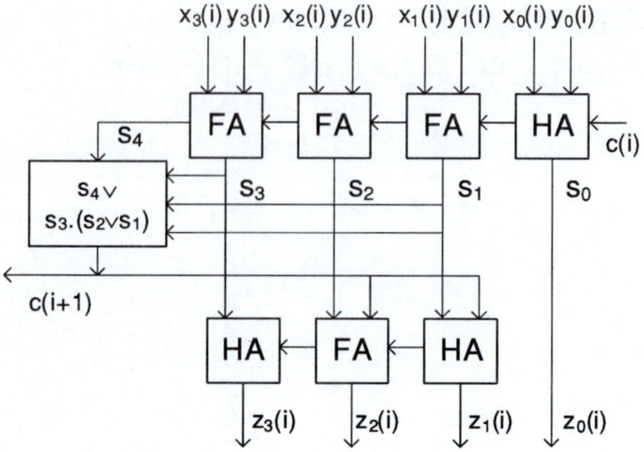

Fig. 11.1 Decimal ripple-carry adder cell

Algorithm 11.1: Classic addition (ripple carry)

```
c(0) := c_in;
for i in 0 .. n-1 loop
    if x(i) + y(i) + c(i) > B-1 then c(i+1) := 1;
    else c(i+1) := 0; end if;
    z(i) := (x(i) + y(i) + c(i)) mod B;
end loop;
z(n) := c(n);
```

For $B = 10$, the classic ripple-carry for a *BCD* decimal adder cell can be implemented as suggested in Fig. 11.1 The mod-10 addition is performed adding 6 to the binary sum of the digits, when a carry for the next digit is generated. The VHDL model *ripple_carry_adder_BCD.vhd* is available at the Authors' web page.

As described in Chap. 7, the na implementation of an adder (ripple-carry, Fig. 7.1) has a meaningful critical path. In order to reduce the execution time of each iteration step, Algorithm 11.1 can be modified as shown in the next section.

11.1.2 Base-B Carry-Chain Adders

In order to improve the ripple-carry adder, a better solution is the use of two binary functions of two *B*-valued variables, namely the *propagate* (*P*) and *generate* (*G*) functions.

$p_i = p(x_i, y_i) = 1 \text{ if } x_i + y_i = B - 1, p(x_i, y_i) = 0$ otherwise;
$g_i = g(x_i, y_i) = 1 \text{ if } x_i + y_i \geq B,\ g(x_i, y_i) = 0 \text{ if } x_i + y_i \leq B - 2,$ otherwise, any value.

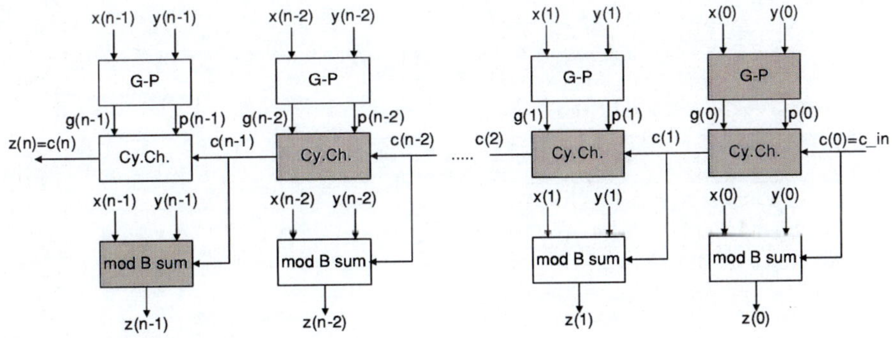

Fig. 11.2 n-digits carry-chain adder

So, c_{i+1} can be expressed under the following way:

$$c_{i+1} = p_i \cdot c_i + \text{not}\,(p_i) \cdot g_i.$$

The corresponding modified Algorithm 11.2 is the following one.

Algorithm 11.2: Carry-chain addition

```
--computation of the generation and propagation conditions:
for i in 0 .. n-1 loop
    g(i) := p(x(i), y(i)); p(i) := p(x(i) ,y(i));
end loop;
c(0) := c_in;  --carry computation:
for i in 0 .. n-1 loop
     if p(i) = 1 then c(i+1) := c(i); else c(i+1) := g(i); end if;
end loop;
for i in 0 .. n-1 loop --sum computation
    z(i) := (x(i) + y(i) + c(i)) mod B;
end loop;
z(n) := c(n);
```

The use of *propagate* and *generate* functions allow generating a *n*-digit adder carry-chain array of Fig. 11.1. It is based on the Algorithm 11.2. The *Generate-Propagate* (*G-P*) cell calculates the *Generate* and *Propagate* functions; and the *carry-chain* (*Cy.Ch*) cell computes the next carry. Observe that the carry-chain cells are binary circuits, whereas the *generate-propagate* and the *mod B sum* cells are *B*-ary ones. As regards the computation time, the critical path is shaded in Fig. 11.2. (It has been assumed that $T_{sum} > T_{Cy.\ Ch}$)

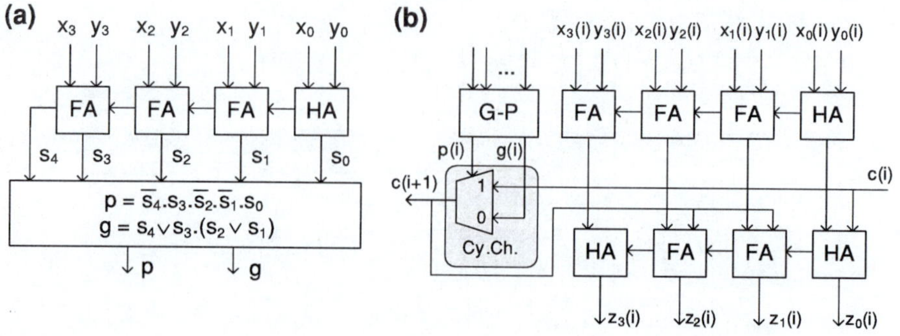

Fig. 11.3 **a** Simple *G-P* cell for *BCD* adder. **b** Carry-chain *BCD* adder *i*th digit computation

11.1.3 Base-10 Carry-Chain Adders

If $B = 10$, the carry-chain circuit remains unchanged but the *P* and *G* functions as well as the modulo-10 sums are somewhat more complex. In base 2 (Chap. 7), the *P* and *G* cells are respectively synthesized by *XOR* and *AND* functions, while in base 10, *P and G* are now defined as follows:

$$p_i = 1 \text{ if } x_i + y_i = 9,\ p_i = 0 \text{ otherwise;} \tag{11.1}$$

$$g_i = 1 \text{ if } x_i + y_i > 9,\ g_i = 0 \text{ otherwise;} \tag{11.2}$$

A straightforward way to synthesize *P* and *G* is shown at Fig. 11.3a. That is add the BCD numbers and detects if the sum is equal to nine and greater than nine respectively. Nevertheless, functions *P* and *G* may be directly computed from x_i, y_i inputs. The following formulas (11.3) and (11.4) are Boolean expressions of conditions (11.1) and (11.2).

$$p_i = P_0 \cdot [K_1 \cdot (P_3 \cdot K_2 \vee K_3 \cdot G_2) \vee G_1 \cdot K_3 \cdot P_2] \tag{11.3}$$

$$g_i = G_0 \cdot [P_3 \vee G_2 \vee P_2 \cdot G_1] \vee G_3 \vee P_3 \cdot P_2 \vee P_3 \cdot P_1 \vee G_2 \cdot P_1 \vee G_2 \cdot G_1 \tag{11.4}$$

where $P_j = x_j \oplus y_j$, $G_j = x_j \cdot y_j$ and $K_j = x_j' \cdot y_j'$ are the binary propagator, generator and carry-kill for the *j*th components of the *BCD* digits *x*(*i*) and *y*(*i*).

The *BCD* carry-chain adder *i*th digit computation is shown at Fig. 11.3b and it is made of a first binary *mod* 16 adder stage, a carry-chain cell driven by the *G-P* functions, and an output adder stage performing a correction (adding 6) whenever the carry-out is one. Actually, a zero carry-out $c(i + 1)$ identifies that the *mod* 16 sum does not exceed 9, so no corrections are needed. Otherwise, the *add*-6 correction applies. Naturally, the *G-P* functions may be computed according to Fig. 11.3, using the outputs of the *mod* 16 stage, including the carry-out s_4.

The VHDL model *cych_adder_BCD_v1.vhd* that implements a behavioral model of the decimal carry-chain adder is available at the Authors' web page.

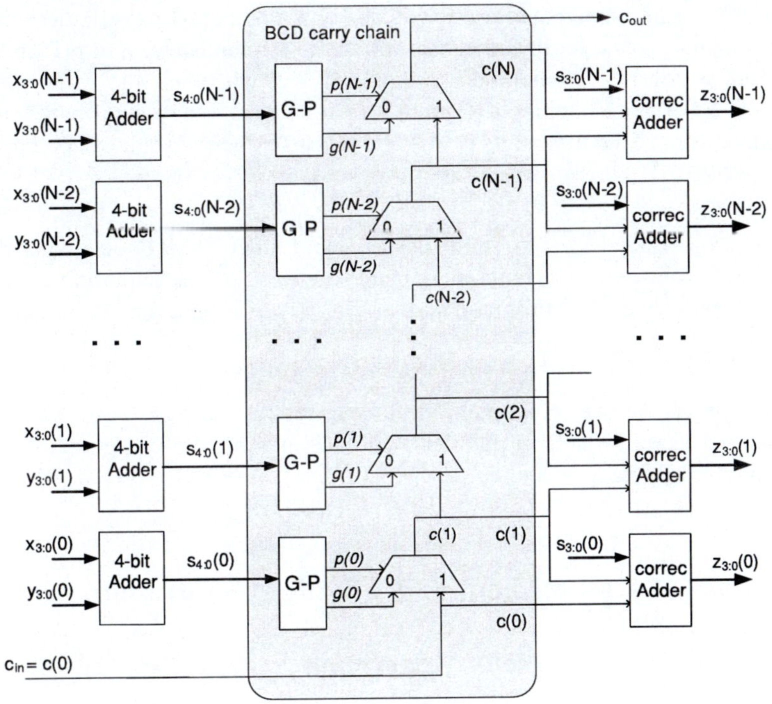

Fig. 11.4 *FPGA* implementation of an *N*-digit *BCD* Adder

11.1.4 FPGA Implementation of the Base-10 Carry-Chain Adders

The FPGA vendors provides dedicated resources to implement binary carry-chain adders [1, 2]. As mentioned in Chap. 7 a simple HDL description of a binary adder is implemented efficiently using the carry-logic. Nevertheless in order to use this resources in other designs it is necessary to instantiate the components manually.

Figure 11.4 shows a Xilinx implementation of the decimal carry-chain adder. The VHDL model *cych_adder_BCD_v2.vhd* that implements a low level component instantiation model of the decimal carry-chain adder is available at the Authors' web page.

Observe that the critical path includes a 4-bit adder, the G-P computation; the n-digits carry propagation and a final 3-bit correction adder. Xilinx 6-input/2-output *LUT* is built as two 5-input functions while the sixth input controls a 2-1 multiplexor allowing to implement either two 5-input functions or a single 6-input one; so G and P functions fit in a single *LUT* as shown at Fig. 11.5a.

Other alternatives to implement in FPGA the decimal carry-chain adders, include computing the functions P and G directly from the decimal digits ($x(i)$, $y(i)$ inputs) using the Boolean expressions, instead of the intermediate sum bits s_k [3].

For this purpose one could use formulas (11.3) and (11.4), nevertheless, in order to minimize time and hardware consumption the implementation of $p(i)$ and $g(i)$ is revisited as follows. Remembering that $p(i) = 1$ whenever the arithmetic sum $x(i) + y(i) = 9$, one defines a 6-input function $pp(i)$ set to be 1 whenever the arithmetic sum of the first 3 bits of $x(i)$ and $y(i)$ is 4. Then $p(i)$ may be computed as:

$$p(i) = (x_0(i) \oplus y_0(i)) \cdot pp(i). \tag{11.5}$$

On the other hand, $gg(i)$ is defined as a 6-input function set to be 1 whenever the arithmetic sum of the first 3 bits of $x(i)$ and $y(i)$ is 5 or more. So, remembering that $g(i) = 1$, whenever the arithmetic sum $x(i) + y(i) > 9$, $g(i)$, may be computed as:

$$g(i) = gg(i) \vee (pp(i) \cdot x_0(i) \cdot y_0(i)). \tag{11.6}$$

As Xilinx *LUTs* may compute 6-variable functions, then $gg(i)$ and $pp(i)$ may be synthesized using 2 *LUTs* in parallel while $g(i)$ and $p(i)$ are computed through an additional single *LUT* as shown at Fig. 11.5b.

11.2 Base-10 Complement and Addition: Subtration

11.2.1 Ten's Complement Numeration System

B's complement representation general principles are available in the literature. One restricts to 10's complement system to cope with the needs of this section. A one-to-one function $R(x)$, associating a natural number to x is defined as follows.

Every integer x belonging to the range $-10^n/2 \le x < 10^n/2$, is represented by $R(x) = x \bmod 10^n$, so that the integer represented in the form $x_{n-1}x_{n-2}\cdots x_1x_0$ is

$$x_{n-1} \cdot 10^{n-1} + x_{n-2} \cdot 10^{n-2} + \cdots + x_0 \text{ if } x_{n-1} \cdot 10^{n-1} + x_{n-2} \cdot 10^{n-2} + \cdots + x_0 < 10^n/2, \tag{11.7}$$

$$x_{n-1} \cdot 10^{n-1} + x_{n-2} \cdot 10^{n-2} + \cdots + x_0 - 10^n \text{ if } x_{n-1} \cdot 10^{n-1} + x_{n-2} \cdot 10^{n-2} + \cdots + x_0 \ge 10^n/2. \tag{11.8}$$

The conditions (11.7) and (11.8) may be more simply expressed as

$$x_{n-1} \cdot 10^{n-1} + x_{n-2} \cdot 10^{n-2} + \cdots + x_0 \text{ if } x_{n-1} < 5, \tag{11.9}$$

$$x_{n-1} \cdot 10^{n-1} + x_{n-2} \cdot 10^{n-2} + \cdots + x_0 - 10^n \text{ if } x_{n-1} \ge 5. \tag{11.10}$$

Another way to express a 10's complement number is:

$$x'_{n-1} \cdot 10^{n-1} + x_{n-2} \cdot 10^{n-2} + \cdots + x_0 \tag{11.11}$$

where $x'_{n-1} = x_{n-1} - -10$ if $x_{n-1} \geq 5$ and $x'_{n-1} = x_{n-1}$ if $x_{n-1} < 5$, while the sign definition rule is the following one: if x is negative then $x_{n-1} \geq 5$; otherwise $x_{n-1} < 5$.

11.2.2 Ten's Complement Sign Change

Given an n-digit 10's complement integer x, the inverse $z = -x$ of x, is an n-digit 10's complement integer. Actually the only case $-x$ cannot be represented with n digits is when $x = -10^n/2$, so $-x = 10^n/2$. The computation of the representation of $-x$ is based on the following property. Assuming x to be represented as an n-digit 10's complement number $R(x)$, $-x$ may be readily computed as

$$-x = 10^{n+1} - R(x). \tag{11.12}$$

A straightforward inversion algorithm then consists of representing x with $n + 1$ digits, complementing every digit to 9, then adding 1. Observe that sign extension is obtained by adding a digit 0 to the left of a positive number, or 9 for a negative number, respectively.

11.2.3 10's Complement BCD Carry-Chain Adder-Subtractor

To compute $X + Y$ similar algorithm as in Algorithm 11.2 (Sect. 11.1.2) can be used. In order to compute $X - Y$, 10's complement subtraction algorithm actually adds $(-Y)$ to X.

10's complement sign change algorithm may be implemented through a digitwise 9's complement stage followed by an add-1 operation. It can be shown that the 9's complement binary components w_3, w_2, w_1, w_0 of a given *BCD* digit y_3, y_2, y_1, y_0 are expressed as

$$w_3 = y'_3 \cdot y'_2 \cdot y'_1; \quad w_2 = y_2 \oplus y_1; \quad w_1 = y_1; \quad w_0 = y'_0 \tag{11.13}$$

An improvement to the adder stage could be carried out by avoiding the delay produced by the 9's complement step. Thus, this operation may be carried out within the first binary adder stage, where $p(i)$ and $g(i)$ are computed as

$$p_0(i) = x_0(i) \oplus y_0(i) \oplus \left(A'/S\right); \; p_1(i) = x_1(i) \oplus y_1(i);$$

$$p_2(i) = x_2(i) \oplus y_2(i) \oplus y_1(i).\left(A'/S\right)$$

$$p_3(i) = x_3(i) \oplus \left(y_3(i)' \cdot y_2(i)' \cdot y_1(i)'\right) \cdot \left(A'/S\right) \oplus y_3(i) \cdot \left(A'/S\right)' \tag{11.14}$$

$$g_k(i) = x_k(i), \; \forall k. \tag{11.15}$$

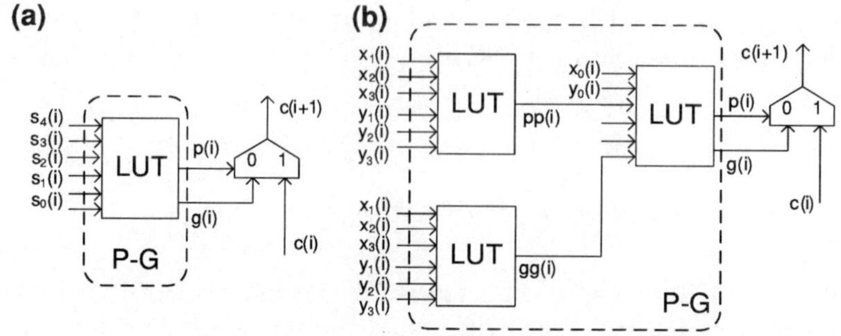

Fig. 11.5 *FPGA* carry-chain for decimal addition. **a**. P-G calculation using an intermediate addition. **b**. P-G calculation directly from BCD digits

Fig. 11.6 *FPGA* 9's complement circuit for *BCD* digit

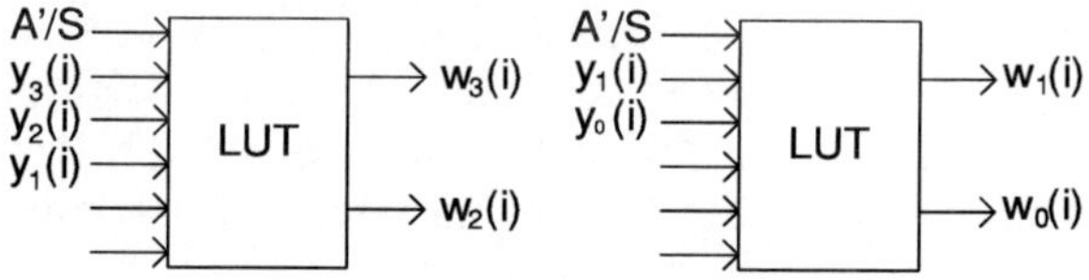

A third alternative is computing G and P directly from the input data. As far as addition is concerned, the P and G functions may be implemented according to formulas (11.4) and (11.5). The idea is computing the corresponding functions in the subtract mode and then multiplexing according to the add/subtract control signal A'/S.

11.2.4 FPGA Implementations of Adder Subtractors

To compute $X - Y$, 10's complement subtraction algorithm actually adds $(-Y)$ to X. So for a first implementation, Fig. 11.6 presents a 9's complement implementation using 6-input/2-output *LUT*s, available in the Xilinx (Virtex-5, Virtex-6, spatan6, 7-series) technology. A'/S is the add/subtract control signal; if $A'/S = 1$ (subtract), formulas (11.13) apply, otherwise $A'/S = 0$ and $w_j(i) = y_j(i)\forall i,j$.

The complete circuit is similar to the circuit of Fig. 11.4, but instead of input y, the input w as produced by the circuit of Fig. 11.5.

The better alternative intended to avoid the delay produced by the 9's complement step, embeds the 9's complementation within the first binary adder stage, as depicted in Fig. 11.7a, where $p(i)$ and $g(i)$ are computed as explained in (11.14) and (11.15).

The VHDL model *addsub_1BDC.vhd* that implements the circuit of Fig. 11.7a is available at the Authors' web page. The VHDL model has two architectures a behavioral and a low level that instantiates components (Luts, muxcy, xorcy, etc.).

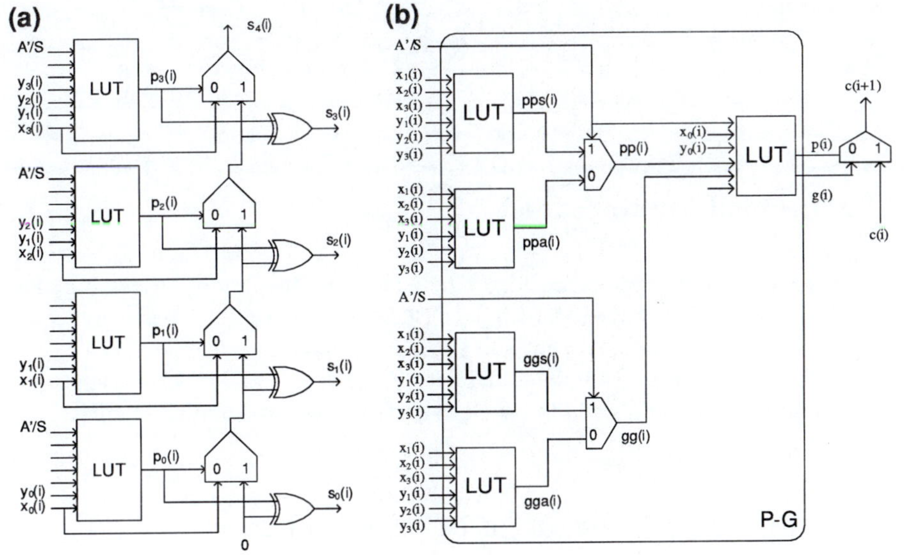

Fig. 11.7 *FPGA* implementation of adder-subtractor. **a** Adder-subtractor for one BCD digit. **b** Direct computation of P-G function

Then we can use the circuit of Fig. 11.6a to compute the P-G function using the previous computed addition of BDC digits. The VHDL model *addsub_BDC_v1.vhd* that implements a complete decimal adder-subtractor and it is available at the Authors' web page. To complete the circuit, a final correction adder (correct_add.vhd) corrects the decimal digit as a function of the carries.

The third alternative is computing G and P directly from the input data. For this reason, assuming that the operation at hand is $X + (\pm Y)$, one defines on one hand $ppa(i)$ and $gga(i)$ according to (11.4) and (11.5) (Sect. 11.1.4), i.e. using the straight values of Y's *BCD* components. On the other hand, $pps(i)$ and $ggs(i)$ are defined using $w_k(i)$ as computed by the 9's complement circuit (11.13). As $w_k(i)$ are expressed from the $y_k(i)$ both $pps(i)$ and $ggs(i)$ may be computed directly from $x_k(i)$ and $y_k(i)$. Then the correct $pp(i)$ and $gg(i)$ signal is selected according to the add/subtract control signal A'/S. Finally, the *propagate* and *generate* function are computed as:

$$p(i) = (x_0(i) \oplus y_0(i) \oplus (A'/S)) \cdot pp(i), \tag{11.16}$$

$$g(i) = gg(i) \vee (pp(i) \cdot x_0(i) \cdot (y_0(i) \oplus (A'/S))). \tag{11.17}$$

Figure 11.7b shows the Xilinx LUT based implementation. The multiplexers are implemented using dedicated muxF7 resources.

The VHDL model *addsub_BDC_v2.vhd* that implements the adder-subtractor using for the P-G computation from the direct inputs (*carry-chain_v2.vhd*) is available at the Authors' web page.

Fig. 11.8 Binary to BCD arithmetic reduction

	80	40	20	10	8	4	2	1
		p_6	p_5	p_4	p_3	p_2	p_1	p_0
+			p_6	p_5		p_4	p_4	
						p_6	p_5	
	d_3	d_2	d_1	d_0	c_3	c_2	c_1	c_0

11.3 Decimal Multiplication

This section presents approaches for *BCD* multipliers. It starts by discussing algorithmic alternatives to implement a 1×1 *BCD* digit multiplier. Then, this section proposes an implementation for an $N \times 1$ *BCD* multiplier. This operation can be used to carried out an $N \times M$ multiplier, but by itself, the $N \times 1$ multiplication appears to be a primitive for other algorithms, such as logarithm or exponential functions.

11.3.1 One-Digit by One-Digit BCD Multiplication

11.3.1.1 Binary Arithmetic with Correction

The decimal product can be obtained through a binary product and a post correction stage [4, 5]. Let *A* and *B* be two *BCD* digits ($a_3\, a_2\, a_1\, a_0$) and ($b_3\, b_2\, b_1\, b_0$) respectively. The *BCD* coded product consists of two *BCD* digits *D* and *C* such that:

$$A{*}B = D{*}10 + \mathrm{C} \tag{11.18}$$

$A * B$ is first computed as a 7-bit binary number $P(6{:}0)$ such that

$$A{*}B = P = p_6\, p_5\, p_4\, p_3\, p_2\, p_1\, p_0 \tag{11.19}$$

Although a classic binary-to-*BCD* decoding algorithm can be used, it can be shown that the *BCD* code for *P* can be computed through binary sums of correcting terms described in Fig. 11.8. The first row in Figure shows the *BCD* weights. The weights of p_3, p_2, p_1 and p_0 are the same as those of the original binary number "$p_6\, p_5\, p_4\, p_3\, p_2\, p_1\, p_0$". But weights 16, 32 and 64 of p_4, p_5, and p_6 have been respectively decomposed as (10, 4, 2), (20, 10, 2) and (40, 20, 4). Observe that "p_3 p_2 p_1 p_0" could violate the interval [0, 9], then an additional adjust could be necessary.

First the additions of Row 1, 2, 3, and correction of "$p_3\, p_2\, p_1\, p_0$" are completed (least significant bit p_0 is not necessary in computations). Then the final correction is computed.

One defines (Arithmetic I):

1. the binary product $A{*}B = P = p_6\, p_5\, p_4\, p_3\, p_2\, p_1\, p_0$
2. the Boolean expression: $adj_1 = p_3 \wedge (p_2 \vee p_1)$,

3. the arithmetic sum: $dcp = p_6\,p_5\,p_4\,p_3\,p_2\,p_1\,p_0 + 0\,p_6\,p_5\,0\,p_4\,p_3 + 0\,0\,0\,0 p_6\,p_5\,0 + 0\,0\,0\,0\,adj_i\ adj_1\,0$,

4. the Boolean expression: $adj_2 = (dcp_3 \wedge (dcp_2 \vee dcp_1)) \vee (p_5 \wedge p_4 \wedge p_3)$.

   One computes

$$dc = dcp + 0\,0\,0\,0\,adj_2\,adj_2\,0. \tag{11.20}$$

Then

$$D = dc_7\,dc_6\,dc_5\,dc_4 \text{ and } C = dc_3\,dc_2\,dc_1\,dc_0$$

A better implementation can be achieved using the following relations (Arithmetic II):

1. the product $A{*}B = P = p_6\,p_5\,p_4\,p_3\,p_2\,p_1\,p_0$
2. compute:

$$cc = p_3\,p_2\,p_1\,p_0 + 0\,p_4\,p_4\,0 + 0\,p_6\,p_5\,0,$$
$$dd = p_6\,p_5\,p_4 + 0\,p_6\,p_5$$

   (cc is 5 bits, dd has 4 bits, computed in parallel)

3. define:

$$cy1 = 1 \text{ iff } cc > 19,\ cy0 = 1 \text{ iff } 9 < cc < 20$$

   (*cy1* y *cy2* are function of $cc_3\ cc_2\ cc_1\ cc_0$, and can be computed in parallel)

4. compute:

$$c = cc_3\,cc_2\,cc_1\,cc_0\ + cy1(cy1 \text{ or } cy0)\,cy\,0\,0, \tag{11.21}$$

$$d = dd_3\,dd_2\,dd_1\,dd_0 + 0\,0\,cy_1\,cy_0$$

(c and d calculated in parallel)

Compared with the first approach, the second one requires smaller adders (5 and 4-bit vs. 8-bit) and the adders can operate in parallel as well. The VHDL models *bcd_mul_arith1.vhd* and *bcd_mul_arith2.vhd* that described the previous method for digit by digit multiplication are available at the Authors' web page.

11.3.1.2 Using ROM

Actually a (100 × 8)-bit *ROM* can fit to store all the *A** *B* multiplications. However, as *A* and *B* are two 4-bit operands (BCD digits) the product can be mapped into a $2^8 \times 8$-bit *ROM*. In FPGA devices there mainly two main memory recourses: Block and distributed RAMS. For the examples in Xilinx devices two main possibilities are considered: Block RAMs *BRAM*'s or distributed *RAM*'s (LUT based implementation of *RAM*'s).

***BRAM*-based implementation**: Xilinx Block *RAM*'s are 18-kbit (or 36-kbits) configurable and synchronous dual-port *RAM* (or *ROM*) blocks. They can be configured into different layouts. The $2^{11} \times 8$-bit configuration option has been selected, though wasting some memory capacity. As *BRAM* is dual-port, two one-by-one digit multiplications can be implemented in a single *BRAM* and in a single clock cycle. However the main characteristic to cope with is that *BRAM*'s are synchronous, so either the address or the output should be registered.

Distributed *RAM* (*LUT*-based): 6-input *LUT*'s can be configured as $64 \times$ 1-bit RAM's. Moreover the four 6-LUTs in a slice has additional multiplexers to implement 256×1 bit RAMs. Then the $2^8 \times 8$-bit ROM can be implemented using 8 slices (32 *LUT*'s).

A trivial optimization reduces the area considering the computation of the *BCD* final result components c_0 and d_3 straightforward. Actually $c_0 = a_0 \wedge b_0$ and $d_3 = a_0 \wedge b_0 \wedge a_3 \wedge b_3$. That is, c_0 is related to the parity while d_3 emphasizes that most significant bit is set for one in only one case (decimal $9 \times 9 = 81$). It is thus possible to reduce the required memory size to a $2^8 \times 6$-bit *ROM* only plus two LUTs to implement c_0 and d_3.

Comments

1. *BRAM*-based design is fast but synchronous. It is useless for combinational implementations, but suitable for sequential and pipelined ones.
2. The existence of "do-not-care" conditions in the memory definition allows the synthesizer to reduce the effective memory requirement.

The VHDL model *bcd_mul_bram.vhd* implements the BRAM based implementation for the digit by digit multiplication. Additionally, and *bcd_mul_-mem1.vhd* and *bcd_mul_mem2.vhd* provides the LUT based implementation of decimal BCD multiplication. These models with the corresponding test bench (*test_mul_1by1BCB.vhd*) are available at the Authors' web page.

11.3.2 N by One BCD Digit Multiplier

A $N \times 1$ *BCD* digit multiplier is readily achieved through N 1×1-digit multiplications followed by a *BCD* decimal addition. Fig. 11.9 shows how the partial products are arranged to feed the *BCD* *N*-digit adder stage. The carry-chain adder of Sect. 13.1 can be used.

The VHDL model *mult_Nx1_BCD.vhd* and *mult_Nx1_BCD_bram.vhd* that describe the N by one decimal multiplier are available at the Authors' web page.

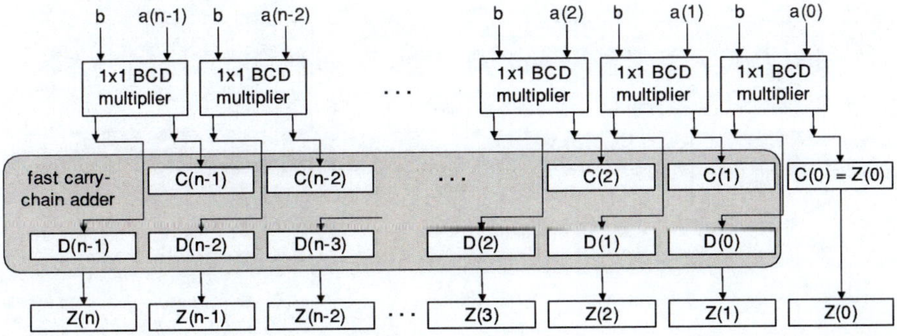

Fig. 11.9 N by one digit circuit multiplication. It includes *N* one by one digit multipliers and a fast carry-chain adder

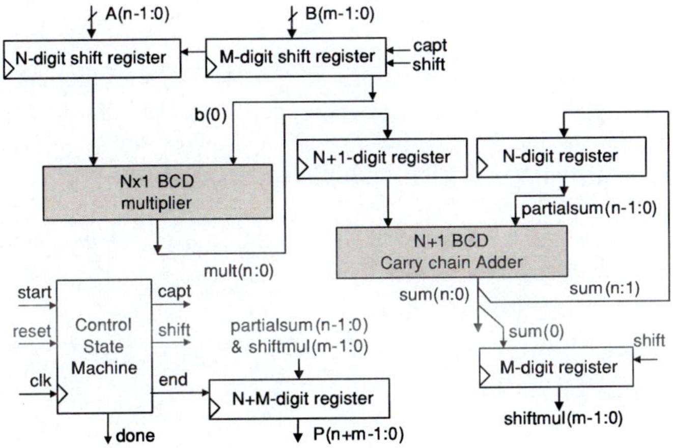

Fig. 11.10 A sequential $N \times M$ multiplier

11.3.3 N by M Digits Multiplier

Using the previously designed *N* by one digit multiplier it is possible to perform the $N \times M$ digits multiplication. The best area compromise is obtained using a sequential implementation that uses an $N \times 1$ digits multiplier and an $N + 1$ digit adder. Figure 11.10 show the scheme of the Least Significant Digit (LSD) first algorithm implemented. The *B* operand is shifted right at each clock cycle. The LSD digit of *B* is multiplied by the *A* operand and accumulated in the next cycle to shorten the critical path. After $M + 1$ cycles, the $N + M$ digit result is available.

The VHDL models *mult_BCD_seq.vhd* and *mult_BCD_bram_seq.vhd* that describes the *N* by *M* digits decimal multiplier and the matching test bench (*test_mult_BCB_seq.vhd*) is available at the Authors' web page.

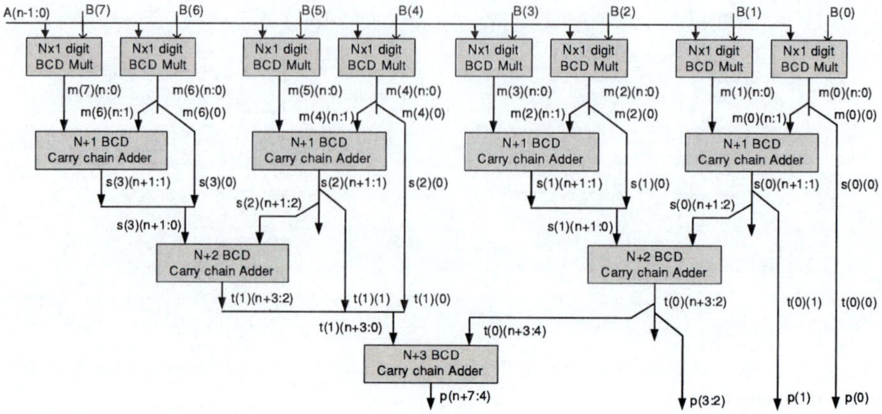

Fig. 11.11 An example of a paralell $N \times M$ multiplier ($N \times 8$)

Fully combinational implementations of $N \times M$ digits multipliers are possible also based on the $N \times 1$ digit multiplication cell. After the first multiplication, an adder tree sums up the intermediate result and gives the $N + M$ digit result. Figure 11.11 show an $N \times 8$ digit multiplication example.

The VHDL model *mult_BCD_comb.vhd* that describes the N by M digits decimal multiplier and the test bench (*test_mult_BCB_comb.vhd*) are available at the Authors' web page.

11.4 Decimal Division

As described in Chap. 9, the basic algorithms are based on digit recurrence. Using Eqs. (9.1), (9.2) and (9.3) at each step of (9.2) $q_{-(i+1)}$ and r_{i+1} are computed in function of r_i and y in such a way that $10 \cdot r_i = q_{-(i+1)}y + r_{i+1}$, with $-y \leq r_{i+1} < y$, that is

$$r_{i+1} = 10 \cdot r_i - q_{-(i+1)}y, \text{ with } -y \leq r_{i+1} < y. \tag{11.22}$$

The *Robertson diagram* applied to radix-10 ($B = 10$) is depicted at Fig. 11.12. It defines the set of possible solutions: the dotted lines define the domain $\{(10 \cdot r_i, r_{i+1}) | -10 \cdot y \leq 10 \cdot r_i < 10 \cdot y \text{ and } -y \leq r_{i+1} < y\}$, and the diagonals correspond to the equations $r_{i+1} = 10 \cdot r_i - ky$ with $k \in \{-10, -9, \ldots, -1, 0, 1, \ldots, 9, 10\}$. If $ky \leq 10 \cdot r_i < (k+1)y$, there are two possible solutions for $q_{-(i+1)}$, namely k and $k + 1$. To the first one corresponds a non-negative value of r_{i+1}, and to the second one a negative value.

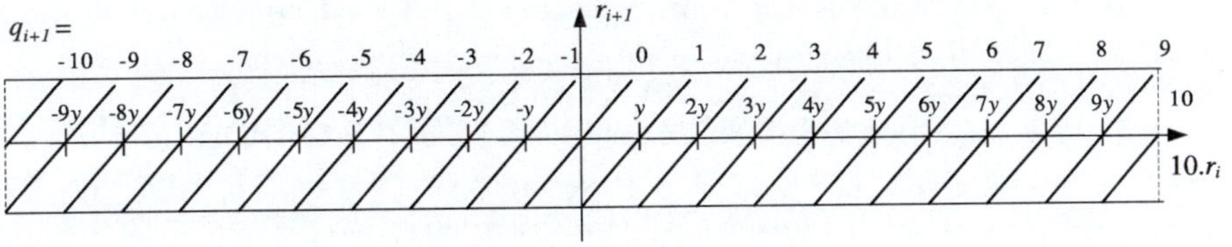

Fig. 11.12 Robertson diagram for radix 10

11.4.1 Non-Restoring Division Algorithm

A slightly modified version of the base-10 non-restoring algorithm (Sect. 9.2.1) is used. For that y must be a normalized n-digit natural, that is $10^{n-1} \le y < 10^n$. The remainders r_i satisfy the condition $-y \le r_i < y$ and belong to the range $-10^n < r_i < 10^n$. Define $w_i = 10 \cdot r_i$, so that

$$-10^{n+1} < w_i < 10^{n+1}. \tag{11.23}$$

Thus, w_i is an $(n + 2)$-digit 10's complement number. The selection of every digit $q_{-(i+1)}$ is based on a truncated value of w_i, namely $w' = \lfloor w_i/10^\alpha \rfloor$, for some α that will be defined later, so that $w_i - 10^\alpha < w' \cdot 10^\alpha \le w_i$ and

$$w' \cdot 10^\alpha \le w_i < w' \cdot 10^\alpha + 10^\alpha. \tag{11.24}$$

According to (11.23) and (11.24), $-10^{n+1} - 10^\alpha < w' \cdot 10^\alpha < 10^{n+1}$, so that

$$-10^{n+1-\alpha} \le w' < 10^{n+1-\alpha}. \tag{11.25}$$

Thus, w' is an $(n + 2 - \alpha)$-digit 10's complement number. Assume that a set of integer-valued functions $m_k(y)$, for k in $\{-10, -9,\ldots, -1, 0, 1,\ldots, 8, 9\}$, satisfying

$$k \cdot y \le m_k(y) \cdot 10^\alpha < (k + 1) \cdot y - 10^\alpha \tag{11.26}$$

has been defined. The interval $[k \cdot y, (k + 1) \cdot y - 10^\alpha]$ must include a multiple of 10^α. Thus, y must be greater than or equal to $2 \cdot 10^\alpha$. Taking into account that $y \ge 10^{n-1}$, the condition is satisfied if $\alpha \le n - 2$.

The following property is a straightforward consequence of (11.24) and (11.26):

Property 11.1 If $m_k(y) \le w' < m_{k+1}(y)$, then $k \cdot y \le w_i < (k + 2) \cdot y$.

According to the Robertson diagram of Fig. 11.12, a solution $q_{-(i+1)}$ can be chosen as follows:

if $w' < m_{-9}(y)$ then $q_{-(i+1)} = -9$,

if $w' \ge m_8(y)$ then $q_{-(i+1)} = 9$,

if $m_k(y) \le w' < m_{k+1}(y)$ for some k in $\{-9, -8, \ldots, -1, 0, 1, \ldots, 7\}$, then $q_{-(i+1)} = k + 1$.

Thus, this non-restoring algorithm generates a p-digit decimal quotient $0.q_{-1}q_{-2}\ldots q_{-p}$ and a remainder r_p satisfying

$$x = (0 \cdot q_{-1}q_{-2}\ldots q_{-p})y + r_p \cdot 10^{-p}, \text{ with } -y \cdot 10^{-p} \leq r_p \cdot 10^{-p} < y \cdot 10^{-p}, \tag{11.27}$$

where every q_{-i} is a signed decimal digit. It can be converted to a decimal number by computing the difference between two decimal numbers

$$pos = q'_{-1} \cdot 10^{-1} + q'_{-2} \cdot 10^{-2} + \cdots + q'_{-p} \cdot 10^{-p} \text{ and}$$
$$neg = q''_{-1} \cdot 10^{-1} + q''_{-2} \cdot 10^{-2} + \cdots + q''_{-p} \cdot 10^{-p},$$

with $q'_{-i} = q_i$ if $q_i > 0$, $q'_{-i} = 0$ if $q_i < 0$, $q''_{-1} = q_i$ if $q_i < 0$, $q''_{-i} = 0$ if $q_i > 0$.

It remains to define a set of integer-valued functions $m_k(y)$ satisfying (11.26). In order to simplify their computation, they should only depend on the most significant digits of y. The following definition satisfies (11.26):

$$m_k(y) = \lfloor k \cdot y'/10 \rfloor + bias \text{ where } y' = \lfloor y/10^{\alpha-1} \rfloor \text{ if } k \geq 0 \text{ and}$$
$$m_k(y) = -\lfloor -k \cdot y'/10 \rfloor + bias \text{ if } k < 0,$$

where $bias$ is any natural belonging to the range $2 \leq bias \leq 6$. With $\alpha = n$-2, y' as a 3-digit natural, and w' and $m_k(y)$ are 4-digit 10's complement numbers. In the following Algorithm 1 $m_k(y)$ is computed without adding up $bias$ to $\lfloor k \cdot y'/10 \rfloor$ or $-\lfloor -k \cdot y'/10 \rfloor$, and w' is substituted by w'—$bias$.

Algorithm 11.3: Non-restoring algorithm for decimal numbers

```
yy := y / (10**(n-3));
plus_0_y := 0;  plus_1_y := yy / 10; plus_2_y := 2*yy / 10;
...; plus_9_y := 9*yy  / 10;
r := x; power := 10**(p-1);
quotient_pos := 0; quotient_neg := 0;
for i in 0 .. p-1 loop
  ww := (10*r)/(10**(n-2));
  ww := ww - bias;
  if ww >= 0 then
    sum1 := ww - plus_1_y;
    ...
    sum8 := ww - plus_8_y;
    if sum1 < 0 then q := 1;
    elsif sum2 < 0 then q := 2;
    ...
    elsif sum8 < 0 then q := 8;
    else q := 9;
    end if;
```

```
    q_y := q*y;
    quotient_pos := quotient_pos + q*power;
  else --ww < 0
    sum1 := ww + plus_1_y;
    ...
    sum9 := ww + plus_9_y;
    if sum1 >= 0 then q := 0;
    elsif sum2 >= 0 then q := 1;
    ...
    elsif sum9 >= 0 then q := 8;
    else q := 9;
    end if;
    q_y := -q*y;
    quotient_neg := quotient_neg + q*power;
  end if;
  r := 10*r - q_y;
  power := power / 10;
end loop;
quotient := quotient_pos - quotient_neg;
```

The structure of the data path corresponding to Algorithm 11.3 is shown in Fig. 11.13. Additionally, two decimal shift registers storing *pos* and *neg* and an output subtractor are necessary. Alternatively the conversion could be done *on-the fly*. The computation time and complexity of the *k_y's generation* and *range detection* components are independent of *n*. The execution time of the iteration step is determined by the *n*-digit by 1-digit multiplier and by the $(n + 1)$-digit adder. The total computation time is $O(p \cdot n)$.

The simplified VHDL model *decimal_divider_nr.vhd* that describes the *N* by *N* digits divider that produces *P* digits of decimal result and the test bench (*test_dec_div_seq.vhd*) are available at the Authors' web page.

11.4.2 An SRT-Like Division Algorithm

In order to make the computation time linear, a classical method consists of using carry-save adders and multipliers instead of ripple-carry ones. In this way, the iteration step execution time can be made independent of the number *n* of digits. Nevertheless, this poses the following problem: a solution $q_{-(i+1)}$ of Eq. (9.2) must be determined in function of a carry-stored representation (s_i, c_i) of r_i, without actually computing $r_i = s_i + c_i$. An algorithm similar to the SRT-algorithms for radix-2^K dividers can be defined for decimal numbers (Sect. 9.2.3). Once again the divisor *y* must be assumed to be a normalized *n*-digit natural.

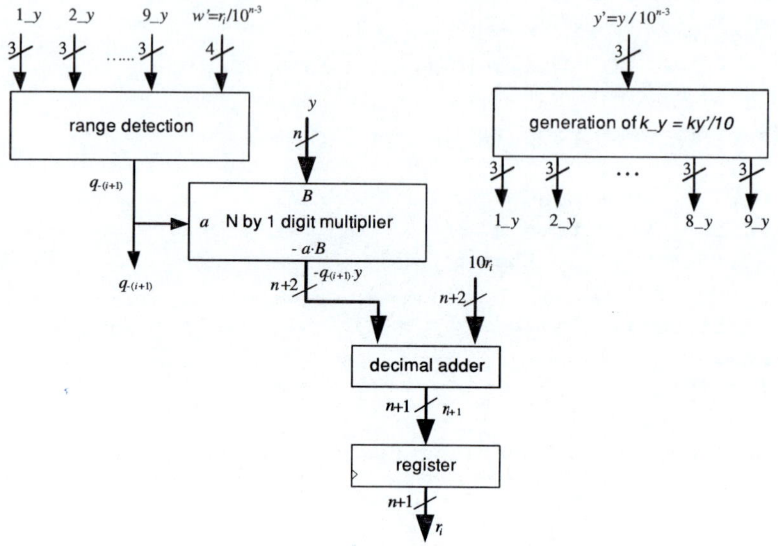

Fig. 11.13 Non-restoring radix 10 algorithm data path

All along the algorithm execution, r_i will be encoded in the form $w_i = s_i + c_i$ (stored-carry encoding) where s_i and c_i are 10's-complement numbers. Define w_i and w' as in previous section, that is $w_i = 10 \cdot r_i$, and $w' = \lfloor w_i/10^\alpha \rfloor$. Thus, $w_i = 10 \cdot s_i + 10 \cdot c_i$ and $w' = \lfloor (10 \cdot s_i + 10 \cdot c_i)/10^\alpha \rfloor$. Define also truncated values s_t and c_t of $10{\cdot}s_i$ and $10{\cdot}c_i$, that is $s_t = \lfloor 10 \cdot s_i/10^\alpha \rfloor$ and $c_t = \lfloor 10 \cdot c_i/10^\alpha \rfloor$, and let w'' be the result of adding s_t and c_t. The difference between $w' = \lfloor (10 \cdot s_i + 10 \cdot c_i)/10^\alpha \rfloor$ and $w'' = \lfloor 10 \cdot s_i/10^\alpha \rfloor + \lfloor 10 \cdot c_i/10^\alpha \rfloor$ is the possible carry from the rightmost positions, so that $w' - 1 \leq w'' \leq w'$, and thus

$$w'' \leq w' \leq w'' + 1. \tag{11.28}$$

According to (11.25) and (11.28),

$$-10^{n+1-\alpha} - 1 \leq w'' < 10^{n+1-\alpha}, \tag{11.29}$$

so that w'' is an (n + 3—α)-digit 10's-complement number. The relation between w_i and the estimate $w''{\cdot}10^\alpha$ of w_i is deduced from (11.24) and (11.28):

$$w'' \cdot 10^\alpha \leq w_i < w'' \cdot 10^\alpha + 2 \cdot 10^\alpha. \tag{11.30}$$

Assume that a set of integer-valued functions $M_k(y)$, for k in {−10, −9,..., −1, 0, 1,..., 8, 9}, satisfying

$$k \cdot y \leq M_k(y) \cdot 10^\alpha < (k+1) \cdot y - 2 \cdot 10^\alpha, \tag{11.31}$$

have been defined. The interval $[k{\cdot}y, (k + 1){\cdot}y - 2{\cdot}10^\alpha[$ must include a multiple of 10^α. Thus y must be greater than or equal to $3{\cdot}10^\alpha$. Taking into account that

$y \geq 10^{n-1}$, once again the condition is satisfied if $\alpha \leq n - 2$. The following property is a straightforward consequence of (11.30) and (11.31).

Property If $M_k(y) \leq w'' < M_{k+1}(y)$, then $k \cdot y \leq w_i < (k+2) \cdot y$.

Thus, according to the Robertson diagram of Fig. 11.12, a solution $q_{-(i+1)}$ can be chosen as follows:

if $w'' < M_{-9}(y)$ then $q_{-(i+1)} = -9$,
if $w'' \geq M_8(y)$ then $q_{-(i+1)} = 9$,
if $M_k(y) \leq w'' < M_{k+1}(y)$ for some k in $\{-9, -8, \ldots, -1, 0, 1, \ldots, 7\}$, then $q_{-(i+1)} = k + 1$.

This SRT-like algorithm generates a p-digit decimal quotient $0.q_{-1}\, q_{-2} \ldots q_{-p}$ and a remainder r_p satisfying (11.27), and can be converted to a decimal number by computing the difference between two decimal numbers as in non-restoring algorithm.

It remains to define a set of integer-valued functions $M_k(y)$ satisfying (11.31). In order to simplify their computation, they should only depend on the most significant digits of y. Actually, the same definition as in Sect. 11.3.1 can be used, that is

$$M_k(y) = \lfloor k \cdot y'/10 \rfloor + bias \text{ where } y' = \lfloor y/10^{\alpha-1} \rfloor \text{ if } k \geq 0 \text{ and}$$
$$M_k(y) = -\lfloor -k \cdot y'/10 \rfloor + bias \text{ if } k < 0.$$

In this case the range of *bias* is $3 \leq bias \leq 6$. With $\alpha = n - 2$, y' as a 3-digit natural, w' and $m_k(y)$ are 4-digit 10's complement numbers, and w'' is a 5-digit 10's complement number. In the following Algorithm 2 $M_k(y)$ is computed without adding up *bias* to $\lfloor k \cdot y'/10 \rfloor$ or $-\lfloor -k \cdot y'/10 \rfloor$, and w'' is substituted by w''—*bias*.

Algorithm 11.4: SRT like algorithm for decimal numbers

```
--computation of k_y for k in 0 to 9:
yy := y / (10**(n-3));
plus_0_y := 0; plus_1_y := yy/10; ...; plus_9_y := 9*yy/10;
--digit-recurrence algorithm:
s := x; c := 0; quotient_pos := 0; quotient_neg := 0;
power := 10**(p-1);
for i in 0 .. p-1 loop
  --estimate of w:
  ww := (10*s)/(10**(n-2)) + (10*c)/(10**(n-2));
  --detection of the range of ww:
  ww := ww - bias;
  if ww >= 0 then
    sum1 := ww - plus_1_y;
    sum2 := ww - plus_2_y;
    ...
```

```
      sum8 := ww - plus_8_y;
      if sum1 < 0 then q := 1; q_y := - y;
      elsif sum2 < 0 then q := 2; q_y := - 2*y;
      ...
      elsif sum8 < 0 then q := 8; q_y := - 8*y;
      else q := 9; q_y := - 9*y;
      end if;
      quotient_pos := quotient_pos + q*power;
    else
      sum1 := ww + plus_1_y;
      sum2 := ww + plus_2_y;
      ...
      sum9 := ww + plus_9_y;
      if sum1 >= 0 then q := 0; q_y := 0;
      elsif sum2 >= 0 then q := 1; q_y := y;
      ...
      elsif sum9 >= 0 then q := 8; q_y := 8*y;
      else q := 9; q_y := 9*y;
      end if;
      quotient_neg := quotient_neg + q*power;
    end if;
    --carry-save addition of 10s, 10c and -qy:
    csa (10*s, q_y, 10*c, next_s, next_c);
    s := next_s; c := next_c;
    power := power / 10;
  end loop;
  quotient := quotient_pos - quotient_neg;
  --stored-carry decoding:
  r := s + c;
```

The structure of the data path corresponding to Algorithm 11.4 is shown in Fig. 11.4. The carry-free multiplier is a set of 1-digit by 1-digit multipliers working in parallel. Each of them generates two digits $p_{1,\,j\,+\,1}$ and $p_{0,\,j}$ such that $q_{-(i+1)} \cdot y_j = 10 \cdot p_{1\,j+1} + p_{0,j}$, and the product $q_{-(i+1)} \cdot y$ is obtained under the form $p_1 + p_0$. The 4-to-2 counter computes $10{\cdot}s_i + 10{\cdot}c_i - (p_1 + p_0)$, that is $10{\cdot}r_i - q_{-(i\,+\,1)}{\cdot}y$, under the form $s_{i\,+\,1} + c_{i\,+\,1}$. Two decimal shift registers storing *pos* and *neg* and an output ripple-carry subtractor are necessary. Another output ripple-carry adder is necessary for computing the remainder $r_p = s_p + c_p$. Thus, all the components, but the output ripple-carry components, have computation times independent of *n* and *p*. The total computation time is $O(p + n)$.

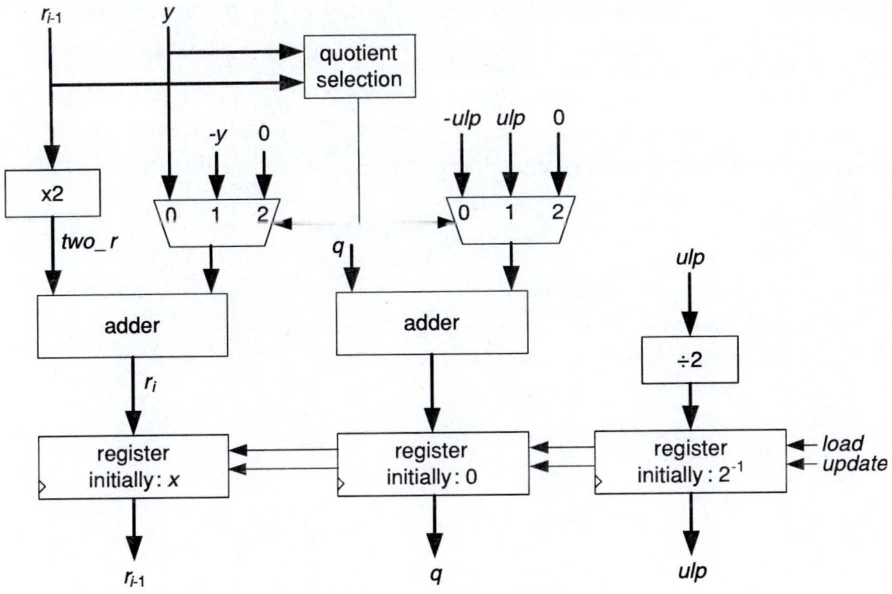

Fig. 11.14 Binary digit-recurrence data path for decimal division

The VHDL model *decimal_divider_SRT_like.vhd* that describes the *N* by *N* digits divider that produces *P* digits decimal result with several modules (*decimal_shift_register.vhd, mult_Nx1_BCD_carrysave.vhd, special_5digit_adder.vhd, range_detection3.vhd, bcd_csa_addsub_4to2.vhd*) and the test bench (*test_dec_div_seq.vhd*) are available at the Authors' web page.

11.4.3 Other Methods for Decimal Division

Other methods of division could be considered such as the use digit recoding in dividend and or divisor and also use extra degree of pre-normalization for the operands. The idea behind these methods is to ease the digit selection process.

Another idea is the use of the binary digit recurrence algorithm of describe in Chap. 9 but using decimal operands.

Observe that the Algorithm 9.1 can be executed whatever the representation of the numbers. If *B*'s complement radix-*B* representation is used, then the following operations must be available: radix-*B* doubling, adding, subtracting and halving.

Algorithm 11.5: Binary digit-recurrence algorithm for decimal division

```
r0 := x; q := 0; ulp := 2^-1;
for i in 1 .. p loop
  two_r := doubling(ri-1);
  sum := adding(two_r, y);
  dif := subtracting(two_r, y);
  if quotient_selection(ri-1, y) = 1 then
    ri := dif;
    q := adding(q, ulp);
  elsif quotient_selection(ri-1, y) = -1 then
    ri := sum;
    q := subtracting(q, ulp);
  else
    ri := two_r;
  end if;
  ulp := halving(ulp);
end loop ;
r := rp;
```

In order to obtain m decimal digits results, you need to choose p so that $2^{-p} \cong 10^{-m}$, that is $p \cong m \cdot log_2 10 \cong 3.3 \cdot m$. A possible data path is shown in Fig. 11.14. The implementation of this architecture leads to a smaller circuit, but slower than the decimal digit recurrence due to the difference in the amount of cycles to be executed.

11.5 FPGA Implementation Results

The circuits have been implemented on Xilinx Virtex-5 family with speed grade -2 [6]. The Synthesis and implementation have been carried out on *XST* (Xilinx Synthesis Technology) [7] and Xilinx Integrated System environment (*ISE*) version 13.1 [2]. The critical parts were designed using low level components instantiation (lut6_2, muxcy, xorcy, etc.) in order to obtain the desired behavior.

11.5.1 Adder-Subtractor Implementations

The adder and adder-subtractor implementation results are presented in this section. Performances of different N-digit *BCD* adders have been compared to those of an M-bit binary carry chain adder (implemented by *XST*) covering the same range, i.e. such that $M = \lfloor N.log_2(10) \rfloor \cong 3.322N$.

Table 11.1 Delays in *ns* for decimal and binary adders and adder-subtractor

N (digits)	RpCy add	CyCh add	AddSub V1	AddSub V2	M (bits)	Binary add-sub
8	12.4	3.5	3.5	3.4	27	2.1
16	24.4	3.8	3.8	3.7	54	2.6
32	48.5	4.5	4.6	4.8	107	3.8
48	72.3	5.1	5.2	5.3	160	5.2
64	95.9	5.2	5.5	5.5	213	6.6
96	–	5.9	6.1	6.1	319	8.8

Table 11.2 Area in 6-input *LUTs* for different decimal adders and adders-subtractors

Circuit	# LUTs
Ripple carry adder	$7 \times N$
Carry chain adder	$9 \times N$
Binary	$\lceil 3.32 \times N \rceil$
Adder-subtractor V2 (PG from binary addition)	$9 \times N$
Adder-subtractor V2 (PG direct form inputs)	$13 \times N$
Binary adder and adder-subtractor	$\lceil 3.32 \times N \rceil$

Table 11.1 exhibits the post placement and routing delays in *ns* for the decimal adder implementations *Ad*-I and *Ad*-II of Sect. 7.1; and the delays in *ns* for the decimal adder-subtractor implementations *AS*-I and *AS*-II of Sect. 7.2. Table 11.2 lists the consumed areas expressed in terms of 6-input look-up tables (6-input *LUT*s). The estimated area presented in Table 11.2 was empirically confirmed.

Comments

1. Observe that for large operands, the decimal operations are faster than the binary ones.
2. The delay for the carry-chain adder and the adder-subtractor are similar in theory. The small difference is due to the placement and routing algorithm.
3. The overall area with respect to binary computation is not negligible. In Xilinx 6-input LUT family an adder-subtractor is between 3 and 4 times bigger.

11.5.2 Multiplier Implementations

The decimal multipliers use the one by one digit multiplication described in Sect. 11.2.1 and the decimal adders of Sect. 11.1. The results are for the same Virtex 5 device speed grade −2.

Table 11.3 Results of BCD N × 1 multipliers using LUTs cells and BRAM cells

N	Mem in LUTs cells		BRAM-based cells		
	T (ns)	# LUT	T (ns)	# LUT	# BRAM
4	5.0	118	5.0	41	1
8	5.1	242	5.1	81	2
16	5.3	490	5.4	169	4
32	6.1	986	6.1	345	8

Table 11.4 Results of sequential implementations of $N \times M$ multipliers using one by one digit multiplication in LUTs

N	M	T (ns)	# FF	# LUT	# cycles	Delay (ns)
4	4	5.0	122	243	4	25.0
8	4	5.1	186	451	5	25.5
8	8	5.1	235	484	9	45.9
8	16	5.1	332	553	17	86.7
16	8	5.3	363	921	9	47.7
16	16	5.3	460	986	17	90.1
32	16	5.7	716	1,764	17	96.9
16	32	5.3	653	1,120	33	174.9
32	32	5.7	909	1,894	33	188.1

11.5.2.1 Decimal N × 1 Digits Implementation Results

Comparative figures of merit (minimum clock period (T) in ns, and area in LUTs and BRAMS) of the $N \times 1$ multiplier are shown in Table 11.3, for several values of N; the result are shown for circuits using LUTs, and BRAMS to implement the digit by digit multiplication respectively.

11.5.2.2 Sequential Implementations

The sequential circuit multiplies N digits by 1 digit per clock cycle, giving the result $M + 1$ cycles later. In order to speed up the computation, the addition of partial results is processed in parallel, but one cycle later (Fig. 11.10). Results for sequential implementation using LUT based cells for 1 × 1 *BCD* digit multiplication are given in Table 11.4. If the *BRAM*-based cell is used, similar periods (T) can be achieved, by means of less LUTs but using BRAMs blocks.

11.5.2.3 Combinational Implementations of N by M Multipliers

For the implementation of the $N \times M$-digit multiplier, the $N \times 1$ *mux*-based multiplication stage has been replicated M times: it is the best choice because *BRAM*-

Table 11.5 Results of combinational implementations of $N \times M$ multipliers

N	M	Delay (ns)	# LUT
4	4	10.2	719
8	4	10.7	1,368
8	8	13.4	2,911
8	16	15.7	6,020
16	8	13.6	5,924
16	16	16.3	12,165

based multipliers are synchronous. Partial products are inputs to an addition tree. For all *BCD* additions the fast carry-chain adder of Sect. 11.1.4 has been used.

Input and output registers have been included in the design. Delays include *FF* propagation and connections. The amount of *FF*'s actually used is greater than $8{*}(M + N)$ because the ISE tools [2] replicate the input register in order to reduce fan-outs. The most useful data for area evaluation is the number of *LUT*'s (Table 11.5).

Comments

1. Observe that computation time and the required area are different for N by M than M by N. That is due mainly by the use of carry logic in the adder tree.

11.5.3 Decimal Division Implementations

The algorithms have been implemented in the same Xilinx Virtex-5 device as previous circuits in the chapter. The adders and multipliers used in division are the ones described in previous sections. The non-restoring like division circuit correspond to Algorithm 11.4 (Fig. 11.13), meanwhile the SRT-like to Algorithm 11.5 (Fig. 11.15). In tables the number of decimal digits of dividend and divider is expressed as *N* and the number of digits of quotient as *P*, meanwhile the period and latency in *ns* (Table 11.6, 11.7).

Comments

1. Both types of dividers have approximately the same costs and similar delay. They are also faster than equivalent binary dividers. As an example, the computation time of an SRT-like 48-digit divider ($n = p = 48$) is about $50{\cdot}10.9 = 545$ ns, while the computation time of an equivalent binary non-restoring divider, that is a 160-bit one ($48/\log 102 \cong 160$), is more than 900 ns.
2. On the other hand, the area of the 48-digit divider (4,607 LUTs) is about five times greater than that of the 160-bit binary divider (970 LUTs). Why so a great difference? On the one hand it has been observed that decimal computation resources (adders and subtractors) need about three times more slices than binary resources (Sects. 11.4.1 and 11.4.2), mainly due to the more complex definition of the carry propagate and carry generate functions, and to the final

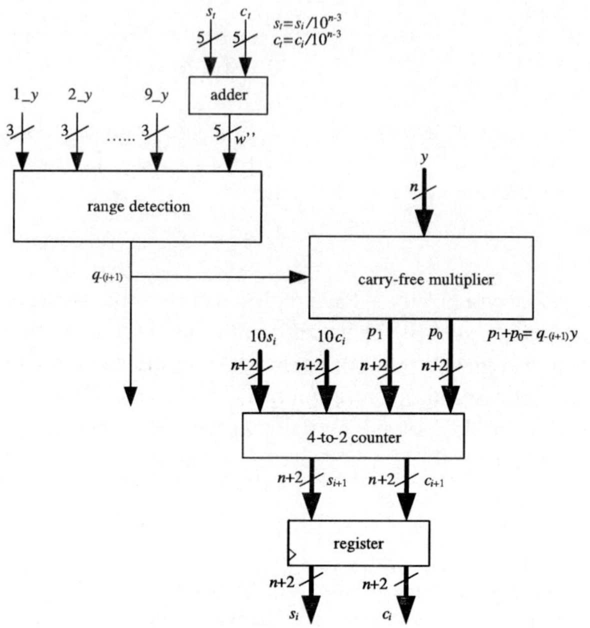

Fig. 11.15 SRT-like radix-10 algorithm data path

Table 11.6 Result for non-restoring division algorithm

N, P	FF	LUTS	Period	Latency
8	106	1,082	11.0	110.0
16	203	1,589	11.3	203.4
32	396	2,543	11.6	394.4
48	589	3,552	12.1	605.0

Table 11.7 Result for SRT-like division algorithm

N, P	FF	LUTS	Period	Latency
8	233	1,445	10.9	109.0
16	345	2,203	10.9	196.2
32	571	3,475	10.9	370.6
48	795	4,627	10.9	545.0

mod 10 reduction. On the other hand, the computation of the next quotient digit is much more complex than in the binary case.

11.6 Exercises

1. Implement a decimal comparator. Based on the decimal adder subtractor architecture modify it to implement a comparator.
2. Implement a decimal "greater than" circuit that returns '1' if A $\geq$ B else '0'. Tip: Base your design on the decimal adder subtractor.
3. Implement a $N \times 2$ digits circuit. In order to speed up computation analyze the use of a 4 to 2 decimal reducer.
4. Implement a $N \times 4$ digits circuit using a 8 to 2 decimal reducer and only one carry save adder.
5. Design a N by M digit multiplier using the $N \times 2$ or the $N \times 4$ digit multiplier. Do you improve the multiplication time? What is the area penalty with respect to the use of a $N \times 1$ multiplier?
6. Implement the binary digit-recurrence algorithm for decimal division (Algorithm 11.5). The key point is an efficient implementation of radix-B doubling, adding, subtracting and halving.

References

1. Altera Corp (2011) Advanced synthesis cookbook. http://www.altera.com
2. Xilinx Inc (2011b) ISE Design Suite Software Manuals (v 13.1). http://www.xilinx.com
3. Bioul G, Vazquez M, Deschamps J-P, Sutter G (2010) High speed FPGA 10's complement adders-subtractors. Int J Reconfigurable Comput 2010:14, Article ID 219764
4. Jaberipur G, Kaivani A (2007) Binary-coded decimal digit multiplier. IET Comput Digit Tech 1(4):377–381
5. Sutter G, Todorovich E, Bioul G, Vazquez M, Deschamps J-P (2009) FPGA implementations of BCD multipliers. V International Conference on ReConFigurable Computing and FPGAs (ReConFig 09), Mexico
6. Xilinx Inc (2010) Virtex-5 FPGA Data Sheet, DS202 (v5.3). http://www.xilinx.com
7. Xilinx Inc (2011) XST User Guide UG687 (v 13.1). http://www.xilinx.com

Chapter 12
Floating Point Arithmetic

There are many data processing applications (e.g. image and voice processing), which use a large range of values and that need a relatively high precision. In such cases, instead of encoding the information in the form of integers or fixed-point numbers, an alternative solution is a floating-point representation. In the first section of this chapter, the IEEE standard for floating point is described. The next section is devoted to the algorithms for executing the basic arithmetic operations. The two following sections define the main rounding methods and introduce the concept of guard digit. Finally, the last few sections propose basic implementations of the arithmetic operations, namely addition and subtraction, multiplication, division and square root.

12.1 IEEE 754-2008 Standard

The IEEE-754 Standard is a technical standard established by the Institute of Electrical and Electronics Engineers for floating-point operations. There are numerous CPU, FPU and software implementations of this standard. The current version is IEEE 754-2008 [1], which was published in August 2008. It includes nearly all the definitions of the original IEEE 754-1985 and IEEE 854-1987 standards. The main enhancement in the new standard is the definition of decimal floating point representations and operations. The standard defines the arithmetic and interchange formats, rounding algorithms, arithmetic operations and exception handling.

J.-P. Deschamps et al., *Guide to FPGA Implementation of Arithmetic Functions*, Lecture Notes in Electrical Engineering 149, DOI: 10.1007/978-94-007-2987-2_12,

12.1.1 Formats

Formats in IEEE 754 describe sets of floating-point data and encodings for interchanging them. This format allows representing a finite subset of real numbers. The floating-point numbers are represented using a triplet of natural numbers (positive integers). The finite numbers may be expressed either in base 2 (binary) or in base 10 (decimal). Each finite number is described by three integers: the *sign* (zero or one), the significand s (also known as coefficient or mantissa), and the exponent e. The numerical value of the represented number is $(-1)^{sign} \times s \times B^e$, where B is the base (2 or 10).

For example, if $sign = 1$, $s = 123456$, $e = -3$ and $B = 10$, then the represented number is -123.456.

The format also allows the representation of infinite numbers ($+\infty$ and $-\infty$), and of special values, called *Not a Number* (NaN), to represent invalid values. In fact there are two kinds of NaN: qNaN (quiet) and sNaN (signaling). The latter, used for diagnostic purposes, indicates the source of the NaN.

The values that can be represented are determined by the base (B), the number of digits of the significand (precision p), and the maximum and minimum values e_{min} and e_{max} of e. Hence, s is an integer belonging to the range 0 to B^{p-1}, and e is an integer such that $e_{min} \leq e \leq e_{max}$.

For example if $B = 10$ and $p = 7$ then s is included between 0 and 9999999. If $e_{min} = -96$ and $e_{max} = 96$, then the smallest non-zero positive number that can be represented is 1×10^{-101}, the largest is 9999999×10^{90} (9.999999×10^{96}), and the full range of numbers is from -9.999999×10^{96} to 9.999999×10^{96}. The numbers closest to the inverse of these bounds (-1×10^{-95} and 1×10^{-95}) are considered to be the smallest (in magnitude) normal numbers. Non-zero numbers between these smallest numbers are called subnormal (also denormalized) numbers.

Zero values are finite values whose significand is 0. The sign bit specifies if a zero is +0 (positive zero) or −0 (negative zero).

12.1.2 Arithmetic and Interchange Formats

The arithmetic format, based on the four parameters B, p, e_{min} and e_{max}, defines the set of represented numbers, independently of the encoding that will be chosen for storing and interchanging them (Table 12.1). The interchange formats define fixed-length bit-strings intended for the exchange of floating-point data. There are some differences between binary and decimal interchange formats. Only the binary format will be considered in this chapter. A complete description of both the binary and the decimal format can be found in the document of the IEEE 754-2008 Standard [1].

For the interchange of binary floating-point numbers, formats of lengths equal to 16, 32, 64, 128, and any multiple of 32 bits for lengths bigger than 128, are

Table 12.1 Binary and decimal floating point format in IEEE 754-2008

	Binary formats ($B = 2$)				Decimal formats ($B = 10$)		
Parameter	Binary 16	Binary 32	Binary 64	Binary 128	Decimal 132	Decimal 164	Decimal 128
p, digits	$10 + 1$	$23 + 1$	$52 + 1$	$112 + 1$	7	16	34
e_{max}	+15	+127	+1023	+16383	+96	+384	+16,383
e_{min}	−14	−126	−1022	−16382	−95	−383	−16,382
Common name	Half precision	Single precision	Double precision	Quadruple precision			

Table 12.2 Binary interchange format parameters

Parameter	Binary16	Binary32	Binary64	Binary128	Binary{k} ($k \geq 128$)
k, storage width in bits	16	32	64	128	Multiple of 32
p, precision in bits	11	24	53	113	$k - w$
e_{max}	15	127	1,023	16,383	$2^{(k-p-1)} - 1$
$bias$, $E - e$	15	127	1,023	16,383	e_{max}
w, exponent field width	5	8	11	15	Round($4{\cdot}\log_2 k$) − 13
t, trailing significand bits	10	23	52	112	$k - w - 1$

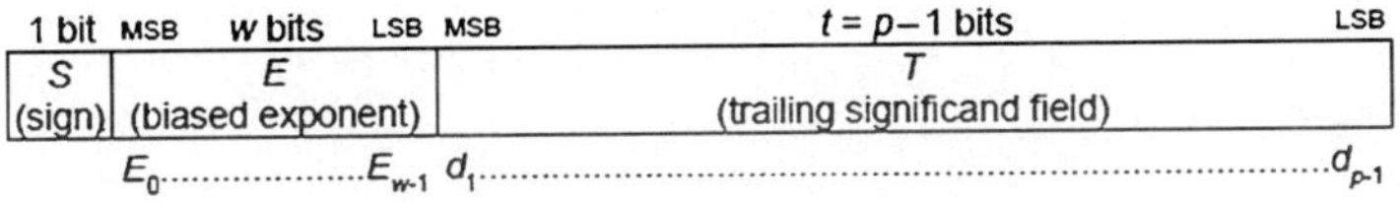

Fig. 12.1 Binary interchange format

defined (Table 12.2). The 16-bit format is only for the exchange or storage of small numbers.

The binary interchange encoding scheme is the same as in the IEEE 754-1985 standard. The k-bit strings are made up of three fields (Fig. 12.1):

- a 1-bit sign S,
- a w-bit biased exponent $E = e + bias$,
- the $p - 1$ trailing bits of the significand; the missing bit is encoded in the exponent (hidden first bit).

Each binary floating-point number has just one encoding. In the following description, the significand s is expressed in scientific notation, with the radix point immediately following the first digit. To make the encoding unique, the value of the significand s is maximized by decreasing e until either $e = e_{min}$ or $s \geq 1$ (normalization). After normalization, there are two possibilities regarding the significand:

- If $s \geq 1$ and $e \geq e_{min}$ then a normalized number of the form $1.d_1\, d_2 \ldots d_{p-1}$ is obtained. The first "1" is not stored (implicit leading 1).

- If $e = e_{min}$ and $0 < s < 1$, the floating-point number is called subnormal. Subnormal numbers (and zero) are encoded with a reserved biased exponent value. They have an implicit leading significand bit = 0 (Table 12.2).

The minimum exponent value is $e_{min} = 1 - e_{max}$. The range of the biased exponent E is 1 to $2^w - 2$ to encode normal numbers. The reserved value 0 is used to encode ± 0 and subnormal numbers. The value $2^w - 1$ is reserved to encode $\pm\infty$ and NaNs.

The value of a binary floating-point data is inferred from the constituent fields as follows:

- If $E = 2^w - 1$ (all 1's in E), then the data is an NaN or infinity. If $T \neq 0$, then it is a qNaN or an sNaN. If the first bit of T is 1, it is an sNaN. If $T = 0$, then the value is $(-1)^{sign} \times (+\infty)$.
- If $E = 0$ (all 0's in E), then the data is 0 or a subnormal number. If $T = 0$ it is a signed 0. Otherwise ($T \neq 0$), the value of the corresponding floating-point number is $(-1)^{Sign} \times 2^{emin} \times (0 + 2^{1-p} \times T)$.
- If $1 \leq E \leq 2^w - 2$, then the data is $(-1)^{sign} \times 2^{(E-bias)} \times (1 + 2^{1-p} \times T)$. Remember that the significand of a normal number has an implicit leading 1.

Example 12.1
Convert the decimal number −9.6875 to its binary32 representation.

- First convert the absolute value of the number to binary (Chap. 10): $9.6875_{10} = 1001.1011_2$.
- Normalize: $1001.1011 = 1.00110011 \times 2^3$. Hence $e = 3$, $s = 1.0011\,0011$.
- Hide the first bit and complete with 0's up to 23 bits: 00110011000000000000000.
- Add *bias* to the exponent. In this case, $w = 8$, $bias = 2^8 - 1 = 127$ and thus $E = e + bias = 3 + 127 = 130_{10} = 10000010_2$.
- Compose the final 32-bit representation:
- $1\,10000010\,00110011000000000000000_2 = C1198000_{16}$.

Example 12.2
Convert the following binary32 numbers to their decimal representation.

- $7FC00000_{16}$: $sig\,n = 0$, $E = FF_{16}$, $T \neq 0$, hence it is an NaN. Since the first bit of T is 1, it is a quiet NaN.
- $FF800000_{16}$: $sig\,n = 0$, $E = FF_{16}$, $T = 0$, hence it is $-\infty$.
- $6545AB78_{16}$: $sign = 0, E = CA_{16}, = 202_{10}, e = E - bias = 202 - 127 = 75_{10}$,

  $T = 10001011010101101111000_2$,

  $s = 1.10001011010101101111000_2 = 1.5442953_{10}$.

  The number is $1.10001011010101101111_2 \times 2^{75} = 5.8341827 \times 10^{22}$.

- 12345678_{16} : $sign = 0$, $E = 24_{16} = 36_{10}$, $e = E - bias = 36 - 127 = -91_{10}$,

  $T = 01101000101011001111000_2$,

  $s = 1.01101000101011001111000_2 = 1.408888_{10}$.

  The number is 1.01101000101011001111

  $\times 2^{-91} = 5.69045661 \times 10^{-28}$.8000000016: sign = 1, E = 0016, T = 0. The number is −0.0.
- 8000000016: sign = 1, E = 0016, T = 0. The number is −0.0.
- 00012345_{16}:$sign = 0$, $E = 00_{16}$, $T \neq 0$, hence it is a subnormal number, $e = E - bias = -127$, $T = 00000100100011010001012_2$,

  $s = 0.00000100100011010001012_2 = 0.0177777_{10}$.

  The number is $0.00000100100011010000101 \times 2^{-127} = 1.0448782 \times 10^{-40}$.

12.2 Arithmetic Operations

First analyze the main arithmetic operations and generate the corresponding computation algorithms. In what follows it will be assumed that the significand s is represented in base B (in binary if $B = 2$, in decimal if $B = 10$) and that it belongs to the interval $1 \leq s \leq B - ulp$, where ulp stands for the unit in the last position or unit of least precision. Thus s is expressed in the form $(s_0 \cdot s_{-1} \cdot s_{-2} \ldots \cdot s_{-p}) \cdot B^e$ where $e_{min} \leq e \leq e_{max}$ and $1 \leq s_0 \leq B - 1$.

Comment 12.1
The binary subnormals and the decimal floating point are not normalized numbers and are not included in the following analysis. This situation deserves some special treatment and is out of the scope of this section.

12.2.1 Addition of Positive Numbers

Given two positive floating-point numbers $s_1 \cdot B^{e1}$ and $s_2 \cdot B^{e2}$ their sum $s \cdot B^e$ is computed as follows: assume that e_1 is greater than or equal to e_2; then (*alignment*) the sum of $s_1 \cdot B^{e1}$ and $s_2 \cdot B^{e2}$ can be expressed in the form $s \cdot B^e$ where

$$s = s_1 + s_2/\left(B^{e1-e2}\right) \text{ and } e = e_1. \tag{12.1}$$

The value of s belongs to the interval

$$1 \leq s \leq 2 \cdot B - 2 \cdot ulp, \tag{12.2}$$

so that s could be greater than or equal to B. If it is the case, that is if

$$B \leq s \leq 2 \cdot B - 2 \cdot ulp, \tag{12.3}$$

then (*normalization*) substitute s by s/B, and e by $e + 1$, so that the value of $s \cdot B^e$ is the same as before, and the new value of s satisfies

$$1 \leq s \leq 2 - (2/B) \cdot ulp \leq B - ulp. \tag{12.4}$$

The significands s_1 and s_2 of the operands are multiples of *ulp*. If e_1 is greater than e_2, the value of s could no longer be a multiple of *ulp* and some rounding function should be applied to s. Assume that $s' < s < s'' = s' + ulp$, s' and s'' being two successive multiples of *ulp*. Then the *rounding* function associates to s either s' or s'', according to some rounding strategy. According to (12.4) and to the fact that 1 and B—*ulp* are multiples of *ulp*, it is obvious that $1 \leq s' < s'' \leq B—ulp$. Nevertheless, if the condition (12.3) does not hold, that is if $1 \leq s < B$, s could belong to the interval

$$B - ulp < s < B, \tag{12.5}$$

so that *rounding*(s) could be equal to B, then a new *normalization step* would be necessary, i.e. substitution of $s = B$ by $s = 1$ and e by $e + 1$.

Algorithm 12.1: Sum of positive numbers

```
if e1 ≥ e2 then e := e1; s := s1 + s2/B^(e1-e2);
else e := e2; s := s1/B^(e2-e1) + s2; end if;
if s ≥ B then e := e+1; s := s/B; end if; --1st normalization
s := round(s);
if s ≥ B then e := e+1; s := s/B; end if; --2nd normalization
```

Examples 12.3

Assume that $B = 2$ and $ulp = 2^{-5}$, so that the numbers are represented in the form $s \cdot 2^e$ where $1 \leq s \leq 1.11111_2$. For simplicity e is written in decimal (base 10).

1. Compute $z = (1.10101 \times 2^3) + (1.00010 \times 2^{-1})$.

   Alignment: $z = (1.10101 + 0.000100010) \times 2^3 = 1.101110010 \times 2^3$.

   Rounding: $s \cong 1.10111$.

   Final result: $z \cong 1.10111 \times 2^3$.

2. Compute $z = (1.11010 \times 2^3) + (1.00110 \times 2^2)$.

   Alignment:$z = (1.11010 + 0.100110) \times 2^3 = 10.011010 \times 2^3$.

   Normalization : $s = 1.0011010,\ e = 4$.

   Rounding: $s \cong 1.00110$.

   Final result: $z \cong 1.00110 \times 2^4$.

3. Compute $z = (1.10101 \times 2^3) + (1.10101 \times 2^1)$.

   Alignment: $z = (1.10010 + 0.0110101) \times 2^3 = 1.1111101 \times 2^3$.

   Rounding: $s \cong 10.00000$.

   Normalization: $s \cong 1.00000, e = 4$.

   Final result : $z \cong 1.00000 \times 2^4$.

Comments 12.2

1. The addition of two positive numbers could produce an *overflow* as the final value of e could be greater than e_{max}.
2. Observe in the previous examples the lack of precision due to the small number of bits (6 bits) used in the significand s.

12.2.2 Difference of Positive Numbers

Given two positive floating-point numbers $s_1 \cdot B^{e1}$ and $s_2 \cdot B^{e2}$ their difference $s \cdot B^e$ is computed as follows: assume that e_1 is greater than or equal to e_2; then (*for alignment*) the difference between $s_1 \cdot B^{e1}$ and $s_2 \cdot B^{e2}$ can be expressed in the form $s \cdot B^e$ where

$$s = s_1 - s_2/\left(B^{e1-e2}\right) \text{ and } e = e_1. \tag{12.6}$$

The value of s belongs to the interval $-s \cdot -(B - ulp) \leq s \leq B - ulp$. If s is negative, then it is substituted by and the sign of the final result will be modified accordingly. If s is equal to 0, then an exception *equal_zero* could be raised. It remains to consider the case where $0 < s \leq B - ulp$. The value of s could be smaller than 1. In order to normalize the significand, s is substituted by $s \cdot B^k$ and e by $e - k$, where k is the minimum exponent k such that $s \cdot B^k \geq 1$. Thus, the relation $1 \leq s \leq B$ holds. It remains to round (up or down) the significand and to normalize it if necessary.

In the following algorithm, the function *leading_zeroes(s)* computes the smallest k such that $s \cdot B^k \geq 1$.

Algorithm 12.2: Difference of positive numbers

```
if e1 ≥ e2 then e := e1; s := s1 - s2/B^(e1-e2);
else e := e2; s := s1/B^(e2-e1) - s2; end if;
if s < 0 then s := -s; sign := 1; end if;
k := leading_zeroes(s);
s := s·B^k; e := e-k; --1st norm.
s := round(s);
if s ≥ B then e := e+1; s := s/B; end if; --2nd norm.
```

Examples 12.4
Assume again that $B = 2$ and $ulp = 2^{-5}$, so that the numbers are represented in the form $s \cdot 2^e$ where $1 \leq s \leq 1.11111_2$. For computing the difference, the 2's complement representation is used (one extra bit is used).

1. Compute $z = \left(1.10101 \times 2^{-2}\right) - \left(1.01010 \times 2^1\right)$.

   Alignment: $z = (0.00110101 - 1.01010) \times 2^1$.

   2's complement addition: $(00.00110101 + 10.10101 + 00.00001) \times 2^1 =$

   10.11100101×2^1.

   Change of sign: $-s = 01.00011010 + 00.00000001 = 01.00011011$.

   Rounding: $-s \cong 1.00011$.

   Final result: $z \cong -1.00011 \times 2^1$.

2. Compute $z = \left(1.00010 \times 2^3\right) - \left(1.10110 \times 2^2\right)$.

   Alignment: $z = (1.00010 - -0.110110) \times 2^3$.

   2's complement addition: $(01.00010 + 11.001001 + 00.000001) \times 2^3 =$

   00.001110×2^3.

   Leading zeroes: $k = 3,\ s = 1.11000,\ e = 0$.

   Final result: $z = 1.11000 \times 2^0$.

3. Compute $z = \left(1.01010 \times 2^3\right) - \left(1.01001 \times 2^1\right)$.

   Alignment: $z = (1.01010 - -0.0101001) \times 2^3 = 0.1111111 \times 2^3$.

   Leading zeroes: $k = 1,\ s = 1.111111,\ e = 2$.

   Rounding: $s \cong 10.00000$.

   Normalization: $s \cong 1.0000,\ e = 3$.

   Final result: $z \cong 1.00000 \times 2^3$.

Comment 12.3
The difference of two positive numbers could produce an *underflow* as the final value of e could be smaller than e_{min}.

Table 12.3 Effective operation in floating point adder–subtractor

Operation	$Sign_1$	$Sign_2$	Actual operation
0	0	0	$s_1 + s_2$
0	0	1	$s_1 - s_2$
0	1	0	$-(s_1 - s_2)$
0	1	1	$-(s_1 - s_2)$
1	0	0	$s_1 - s_2$
1	0	1	$s_1 + s_2$
1	1	0	$-(s_1 + s_2)$
1	1	1	$-(s_1 - s_2)$

12.2.3 Addition and Subtraction

Given two floating-point numbers $(-1)^{sign1}\cdot s_1 \cdot B^{e1}$ and $(-1)^{sign2}\cdot s_2 \cdot B^{e2}$, and a control variable *operation*, an algorithm is defined for computing

$$z = (-1)^{sign}\cdot s \cdot B^e = (-1)^{sign1}\cdot s_1 \cdot B^{e1} + (-1)^{sign2}\cdot s_2 \cdot B^{e2}, \quad \text{if } operation = 0,$$
$$z = (-1)^{sign}\cdot s \cdot B^e = (-1)^{sign1}\cdot s_1 \cdot B^{e1} - (-1)^{sign2}\cdot s_2 \cdot B^{e2}, \quad \text{if } operation = 1.$$

Once the significands have been aligned, the actual operation (addition or subtraction of the significands) depends on the values of *operation*, $sign_1$ and $sign_2$ (Table 12.3). The following algorithm computes z. The procedure *swap (a, b)* interchanges *a* and *b*.

Algorithm 12.3: Addition and subtraction

```
if operation = 1 then sign2 := 1 - sign2; end if;
if e1 < e2 then
  swap(sign1, sign2); swap(s1, s2); swap (e1, e2);
end if;
e := e1; s2 := s2/B^(e1-e2); sign := sign1;
if sign1 xor sign2 = 0 then --addition
  s := s1 + s2;
  if s ≥ B then e := e+1; s := s/B; end if;
else --subtraction
  if (e1 = e2) and (s1 < s2) then
    swap(s1, s2); sign := 1 - sign;
  end if;
  s := s1 - s2; k := leading_zeroes(s);
  s := s·B^k; e := e - k;
end if;
s := round(s);
if s ≥ B then e := e+1; s := s/B; end if;
```

12.2.4 Multiplication

Given two floating-point numbers $(-1)^{sign1} \cdot s_1 \cdot B^{e1}$ and $(-1)^{sign2} \cdot s_2 \cdot B^{e2}$ their product $(-1)^{sign} \cdot s \cdot B^{e}$ is computed as follows:

$$sign = sign_1 \, xor \, sign_2, \quad s = s_1 \cdot s_2, \quad e = e_1 + e_2. \tag{12.7}$$

The value of s belongs to the interval $1 \leq s \leq (B - ulp)^2$, and could be greater than or equal to B. If it is the case, that is if $B \leq s \leq (B - ulp)^2$, then (*normalization*) substitute s by s/B, and e by $e + 1$. The new value of s satisfies

$$1 \leq s \leq (B - ulp)^2/B = B - 2.ulp + (ulp)^2/B < B - -ulp \tag{12.8}$$

($ulp < B$ so that $2 - ulp/B > 1$). It remains to round the significand and to normalize if necessary.

Algorithm 12.4: Multiplication

```
sign := sign1 xor sign2; s := s1·s2; e := e1 + e2;
if s ≥ B then e := e+1; s := s/B; end if; --1st normalization
s := round(s);
if s ≥ B then e := e+1; s := s/B; end if; --2nd normalization
```

Examples 12.5
Assume again that $B = 2$ and $ulp = 2^{-5}$, so that the numbers are represented in the form $s \cdot 2^e$ where $1 \leq s \leq 1.11111_2$. The exponent e is represented in decimal.

1. Compute $z = (1.11101 \times 2^{-2}) \times (1.00010 \times 2^{5})$.

   Multiplication : $z = 01.1100101100 \times 2^3$.

   Rounding : $s \cong 1.11001$.

   Final result : $z \cong 1.11001 \times 2^3$.

2. Compute $z = (1.11101 \times 2^{3}) \times (1.00011 \times 2^{-1})$.

   Multiplication : $z = 10.0001010111_2 \times 2^2$.

   Normalization : $s = 1.00001010111_2, \quad e = 3$.

   Rounding : $s \cong 1.00001$.

   Final result : $z \cong 1.00001 \times 2^3$.

3. Compute $z = (1.01000 \text{ x } 2^1) \times (1.10011 \times 2^2)$.

   Multiplication : $z = 01.111111000 \times 2^2$.

   Normalization : $s = 1.11111, \quad e = 3$.

   Rounding : $s \cong 10.00000$.

   Normalization : $s \cong 1, \quad e = 4$.

   Final result : $z \cong 1.00000_2 \times 2^4$.

Comment 12.4
The product of two real numbers could produce an *overflow* or an *underflow* as the final value of e could be greater than e_{max} or smaller than e_{min} (addition of two negative exponents).

12.2.5 Division

Given two floating-point numbers $(-1)^{sign1} \cdot s_1 \cdot B^{e1}$ and $(-1)^{sign2} \cdot s_2 \cdot B^{e2}$ their quotient
$(-1)^{sign} \cdot s \cdot B^e$ is computed as follows:

$$sign = sign_1 \ xor \ sign_2, \quad s = s_1/s_2, \quad e = e_1 - e_2. \tag{12.9}$$

The value of s belongs to the interval $1/B < s \leq B - ulp$, and could be smaller than 1. If that is the case, that is if $s = s_1/s_2 < 1$, then $s_1 < s_2, s_1 \leq s_2 - ulp$, $s_1/s_2 \leq 1 - ulp/s_2 < 1 - ulp/B$, and $1/B < s < 1 - ulp/B$.

Then (*normalization*) substitute s by $s \cdot B$, and e by $e - 1$. The new value of s satisfies $1 < s < B - ulp$. It remains to round the significand.

Algorithm 12.5: Division

```
sign := sign1 xor sign2; s := s1/s2; e := e1 - e2;
if s < 1 then e := e-1; s := s·B; end if;
s := round(s);
```

Examples 12.6
Assume again that $B = 2$ and $ulp = 2^{-5}$, so that the numbers are represented in the form $s \bullet 2^e$ where $1 \leq s \leq 1.11111_2$. The exponent e is represented in decimal.

1. Compute $z = (1.11101 \times 2^3)/(1.00011 \times 2^{-1})$.

   Division: $z = 1.1011111000 \times 2^4$.

   Rounding: $s \cong 1.10111$.

   Final result: $z \cong 1.00001 \times 2^3$.

2. Compute $z = (1.01000 \times 2^1)/(1.10011 \times 2^2)$.

   Division: $z = 0.1100100011 \times 2^{-1}$.

   Normalization: $s \cong 1.100100011, \quad e = -2$.

   Rounding: $s \cong 1.10010$.

   Final result: $z \cong 1.10010 \times 2^{-2}$.

Comment 12.5
The quotient of two real numbers could produce an *underflow* or an *overflow* as the final value of e could be smaller than e_{min} or bigger than e_{max}. Observe that a second normalization is not necessary as in the case of addition, subtraction and multiplication.

12.2.6 Square Root

Given a positive floating-point number $s_1 \cdot B^{e1}$, its square root $s \cdot B^e$ is computed as follows:

$$\text{if } e_1 \text{ is even,} \quad s = (s_1)^{1/2}, \quad e = e_1/2, \tag{12.10}$$

$$\text{if } e_1 \text{ is odd,} \quad s = (s_1/B)^{1/2}, \quad e = (e_1 + 1)/2. \tag{12.11}$$

In the first case (12.10), $1 \le s \le (B - ulp)^{1/2} < B - ulp$. In the second case $(1/B)^{1/2} \le s < 1$. Hence (*normalization*) s must be substituted by $s \cdot B$ and e by $e - 1$, so that $1 \le s < B$. It remains to round the significand and to normalize if necessary.

Algorithm 12.6: Square root

```
if (e1  mod 2) = 1 then s1 := s1/B; e1 := e1+1; end if;
s := square_root(s1); e := e1/2;
if s < 1 then e := e-1; s := s·B; end if; --1st norm.
s := round(s);
if s ≥ B then e := e+1; s := s/B; end if; --2nd norm.
```

An alternative is to replace (12.11) by:

$$\text{if } e_1 \text{ is odd,} \quad s = (s_1 \cdot B)^{1/2}, \quad e = (e_1 - 1)/2. \tag{12.12}$$

In this case $B^{1/2} \leq s \leq (B^2 - ulp \cdot B)^{1/2} < B$, then the first normalization is not necessary. Nevertheless, s could be $B - ulp < s < B$, and then depending on the rounding strategy, normalization after rounding could be still necessary.

Algorithm 12.7: Square root, second version

```
if (e1 mod 2) = 1 then s1 := s1 * B; e1 := e1 - 1; end if;
s := square_root(s1); e := e1/2;
s := round(s);
if s ≥ B then e := e+1; s := s/B; end if; --2nd normalization
```

Note that the "round to nearest" (default rounding in IEEE 754-2008) and the "truncation" rounding schemes allow avoiding the second normalization.

Examples 12.7
Assume again that $B = 2$ and $ulp = 2^{-5}$, so that the numbers are represented in the form $s \cdot 2^e$ where $1 \leq s \leq 1.11111_2$. The exponent e is represented in decimal form.

1. Compute $z = \left(1.11101 \times 2^4\right)^{1/2}$.

   Square rooting : $z = 1.01100001 \times 2^2$.

   Rounding : $s \cong 1.01100$.

   Final result : $z \cong 1.01100 \times 2^2$.

2. Compute $z = (1.00101 \times 2^{-1})^{1/2}$.

   Even exponent : $s = 10.0101, \quad e = -2$.

   Square rooting : $z = 1.10000101 \times 2^{-1}$.

   Rounding : $s \cong 1.10000$

   Final result : $z \cong 1.10000 \times 2^{-1}$.

3. Compute $z = \left(1.11111 \times 2^3\right)^{1/2}$.

   Even exponent; $s = 11.1111, \quad e = 2$.

   Square rooting : $z = 1.11111011 \times 2^1$.

   Rounding : $s \cong 1.11111$(round to nearest).

   Final result : $z \cong 1.11111_2 \times 2^1$.

However, some rounding schemes (e.g. toward infinite) generate $s \cong 10.00000$. Then, the result after normalization is $s \cong 1.00000$, $e = 2$, and the final result $z \cong 1.00000 \times 2^2$.

Comment 12.6

The square rooting of a real number could produce an *underflow* as the final value of e could be smaller than e_{min}.

12.3 Rounding Schemes

Given a real number x and a floating-point representation system, the following situations could happen:

$|x| < s_{min} \cdot B^{e\_min}$, that is, an *underflow* situation,
$|x| > s_{max} \cdot B^{e\_max}$, that is, an *overflow* situation,
$|x| = s \cdot B^e$, where $e_{min} \leq e \leq e_{max}$ and $s_{min} \leq s \leq s_{max}$.

In the third case, either s is a multiple of *ulp*, in which case a rounding operation is not necessary, or it is included between two multiples s' and s'' of *ulp*: $s' < s < s''$.

The *rounding* operation associates to s either s' or s'', according to some rounding strategy. The most common are the following ones:

- The *truncation* method (also called *round toward* 0 or *chopping*) is accomplished by dropping the extra digits, i.e. $round(s) = s'$ if s is positive, $round(-s) = s''$ if s is negative.
- The round toward plus infinity is defined by round(s) $= s''$.
- The round toward minus infinity is defined by round(s) $= s'$.
- The *round to nearest* method associates s with the closest value, that is, if $s < s' + ulp/2$, $round(s) = s'$, and if $s > s' + ulp/2$, $round(s) = s''$.

If the distances to s' and s'' are the same, that is, if $s = s' + ulp/2$, there are several options. For instance:

- $round(s) = s'$;
- $round(s) = s''$;
- $round(s) = s'$ if s is positive, $round(s) = s''$ if s is negative. It is the *round to nearest, ties to zero* scheme.
- $round(s) = s''$ if s is positive, $round(s) = s'$if s is negative. It is the *round to nearest, ties away from zero* scheme.
- $round(s) = s'$ if s' is an even multiple of *ulp*, $round(s) = s''$ if s'' is an even multiple of *ulp*. It is the default scheme in the IEEE 754 standard.
- $round(s) = s'$ if s' is an odd multiple of *ulp*, $round(s) = s''$ if s'' is an odd multiple of *ulp*.

The preceding schemes (*round to the nearest*) produce the smallest absolute error, and the two last (*tie to even, tie to odd*) also produce the smallest average absolute error (*unbiased* or 0-*bias* representation systems).

Assume now that the exact result of an operation, after normalization, is

$$s = 1.s_{-1}\,s_{-2}\,s_{-3}\ldots s_{-p}|s_{-(p+1)}\,s_{-(p+2)}\,s_{-(p+3)}\cdots$$

where *ulp* is equal to B^{-p} (the | symbol indicates the separation between the digit which corresponds to the *ulp* and the following). Whatever the chosen rounding scheme, it is not necessary to have previously computed all the digits $s_{-(p+1)}\,s_{-(p+2)}\ldots$; it is sufficient to know whether all the digits $s_{-(p+1)}\,s_{-(p+2)}\ldots$ are equal to 0, or not. For example the following algorithm computes *round*(*s*) if the *round to the nearest, tie to even* scheme is used.

Algorithm 12.8: Round to the nearest, tie to even

```
s1 := 1.s-1 s-2 ... s-p;
s2 := s - s1 - s-(p+1)·ulp/B;
--s2 = 0.00 ... 0|0 s-(p+2) s-(p+3) ...
if s-(p+1) < B/2 then round := s1;
elsif s-(p+1) > B/2 then round := s1 + ulp;
elsif s-(p+1) = B/2 and s2 > 0 then round := s1 + ulp;
elsif s-(p+1) = B/2 and s2 = 0 and (s-p mod 2) = 0 then
      round := s1;
elsif s-(p+1) = B/2 and s2 = 0 and (s-p mod 2) = 1 then
      round := s1 + ulp;
end if;
```

In order to execute the preceding algorithm it is sufficient to know the value of $s_1 = 1\cdot s_{-1}\,s_{-2}\,s_{-3}\ldots s_{-p}$, the value of $s_{-(p+1)}$, and whether $s_2 = 0.00\ldots 0|0\,s_{-(p+2)}s_{-(p+3)}\cdots$ is equal to 0, or not.

12.3.1 Rounding Schemes in IEEE 754

From the previous description, the IEEE 754-2008 standard defines five rounding algorithms. The two first round to a nearest value; the others are called *directed roundings*:

- *Round to nearest, ties to even*; this is the default rounding for binary floating-point and the recommended default for decimal.
- *Round to nearest, ties away from zero.*
- *Round toward* 0—directed rounding towards zero.
- *Round toward* $+\infty$—directed rounding towards positive infinity
- *Round toward* $-\infty$—directed rounding towards negative infinity.

12.4 Guard Digits

Consider the exact result r of an operation, before normalization. According to the preceding paragraph:

$$r<B^2,\ \text{i.e.}\ r = r_1\, r_0 \cdot r_{-1}\, r_{-2}\, r_{-3} \ldots r_{-p}|r_{-(p+1)}\, r_{-(p+2)}\, r_{-(p+3)} \ldots$$

The normalization operation (if necessary) is accomplished by

- dividing the result by B (sum of positive numbers, multiplication),
- multiplying the result by B (division),
- multiplying the result by B^k (difference of positive numbers).

Furthermore, if the operation is a difference of positive numbers (Algorithm 12.2), consider two cases:

- if $e_1 - e_2 \geq 2$, then $r = s_1 - s_2/(B^{e1-e2}) > 1 - B/B^2 = 1 - 1/B \geq 1/B(\text{as}\ B \geq 2)$, so that the number k of leading zeroes is equal to 0 or 1, and the normalization operation (if necessary i.e. $k = 1$) is accomplished by multiplying the result by B;
- if $e_1 - e_2 \leq 1$, then the result before normalization is either

$$r_0 \cdot r_{-1}\, r_{-2}\, r_{-3} \ldots r_{-p}|r_{-(p+1)}0\,0 \ldots (e_1 - e_2 = 1),\ \text{or}$$

$$r_0 \cdot r_{-1}\, r_{-2}\, r_{-3} \ldots r_{-p}|0\,0\,0 \ldots (e_1 - e_2 = 0).$$

A consequence of the preceding analysis is that the result after normalization can be either

$$r_0 \cdot r_{-1}\, r_{-2}\, r_{-3} \ldots r_{-p}|r_{-(p+1)}\, r_{-(p+2)}\, r_{-(p+3)} \ldots (\text{no normalization operation}), \tag{12.13}$$

or

$$r_1 \cdot r_0\, r_{-1}\, r_{-2} \ldots r_{-p+1}|r_{-p}\, r_{-(p+1)}\, r_{-(p+2)} \ldots (\text{divide by}\ B), \tag{12.14}$$

or

$$r_{-1} \cdot r_{-2}\, r_{-3}\, r_{-4} \ldots r_{-(p+1)}|r_{-(p+2)}\, r_{-(p+3)}\, r_{-(p+4)} \ldots (\text{multiply by}\ B), \tag{12.15}$$

or

$$r_{-k} \cdot r_{-(k+1)}\, r_{-(k+2)} \ldots r_{-p}\, r_{-(p+1)}0 \ldots 0|0\,0 \ldots (\text{multiply by}\ B^k \text{where}\ k > 1). \tag{12.16}$$

For executing a rounding operation, the worst case is (12.15). In particular, for executing Algorithm 12.8, it is necessary to know

- the value of $s_1 = r_{-1} \cdot r_{-2}\, r_{-3}\, r_{-4} \ldots r_{-(p+1)}$,
- the value of $r_{-(p+2)}$,
- whether $s_2 = 0.00\ldots0|0\, r_{-(p+3)}\, r_{-(p+4)} \ldots$ is equal to 0, or not.

The conclusion is that the result r of an operation, before normalization, must be computed in the form

$$r \cong r_1\, r_0 \cdot r_{-1}\, r_{-2}\, r_{-3} \ldots r_{-p} | r_{-(p+1)}\, r_{-(p+2)}\, T,$$

that is, with two *guard digits* $r_{-(p+1)}$ and $r_{-(p+2)}$, and an additional *sticky digit* T equal to 0 if all the other digits $(r_{-(p+3)}, r_{-(p+4)}, \ldots)$ are equal to 0, and equal to any positive value otherwise.

After normalization, the significand will be obtained in the following general form:

$$s \cong 1.s_{-1}\, s_{-2}\, s_{-3} \ldots s_{-p} | s_{-(p+1)}\, s_{-(p+2)}\, s_{-(p+3)}.$$

The new version of Algorithm 12.8 is the following:

Algorithm 12.9: Round to the nearest, tie to even, second version

```
s1 := 1.s-1 s-2 ... s-p;
if s-(p+1) < B/2 then round := s1;
elsif s-(p+1) > B/2 then round := s1 + ulp;
elsif s-(p+2) > 0 or s-(p+3) >0 then round := s1 + ulp;
elsif (s-p mod 2) = 0 then round := s1;
else round := s1 + ulp;
end if;
```

Observe that in binary representation, the following algorithm is even simpler.

Algorithm 12.10: Round to the nearest, tie to even, third version

```
s1 := 1.s-1 s-2 ... s-p;
if s-(p+1) = '0' then round := s1;
elsif s-p-1 = '1' and (s-p-2 = '1' or s-p-3 = '1') then
  round := s1 + ulp;
elsif s-p='0' then round := s1; --s-p-1:-p-3 = "100"
else round := s1 + ulp;
end if;
```

12.5 Arithmetic Circuits

This section proposes basic implementations of the arithmetic operations, namely addition and subtraction, multiplication, division and square root. The implementation is based on the previous section devoted to the algorithms, rounding and guard digit.

The circuits support normalized binary IEEE 754-2008 operands. Regarding the binary subnormals the associated hardware to manage this situation is complex. Some floating point implementations solve operations with subnormals via software routines. In the FPGA arena, most cores do not support denormalized numbers. The dynamic range can be increased using fewer resources by increasing the size of the exponent (a 1-bit increase in exponent, roughly doubles the dynamic range) and is typically the solution adopted.

12.5.1 Adder–Subtractor

An adder-subtractor based on Algorithm 12.3 will now be synthesized. The operands are supposed to be in IEEE 754 binary encoding. It is made up of five parts, namely unpacking, alignment, addition, normalization and rounding, and packing. The following implementation does not support subnormal numbers; they are interpreted as zero.

12.5.1.1 Unpacking

The *unpacking* separates the constitutive parts of the Floating Points and additionally detects the special numbers (infinite, zeros and NaNs). The special number detection is implemented using simple comparators. The following VHDL process defines the unpacking of a floating point operand *FP*; *k* is the number of bits of *FP*, *w* is the number of bits of the exponent, and *p* is the significand precision.

```
sign := FP(k-1); e := FP(k-2..k-w-1); s := '1' & FP(p-2..0);
exp_FF := '1' when e = ALL_ONES else '0';
exp_Z := '1' when e = ALL_ZEROS else '0';
sig_Z := '1' when FP(p-2..0) = ALL_ZEROS;
isNaN := exp_FF and (not sig_Z);
isInf := exp_FF and sig_Z;
isZero := exp_Z and sig_Z;
isSubn := exp_Z and (not sig_Z);
```

The previous is implemented using two *w* bits comparators, one *p* bits comparator and some additional gates for the rest of the conditions.

12.5.1.2 Alignment

The alignment circuit implements the three first lines of the Algorithm 12.3, i.e.

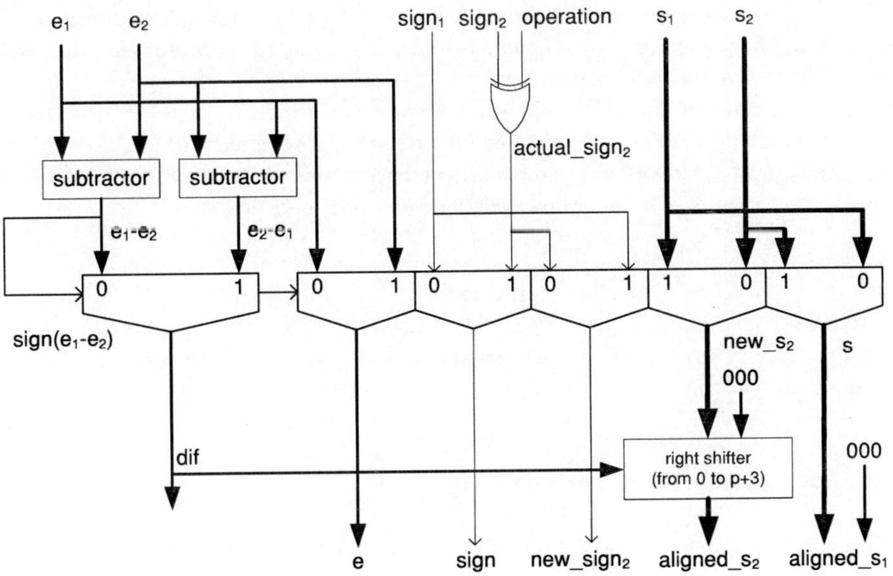

Fig. 12.2 Alignment circuit in floating point addition/subtraction

```
if operation = 1 then sign2 := 1 - sign2; end if;
if e1 < e2 then
   swap(sign1, sign2); swap(s1, s2); swap (e1, e2);
end if;
e := e1; s2 := (s2 / B**(e1-e2); sign := sign1;
```

An example of implementation is shown in Fig. 12.2. The principal and more complex component is the right shifter.

Given a $(p + 3)$-bit input vector $[1\, a_{k-1}\, a_{k-2} \ldots a_1\, a_0\, 0\, 0\, 0]$, the shifter generates a $(p + 3)$-bit output vector. The *lsb* (least significand bit) of the output is the sticky bit indicating if the shifted bits are equal to zero or not. If $B = 2$, the sticky-digit circuit is an OR circuit.

Observe that if $e_1 - e_2 \geq p + 3$, then the shifter output is equal to [0 0... 0 1], since the last bit is the sticky bit and the input number is a non-zero. The operation with zero is treated differently.

12.5.1.3 Addition and Subtraction

Depending on the respective signs of the aligned operands, one of the following operations must be executed: if they have the same sign, the sum *aligned_-*s_1 + *aligned_*s_2 must be computed; if they have different signs, the difference *aligned_s$_1$—aligned_s$_2$* is computed. If the result of the significand difference is negative, then *aligned_s$_2$—aligned_s$_1$* must be computed. Moreover, the only situation in which the final result could be negative is when the $e_1 = e_2$.

In the circuit of Fig. 12.3 two additions are performed in parallel: *result* = *aligned_s$_1$* ± *aligned_s$_2$*, where the actual operation is selected with the signs of the operands, and *alt_result* = $s_2 - s_1$. At the end of this stage a multiplexer selects the correct results. The operation selection is done as follows:

```
significand :=
  '0' & s1 & "000" when isZero2
  else '0' & s2 & "000" when isZero1
  else alt_result when alt_result > 0 and e1 = e2 and isSUB
  else result;
```

12.5.1.4 Normalization and Rounding

The normalization circuit executes the following part of Algorithm 12.3:

```
if sign1 xor sign2 = 0 then --addition
  s := s1 + s2;
  if s ≥ B then e := e+1; s := s/B; end if;
else --subtraction
  if (e1 = e2) and (s1 < s2) then
    swap(s1, s2); sign := 1 - sign;
  end if;
  s := s1 - s2; k := leading_zeroes(s);
  s := s·B^k; e := e - k;
end if;
```

If the number of leading zeroes is greater than $p + 3$, i.e. $s_1 - s_2 > B^{-(p+2)}$, then $s_2 > s_1 - B^{-(p+2)}$. If e_1 were greater than e_2 then $s_2 \leq (B - ulp)/B = 1 - B^{-(p+1)}$ so that $1 - B^{-(p+1)} \geq s_2 < s_1 - B^{-(p+2)} \geq 1 - B^{-(p+2)}$, which is impossible. Thus, the only case where the number of leading zeroes can be greater than $p + 3$ is when $e_1 = e_2$ and $s_1 = s_2$. If more than $p + 3$ leading 0's are detected in the circuit of Fig. 12.4, a zero_flag is raised.

It remains to execute the following algorithm where *operation* is the internal operation computed in Fig. 12.3:

```
if operation = 0 then
  if s ≥ B then e := e+1; s := s/B; end if;
else
  k := leading_zeroes(s); s := s·B^k; e := e - k;
end if;
```

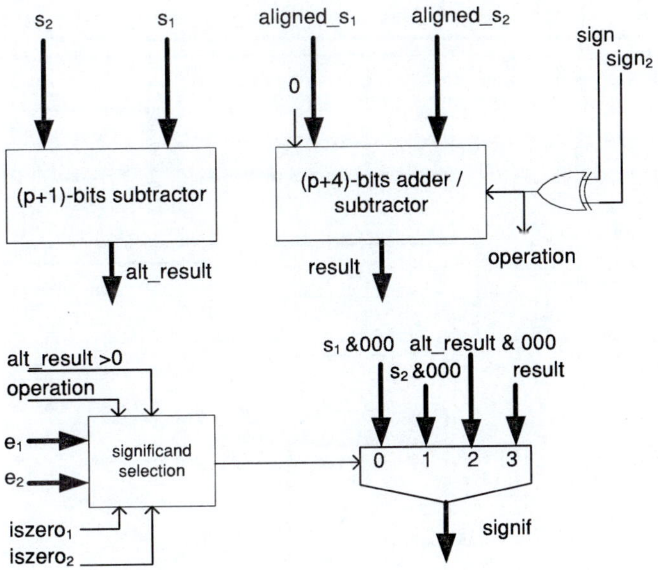

Fig. 12.3 Effective addition and subtraction

The rounding depends on the selected rounding strategy. An example of *rounding* circuit implementation is shown in Fig. 12.4. If the *round to the nearest, tie to even* method is used (Algorithm 12.10), the block named *rounding decision* computes the following Boolean function *decision*:

```
isRoundUp := '1' when ((s2 = '1' and (s1 = '1' or s0 =
'1')) or s3..0 = "1100") else '0';
```

The second rounding is avoided in the following way. The only possibility to need a second rounding is when, as the result of an addition, the significand is 1.111...1111xx. This situation is detected in a combinational block that generates the signal "isTwo" and adds one to the exponent. After rounding, the resulting number is 10.000...000, but the two most significand bits are discarded and the hidden 1 is appended.

12.5.1.5 Packing

The *packing* joins the constitutive parts of the floating point result. Additionally depending on special cases (infinite, zeros and NaNs), generates the corresponding codification.

Example 12.8 (Complete VHDL code available)
Generate the VHDL model of an IEEE decimal floating-point adder-subtractor. It is made up of five previously described blocks. Fig. 12.5 summarizes the interconnections. For clearness and reusability the code is written using parameters,

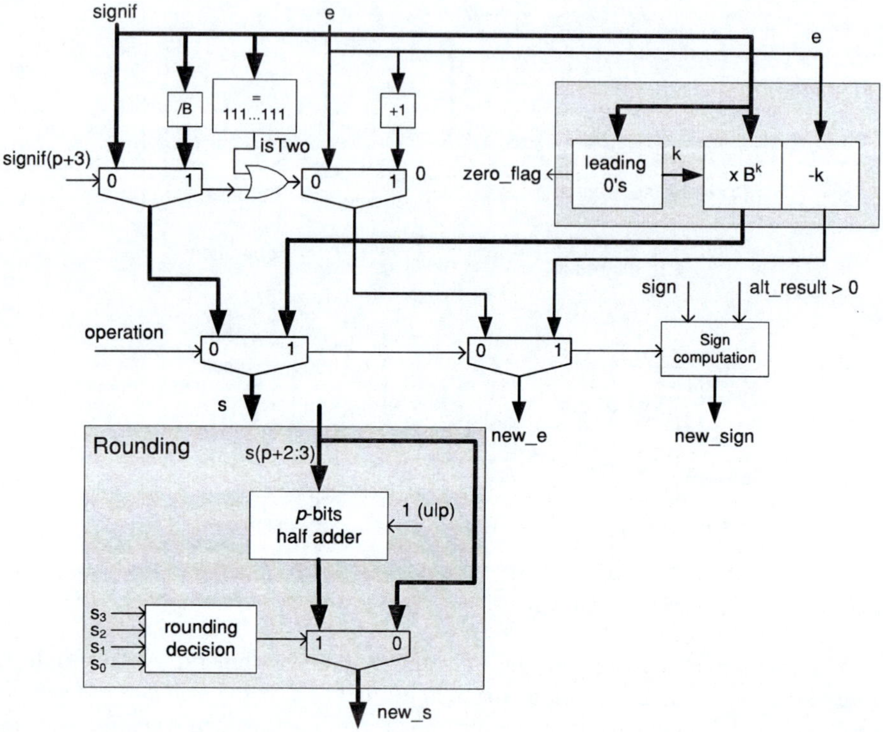

Fig. 12.4 Normalization and rounding

where *K* is the size of the floating point numbers (sign, exponent, significand), *E* is the size of the exponent and *P* is the size of the significand (including the hidden 1). The entity declaration of the circuit is:

```
entity FP_Add is
   generic(K:natural; P:natural; E:natural);
Port (
  FP_A : in  std_logic_vector (K-1 downto 0);
  FP_B : in  std_logic_vector (K-1 downto 0);
  add_sub: in std_logic;
  clk  : in  std_logic;
  ce   : in  std_logic;
  FP_Z : out  std_logic_vector (K-1 downto 0));
end FP_Add;
```

For simplicity the code is written as a single VHDL code except additional files to describe the right shifter of Fig. 12.2 and the leading zero detection and shifting of Fig. 12.4. The code is available at the book home page.

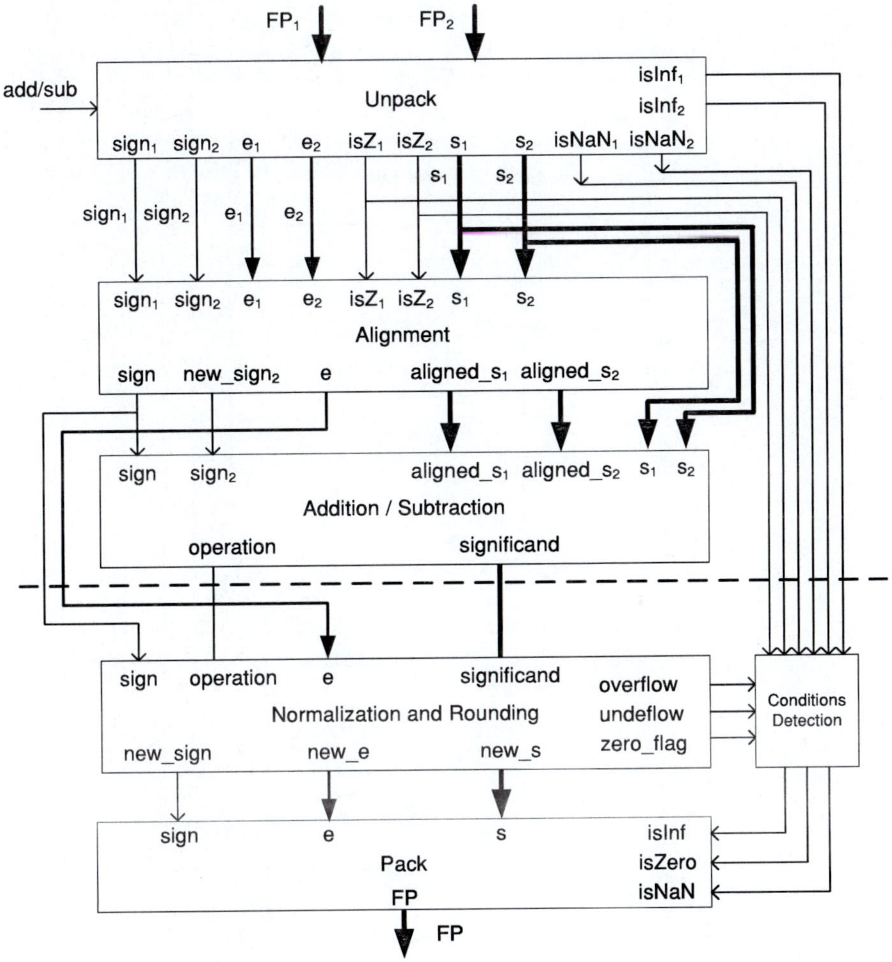

Fig. 12.5 General structure of a floating point adder-subtractor

A two stage pipeline could be achieved dividing the data path between addtion/ subtraction and normalization and rounding stages (dotted line in Fig. 12.5).

12.5.2 Multiplier

A basic multiplier deduced from Algorithm 12.4 is shown in Fig. 12.6. The unpacking and packing circuits are the same as in the case of the adder-subtractor (Fig. 12.5, Sects. 12.5.1.1 and 12.5.1.5) and for simplicity, are not drawn. The "normalization and rounding" is a simplified version of Fig. 12.4, where the part related to the subtraction is not necessary.

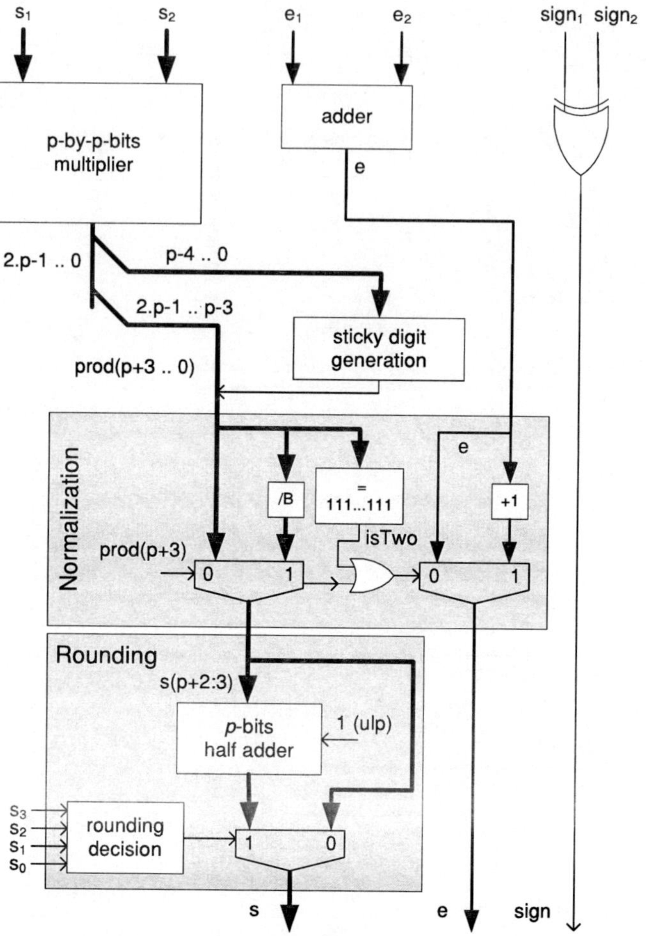

Fig. 12.6 General structure of a floating point multiplier

The obvious method of computing the sticky bit is with a large fan-in OR gate on the low order bits of the product. Observe, in this case, that the critical path includes the p by p bits multiplication and the sticky digit generation.

An alternative method consists of determining the number of trailing zeros in the two inputs of the multiplier. It is easy to demonstrate that the number of trailing zeros in the product is equal to the sum of the number of trailing zeros in each input operand. Notice that this method does not require the actual low order product bits, just the input operands, so the computation can occur in parallel with the actual multiply operation, removing the sticky computation from the critical path.

The drawback of this method is that significant extra hardware is required. This hardware includes two long length priority encoders to count the number of

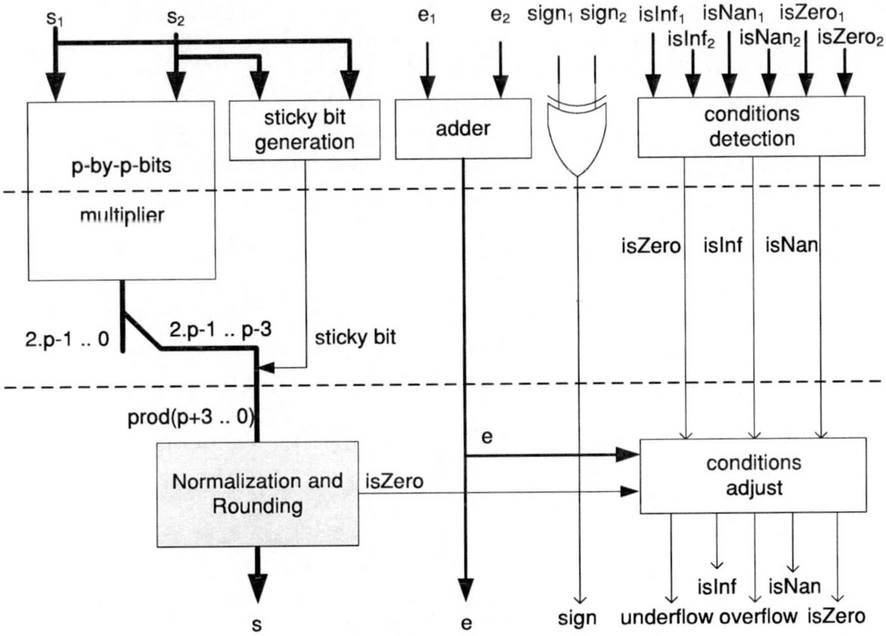

Fig. 12.7 A better general structure of a floating point multiplier

trailing zeros in the input operands, a small length adder, and a small length comparator. On the other hand, some hardware is eliminated, since the actual low order bits of the product are no longer needed.

A faster floating point multiplier architecture that computes the p by p multiplication and the sticky bit in parallel is presented in Fig. 12.7. The dotted lines suggest a three stage pipeline implementation using a two stage p by p multiplication. The two extra blocks are shown to indicate the special conditions detections. In the second block, the range of the exponent is controlled to detect *overflow* and *underflow* conditions. In this figure the packing and unpacking process are omitted for simplicity.

Example 12.9 (complete VHDL code available)
Generate the VHDL model of a generic floating-point multiplier. It is made up of the blocks depicted in Fig. 12.7 described in a single VHDL file. For clearness and reusability the code is written using parameters, where K is the size of the floating point numbers (sign, exponent, significand), E is the size of the exponent and P is the size of the significand (including the hidden 1). The entity declaration of the circuit is:

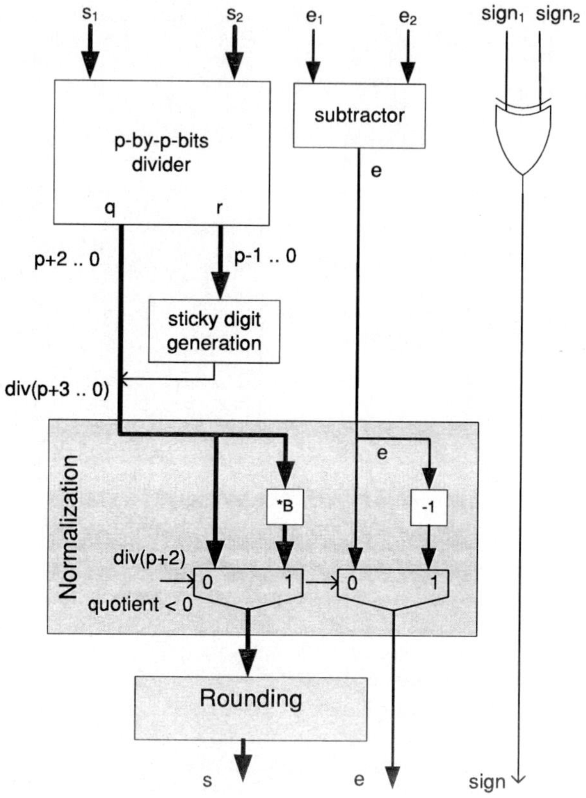

Fig. 12.8 A general structure of a floating point divider

```
entity FP_mul is
  generic(K: natural; P: natural; E: natural);
port (
  FP_A : in  std_logic_vector (K-1 downto 0);
  FP_B: in  std_logic_vector (K-1 downto 0);
  clk : in  std_logic;
  ce   : in  std_logic;
  FP_Z: out std_logic_vector (K-1 downto 0));
end FP_mul ;
```

The code is available at the home page of this book. The combinational circuit registers inputs and outputs to ease the synchronization. A two or three stage pipeline is easily achievable adding the intermediate registers as suggested in Fig. 12.7. In order to increase the clock frequency, more pipeline registers can be inserted into the integer multiplier.

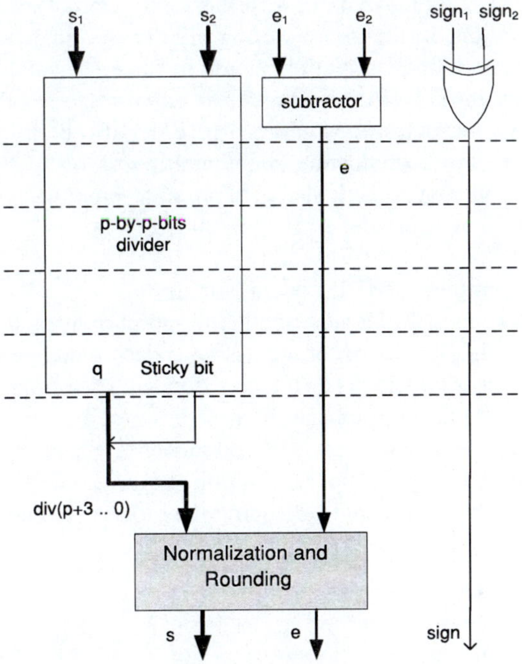

Fig. 12.9 A pipelined structure of a floating point divider

12.5.3 Divider

A basic divider, deduced from Algorithm 12.6, is shown in Fig. 12.9. The unpacking and packing circuits are similar to those of the adder-subtractor or multiplier. The "normalize and rounding" is a simplified version of Fig. 12.4, where the part related to the subtraction is not necessary.

The inputs of the p-bit divider are s_1 and s_2. The first operator s_1 is internally divided by two (s_1/B, i.e. right shifted) so that the dividend is smaller than the divisor. The precision is chosen equal to $p + 3$ digits. Thus, the outputs quotient and remainder satisfy the relation $(s_1/B).B^{p+3} = s_2.q + r$ where $r < s_2$, that is,

$$s_1/s_2 = q.B^{-(p+2)} + (r/s_2) \cdot B^{-(p+2)} \text{ where } (r/s_2) \cdot B^{-(p+2)} < B^{-(p+2)}.$$

The sticky digit is equal to 1 if $r < 0$ and to 0 if $r = 0$. The final approximation of the exact result is

$$quotient = q.B^{-(p+2)} + sticky\_digit.\, B^{-(p+3)}.$$

If a non-restoring divider is used, a further optimization could be done in the sticky bit computation. In the non-restoring divider the final remainder could be negative. In this case, a final correction should be done. This final operation can be avoided: the sticky bit is 0 if the remainder is equal to zero, otherwise it is 1.

A divider is a time consuming circuit. In order to obtain circuits, with frequencies similar to those of floating point multipliers or adders, more pipeline stages are necessary. Figure 12.9 shows a possible pipeline for a floating point divider where the integer divider is also pipelined.

Example 12.10 (complete VHDL code available)
Generate the VHDL model of a generic floating-point divider. It is made up of the blocks depicted in Fig. 12.9. Most of the design is described in a single VHDL file, but for the integer divider. The integer divider is a non-restoring divider (*div_nr_wsticky.vhd*) that uses a basic non-restoring cell (*a_s.vhd*). For clearness and reusability the code is written using parameters, where K is the size of the floating point numbers (sign, exponent, significand), E is the size of the exponent and P is the size of the significand (including the hidden 1). The entity declaration of the circuit is:

```
entity FP_div is
   generic(K: natural; P: natural; E: natural);
port ( FP_A : in  std_logic_vector (K-1 downto 0);
   FP_B: in  std_logic_vector (K-1 downto 0);
   clk : in  std_logic;
   ce  : in  std_logic;
   FP_Z: out std_logic_vector (K-1 downto 0));
end FP_div;
```

12.5.4 Square Root

A basic square rooter deduced from Algorithm 12.7 is shown in Fig. 12.10. The unpacking and packing circuits are the same as in previous operations and, for simplicity, are not drawn. Remember that the first normalization is not necessary, and for most rounding strategies the second normalization is not necessary either.

The exponent is calculated as follows:

$$E = e_1/2 + bias = (E_1 - bias)/2 + bias = E_1/2 + bias/2.$$

- If e_1 is even, then both E_1 and $bias$ are odd ($bias$ is always odd). Thus, $E = \lfloor E_1/2 \rfloor + \lfloor bias/2 \rfloor + 1$ where $\lfloor E_1/2 \rfloor$ and $\lfloor bias/2 \rfloor$ amount to right shifts.
- If e_1 is odd, then E_1 is even. The significand is multiplied by 2 and the exponent reduced by one unit. The biased exponent $E = (E_1 - 1)/2 + bias/2 = E_1/2 + \lfloor bias/2 \rfloor$.

To summarize, the biased exponent $E = \lfloor E_1/2 \rfloor + \lfloor bias/2 \rfloor + parity(E_1)$.

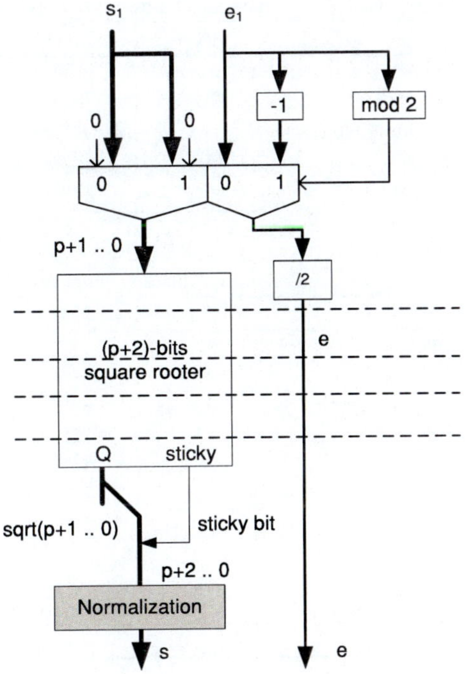

Fig. 12.10 Structure of a floating point squarer

Since the integer square root is a complex operation, a pipelined floating-point square rooter should be based on a pipelined integer square root. The dotted line of Fig. 12.10 shows a possible five stage pipeline circuit.

Example 12.11 (complete VHDL code available)
Generate the VHDL model of a generic floating-point square rooter. It is made up of the blocks depicted in Fig. 12.10. The design is described in a single VHDL file, but for the integer square root. The integer square root is based in a non-restoring algorithm (*sqrt_wsticky.vhd*) that uses two basic non-restoring cells (*sqrt_cell.vhd* and *sqrt_cell_00.vhd*). For clearness and reusability, the code is written using parameters, where *K* is the size of the floating point numbers (sign, exponent, significand), *E* is the size of the exponent and *P* is the size of the significand (including the hidden 1). The entity declaration of the circuit is:

```
entity FP_sqrt is
  generic(K: natural; P: natural; E: natural);
port ( FP_A : in  std_logic_vector (K-1 downto 0);
  clk : in  std_logic;
  ce  : in  std_logic;
  FP_Z: out std_logic_vector (K-1 downto 0));
end FP_sqrt;
```

Table 12.4 Combinational floating point operators in binary32 format

	FF	LUTs	DSP48	Slices	Delay
FP_add	96	699	–	275	11.7
FP_mult	96	189	2	105	8.4
FP_mult_luts	98	802	–	234	9.7
FP_div	119	789	–	262	46.6
FP_sqrt	64	409	–	123	38.0

Table 12.5 Combinational floating point operators in binary64 format

	FF	LUTs	DSP48	Slices	Delay
FP_add	192	1372	–	585	15.4
FP_mult	192	495	15	199	15.1
FP_mult_luts	192	3325	–	907	12.5
FP_div	244	3291	–	903	136.9
FP_sqrt	128	1651	–	447	97.6

Table 12.6 Pipelined floating point operators in binary32 format

	FF	LUTs	DSP48	Slices	Period	Latency
FP_add	137	637	–	247	6.4	2
FP_mult	138	145	2	76	5.6	2
FP_mult_luts	142	798	–	235	7.1	2
FP_mult	144	178	2	89	3.8	3
FP_mult_luts	252	831	–	272	5.0	3
FP_sqrt	384	815	–	266	9.1	6
FP_div	212	455	–0	141	9.2	5

12.5.5 Implementation Results

Several implementation results are now presented. The circuits were implemented in a Virtex 5, speed grade 2, device, using ISE 13.1 and XST for synthesis. Tables 12.4 and 12.5 show combinational circuit implementations for binary32 and binary64 floating point operators. The inputs and outputs are registered. When the number of registers (FF) is greater than the number of inputs and outputs, this is due to the register duplication made by synthesizer. The multiplier is implemented using the embedded multiplier (DSP48) and general purpose logic.

Table 12.6 shows the results for pipelined versions in the case of decimal32 data. The circuits include input and output registers. The adder is pipelined in two stages (Fig. 12.5). The multiplier is segmented using three pipeline stages (Fig. 12.7). The divider latency is equal to 6 cycles and the square root latency is equal to five cycles (Figs. 12.9 and 12.10).

12.6 Exercises

1. How many bits are there in the exponent and the significand of a 256-bit binary floating point number? What are the ranges of the exponent and the bias?
2. Convert the following decimal numbers to the binary32 and binary64 floating-point format. (a) 123.45; (b) -1.0; (c) 673.498e10; (d) qNAN; (e) -1.345e$-$129; (f) ∞; (g) 0.1; (h) 5.1e5
3. Convert the following binary32 number to the corresponding decimal number. (a) 08F05352; (b) 7FC00000; (c) AAD2CBC4; (d) FF800000; (e) 484B0173; (f) E9E55838; (g) E9E55838.
4. Add, subtract, multiply and divide the following binary floating point numbers with $B = 2$ and $ulp = 2^{-5}$, so that the numbers are represented in the form $s \cdot 2^e$ where $1 \leq s \leq 1.11111_2$. For simplicity e is written in decimal (base 10).

   (a) 1.10101×2^3 op 1.10101×2^1
   (b) 1.00010×2^{-1} op 1.00010×2^{-1}
   (c) 1.00010×2^{-3} op 1.10110×2^2
   (d) 1.10101×2^3 op 1.00000×2^4

5. Add, subtract, multiply and divide the following decimal floating point numbers using $B = 10$ and $ulp = 10^{-4}$, so that the numbers are represented in the form $s \cdot 10^e$ where $1 \leq s \leq 9.9999$ (normalized decimal numbers).

   (a) 9.4375×10^3 op 8.6247×10^2
   (b) 1.0014×10^3 op 9.9491×10^2
   (c) 1.0714×10^4 op 7.1403×10^2
   (d) 3.4518×10^{-1} op 7.2471×10^3

6. Analyze the consequences and implication to support denormalized (subnormal in IEEE 754-2008) numbers in the basic operations.
7. Analyze the hardware implication to deal with no-normalized significands (s) instead of normalized as in the binary standard.
8. Generate VHDL models adding a pipeline stage to the binary floating point adder of Sect. 12.5.1.
9. Add a pipeline stage to the binary floating point multiplier of Sect. 12.5.2.
10. Generate VHDL models adding two pipeline stages to the binary floating point multiplier of Sect. 12.5.2.
11. Generate VHDL models adding several pipeline stages to the binary floating point divider of Sect. 12.5.3.
12. Generate VHDL models adding several pipeline stages to the binary floating point square root of Sect. 12.5.4.

Reference

1. IEEE (2008) IEEE standard for floating-point arithmetic, 29 Aug 2008

Chapter 13
Finite-Field Arithmetic

Finite fields are used in different types of computers and digital communication systems. Two well-known examples are error-correction codes and cryptography. The traditional way of implementing the corresponding algorithms is software, running on general-purpose processors or on digital-signal processors. Nevertheless, in some cases the time constraints cannot be met with instruction-set processors, and specific hardware must be considered.

The operations over the finite ring Z_m are described in Sect. 13.1. Two multiplication algorithms are considered: "multiply and reduce" (Sect. 13.1.2.1) and "interleaved multiplication" (Sect. 13.1.2.2). The Montgomery multiplication, and its application to exponentiation algorithms, are the topics of Sect. 13.1.2.3. Section 13.2 is dedicated to the division over Z_p, where p is a prime. The proposed method is the "binary algorithm", an extension of an algorithm that computes the greatest common divider of two naturals. The operations over the polynomial ring $Z_2[x]/f(x)$ are described in Sect. 13.3. Two multiplication algorithms are considered: "multiply and reduce" (Sect. 13.3.2.1) and "interleaved multiplication" (Sect. 13.3.2.2). Squaring is the topic of Sect. 13.3.2.3. Finally, Sect. 13.4 is dedicated to the division over $GF(2^n)$.

As a matter of fact, only some of the most important algorithms have been considered. According to the Authors' experience they generate efficient FPGA implementations (Sect. 13.5). Furthermore, non-binary extension fields $GF(p^n)$ are not considered. A much more complete presentation of finite field arithmetic can be found in [1] and [2].

13.1 Operations Modulo *m*

Given a natural m, the set $Z_m = \{0, 1, \ldots, m-1\}$ is a ring whose operations are defined as modulo m.

J.-P. Deschamps et al., *Guide to FPGA Implementation of Arithmetic Functions*, Lecture Notes in Electrical Engineering 149, DOI: 10.1007/978-94-007-2987-2_13,

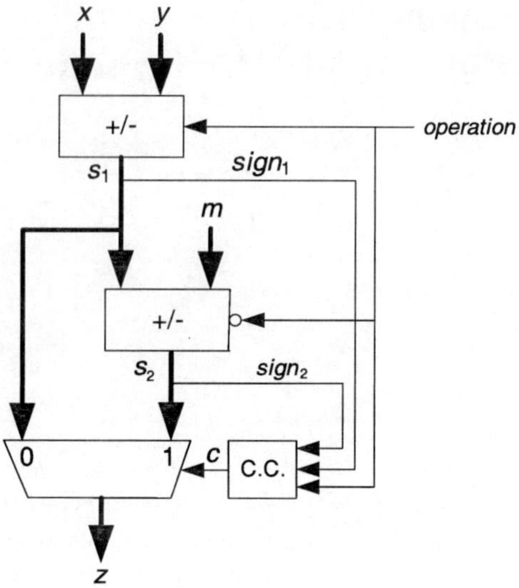

Fig. 13.1 Adder–subtractor modulo *m*

13.1.1 Addition and Subtraction Mod m

Given two natural numbers x and y belonging to Z_m, compute $z = (x + y)$ mod m. Taking into account that $0 \leq x + y < 2 \cdot m$, z must be equal to either $x + y$ or $x + y - m$, the following algorithm computes z.

Algorithm 13.1: Mod *m* addition

```
z1 := x + y; z2 := z1 - m;
if z2 >= 0 then z := z2; else z := z1; end if;
```

As regards the computation of $z = (x - y)$ mod m, take into account that $-m < x - y < m$, so that z must be equal to either $x - y$ or $x - y + m$. The corresponding algorithm is the following.

Algorithm 13.2: Mod *m* subtraction

```
z1 := x - y; z2 := z1 + m;
if z1 < 0 then z := z2; else z := z1; end if;
```

The circuit of Fig. 13.1 is an adder-subtractor, which computes $z = (x + y)$ mod m if *operation* $= 0$ and $z = (x - y)$ mod m if *operation* $= 1$. It is described by the following VHDL model in which k is the number of bits of m.

```
WITH operation SELECT s1 <=
  ('0'&x) + y WHEN '0', ('0'&x) - y WHEN OTHERS;
WITH operation SELECT s2 <=
  s1 + m WHEN '1', s1 - m WHEN OTHERS;
c <=
  (NOT(operation) AND s2(k)) OR (operation AND NOT(s1(k)));
WITH c SELECT z <=
  s1(k-1 DOWNTO 0) WHEN '1', s2(k-1 DOWNTO 0) WHEN OTHERS;
```

A complete VHDL model is *mod_m_AS.vhd* is available at the Authors' web page.

13.1.2 Multiplication Mod m

Given x and $y \in Z_m$, compute $z = x \cdot y \bmod m$, where m is a k-bit natural.

13.1.2.1 Multiply and Reduce

A straightforward method consists of multiplying x by y, so that a $2k$-bit result *product* is obtained, and then reducing *product* mod m. For that, any combination of multiplier and mod m reducer can be used. For fixed values of m, specific combinational mod m reducers can be considered.

As an example, synthesize a mod m multiplier with $m = 2^{192} - 2^{64} - 1$. Any 192-by-192 multiplier can be used. A 384-bit to 192-bit mod m reducer can be synthesized as follows: given $x = x_{383} \cdot 2^{383} + x_{382} \cdot 2^{382} + \ldots + x_0 \cdot 2^0$, it can be divided up under the form

$$(x_{383} \cdot 2^{63} + x_{382} \cdot 2^{62} + \ldots + x_{320} \cdot 2^0)2^{320} + (x_{319} \cdot 2^{63} + x_{318} \cdot 2^{62} + \ldots + x_{256} \cdot 2^0)2^{256}$$
$$+ (x_{255} \cdot 2^{63} + x_{254} \cdot 2^{62} + \ldots + x_{192} \cdot 2^0)2^{192} + x_{191} \cdot 2^{191} + x_{190} \cdot 2^{190} + \ldots + x_0 \cdot 2^0.$$

Then, substitute 2^{320} by $2^{128} + 2^{64} + 1 \equiv 2^{320} \bmod m$, 2^{256} by $2^{128} + 2^{64} \equiv 2^{256} \bmod m$, and 2^{192} by $2^{64} + 1 \equiv 2^{192} \bmod m$. So, $x \equiv x' + x'' + x''' + x''''$, where

$$\begin{aligned}
x' &= (x_{383} \cdot 2^{191} + x_{382} \cdot 2^{190} + \ldots + x_{320} \cdot 2^{128}) \\
&\quad + (x_{383} \cdot 2^{127} + x_{382} \cdot 2^{126} + \ldots + x_{320} \cdot 2^{64}) \\
&\quad + (x_{383} \cdot 2^{63} + x_{382} \cdot 2^{62} + \ldots + x_{320} \cdot 2^{0}), \\
x'' &= (x_{319} \cdot 2^{191} + x_{318} \cdot 2^{190} + \ldots + x_{256} \cdot 2^{128}) \\
&\quad + (x_{319} \cdot 2^{127} + x_{318} \cdot 2^{126} + \ldots + x_{256} \cdot 2^{64}), \\
x''' &= (x_{255} \cdot 2^{127} + x_{254} \cdot 2^{126} + \ldots + x_{192} \cdot 2^{64}) \\
&\quad + (x_{255} \cdot 2^{63} + x_{254} \cdot 2^{62} + \ldots + x_{192} \cdot 2^{0}), \\
x'''' &= x_{191} \cdot 2^{191} + x_{190} \cdot 2^{190} + \ldots + x_0 \cdot 2^0.
\end{aligned}$$

The sum $s = x' + x'' + x''' + x''''$ is smaller than $4 \times (2^{192} - 2^{64} - 1) = 4m$, so that x mod m is either s, $s - m$, $s - 2m$ or $s - 3m$.

The corresponding circuit is shown in Fig. 13.2 and is described by the following VHDL model.

```
s1 <= ('0' & x(383 DOWNTO 320) & x(383 DOWNTO 320) &
  x(383 DOWNTO 320)) + x(191 DOWNTO 0);
s2(192 DOWNTO 64)  <= ('0' & x(319 DOWNTO 256) &
  x(319 DOWNTO 256)) + x(255 DOWNTO 192);
s2(63 DOWNTO 0) <= x(255 DOWNTO 192);
s <= ('0'&s1) + s2;
z1 <= ('0'&s) + ("11"&minus_m);
z2 <= s + minus_2m;
z3 <= s(192 DOWNTO 0) + minus_3m;
PROCESS(z1, z2, z3, s)
BEGIN
  IF z1(194) = '1' THEN z <= s(191 DOWNTO 0);
  ELSIF z2(193) = '1' THEN z <= z1(191 DOWNTO 0);
  ELSIF z3(192) = '1' THEN z <= z2(191 DOWNTO 0);
  ELSE z <= z3(191 DOWNTO 0);
  END IF;
END PROCESS;
```

A complete VHDL model is *mod_p192_reducer2.vhd* is available at the Authors' web page.

In order to complete the multiplier design, any 192-bit by 192-bit multiplier can be used (Chap. 8). This is left as an exercise.

13.1.2.2 Interleaved Multiplier

Another option is to modify a classical left-to-right multiplication algorithm based on the following computation scheme

$$x \cdot y = (\ldots((0 \cdot 2 + x_{n-1} \cdot y) \cdot 2 + x_{n-2} \cdot y) \cdot 2 + \ldots + x_1 \cdot y) \cdot 2 + x_0 \cdot y.$$

Algorithm 13.3: Mod *m* multiplication, left-to-right algorithm

```
accumulator := 0;
for i in 0 .. k-1 loop
  accumulator := (accumulator·2 + x_{k-i-1}·y) mod m;
end loop;
product := accumulator;
```

The data path corresponding to Algorithm 13.3 is shown in Fig. 13.3. It is described by the following VHDL model.

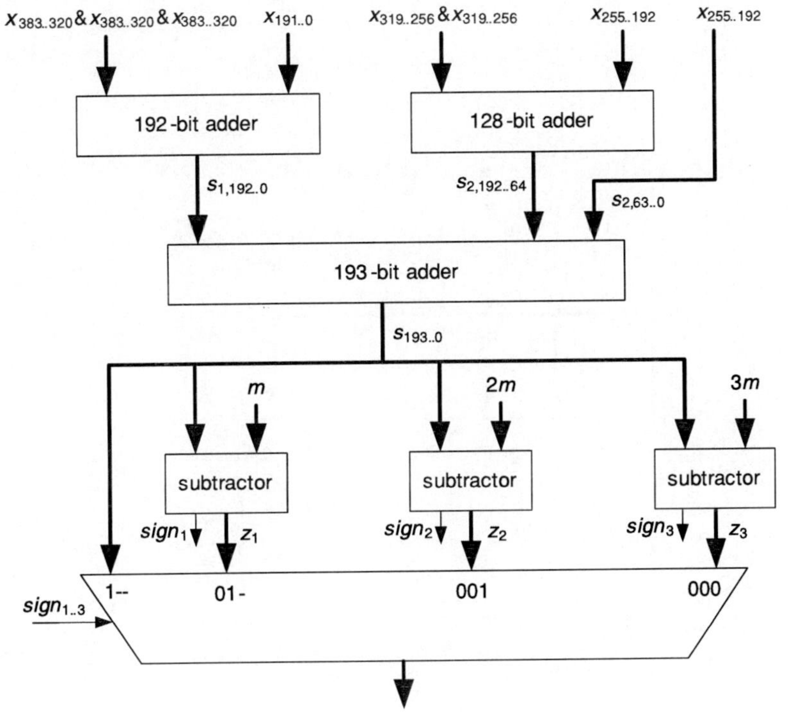

Fig. 13.2 Mod $2^{192} - 2^{64} - 1$ reducer

```
vector_xi <= (OTHERS => xi); xi_by_y <= y AND vector_xi;
s <= '0'&acc&'0' + xi_by_y;
z1 <= ('0'&s) + ("11"&minus_m);
z2 <= ('0'&s) + ('1'&minus_2m);
PROCESS(z1, z2, s)
BEGIN
  IF z1(k+2) = '1' THEN next_acc <= s(k-1 DOWNTO 0);
  ELSIF z2(k+2) = '1' THEN next_acc <= z1(k-1 DOWNTO 0);
  ELSE next_acc <= z2(k-1 DOWNTO 0);
  END IF;
END PROCESS;
register_acc: PROCESS(clk) ...
shift_register: PROCESS(clk) ...
xi <= int_x(k-1);
```

A complete VHDL model *mod-_m_multiplier.vhd-*, including a *k*-state counter and a control unit, is available at the Authors' web page.

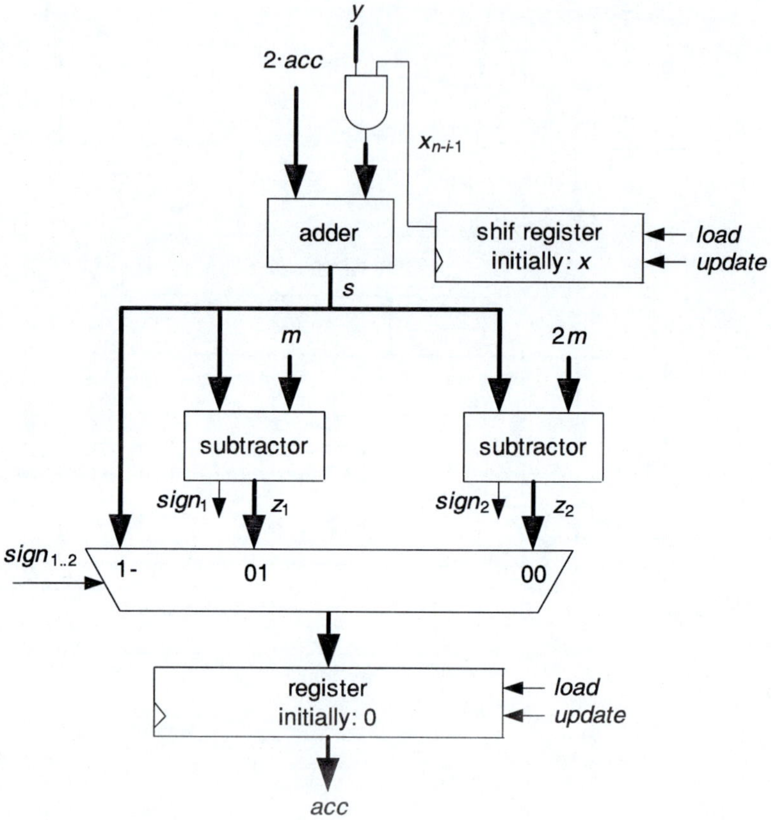

Fig. 13.3 Interleaved mod *m* multiplier

If ripple-carry adders are used, the minimum clock period is about $k \cdot T_{FA}$, so that the total computation time is approximately equal to $k^2 \cdot T_{FA}$. In order to reduce the computation time, the stored-carry encoding principle could be used [2]. For that, Algorithm 13.3 is modified: *accumulator* is represented under the form $acc_s + acc_c$; the conditional sum $(acc_s + acc_c) \cdot 2 + x_{n-k-i} \cdot y$ is computed in stored-carry form, and every sum is followed by a division by m, also in stored-carry form (Sect. 9.2.4), without on-the-fly conversion as only the remainder must be computed. The corresponding computation time is proportional to k instead of k^2. The design of the circuit is left as an exercise.

13.1.2.3 Montgomery Multiplication

Assume that m is an odd k-bit number. As m is odd, then $gcd(m, 2^k) = 1$, and there exists an element 2^{-k} of Z_m such that $2^k \cdot 2^{-k-} \equiv 1 \bmod m$. Define a one-to-one and onto application T from Z_m to Z_m:

$$T(x) = x.2^k \bmod m \text{ and } T^{-1}(y) = y.2^{-k} \bmod m.$$

The following properties hold true: $T((x + y) \bmod m) = (T(x) + T(y)) \bmod m$, $T((x - y) \bmod m) = (T(x) - T(y)) \bmod m$, $T(x{\cdot}y \bmod m) = T(x){\cdot}T(y).2^{-k} \bmod m$. The latter suggests the definition of a new operation on Z_m, the so-called *Montgomery product MP* [3]:

$$MP(x, y) = x \cdot y \cdot 2^{-k} \bmod m.$$

Assume that the value $2^{2k} \bmod m$ has been previously computed. Then

$$T(x) = MP\left(x, 2^{2k} \bmod m\right) \text{ and } T^{-1}(y) = MP(y, 1).$$

The main point is that the Montgomery product MP is easier to compute than the mod m product. The following algorithm computes $MP(x, y)$.

Algorithm 13.4: Montgomery product

```
p := 0;
for i in 0 .. k-1 loop
  qi := (p0 + xi·y0) mod 2;
  p := (p + xi·y + qi·m)/2;
end loop;
if p ≥ m then z := p-m; else z := p; end if;
```

The data path corresponding to Algorithm 13.4 (without the final correction) is shown in Fig. 13.4. It is described by the following VHDL model.

```
q <= p(0) XOR (xi AND y(0));
vector_xi <= (OTHERS => xi); vector_q <= (OTHERS => q);
two_p <= ('0'&p) + (vector_xi AND y) +  (vector_q AND m);
next_p <= two_p(k+1 DOWNTO 1);
p_minus_m <= p + minus_m;
register_p: PROCESS(clk) ...
shift_register: PROCESS(clk) ...
xi <= int_x(0);
```

If ripple-carry adders are used, the total computation time is approximately equal to $k^2{\cdot}T_{FA}$. In order to reduce the computation time, the stored-carry encoding principle could be used [2, 4].

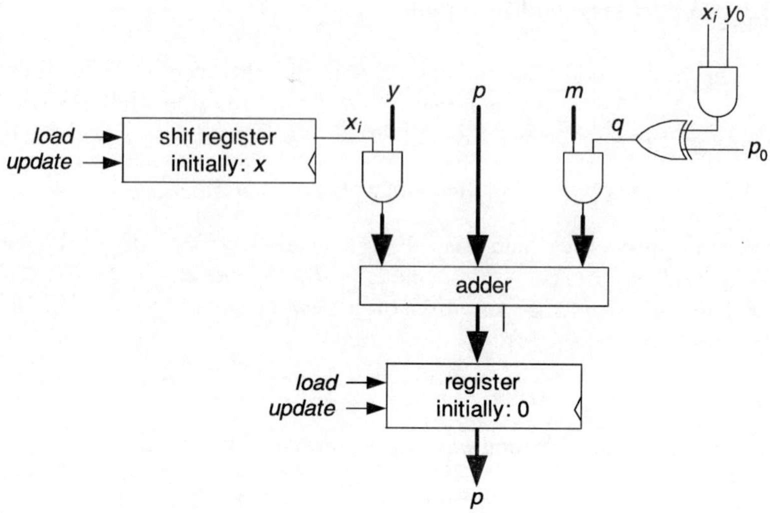

Fig. 13.4 Montgomery product

Algorithm 13.5: Montgomery product, carry-save addition

```
pc := 0; ps := 0;
for i in 0 .. k-1 loop
  qi := (pc0 + ps0 + xi·y0) mod 2;
  (pc, ps) := (pc + ps + xi·y + qi·m)/2;
end loop;
if p ≥ m then z := p-m; else z := p; end if;
```

The corresponding computation time is proportional to k instead of k^2. The design of the circuit is left as an exercise.

In order to compute $z = x{\cdot}y \bmod m$, the Montgomery product concept should be used in the following way: first (initial encoding) substitute x and y by $x' = T(x) = MP(x, 2^{2k} \bmod m)$ and $y' = T(y) = MP(y, 2^{2k} \bmod m)$; then compute $z' = MP(x', y')$; finally (result decoding) compute $z = T^{-1}(z') = MP(z, 1)$. This method is not efficient, unless many operations involving the same initial data are performed, in which case the initial encoding of those data is performed only once. Consider the following modular exponentiation algorithm; it computes $z = y^x \bmod m$ and is based on the following computation scheme:

$$z = y^{x_0 + x_1 \cdot 2 + x_2 \cdot 2^2 + \ldots + x_{k-1} \cdot 2^{k-1}} = y^{x_0} \cdot (y^2)^{x_1} \cdot (y^{2^2})^{x_1} \cdot \ldots \cdot (y^{2^{k-1}})^{x_{k-1}} \bmod m.$$

Algorithm 13.6: Mod *m* exponentiation, LSB-first

```
e := 1;
for i in 0 .. k-1 loop
  if xi = 1 then e := e·y mod m; end if;
  y := y² mod m ;
end loop;
z := e;
```

Mod m operations can be substituted by Montgomery products. For that, 1 is substituted by $T(1) = 2^k$ mod m, and y by $T(y) = MP(y, 2^{2k} \bmod m)$. Thus, assuming that 2^k mod m and 2^{2k} mod m have been previously computed, the following algorithm computes $z = x{\cdot}y$ mod m.

Algorithm 13.7: Modulo *m* exponentiation, Montgomery algorithm, LSB-first

```
te := 2^k mod m; ty := MP(y, 2^2k mod m);
for i in 0 .. k-1 loop
  if xi = 1 then te := MP(te, ty); end if;
  ty := MP(ty, ty);
end loop;
z := MP(te, 1);
```

The corresponding circuit is made up of two *Montgomery multipliers* working in parallel, with some kind of synchronization mechanism. A data-flow VHDL description *mod_m_exponentiation.vhd* is available at the Authors' web page. At each step both multipliers MP_1 and MP_2, with their corresponding $done_1$ and $done_2$ signals, are synchronized with a *wait* instruction:

```
WAIT UNTIL (done1 AND done2) = '1';
```

An MSB-first exponentiation algorithm could also be considered. For that, use the following computation scheme:

$$\begin{aligned} z &= y^{x_0+x_1\cdot 2+x_2\cdot 2^2+\ldots+x_{k-1}\cdot 2^{k-1}} \\ &= ((\ldots(1^2\cdot y^{x_{k-1}})^2\cdot y^{x_{k-2}})^2\cdot\ \ldots\ \cdot y^{x_1})^2\cdot y^{x_0} \bmod m. \end{aligned}$$

Algorithm 13.8: Mod m exponentiation, MSB-first

```
e := 1;
for i in 1 .. k loop
  e := e² mod m;
  if x_{k-i} = 1 then e := e·y mod m; end if;
end loop;
z := e;
```

As before, mod m multiplications can be substituted by Montgomery products. For that, 1 is substituted by $T(1) = 2^k \bmod m$, and y by $T(y) = MP(y, 2^{2k} \bmod m)$. There is a precedence relation between the two main operations $T(e) := T(e^2) = MP(T(e), T(e))$ and $T(e) := T(e \cdot y) = MP(T(e), T(y))$. Thus, a direct implementation includes only one *Montgomery multiplier*, but needs up to $2k$ cycles instead of k in the case of the LSB-first algorithm. The design of the corresponding circuit is left as an exercise.

13.2 Division Modulo p

If p is prime, than all non-zero elements of Z_p have a multiplicative inverse. Thus, given x and $y \neq 0$ in Z_p, there exists an element z of Z_p such that $z = x \cdot y^{-1} \bmod p$.

There are several types of algorithms that compute z. Some of them are generalizations of algorithms that compute the *greatest common divider*: Euclidean algorithm [5, 6], binary algorithm [7], plus-minus algorithm [8–10]. Another option is to substitute division by multiplications: according to the Fermat's little theorem $z = x \cdot y^{p-2} \bmod p$. As an example, the following binary algorithm computes $z = x \cdot y^{-1} \bmod p$. It uses four variables a, b, c and d, initially equal to p, y, 0 and x, respectively. At each step, a and b are updated in such a way that their gcd is unchanged and that b decreases. For that, observe that if b is even and a is odd, then $gcd(a, b) = gcd(a, b/2)$, and if both a and b are odd, then $gcd(a, b) = gcd(a, |b\text{-}a|) = gcd(b, |b\text{-}a|)$. As initially $a = p$ and $b = y$, where p is a prime, after a finite number of steps a is equal to 1. On the other hand, c and d are updated in such a way that $c \cdot y \equiv a \cdot x \bmod p$ and $d \cdot y \equiv b \cdot x \bmod p$. Initially, $c = 0$, $a = p \equiv 0 \bmod p$, $d = x$ and $b = y$, so that the mentioned relations are satisfied. It can be proven that if c and d are updated in the same way as a and b, both relations remain true. In particular, if $a = 1$, then $c \cdot y \equiv x \bmod p$, and $z = c$.

Algorithm 13.9: Mod p division, binary algorithm

```
a := p; b := y; c := 0; d := x;
while a ≠ 1 loop
  if b0 = 0 then b := b/2; d := d·2^-1 mod p;
  elsif b ≥ a then b := b-a; d := (d-c) mod p;
  else (b, a) := (a-b, b); (d, c) := ((c-d) mod p, d);
  end if;
end loop;
z := c;
```

The corresponding circuit is made up of adders, registers and connection resources. A data-flow VHDL description *mod_p_division2.vhd* is available at the Authors' web page.

An upper bound of the number of steps before $a = 1$ is $4k$, k being the number of bits of p. So, if ripple-carry adders are used, the computation time is shorter than $4 \cdot k^2 \cdot T_{FA}$ (a rather pessimistic estimation).

13.3 Operations Over $Z_2[x]/f(x)$

Given a polynomial $f(x) = x^m + f_{m-1}x^{m-1} + \ldots + f_1x + f_0$, whose coefficients f_i belong to the binary field Z_2, the set of polynomials of degree smaller than m over Z_2 is a ring $Z_2[x]/f(x)$ whose operations are defined modulo $f(x)$.

13.3.1 Addition and Subtraction of Polynomials

Given two polynomials $a(x) = a_{m-1}x^{m-1} + \ldots + a_1x + a_0$ and $b(x) = b_{m-1}x^{m-1} + \ldots + b_1x + b_0$, then

$$a(x) + b(x) = a(x) - b(x) = c_{m-1}x^{m-1} + \ldots + c_1x + c_0,$$

where $c_i = (a_i + b_i) \bmod 2, \forall i \text{ in} \{0, 1, \ldots, m-1\}$. In other words, the corresponding circuit is a set of m 2-input XOR gates working in parallel, which can be described by the following VHDL sentence:

```
c <= a XOR b;
```

13.3.2 Multiplication Modulo f(x)

Given two polynomials $a(x) = a_{m-1}x^{m-1} + \ldots + a_1x + a_0$ and $b(x) = b_{m-1}x^{m-1} + \ldots + b_1x + b_0$ of degree smaller than m, and a polynomial $f(x) = x^m + f_{m-1}x^{m-1} + \ldots + f_1x + f_0$, compute $c(x) = a(x) \cdot b(x) \bmod f(x)$.

13.3.2.1 Multiply and Reduce

A straightforward method consists of multiplying $a(x)$ by $b(x)$, so that a polynomial $d(x)$ of degree smaller than $2m$-1 is obtained, and then reducing $d(x)$ mod $f(x)$.

The coefficients d_k of $d(x)$ are the following:

$$d_k = \sum\nolimits_{i=0}^{k} a_i \cdot b_{k-i},\ k = 0, 1, \ldots, m-1,$$
$$d_k = \sum\nolimits_{i=k}^{2m-2} a_{k-i+(m-1)} \cdot b_{i-(m-1)},\ k = m, m+1, \ldots, 2m-2.$$

The preceding equations can be implemented by a combinational circuit made up of 2-input AND gates and XOR gates with up to m inputs.

```
gen_ands: FOR k IN 0 TO m-1 GENERATE
  l1: FOR i IN 0 TO k GENERATE
  a_by_b(k)(i) <= a(i) AND b(k-i); END GENERATE;
END GENERATE;
gen_ands2: FOR k IN m TO 2*m-2 GENERATE
  l2: FOR i IN k TO 2*m-2 GENERATE
  a_by_b(k)(i) <= a(k-i+(m-1)) AND b(i-(m-1)); END GENERATE;
END GENERATE;
d(0) <= a_by_b(0)(0);
gen_xors: FOR k IN 1 TO 2*m-2 GENERATE
  l3: PROCESS(a_by_b(k),c(k))
      VARIABLE aux: STD_LOGIC;
  BEGIN
    IF (k < m) THEN aux := a_by_b(k)(0);
      FOR i IN 1 TO k LOOP aux := a_by_b(k)(i) XOR aux;
      END LOOP;
    ELSE aux := a_by_b(k)(k);
      FOR i IN k+1 TO 2*m-2 LOOP aux := a_by_b(k)(i) XOR aux;
      END LOOP;
    END IF;
    d(k) <= aux;
  END PROCESS;
END GENERATE;
```

It remains to reduce $d(x)$ modulo $f(x)$. Assume that all coefficients $r_{i,j}$, such that

$$x^{m+j} \bmod f(x) = r_{m-1,j}x^{m-1} + r_{m-2,j}x^{m-2} + \cdots + r_{1,j}x^1 + r_{0,j},$$

have been previously computed. Then the coefficients of $c(x) = a(x)\cdot b(x)$ mod $f(x)$ are the following:

$$c_j = d_j + \sum\nolimits_{i=0}^{m-2} r_{j,i} \cdot d_{m+i},\ j = 0, 1, \ldots, m-1. \quad (13.1)$$

The preceding equations can be implemented by a combinational circuit made up of m XOR gates with up to m inputs. The number of gate inputs is determined by the maximum number of 1's within a column of the matrix $[r_{i,j}]$, and this depends on the chosen polynomial $f(x)$.

```
gen_xors: FOR j IN 0 TO m-1 GENERATE
  l1: PROCESS(d)
    VARIABLE aux: STD_LOGIC;
  BEGIN
    aux := d(j);
    FOR i IN 0 TO m-2 LOOP
    aux := aux XOR (d(m+i) and r(j)(i)); END LOOP;
    c(j) <= aux;
  END PROCESS;
END GENERATE;
```

A complete VHDL model *classic_multiplier.vhd*, including both the polynomial multiplier and the polynomial reducer, is available at the Authors' web page.

Every coefficient d_k is the sum of at most m products $a_i \cdot b_{k-i}$, and every coefficient c_j is the sum of, at most, m coefficients d_k. Thus, if tree structures are used, the computation time is proportional to log m. On the other hand, the cost is proportional to m^2. Hence, this type of multiplier is suitable for small values of m. For great values of m, the cost could be excessive and sequential multipliers should be considered.

13.3.2.2 Interleaved Multiplier

The following LSB-first algorithm computes $c(x) = a(x) \cdot b(x) \bmod f(x)$ according to the following computation scheme:

$$c(x) = b_0 \cdot a(x) + b_1 \cdot (a(x) \cdot x) + b_2 \cdot \left(a(x) \cdot x^2\right) + \ldots + b_{m-1} \cdot \left(a(x) \cdot x^{m-1}\right).$$

Algorithm 13.10: Interleaved multiplication, LSB-first

```
c(x) := 0;
for i in 0 .. m-1 loop
  c(x) := c(x) + bi·a(x);
  a(x) := a(x)·x mod f(x);
end loop;
```

The first operation $c(x) + b_i \cdot a(x)$ is executed by the circuit of Fig. 13.5a. In order to compute $a(x) \cdot x \bmod f(x)$, use the fact that $x^m \bmod f(x) = f_{m-1}x^{m-1} + f_{m-2}x^{m-2} + \ldots + f_0$. Thus,

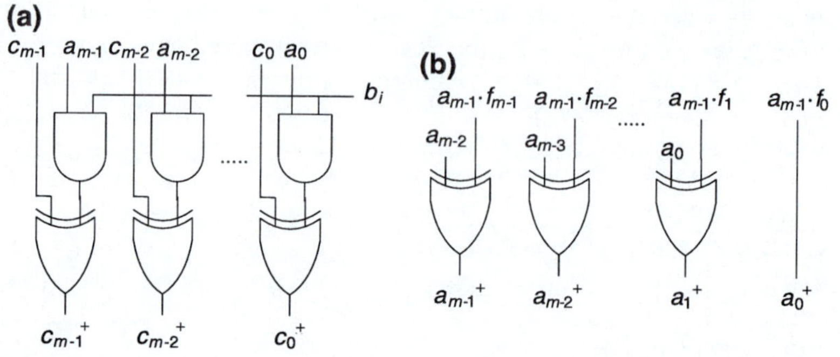

Fig. 13.5 Interleaved multiplier, computation of $c(x) + b_i \cdot a(x)$ and $a(x) \cdot x \bmod f(x)$

$$a(x) \cdot x \bmod f(x) = a_{m-1} \cdot \left(f_{m-1}x^{m-1} + f_{m-2}x^{m-2} + \ldots + f_0\right) + a_{m-2}x^{m-1} + a_{m-3}x^{m-2} + \ldots + a_0 x.$$

The corresponding circuit is shown in Fig. 13.5b. For fixed $f(x)$, the products $a_{m-1} \cdot f_j$ must not be computed: if $f_j = 1$, $a_j^+ = \left(a_{j-1} + a_{m-1}\right) \bmod 2$, and if $f_j = 0$, $a_j^+ = a_{j-1}$.

The circuit also includes two parallel registers a and c, a shift register b, an m-state counter, and a control unit. A complete VHDL model *interleaved_mult.vhd* is available at the Authors' web page.

13.3.2.3 Squaring

Given $a(x) = a_{m-1}x^{m-1} + \ldots + a_1 x + a_0$, the computation of $c(x) = a^2(x) \bmod f(x)$ can be performed with the algorithm of Sect. 13.3.2.1. The first step (multiply) is trivial:

$$d(x) = a^2(x) = (a_{m-1}x^{m-1} + \ldots + a_1 x + a_0)^2$$
$$= a_{m-1}x^{2(m-1)} + a_{m-2}x^{2(m-2)} + \ldots + a_1 x^2 + a_0.$$

Thus, $d_i = a_{i/2}$ if i is even, else $d_i = 0$. According to (13.1),

$$c_j = a_{j/2} + \sum_{\substack{0 \le i \le m-2 \\ m+i \text{ even}}} r_{j,i} \cdot a_{(m+i)/2}, \; j = 0, 2, 4, \ldots$$

$$c_j = \sum_{\substack{0 \le i \le m-2 \\ m+i \text{ even}}} r_{j,i} \cdot a_{(m+i)/2}, \; j = 1, 3, 5, \ldots$$

The cost of the circuit depends on the chosen polynomial $f(x)$, which, in turn, defines the matrix $[r_{i,j}]$. If $f(x)$ has few non-zero coefficients, as is the case of trinomials and pentanomials, then the matrix $[r_{i,j}]$ also has few non-zero coefficients, and the corresponding circuit is very fast and cost-effective. Examples of implementations are given in Sect. 13.5.

13.4 Division Over $GF(2^m)$

If $f(x)$ is irreducible, then all non-zero polynomials in $Z_2[x]/f(x)$ have a multiplicative inverse. Thus, given $g(x)$ and $h(x) \neq 0$ in $Z_2[x]/f(x)$, there exists a polynomial $z(x)$ in $Z_2[x]/f(x)$ such that $z(x) = g(x) \cdot h^{-1}(x) \bmod f(x)$.

There are several types of algorithms for computing $z(x)$. Some of them are generalizations of algorithms that compute the *greatest common divider*, like the Euclidean algorithm and the binary algorithm. Another option is to substitute the division by multiplications: according to the Fermat's theorem $z(x) = g(x) \cdot h^{q-2}(x) \bmod f(x)$ where $q = 2^m$. As an example, the following binary algorithm computes $z(x) = g(x) \cdot h^{-1}(x) \bmod f(x)$. It uses four variables $a(x)$, $b(x)$, $u(x)$ and $v(x)$, initially equal to $f(x)$, $h(x)$, 0 and $g(x)$, respectively. At each step, $a(x)$ and $b(x)$ are updated in such a way that their greatest common divider is unchanged and that the degree of $a(x) + b(x)$ decreases. For that, observe that if $b(x)$ is divisible by x and $a(x)$ is not, then $gcd(a(x), b(x)) = gcd(a(x), b(x)/x)$, and if neither $a(x)$ nor $b(x)$ are divisible by x, then $gcd(a(x), b(x)) = gcd(a(x), (a(x) + b(x))/x) = gcd(b(x), (a(x) + b(x))/x)$. As initially $a(x) = f(x)$ and $b(x) = h(x)$, where $f(x)$ is irreducible, after a finite number of steps $b(x) = 0$ and $a(x) = gcd(f(x), h(x)) = 1$. On the other hand, $u(x)$ and $v(x)$ are updated in such a way that $u(x) \cdot h(x) \equiv a(x) \cdot g(x) \bmod f(x)$ and $v(x) \cdot h(x) \equiv b(x) \cdot g(x) \bmod f(x)$. Initially, $u(x) = 0$, $a(x) = f(x) \equiv 0 \bmod f(x)$, $v(x) = g(x)$ and $b(x) = h(x)$, so that both equivalence relations are satisfied. It can be proven that if $u(x)$ and $v(x)$ are updated in the same way as $a(x)$ and $b(x)$, both relations remain true. In particular, if $a(x) = 1$, then $u(x) \cdot h(x) \equiv g(x) \bmod f(x)$, and $z(x) = u(x)$.

Algorithm 13.11: Mod $f(x)$ division, binary algorithm

```
a(x) := f(x); b(x) := h(x); u(x) := 0; v(x) := g(x);
while b(x) ≠ 0 loop
  if b0 = 0 then b(x) := b(x)/x; v(x) := v(x)·x^-1 mod f(x);
  elsif degree(a(x)) ≥ degree(b(x)) then
    (a(x), b(x), u(x), v(x)) :=
    (b(x), (a(x)+b(x))/x, v(x), (u(x)+v(x))·x^-1 mod f(x));
  else
    b(x) := (a(x)+b(x))/x; v(x) := (u(x)+v(x))·x^-1 mod f(x));
  end if;
end loop;
z(x) := u(x);
```

Given a polynomial $w(x)$, then $w(x) \cdot x^{-1} \bmod f(x) = (w(x) + w_0 \cdot f(x))/x$. A data path for executing the preceding algorithm is shown in Fig. 13.6. The small circles connected to $f_1, f_2, \ldots$, represent programmable connections: if $f_i = 1$, the rightmost input of the corresponding XOR gate is connected to w_0, else it is connected to 0.

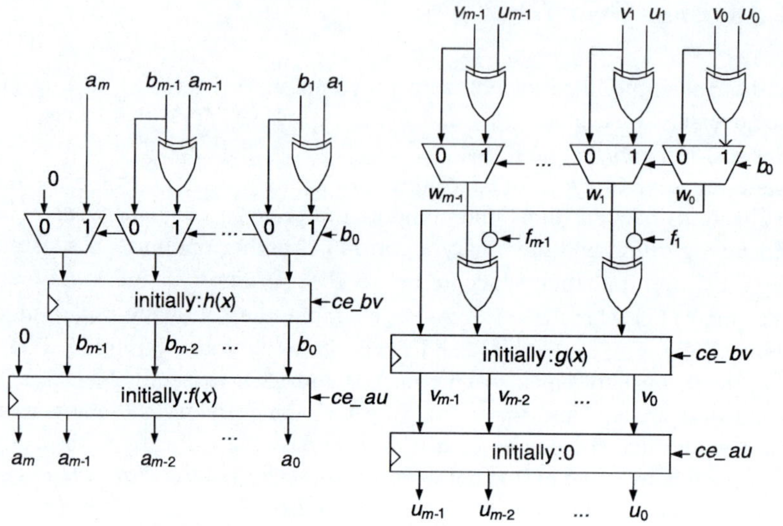

Fig. 13.6 Binary algorithm for polynomials

A drawback of the proposed algorithm is that the degrees of $a(x)$ and $b(x)$ must be computed at each step. A better option is to use upper bounds α and β of the degrees of $a(x)$ and $b(x)$.

Algorithm 13.12: Mod $f(x)$ division, binary algorithm, version 2

```
a(x) := f(x); b(x) := h(x); u(x) := 0; v(x) := g(x);
α := m; β := m-1;
while β ≥ 0 loop
  if b0 = 0 then b(x) := b(x)/x; v(x) := v(x)·x^-1 mod f(x);
    β := β-1;
  elsif α ≥ β then (a(x), b(x), u(x), v(x)) :=
    (b(x), (a(x)+b(x))/x, v(x), (u(x)+v(x))·x^-1 mod f(x));
    (α, β) := (β, α-1);
  else
    b(x) := (a(x)+b(x))/x; v(x) := (u(x)+v(x))·x^-1 mod f(x));
    β := β-1;
  end if;
end loop;
z(x) := u(x);
```

The data path is the same as before (Fig. 13.6). An upper bound of the number of steps is $2m$. As the operations are performed without carry propagation, the computation time is proportional to m. A data-flow VHDL description *mod_f_division2.vhd* is available at the Authors' web page.

In the preceding binary algorithms, the number of steps is not fixed; it depends on the input data values. This is an inconvenience when optimization methods, such as digit-serial processing (Chap. 3), are considered. In the following algorithm [11] α and β are substituted by *count* = $|\alpha\text{-}\beta\text{-}1|$, and a binary variable *state* equal to 0 if $\alpha > \beta$, and equal to 1 if $\alpha \le \beta$.

Algorithm 13.13: Mod *f*(*x*) division, binary algorithm, version 3

```
a(x) := f(x); b(x) := h(x); u(x) := 0; v(x) := g(x);
count := 0; state := 0;
for i in 1 .. 2m loop
  if state = 0 then count := count+1;
    if b0 = 0 then b(x) := b(x)/x; v(x) := v(x)·x^-1 mod f(x);
    else (a(x), b(x), u(x), v(x)) :=
      (b(x), (a(x)+b(x))/x, v(x), (u(x)+v(x))·x^-1 mod f(x));
       state := 1;
    end if;
  else count := count -1;
    if b0 = 0 then b(x) := b(x)/x; v(x) := v(x)·x^-1 mod f(x);
    else b(x) := (a(x)+b(x))/x;
      v(x) := (u(x)+v(x))·x^-1 mod f(x));
    end if;
    if count = 0 then state := 0; end if;
  end if;
```

The data path is still the same as before (Fig. 13.6), and the number of steps is $2m$, independently of the input data values. As the operations are performed without carry propagation, the computation time is proportional to m. A data-flow VHDL description *mod_f_division3.vhd* is available at the Authors' web page. Furthermore, several digit-serial implementations, with different digit definitions, are given in Sect. 13.5.

13.5 FPGA Implementations

Several circuits have been implemented within a Virtex 5–2 device. The times are expressed in *ns* and the costs in numbers of Look Up Tables (LUTs) and flip-flops (FFs). All VHDL models are available at the Authors' web page.

Two combinational multipliers (multiply and reduce) have been implemented (Table 13.1).

For greater values of the degree m of $f(x)$, sequential implementations should be considered. Several interleaved multipliers have been implemented (Table 13.2).

In the case of the squaring operation, combinational circuits can be used, even for great valued of m (Table 13.3).

Table13.1 Classic mod $f(x)$ multipliers

m	LUTs	Delay
8	37	3.2
64	2,125	5.3

Table 13.2 Interleaved mod $f(x)$ multipliers

m	FFs	LUTs	Period	Total time
8	32	34	1.48	13.3
64	201	207	1.80	117.0
163	500	504	1.60	262.4
233	711	714	1.88	439.9

Table 13.3 Mod $f(x)$ squaring (square and reduce)

m	LUTs	Delay
8	8	0.7
64	129	0.7
163	163	0.7
233	153	0.7

Table 13.4 Mod $f(x)$ dividers

m	FFs	LUTs	Period	Total time
8	43	34	2.41	41.0
64	273	157	2.63	339.3
163	673	370	2.96	967.9
233	953	510	2.98	1391.7

Several sequential mod $f(x)$ dividers (Sect. 13.4) have been implemented (Table 13.4).

13.6 Exercises

1. Generate the VHDL model of a reducer modulo 239.
2. Generate the VHDL model of a multiplier modulo $m = 2^{192} - 2^{64} - 1$.
3. Design an interleaved modulo m multiplier using the stored-carry principle.
4. Design a modulo p divider based on Fermat's little theorem.
5. Generate the VHDL model of a multiplier over $Z_2[x]/f(x)$ where $f(x) = x^8 + x^4 + x^3 + x + 1$.

6. Generate the VHDL model of a divider over $Z_2[x]/f(x)$ where $f(x) = x^8 + x^4 + x^3 + x + 1$.
7. Design a squarer over $Z_2[x]/f(x)$ where $f(x) = x^8 + x^4 + x^3 + x + 1$.

References

1. Rodríguez-Henríquez F, Saqib N, Díaz-Pérez A, Koç ÇK (2006) Cryptographic algorithms on reconfigurable hardware. Springer, Heidelberg
2. Deschamps JP, Imaña JL, Sutter G (2009) Hardware implementation of finite-field arithmetic. McGraw-Hill, New York
3. Montgomery PL (1985) Modular multiplication without trial division. Math Comput 44: 519–521
4. Sutter G, Deschamps JP, Imaña JL (2011) Modular multiplication and exponentiation architectures for fast RSA cryptosystem based on digit serial computation. IEEE Trans Industr Electron 58(7):3101–3109
5. Hankerson D, Menezes A, Vanstone S (2004) Guide to elliptic curve cryptography. Springer, Heidelberg
6. Menezes A, van Oorschot PC, Vanstone S (1996) Handbook of applied cryptography. CRC Press, Boca Raton
7. Knuth DE (1981) The art of computer programming, Seminumerical algorithmsvol, vol 2. Addison-Wesley, Reading
8. Deschamps JP, Bioul G, Sutter G (2006) Synthesis of arithmetic circuits. Wiley, New York
9. Meurice de Dormale G, Bulens Ph, Quisquater JJ (2004) Efficient modular division implementation. Lect Notes Comp Sci 3203:231–240
10. Takagi N (1998) A VLSI algorithm for modular division based on the binary GCD algorithm. IEICE Trans Fundam Electron Commun Comp Sci 81-A(5):724–728
11. Kim C, Hong C (2002) High-speed division architecture for GF(2^m). Electron Lett 38: 835–836

Chapter 14
Systems on Chip

This chapter introduces the main concepts related to the implementation of embedded systems on FPGA devices. Many of these concepts will appear during the case studies that are exposed in the next chapter.

14.1 System on Chip

The concept System on Chip (SoC) indicates a technical direction where the increasing capacity of transistors into a chip enables the integration of more complex systems. A SoC integrates all the components of the system, such as processors, memory, analog functions, etc., on a single chip substrate. Some systems may be too complex to be built using a single microelectronic technological process; therefore a system's subset is optimized to be implemented in a single chip. Typical SoC applications are embedded systems that are specific-application computers.

SoCs can be implemented on several microelectronic technologies: full-custom, standard-cells, FPGAs. Advances in FPGA devices permit the mapping of quite complex embedded systems in a single programmable device, thus, leading a growing technology.

14.2 Intellectual Property Cores

Most SoCs are developed from pre-designed hardware cores usually know as Intellectual Property (IP) cores. The IP cores are reusable hardware units designed and verified by third-party companies or open-source developers to perform a specific task. They can be licensed to permit a faster SoC development, since they can be integrated into the design flow as a qualified building block. Software

J.-P. Deschamps et al., *Guide to FPGA Implementation of Arithmetic Functions*, Lecture Notes in Electrical Engineering 149, DOI: 10.1007/978-94-007-2987-2_14,

drivers are enclosed in some IP cores, in order to facilitate the development of applications for embedded systems.

IP cores can be delivered in different abstraction levels. In general terms, a higher level of abstraction provides more flexibility but poorer performances and failure risks due to the following stages of the design flow.

- Soft-core. They are described in files and can be represented in different levels.
  - Register Transfer Logic (RTL). The highest abstraction level where IP cores are generally described in VHDL/Verilog files, therefore, they must be synthesized and implemented on the target device. Open source IPs are suitable on a large number of different devices and they may be modified to match the required needs. By contrast many industrial IP cores, as the IP cores deployed in the Xilinx's EDK suite, may be designed for a particular device and protected in order to avoid their modification.
  - Gate-level. The next lower level are IPs described in netlist files that are obtained after the synthesis for a particular family. They can not be (easily) modified or ported to other devices. The IP is usually optimized to use the most appropriate hardware resources, but the design flow must implement the layout for the target device. The soft macros from Xilinx are pre-defined circuits, composed of library components that have flexible mapping, placement and routing.
  - Layout-level. The next abstraction level is obtained after the IP implementation on a device. The layout is usually optimized to improve area and timing constraints. The Relative Placed Macro (RPMs) from Xilinx are pre-defined layouts that have flexible placement.
- Hard-core. It is the physical IP implementation on the device, therefore, logic and performances are fixed. FPGA devices, usually, provide hard IP cores that are efficiently built in a devoted area. For instance, most FPGAs provide internal memory blocks that can be employed by embedded processors. More advanced devices, such as some of the Xilinx Virtex-4/5 family, integrate PowerPC processors, Digital Signal Processing (DSP) blocks, Ethernet Media Access Control (MAC), etc.

14.3 Embedded Systems

General-purpose computers are designed to execute a wide range of different applications. Usually, they are built with a high-performance microprocessor, a large memory capacity and many different peripherals (USB ports, video card, disk drives, etc.). Applications rely on the Operating System (OS) which manages memory, peripherals, file system, interrupts and task scheduling. By contrast, an embedded system is a computer designed and programmed to meet the requirements of a specific application, such as media players, printers, domotics, etc. Applications may not require OS or may rely on a customized OS. Important aspects concerning these systems are their reduced size, cost and power consumption when they are compared against general-purpose computers. An embedded system is, usually, composed of a low-cost microprocessor, memory and some peripherals. It may include a customized peripheral or coprocessor to speed-up a specific task or computation.

14.3.1 Embedded Microprocessors

Microprocessors are the heart of any computer system since they manage the rest of the system, performing the control tasks and many of the computations. An embedded microprocessor is a general-purpose processor whose cost, size, power consumption and computational performances are adequate for the specific application it executes. The variety of general-purpose microprocessors is huge, but they can be classified according to their architecture. Embedded microprocessors are usually classified according to:

- Data-path width. The basic building blocks of a microprocessor's data-path are the registers and the Arithmetic Logic Unit (ALU) that are of n-bit width. The data-path width indicates the maximum size carried out by a single computation. In general terms, a higher data-path width brings more computational performance but higher cost and power consumption. Most embedded processors are 8-bit to 32-bit width since they offer a good trade-off.
- Instruction set. General-purpose microprocessors can be divided into Reduced Instruction Set Computing (RISC) and Complex Instruction Set Computing (CISC). The machine instructions of a RISC processor are quite simple, such as loading a register from memory, computing an arithmetic/logical operation on registers, and storing a register to memory. A CISC processor improves the efficiency of the instruction memory since a single instruction can perform several of the previously described steps. However, RISC microprocessors are predominant in embedded systems since they benefit from a simpler architecture and increased clock frequency.
- Memory architecture. The von Neumann architecture provides a single bus to access memory which stores instructions and data. The memory architecture is simplified, but it cannot simultaneously read the next instruction while it is reading/writing data, due to the previous instruction. The Harvard architecture provides separated busses (instruction-side and data-side) to memory, in order to improve the execution performance. In the latter case, the system provides an arbiter which multiplexes the accesses to memory, from both sides. The memory devices that store instructions and data can be physically different or not.
- Endianness. The minimum addressable space of memory that a microprocessor, usually, can read/write is a byte (8-bit). A 32-bit processors can also access memory to read/write 16/32-bit words that are stored in a set of 2/4 bytes allocated in consecutive addresses. A little-endian processor stores the least significant byte of a word into the byte of memory which is allocated at the

lower address. By contrast, a big-endian processor stores the most significant byte of the word into the memory byte at the lower address.

- Pipeline architecture. The pipeline architecture is a digital design technique which divides a large combinational path into a set of simpler processing stages that are serially connected through registers. During a time slot, a stage performs a partial computation from the resulting data of the previous stage, which is processing the next data. For instance, a typical 3-stage microprocessor can simultaneously fetch, decode and execute three different instructions. Most microprocessors provide several pipeline stages, since they can significantly increase the instruction throughput and the clock frequency. Their main drawbacks are the increased number of registers and the conditions that flush the pipeline. Some microprocessors can minimize the pipeline flushing, permitting out-of-order execution or the predicting of conditional branches.
- Superscalar architecture. They are microprocessors that have duplicated some of the functional units. For instance, they may provide several ALUs, in order to permit the parallel execution of computations that require several clock cycles. Another possibility is to provide duplicated stages of the pipelined architecture, in order to enhance the execution of conditional branches.
- Single-core or Multi-core. Many embedded microprocessors are single-core since they provide enough computing performance for the application. However, other applications may require a faster microprocessor. Increasing the clock frequency is not always feasible or reliable due to the physical limitations of the technology. Multi-core processors can perform several tasks in parallel in order to increase the performance at the same clock frequency. They are widely applied on general-purpose computers and they are extending their applicability on embedded systems. Usually, they are homogeneous cores since they are composed of a set of microprocessors of the same type. A heterogeneous core is composed of different processor types. For instance, the DM37xx [9] OMAP processors are composed of a general-purpose ARM microprocessor and a Digital Signal Processor (DSP) to accelerate processing of image, video and audio.

FPGAs can implement embedded processors as hard- or soft-cores. Hard-core microprocessors offer better performances. A fixed area of the FPGA implements the microprocessor and the programmable logic fabric is available to implement the rest of the embedded system (busses, peripherals, etc.). On the other hand, soft-core microprocessors on FPGAs offer a greater flexibility since they can be configured to match the required performances on a larger variety of applications. However, they may occupy a large area of the programmable logic. The set of machine instructions can be extended to support some of the optional features of the microprocessor. This way a C/C ++ compiler can optimize the machine code to take benefits from the enhanced architecture of the microprocessor.

Many of the soft-core microprocessors offer configurable features that can be enabled or disabled to match the desired performance/cost trade-off.

- Cache logic. The on-chip memory provides high bandwidth memory, but very limited capacity. By contrast, the external memory is slower, but it provides a higher capacity due to its lower cost per byte. When the on-chip memory does not provide enough capacity for the application, it can be used as cache memory. The cache improves the execution of applications from external memory, storing transparently the most frequently accessed instructions and data. The microprocessor architecture must provide cache logic which implements a policy, in order to try to maximize the number of cache hits.
- Memory Management Unit (MMU). Most sophisticated OS require virtual memory support in order to provide memory protection, and to extend the logical capacity of the physical memory. The address space of the memory is divided in pages and the MMU translates the page numbers from the logical memory to the physical memory. It relies on the Translation Lookaside Buffer (TLB) which stores a page table, in order to provide a fast translation. Many embedded microprocessors may not implement the MMU, since the OS does not support virtual memory.
- Floating-Point Unit (FPU). The unit provides arithmetic operations on real numbers that are usually formatted on the single or double-precision IEEE-754. The FPU may not be adequate on many low-cost applications and it can be emulated by software.
- Parallel multiplier, divider, barrel-shifter, or other additional units. Some applications require frequent operation on integer numbers that can be accelerated using a dedicated hardware unit. For instance, a barrel-shifter can perform an arbitrary n-bit shift in a single clock cycle which is useful in some applications.
- Debug module. Most embedded processors can add logic to support an external debugger tool. This way the external debugger can control the microprocessor which is running an application, by performing actions like stop it, inspect the registers or set breakpoints.

There is a very large collection of embedded microprocessors from different device vendors, third-parties or free sources. Some FPGAs embed hard-core microprocessors that provide the highest performance, but with a lower flexibility since they are physically implemented in the device. The FPGA vendor provides the tool chain (compiler, linker, debugger, etc.) to easily develop the software. Many FPGA vendors provide their own soft-core microprocessors, in order to facilitate the development of embedded systems on their devices. They are highly optimized for the target device, and they can be configured to match the required specifications. However, the flexibility is quite limited since the microprocessor cannot be (easily) modified or implemented in a device from other vendors. The vendor also provides the design flow, which permits the easily development of the hardware and software. The open-source microprocessors usually occupy a larger area and cannot run at the same clock frequencies when they are compared with the previous alternatives. However, they provide the greater flexibility since they can be modified or implemented in any FPGA that provides enough logic capacity. The design flow is also more complicated since the designer uses tools from different parties and they may lack support and warranty.

The Harvard, RISC, 32-bit microprocessors are widely implemented on FPGAs since they can provide a near to optimal performance/cost trade-off for the current technology. Therefore, we introduce some of the most likely used.

Some of the Xilinx Virtex-4/5 FPGAs offer the hard-core PowerPC-405/440 32-bit processors. The PowerPC-405 [10] is a little-endian processor which features a 5-stage pipeline, cache and MMU. The PowerPC-440 [11] is an enhanced version which features superscalar architecture and 7-stage pipeline. The architecture of both PowerPC processors supports an auxiliary FPU or other coprocessors. The newer Zynq-7000 [12] FPGAs from Xilinx integrate a dual-core ARM Cortex-A9 processor. The ARM processors are the de facto standard for embedded 32-bit microprocessors on Application Specific Integrated Circuit (ASICs).

Xilinx provides the MicroBlaze [13], a soft-core 32-bit microprocessor suitable for embedded systems implemented on their FPGA families. It can be configured to a 3/5-stage pipeline, and it can enable cache logic, MMU, single-precision FPU, divider unit, barrel-shifter, etc. Altera provides the Nios II [2] 32-bit microprocessor, a competing soft-core, licensable for their FPGAs or third-party standard-cells. The byte ordering is little-endian, with 1/5/6-stage pipeline, and it provides many of the configurable features available as in MicroBlaze. The LatticeMico32 [6] is an open source microprocessor which can be embedded in any FPGA device or ASIC technology. It features a 6-stage pipeline and big-endian ordering. The core can be configured to enable cache logic, parallel multiplier, divider or barrel shifter, but not MMU nor FPU. Actel FPGAs can license the Cortex-M1 [1] microprocessor which is an implementation ARMv6 processor specially designed for major FPGA devices. The processor provides a 3-stage pipeline, but the configurable options are not as extended as in the previous microprocessors.

Leon-3/4 [4, 5] are synthesizable VHDL models of SPARC v8 microprocessors that can be licensed for commercial applications. They feature a 7-stage pipeline, cache, MMU, single or double-precision FPU, etc. The microprocessors can be implemented on FPGAs, or attached to build a multi-core embedded system. The OpenRISC-1200 [7] is delivered in Verilog files and it is suitable for many FPGAs. It provides a 5-stage pipeline and optionally cache logic, MMU and single-precision FPU.

14.3.2 Peripherals

A general-purpose microprocessor is attached to other hardware components to add functionalities to the computer system. Peripherals extend the capabilities of the computer and they are dependent on it. The peripheral provides a bus interface which transforms signals and the handshake from the core to the system bus, in order to permit the microprocessor to control/monitor it. Many peripherals can generate an interrupt request which triggers the execution of a service routine on the microprocessor when the peripheral detects an event.

There are many types of peripherals. Some examples of typical peripherals are:

- General Purpose Input/Output (GPIO), a peripheral which can be used to get or set the state of I/O pins that are externally connected to other device. Typical applications are to drive led diodes or to read switches. However, they can control more complex devices when the microprocessor is programmed accordingly.
- Universal Asynchronous Receiver-Transmitter (UART). The data transported on a serial communication interface is composed of character frames. The frames on an asynchronous transmission are characterized by the start, stop and parity bits, the character length, as well as the bit speed. The UART peripheral generates and transmits a frame when it receives a new character from the microprocessor. It also performs the reverse transformation when it receives a frame from the serial interface. The transmitted and received characters are temporally stored in First Input First Output (FIFOs) in order to permit the microprocessor to not attend the UART immediately.
- Counter/Timer. The peripheral implements a register which counts an event from an external signal or from an internal clock. In the second case the elapsed time is deducted from the number of counts and the clock period. Another typical application is to generate a Programmable Interval Timer (PIT) which periodically generates an interrupt request.
- Memory controllers. They assign an address space to the memory and they control efficiently the read/write accesses from the microprocessor, in order to maximize the memory bandwidth. They can also match the data width between the memory and system bus. Depending on the memory they may permit burst accesses and refresh the dynamic memory.
- Interrupt controller. Combines interrupt requests from several sources to the microprocessor's interrupt input. They can mask interrupts to enable/disable them or assign priorities.
- Direct Memory Access (DMA). Some peripherals, such as video and disk controllers, require very frequent accesses to memory. The microprocessor can move data from memory to the peripheral (or vice versa) through the system bus, although, it is quite inefficient. A DMA is connected as a master of the system bus in order to permit to copy data between memory and peripherals while the microprocessor performs any other action. The microprocessor task can be reduced to program the DMA with the source and destination addresses, the data length and the burst size. These systems require a bus arbiter since they are several bus masters that can concurrently initiate a new access. The DMA can be a centralized peripheral or be a part of a peripheral.
- Video controller. A peripheral which generates the video signals required to display an image. They require a DMA in order to read pixels from the video memory (frame buffer) to generate the images at the required frequency.

Peripherals implement a set of internal registers in order to control or monitor them. Each register provides different functionalities that depend on the peripheral type and designer. A basic classification of peripheral registers is:

- Control register. A write-only register which configures and controls the peripheral. For instance, it sets the communication parameters in an UART.
- Status register. A read-only register used to get the status of the peripheral. For instance, it gets the detected error conditions and the state of the FIFOs in a UART.
- Data register. A read/write register which sets or gets data through the peripheral. For instance, it is the register used in the UART to read/write the characters stored in the reception/transmission FIFOs.

Each register is mapped on an address, in order to be accessible from the microprocessor. Two registers may be assigned to the same address. This is a quite common situation since the control and the status registers are, usually, mapped on the same address. In this way, a read instruction on the assigned address will get the status register and a write instruction will set the control register.

Peripherals attach to the system in order to permit the microprocessor to access their internal registers. Depending on the attachment, peripherals are classified as:

- Port-mapped peripherals are attached to a microprocessor's dedicated ports that are physically isolated from memory. The peripheral's registers are accessed through specific machine instructions that are different from the memory instructions.
- Memory-mapped peripherals share the system bus with memory. The peripheral's registers are accessed with the same machine instructions used to read/write memory. The bus provides handshake signals to introduce wait states, in order to adapt it to the different latencies of the memory and peripherals. Usually, the system reserves an address range for peripherals to easily differentiate them from the regular memory space, simplifying the address decoding logic.

The address decoder enables the selected register according to the address bus which attaches to the peripheral. In a memory-mapped peripheral, the address bus is much larger than required, due to the number of registers. For instance, the address decoder of a peripheral, which implements 4 registers, requires only 2 of the bits. The address decoding is divided into:

- Complete address decoding. The decoder checks all the arriving bits from the address bus. Each register is mapped on a single address.
- Incomplete address decoding. It does not check some bits of the address bus to simplify the logic. It checks some of the Most Significant Bit (MSBs) to select one of the peripherals and some of the Lower Significant Bit (LSBs) to enable one of its registers. Therefore, a peripheral register can be accessed from different addresses because some bits of the address bus are not decoded.

A software application can directly access the peripheral's registers in order to get the desired behavior. This approach has several drawbacks. Firstly, the programmer must have a deep knowledge about the internal architecture of the peripheral. Secondly, the application may not be portable since the peripheral could change. Finally, it is not adequate on a multi-tasking OS, since several processes could simultaneously control the peripheral. A better approach is to program the application to use the peripheral's driver which is composed of high-level functions that rely on the OS.

14.3.3 Coprocessors

Some algorithms may require a set of specific computations that cannot be re-solved in a general-purpose microprocessor in a reasonable execution time. A co-processor is designed and optimized to quickly perform a specific set of computations to accelerate the system performance. Coprocessors do not fetch instructions or data from memory. The general-purpose microprocessor accesses to memory to read data and enables the coprocessor to accelerate a computation. Coprocessors allow the customization of computer systems, since they can improve computational performances without upgrading the general-purpose microprocessor.

There is a wide range of coprocessors and some of the most typical are:

- FPU coprocessors to accelerate the computations on real numbers. Many embedded microprocessors do not integrate a FPU in order to reduce cost. Others may integrate a single-precision FPU which may be not adequate for some applications. The single and double-precision IEEE-754 may be emulated through software libraries, but the system performance may be greatly affected.
- Cryptoprocessors are designed to accelerate the most time spending computations related to cryptography, such as the encryption and decryption tasks. Their architecture usually implements several parallel processing units to improve the speed-up factor.
- DSPs perform a digital signal processing from a data stream (voice, image, etc.). The internal architecture may provide several pipeline stages to improve the data throughput. The input and output streams are carried from/to the microprocessor which accesses to memory or other components.

The coprocessors can attach to the microprocessor in several ways:

- Data-path extension by custom instructions. For instance the ALU of the soft-core NIOS II [3] is prepared to be connected to custom logic which accepts user's machine instructions that are similar to the native ones. Another example is the APU (Auxiliary Processor Unit) of the PowerPC-405/440 [10, 11], which permits the execution of new instructions on a tightly-coupled coprocessor.
- Dedicated point-to-point attachment, such as the FSL (Fast Simplex Link) for MicroBlaze [13]. The machine instructions set provides some instructions able to transmit/receive data to/from the links attached to the coprocessor.
- System bus. A coprocessor can be also attached to the system bus, as any peripheral. This alternative is not usually implemented since it increases the latency to communicate data, due to the more sophisticated bus handshake.

14.3.4 Memory

A memory circuit is a digital storage device which is commonly used to allocate the program executed by a microprocessor. The memory on embedded systems can be classified in several ways. A first logical division is:

- Instruction memory. Microprocessors execute software applications when reading machine instructions from the instruction memory.
- Data memory. Some of the machine instructions command the microprocessor to read or write data from/to the data memory.

The instruction and data memories may be physically isolated or implemented on the same devices. Moreover, they can be composed of several banks and types of memories. According to the physical placement, memory can be classified as the following:

- Internal memory. It is integrated in the same device as the embedded system. The memory is usually SRAM technology which features fast access time but small capacity (some KB). The internal memory can store a small executable (like a boot loader) or serve as cache memory. Many FPGAs integrate internal blocks of memory. For instance, Xilinx FPGAs implement Block RAM (BRAMs), dual-port synchronous SRAM memories that permit simultaneous access to instructions and data per clock cycle.
- External memory. It is not integrated into the device which implements the embedded system. The external memory can be semiconductor chips or storage devices.
  - The semiconductor technologies that are, usually, applied as external memory are DRAM and FLASH. They provide larger capacity (MB or GB) but slower access time, when compared with SRAM memory. The DRAM can storage instructions and data; however it is a volatile memory. By contrast, the FLASH is non-volatile but it cannot be used as the data memory since it is read-only memory. Many embedded systems make use of both technologies, first copying the program from FLASH to DRAM during the boot-up, and then executing the application from DRAM.
  - Storage devices, such as hard drives or FLASH cards, are more common in general-purpose computers than in embedded systems. They are non-volatile memories that feature very high capacity, but very slow access time. Moreover, data is sequentially accessed in large blocks (some KB) making it inefficient when randomly accessing data. The main purpose is to serve as storage repository of programs and files, although some computer systems can implement virtual memory on them if the microprocessor and OS support it.

Another classification is based on the architecture of the bit cells. The most common memories on embedded systems are as follows:

- Static Random Access Memory (SRAM). The bit cell is larger than the other alternatives; therefore, the memory density is lower and the cost per capacity is increased. The main advantages are the lower access time and lower power consumption when they are in standby mode. Many embedded systems and FPGAs integrate some SRAM blocks, since they are compatible with conventional CMOS technologies to fabricate the integrated circuit.
- Dynamic RAM (DRAM). It is a read/write and volatile memory as is the SRAM. The main advantage is the larger memory density, since the bit cell is very simple. However, the access time is, usually, slower and the bit cells must be periodically refreshed, increasing the power consumption. Moreover, they cannot be integrated efficiently in conventional CMOS technologies; therefore they are usually connected as external memory.
- FLASH. The bit cell is based on a floating-gate transistor which can be erased and programmed providing read-only and non-volatile memory. There are two important variants depending on the interconnections between cells: NAND and NOR. The NAND FLASH memories are employed as storage devices, since they provide higher capacity and bandwidth. However, the disadvantage is that data is sequentially read on pages. By contrast, the NOR FLASHs are, usually, employed as Read-Only Memory (ROMs), since they provide random access to data and similar interface as in SRAM chips.

There are more memory technologies, but most of them are not currently implemented or widely used on commercial applications. The Ferroelectric RAM (FRAM) is an emerging technology on embedded systems. It provides random-access and non-volatile memory which features much better performances on access time, power consumption, and write endurance when compared with FLASH. However, the memory density is much lower. For instance, the MSP430FR57xx family [8] of 16-bit microcontrollers can embed up to 16 KB of FRAM.

Another classification is based in the synchronicity of accesses to data.

- Asynchronous memory. Data is accessed by the address bus, independently of any clock signal. Many external SRAM memories are asynchronous since they provide a very simple interface and low latency when data is randomly accessed. An interesting alternative to external SRAM is the Pseudo Static RAM (PSRAM), which is DRAM surrounded with internal circuitry that permits to control it as a SRAM chip. It offers external memory featured by a high capacity and simple interface, since its internal circuitry hides the specific DRAM operations.
- Synchronous memory. Data is synchronously accessed according to a clock signal. These memories improve the data bandwidth since they implement a pipeline scheme, in order to increment clock frequency and data throughput. However, the latency is increased due to the stage registers. Most of the present Synchronous DRAM (SDRAMs), such as the DDR2 and DDR3, use Double Data Rate (DDR) registers to provide very efficient burst accesses in order to replace the cache pages. Another example of synchronous external memory is

the Zero-Bus Turn-around (ZBT) SRAM which improves the bandwidth of burst accesses and eliminates the idle cycles between read and write operations. Usually, the embedded memory blocks on FPGAs are double-port synchronous SRAMs that can efficiently implement the data and instruction cache of a Harvard microprocessor.

Embedded systems usually provide a memory controller to assign an address space to memory and to adapt the handshake signals and the data width from the system bus. The complexity of the external memory controller depends on the type of memory. The SRAM controllers are quite simple, since data can be accessed randomly. The DRAM controllers are more complex since the address bus is multiplexed and they control the periodical refreshing operation. Moreover, data is usually synchronously accessed on a burst basis; therefore the controller also generates the clock signal and the logic to communicate data between clock domains.

14.3.5 Busses

The interconnection between the microprocessor, peripherals and memory controllers in an embedded system is done through busses. The system bus permits the transfer of data between the building components. It is physically composed of a set of connecting lines that are shared by the components. The component must follow a common communication protocol which is implemented on the bus handshake.

The bus master starts a new access cycle in order to read/write data from/to a slave. An embedded system can provide several masters, but only one of them can take the bus control during a time slot. The bus arbiter assigns the control to one of the masters that are requesting it concurrently. For instance, a typical embedded system is composed of a Harvard microprocessor which connects to an external memory controller through the system bus. The external memory stores the instructions and data of an application program. The bus arbiter assigns the bus control to the microprocessor's instruction-side, or the data-side, according to the arbitration policy, while executing the program from the external memory.

There is a wide variety of busses depending on the embedded processor. Moreover many embedded systems provide several busses; depending on the memory or peripherals that are connected. A typical bus can be divided into the following:

- The address bus is a unidirectional bus, from the masters to the slaves, that carries the address of the memory (or a register in a memory-mapped peripheral) that is going to be accessed.
- The data bus is a bidirectional bus that carries the data which is read from slaves, or written to them. A bidirectional bus requires tri-state buffers that are not, usually, available in the internal FPGA fabric. In FPGAs, the data bus is

replaced by two unidirectional busses: the write data bus (from masters to slaves) and the read data bus (from slaves to masters).

- The control bus is the set of signals which implements the bus handshake, such as the read/write signal, or signals that indicate the access start and to acknowledge the completion. Some advanced busses permit burst accesses, in order to improve the throughput of sequential data transfers, and error signals, to permit the microprocessor to detect error conditions on the bus.

References

1. Actel (2010) Cortex-M1 v3.1 Handbook
2. Altera (2011a) NiosII processor reference handbook (NII5V1-11.0)
3. Altera (2011b) Nios II custom instruction User Guide (UG-N2CSTNST-2-0)
4. Gaisler (2010a) LEON3 Multiprocessing CPU core
5. Gaisler (2010b) LEON4 32-bit processor core
6. Lattice (2011) LatticeMico32 processor reference manual
7. OpenCores (2011) OpenRISC 1000 architecture manual
9. Texas Instruments (2011a) DM3730, DM3725 Digital media processors (SPRS685D)
8. Texas Instruments (2011b) Mixed signal microcontroller MSP430FR573x MSP430572x (SLAS639)
10. Xilinx (2008) PowerPC 405 Processor block reference guide (UG018)
11. Xilinx (2010a) PowerPC 440 Embedded processor block reference guide (UG200)
12. Xilinx (2010b) Extensible processing platform. Ideal solution for a wide range of embedded systems (WP369)
13. Xilinx (2011) MicroBlaze processor reference guide (UG081)

Chapter 15
Embedded Systems Development: Case Studies

Embedded systems are computers designed and programmed to meet the requirements of a specific application. Applications may not require an OS (Operating System) or may rely on a customized OS. The system architecture is usually composed of a low-cost microprocessor, memory and peripherals interconnected through busses. It may also include a coprocessor to speed-up a specific computation.

This chapter introduces the design of embedded systems on Xilinx FPGAs through a set of case studies. The studies focus on the hardware development of a peripheral and a coprocessor, as well as their software drivers. It assumes you are familiarized with the VHDL necessary on the hardware implementation, and with C/C++ which is required during the software development. In order to simplify the contents, they only expose the most relevant steps and interesting portions of the source files. They usually make references to specific documentation available in the Xilinx software, to get more details. The full examples can be downloaded from the authors' website.

15.1 Introduction to Xilinx EDK

The FPGA implementation of embedded systems requires a set of tools which permits the building of the hardware and the software (also named firmware). It also provides utilities to simulate and debug the embedded system.

The Xilinx Embedded Development Kit (EDK) suite facilitates the development of those systems. The system's hardware is composed of a general-purpose microprocessor connected to some peripherals through busses. The software development provides the firmware executed by the microprocessor. The EDK generates the bitstream of the system in order to program the FPGA, which can be connected to the debugger tools, in order to test it on a real scenario. EDK also

J.-P. Deschamps et al., *Guide to FPGA Implementation of Arithmetic Functions*, Lecture Notes in Electrical Engineering 149, DOI: 10.1007/978-94-007-2987-2_15,

permits the building of a simulation model in order to check the hardware design through a Hardware Description Language (HDL) simulator.

The EDK is frequently updated and upgraded. We will focus this chapter on the ISE Design Suite 13.1 for Windows since it is probably the most popular operating system for PCs, but there are no significant differences with versions for other operating systems. Although the case studies included in this chapter can be implemented on other upgraded EDK versions, they might require some small changes.

The EDK is composed of a set of tools:

- Xilinx Platform Studio (XPS). A graphical user interface that permits the designing of the hardware of the embedded system from a set of interconnected IP cores. It is the top-level tool which takes care of the necessary files and steps needed to successfully complete the hardware design. The XPS implements the design flow which runs other low-level tools in order to compute the hardware synthesis and implementation (Platgen), the generation of the bitstream (BitGen) and the simulation model (Simgen).
- Software Development Kit (SDK). An integrated environment based on the Eclipse/CDT to manage the development of the software. It launches the C/C++ cross-compiler and linker to build the binary which is executed by the embedded microprocessor. Moreover, it also provides a simple interface with the debugger and profiler tools used by more advanced users. SDK also builds the BSP (Board Support Package) through a low-level tool (Libgen). The BSP contains the set of software drivers used to control the hardware from the executable.
- IP cores. The library of configurable cores (microprocessors, peripherals, busses, etc.) that are used as basic building blocks of the embedded system. Most of these cores are licensed with the EDK, but there are also some cores that must be licensed separately. Many cores include a set of programming functions and drivers that can be used to facilitate the software development.
- GNU tools chain. The set of tools that generate the software libraries and the executable binaries. It includes the GNU C/C++ cross-compiler and linker. The GNU tools are developed for the Linux operating system, but EDK includes ported versions to the Windows OS.

Additionally, the EDK relies on other tools:

- ISE tools. They synthesize and implement the hardware, generate the bitstream and program the device. They also include other tools to generate the simulation model, implement the internal memory, the timing analysis and others. Platform Studio calls the required ISE tools, simplifying the design flow since the user can be abstracted from many specific details.
- Supported HDL simulator. It is recommended if the user designs a new IP, since it permits the simulation of the hardware of the embedded system. Some IP cores deployed with EDK are protected and encrypted, therefore, they can be simulated only on supported simulators. The ISE tools provide the ISim, but the user can choose a third-party simulator.

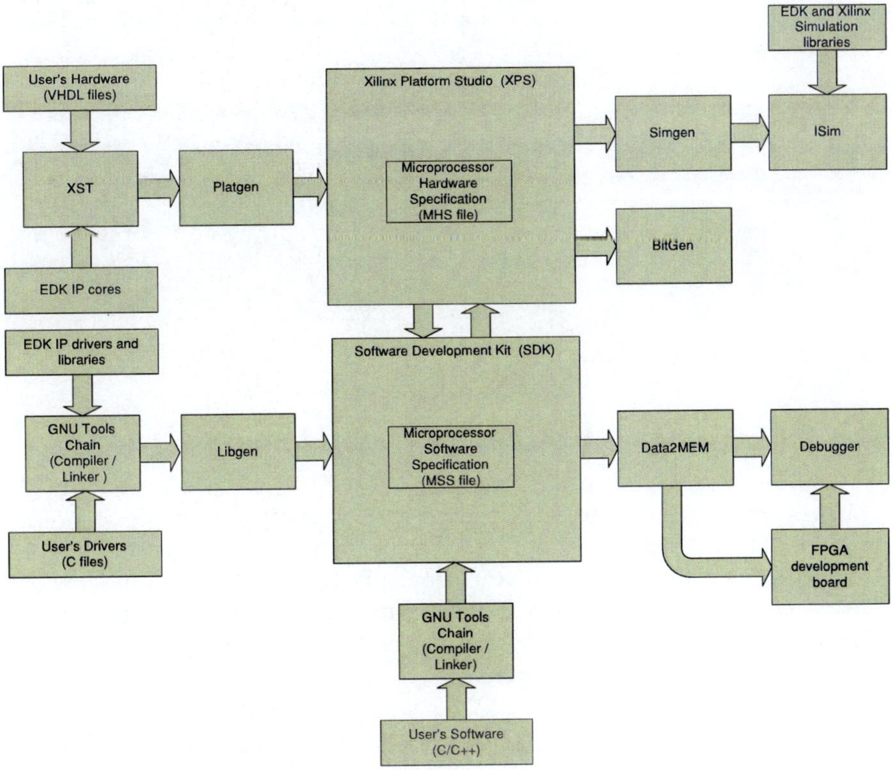

Fig. 15.1 Overview of the EDK design flow

- A development board with a Xilinx FPGA. There is a wide range of boards with different FPGAs (Spartan or Virtex series), memories, displays, communication interfaces and other elements.
- A Xilinx programmer, such as the Parallel III/IV or the Platform Cable USB I/II. Some development boards provide an embedded USB programmer. The programmers can be used to program the FPGA and to debug the executable through the Xilinx Machine Debugger (XMD) low-level tool.

Two files play an important role in the design flow (see Fig. 15.1): the Microprocessor Hardware Specification (MHS) and the Microprocessor Software Specification (MSS). The XPS manages the hardware design flow using a Xilinx Microprocessor Project (XMP) project file. The XPS project relies on the MHS file which configures a set of instances to IP cores that are interconnected as building blocks. The XPS can export the hardware design to SDK in order to generate the Board Support Package (BSP) for the embedded system. The BSP generation relies on the MSS file which configures the drivers and libraries that can be used by the executable to control the hardware.

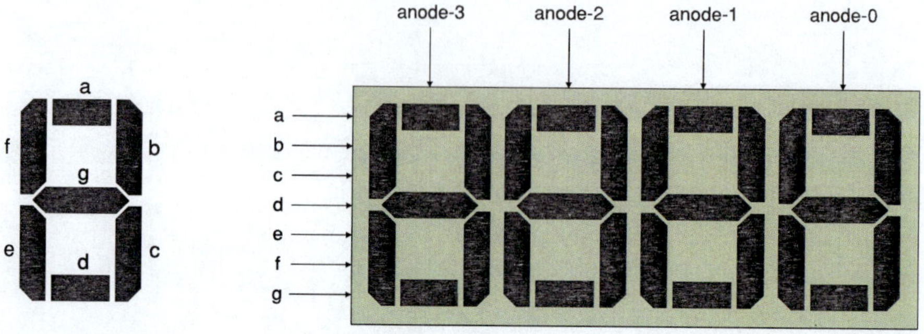

Fig. 15.2 The 4-digit, 7-segment led display

15.1.1 Case Study 1-A: A Basic Embedded System

This case study builds a basic embedded system, composed of the MicroBlaze processor [16], internal memory, and some peripherals. The system controls a 4-digit, 7-segment led display to visualize the hexadecimal content of a variable. The system attaches to two external switches that turn on/off the display and show/hide the left-side zeros. The display can also be controlled by an external computer connected through the serial interface.

The development FPGA board is the Xilinx Spartan-3 Starter Kit [2], a cheap board composed of a XC3S200 FPGA interconnected to a 7-segment display, a serial port and other elements. The four digits of the display share the segment inputs and each digit is enabled with a dedicated anode input, as depicted in Fig. 15.2. The system must continuously perform a time-multiplexed control to refresh the display, driving a single digit during a short time slot (a few milliseconds). Depending on the board, the inputs of the display are asserted to low or high logic levels. The case study can be easily adapted to any FPGA board which provides a similar display.

The display could be continuously controlled by the microprocessor in a loop to enable one digit per refresh cycle. This system is quite inefficient, since most of the execution time would be devoted to wait the refresh of the display, and to poll the serial interface and the state of the switches. A better approach uses two interrupt service routines (ISR) to attend the interrupts from two peripherals. A timer peripheral can, periodically, request an interrupt which triggers an ISR in order to read the switches and refresh the display. The Universal Asynchronous Receiver Transmitter (UART) peripheral requests an interrupt when a new character is received from the serial interface, and its ISR will process it.

The design of an embedded system involves the hardware and software development phases. The output of the hardware phase is a BIT (bitstream) file which configures the hardware resources of the FPGA except the contents of the internal BRAM (Block RAM) used as the microprocessor's local memory. The output of the software phase is the Executable and Linkable Format (ELF) file which must be

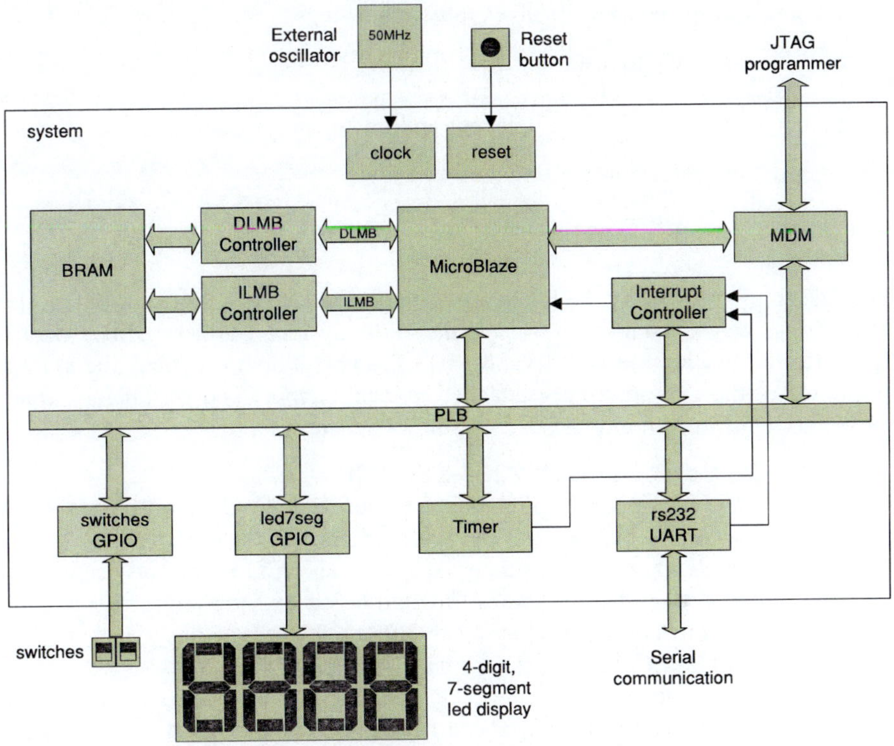

Fig. 15.3 Overview of the system hardware

allocated into the microprocessor's memory. In order to program the FPGA the design flow executes the Data2MEM tool to generate a new bitstream file which configures the FPGA including the BRAM contents to store the executable binary.

15.1.2 Hardware

The system is composed of the MicroBlaze, the internal BRAM and a set of peripherals (see Fig. 15.3). The BRAM implements the microprocessor's local memory and it is connected through two Local Memory Bus (LMBs). The peripherals are connected through a Processor Local Bus (PLB). The MicroBlaze can control the display and the two switches through General Purpose Input Output (GPIO) peripherals. The UART peripheral permits the serial communication with the external PC through the RS-232 interface. The timer and the UART will request the microprocessor to interrupt, therefore, the system includes an interrupt controller. Finally, the Machine Debug Module (MDM) permits the debugging of the application executed by the MicroBlaze through the FPGA programmer and the XMD tool.

The hardware generation involves three main steps:

(1) Hardware specification in the MHS file
(2) Synthesis of the hardware
(3) Implementation of the FPGA layout and bitstream generation

15.1.2.1 Specification

The first step is to specify the hardware of the embedded system in the MHS file. The easiest way to start is by using the Base System Builder (BSB) wizard. It creates the project file and a valid MHS file [14] which describes the system composed of the microprocessor attached to local memory and peripherals. Open the Xilinx Platform Studio (XPS) and follow the steps:

(1) XPS opens a dialog window. Choose the BSB wizard
(2) The next dialog window configures the project file and directory. Change the path to *C:\edk13.1\led7seg* and the project name to *system.xmp*
(3) Next, a new dialog configures the system bus. The AXI is currently supported only on the newer FPGA families (Spartan-6, Virtex-6). Therefore, select the PLB [4] since it is supported by all the FPGA families.
(4) Select the option *Create a new design* in the dialog *Welcome.*
(5) The dialog can configure a pre-built system for a supported FPGA board. Choose *Create a system for a custom board* to setup the system from scratch. Then select the Spartan-3 *xc3s200-ft256-4* device and any polarity for the reset button. These parameters can be easily modified later.
(6) Choose a single processor system since it simplifies the design.
(7) The next dialog configures the frequencies of the reference and system clocks as well as the capacity of the microprocessor's local memory. Leave the default parameters since they are changed later.
(8) The BSB permits the connecting of a set of peripherals to the system. Just continue until the end of the wizard since they are added later. The BSP creates the hardware specification of a basic embedded system.

The XPS permits to display and modify the system in a graphical way using the tab *System Assembly View*, as shown in Fig. 15.4. The view *Bus Interfaces* shows the system composed of instances to IP components that are interconnected through busses. The microprocessor (*microblaze_0*) is connected to the BRAM (*lmb_bram*) through the data and instruction LMBs (*dlmb*, *ilmb*) [17] and their memory controllers (*dlmb_cntrl*, *ilmb_cntrl*). The peripherals attach to the system through the PLB as slaves. The instruction and data PLB sides of the microprocessor are the bus masters. The MDM (*mdm_0*) peripheral attaches to the microprocessor through the PLB (*mb_plb*). The MicroBlaze is master of the busses, meanwhile, the peripheral and memory controllers are the slaves. Finally the last two instances are the generators of the internal clock (*clock_generator_0*) and reset (*proc_sys_reset_0*) that are required by the system.

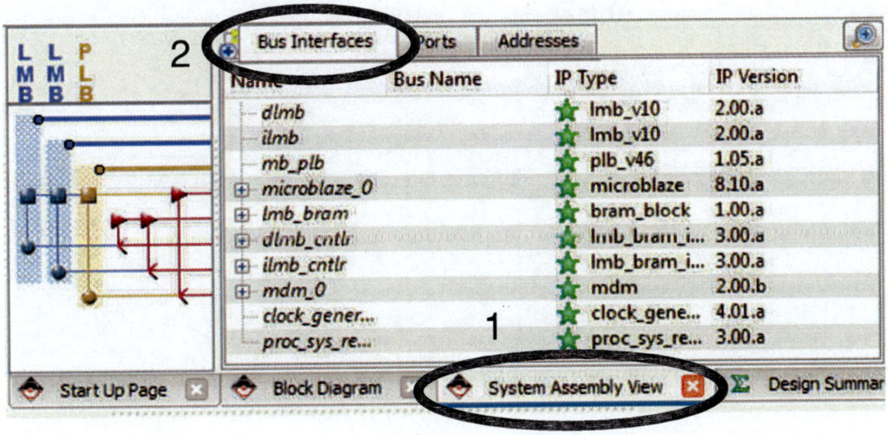

Fig. 15.4 Bus interfaces of the system in EDK

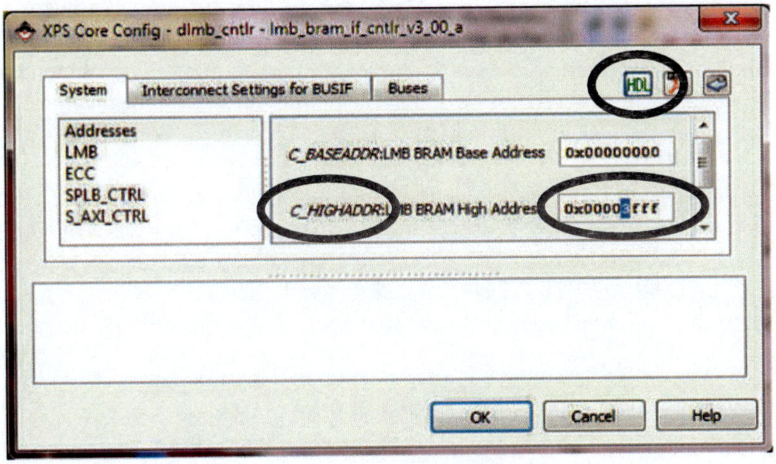

Fig. 15.5 Configuration parameters of the data LMB controller

The IPs provide a set of parameters that can be fixed, auto-computed or configurable. Select an instance and open the contextual menu (click the right button of the mouse) to configure the IP graphically (see Fig. 15.5). The HDL name of a parameter is the same as it appears in the MHS file. The configurable parameters *C_BASEADDR* and *C_HIGHADDR* of the LMB controllers setup the address space of the microprocessor's local memory. Changing the *C_HIGHADDR* to $0 \times 3FFF$ of both LMB controllers increases the auto-computed parameter *C_MEMSIZE* of the BRAM to 16 KB (0×4000). Every IP deployed by EDK provides a PDF file which details the parameters, the input/output ports and internal architecture.

The MHS file is a text file which can be manually edited, as shown in Fig. 15.6. It is synchronized with the *System Assembly View*. Therefore, they both update

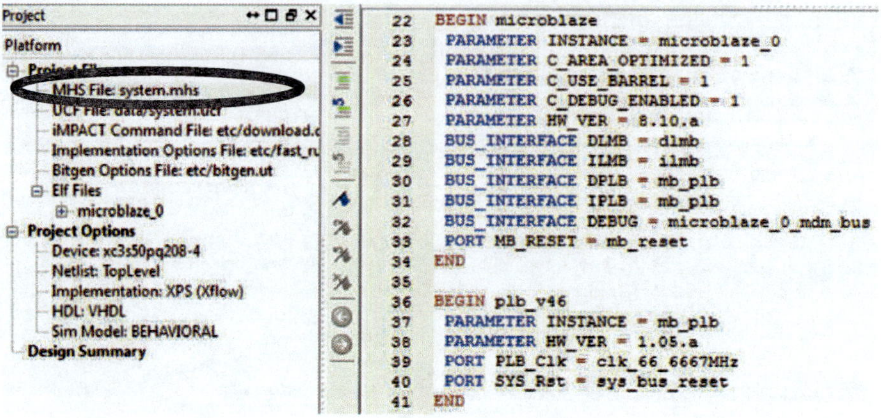

Fig. 15.6 MHS file in the XPS project

when the other one is modified. The MHS specifies the system's hardware as a set of interconnected instances and external FPGA ports. Each instance contains configurable parameters, interface to busses and other ports. The parameters that are not declared in the MHS take a default or the auto-computed value. The bus interfaces or ports that are not declared in the MHS are disconnected.

At the beginning of the MHS file there are the declarations of the external FPGA ports used to input the clock and the reset. The external ports are connected to the internal signals *CLK_S* and *sys_rst_s* that are used by the clock and reset generators. The parameter *CLK_FREQ* declares the frequency of the external oscillator and the *RST_POLARITY* declares the logic level when the reset input asserts. Both parameters must be modified according to the FPGA board.

```
PORT fpga_0_clk_1_sys_clk_pin = CLK_S, DIR = I,
     SIGIS = CLK, CLK_FREQ = 50000000
PORT fpga_0_rst_1_sys_rst_pin = sys_rst_s, DIR = I,
     SIGIS = RST, RST_POLARITY = 1
```

The instance *proc_sys_reset_0* generates the internal reset signals required by the system. The configurable parameter *C_EXT_RESET_HIGH* must be modified according to the reset polarity which was declared in the external port.

```
BEGIN proc_sys_reset
 PARAMETER INSTANCE = proc_sys_reset_0
 PARAMETER C_EXT_RESET_HIGH = 1
 ...
 PORT Ext_Reset_In = sys_rst_s
END
```

The instance *clock_generator_0* infers a Digital Clock Manager (DCM) circuit to generate the system clock from a reference clock (external oscillator). The parameter *C_EXT_RESET_HIGH* must be configured as in the reset generator. The parameter *C_CLKOUT0_FREQ* configures the generated frequency from the reference clock defined by the *C_CLKIN_FREQ*. The BSP generated a signal named *clk_66_6667 MHz* (or similar) which carries out the generated clock, but it can be renamed to *sys_clk*. Also change the name of the signal which drives the clock port of the busses. An incorrect configuration of the clock frequencies will lead to errors during the synthesis or implementation of the hardware.

```
BEGIN clock_generator
 PARAMETER INSTANCE = clock_generator_0
 PARAMETER C_CLKIN_FREQ = 50000000
 PARAMETER C_CLKOUT0_FREQ = 50000000
 ...
 PARAMETER C_EXT_RESET_HIGH = 1
 ...
 PORT CLKIN = CLK_S
 PORT CLKOUT0 = sys_clk
 PORT RST = sys_rst_s
...
END

BEGIN plb_v46
 PARAMETER INSTANCE = mb_plb
 PORT PLB_Clk = sys_clk
 ...
END

BEGIN lmb_v10
 PARAMETER INSTANCE = ilmb
 PORT LMB_Clk = sys_clk
 ...
END

BEGIN lmb_v10
 PARAMETER INSTANCE = dlmb
 PORT LMB_Clk = sys_clk
 ...
END
```

In order to accommodate it to the desired application, the hardware must add new peripherals. The first peripheral is a GPIO [5] which controls the 7 segments and the 4 anodes of the display. As depicted in Fig. 15.7:

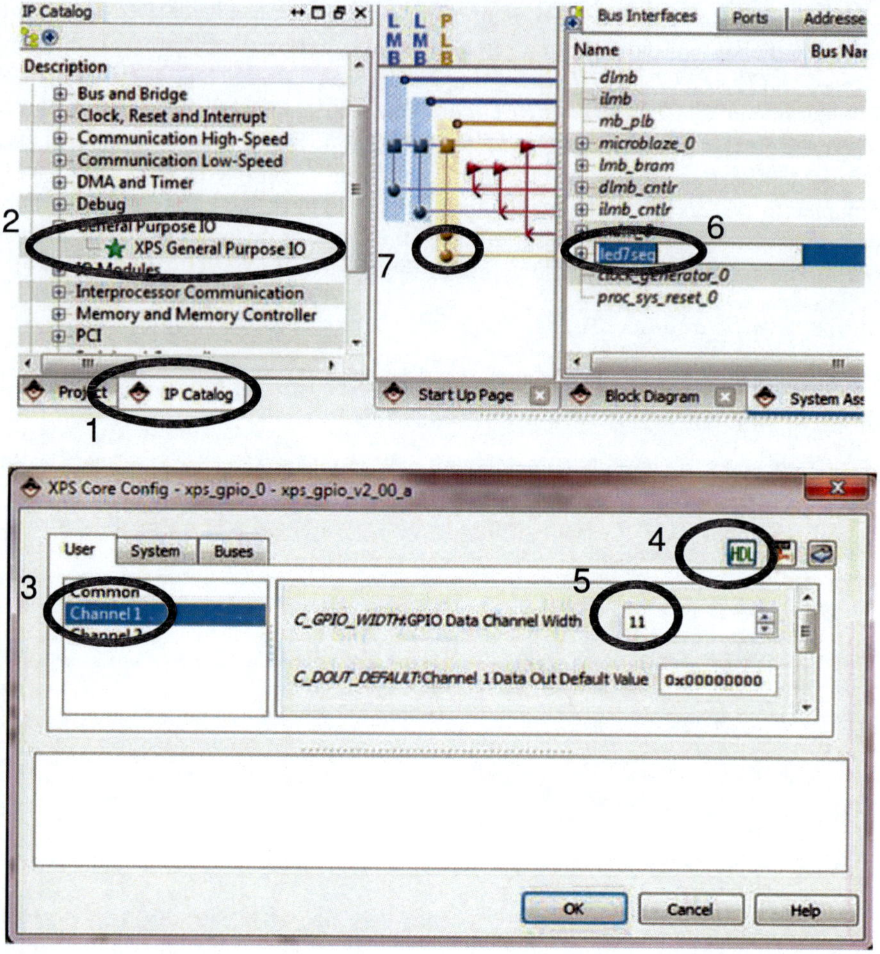

Fig. 15.7 Adding the new GPIO peripheral

(1) Drag the *XPS General Purpose IO* from the *IP Catalog* to the *System Assembly View*.
(2) XPS opens a dialog to configure it. Set the data width of the first channel (parameter *C_GPIO_WIDTH*) to *11* in order to control the display.
(3) Click the created instance and change the name to *led7seg*.
(4) Click the PLB interface to connect the peripheral's SPLB (Slave PLB).

Go to the view *Addresses* and configure automatically the addresses of the peripheral (see Fig. 15.8). The internal registers of the GPIO are accessible from the microprocessor within this addresses range. By default, XPS assigns the addresses range to *64 KB* (*0* × *10000*) from an address above the *0* × *80000000*.

Fig. 15.8 Automatic configuration of the addresses for the GPIO peripheral

The user can change the range, but they must not overlap to the other memory-mapped peripherals.

Finally, the 11-bit width GPIO's output port must be connected to the external FPGA ports that drive the display. Figure 15.9 shows the view *Ports* which permits the external connection of the output port *GPIO_IO_O*. Change the default name to *fpga_0_led7seg_pin* in a similar fashion as the rest of the external FPGA ports and check that the direction is configured as output. Finally, set the range order to *[10:0]* to declare them in descending order. The MSB and the LSB are indexed as 10 and 0, respectively.

Open the MHS file to observe the previous changes. There is a new entry in the section of the external FPGA ports. It also contains the new GPIO instance including its configuration parameters and connections.

```
PORT fpga_0_led7seg_pin=led7seg_GPIO_IO_O, DIR=O, VEC=[10:0]
...
BEGIN xps_gpio
 PARAMETER INSTANCE = led7seg
 PARAMETER HW_VER = 2.00.a
 PARAMETER C_BASEADDR = 0x81400000
 PARAMETER C_HIGHADDR = 0x8140ffff
 PARAMETER C_GPIO_WIDTH = 11
 BUS_INTERFACE SPLB = mb_plb
 PORT GPIO_IO_O = led7seg_GPIO_IO_O
END
```

The next step adds a new GPIO to read the state of the two switches. The hardware can also be modified by editing the MHS file. Copy the previous GPIO instance and change the instance name to *switches* and the data width to *2*. Set the addresses range to *64 KB* and do not overlap it to the other peripherals. Connect the input port *GPIO_IO_I* to a signal which is connected to an external FPGA port named *fpga_0_switches_pin*. Save the MHS file to update the graphical view. The project will close if there is an error in the MHS file, which must be manually corrected.

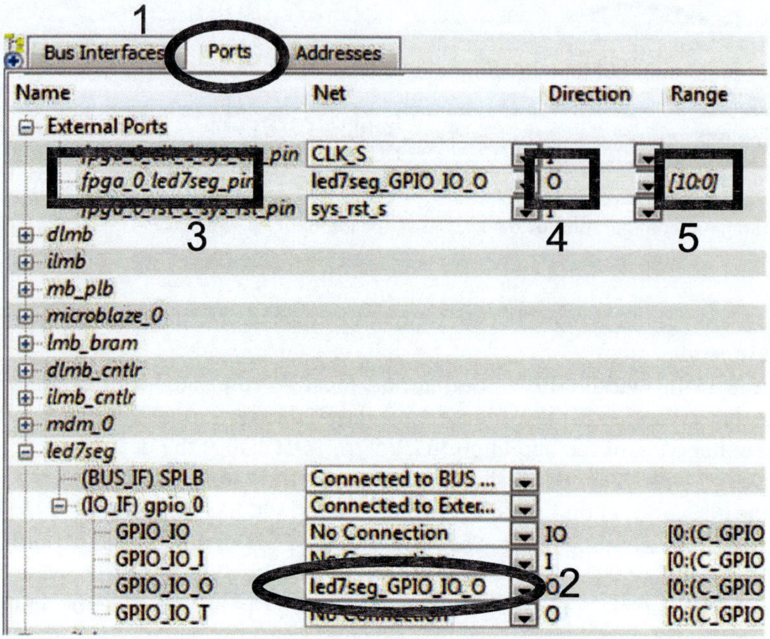

Fig. 15.9 External FPGA connection of the GPIO's output port

```
PORT fpga_0_switches_pin=switches_GPIO_IO_I, DIR=I, VEC=[1:0]
...
BEGIN xps_gpio
 PARAMETER INSTANCE = switches
 PARAMETER HW_VER = 2.00.a
 PARAMETER C_BASEADDR = 0x82400000
 PARAMETER C_HIGHADDR = 0x8240ffff
 PARAMETER C_GPIO_WIDTH = 2
 BUS_INTERFACE SPLB = mb_plb
 PORT GPIO_IO_I = switches_GPIO_IO_I
END
```

The next step adds the two peripherals to the PLB that will request interrupts. The timer [6] will be programmed to request a periodic interrupt. The parameter *C_ ONE_TIMER_ONLY* configures a single timer to minimize the size of the peripheral.

The UART [7] will request an interrupt when it receives a new character from the RS232. It will also transmit messages to the user through the serial communication. Therefore the UART ports transmission (TX) and reception (*RX*) are connected to external FPGA ports. The parameters *C_BAUDRATE* and *C_USE_PARITY* configure the speed and parity of the communication.

Some displays provide an extra input to turn on a dot placed beside the digit. This case study just turns off the dot connecting the *net_gnd* or *net_vcc* to its associated external port.

```
PORT fpga_0_RX_pin = rs232_RX, DIR = I
PORT fpga_0_TX_pin = rs232_TX, DIR = O
PORT fpga_0_led7seg_dot_pin = net_vcc, DIR = O

...
BEGIN xps_timer
 PARAMETER INSTANCE = timer
 PARAMETER HW_VER = 1.02.a
 PARAMETER C_BASEADDR = 0x83c00000
 PARAMETER C_HIGHADDR = 0x83c0ffff
 PARAMETER C_ONE_TIMER_ONLY = 1
 BUS_INTERFACE SPLB = mb_plb
 PORT Interrupt = timer_Interrupt
END

BEGIN xps_uartlite
 PARAMETER INSTANCE = rs232
 PARAMETER HW_VER = 1.01.a
 PARAMETER C_BAUDRATE = 115200
 PARAMETER C_USE_PARITY = 0
 PARAMETER C_BASEADDR = 0x84000000
 PARAMETER C_HIGHADDR = 0x8400ffff
 BUS_INTERFACE SPLB = mb_plb
 PORT RX = rs232_RX
 PORT TX = rs232_TX
 PORT Interrupt = rs232_Interrupt
END
```

The system requires an interrupt controller [8] since MicroBlaze provides a single input port for the interrupt requests. The interrupt controller attaches to the PLB in order to permit the MicroBlaze to enable/disable interrupts or to check the interrupt source. The interrupt controller (*int_control_0*) connects the interrupt requests from the timer and UART peripherals to Microblaze. The controller receives the concatenated signal from the two peripherals and drives the interrupt port of MicroBlaze. The interrupt priority is higher when the interrupt source is concatenated at the right side. The UART is assigned to the lower priority since it provides a receiving First Input First Output (FIFO) memory which temporarily stores the characters. Therefore, a new character received can be processed when the display is not refreshing.

```
BEGIN xps_intc
 PARAMETER INSTANCE = int_control_0
 PARAMETER HW_VER = 2.01.a
 PARAMETER C_BASEADDR = 0x81800000
 PARAMETER C_HIGHADDR = 0x8180ffff
 BUS_INTERFACE SPLB = mb_plb
 PORT Intr = rs232_Interrupt & timer_Interrupt
 PORT Irq = int_control_0_Irq
END
```

MicroBlaze provides configurable parameters to optimize area/performance and to implement optional machine instructions. The parameter *C_AREA_OPTIMIZED* configures a 3-stage pipeline architecture which optimizes the area of the implementation. The parameter *C_USE_BARRELL* implements a barrel shifter and its related machine instructions. Therefore, the C/C++ compiler provides a set of flags to build the BSP and the executable for the configured microprocessor. By default, the MicroBlaze attaches to a MDM instance (*mdm_0*) which permits the debugging of executables.

```
BEGIN microblaze
 PARAMETER C_AREA_OPTIMIZED = 1
 PARAMETER C_USE_BARREL = 1
 PARAMETER C_DEBUG_ENABLED = 1
 ...
 PORT INTERRUPT = int_control_0_Irq
END

BEGIN mdm
 PARAMETER INSTANCE = mdm_0
 ...
END
```

15.1.2.2 Synthesis

The XPS menu *Hardware* → *Generate Netlist* synthesizes the design to generate a set of NGC netlist files. It calls the Platgen tool [13] which starts performing a Design Rule Check (DRC). Then it calls the Xilinx Synthesis Technology (XST) [3] tool to synthesize the IP instances to get their NGC files. The embedded system is finally synthesized and optimized to get the top-level netlist file. A change in the MHS file will force Platgen to synthesise only the required

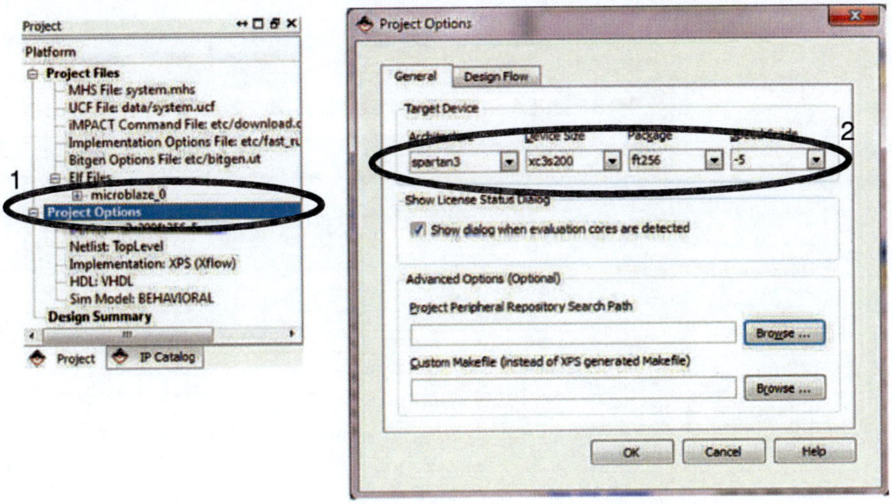

Fig. 15.10 Selection of the device in the XPS project

modules to speed-up the execution. If desired, the XPS can clean up the generated files to start the Platgen from scratch. Synthesis is dependent of the FPGA, therefore, the user must select the correct device in the *Project Options* (see Fig. 15.10) before proceeding.

Figure 15.11 shows the tab *Design Summary* which displays the report files generated by the Platgen and XST in order to check details about the design, such as the occupied FPGA resources or the estimated maximum frequency of the clock.

15.1.2.3 Implementation

The implementation computes the FPGA layout which is stored in a Native Circuit Description (NCD) file. The design flow executes three main tools: NGDBUILD, MAP and PAR [15]. The NGDBUILD translates the NGC files and annotates constraints from a User Constraints File (UCF). The following tools compute the layout based on the annotated constraints. The design flow continues with the MAP and PAR tools to map the netlist into the FPGA resources and to compute their placements and routings.

The BSB wizard generates the UCF for the selected prototyping board. The XPS project refers the UCF which must be edited to specify the attachment of the display and switches to the FPGA board. The UCF also specifies the clock frequency of the external oscillator.

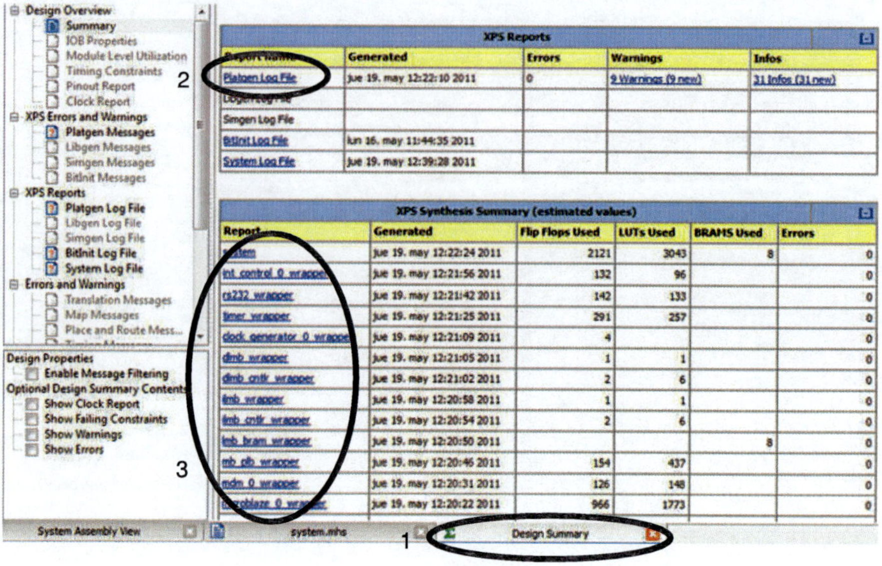

Fig. 15.11 Report files from the Platgen and synthesis

```
Net fpga_0_clk_1_sys_clk_pin TNM_NET = sys_clk_pin;
TIMESPEC TS_sys_clk_pin = PERIOD sys_clk_pin 50000 kHz;
Net fpga_0_rst_1_sys_rst_pin TIG;

Net fpga_0_clk_1_sys_clk_pin   LOC=T9;
Net fpga_0_rst_1_sys_rst_pin   LOC=L14;

Net fpga_0_RX_pin              LOC=T13;
Net fpga_0_TX_pin              LOC=R13;

Net fpga_0_led7seg_pin<0>      LOC=E14;        #segment-A
...
Net fpga_0_led7seg_pin<10>     LOC=E13;        #anode-3
Net fpga_0_led7seg_dot_pin     LOC=P16;        #dot

Net fpga_0_switches_pin<0>     LOC=K13;        #switch-0
Net fpga_0_switches_pin<1>     LOC=K14;        #switch-1
```

The XPS menu *Hardware* → *Generate Bitstream* launches the BitGen tool [15] which generates the bitstream file from the FPGA layout. First, the XPS executes the design flow to implement the FPGA layout, if necessary. Then it generates the BIT (bitstream) file *system.bit* and the BlockRAM Memory Map (BMM) file *system_bd.bmm*. The microprocessor's local memory is implemented on BRAMs,

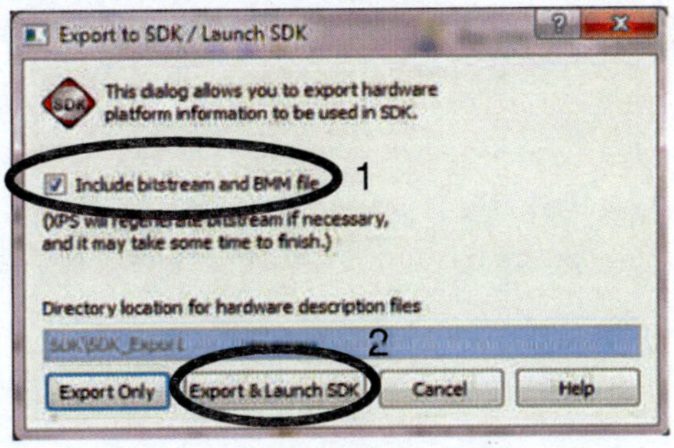

Fig. 15.12 Hardware export from EDK to SDK

but the generated BIT file does not initialize them, since the executable binary is not available at this stage. The file *system_bd.bmm* annotates the physical placement of the BRAMs with the microprocessor's local memory. This file will be required later to update the BRAM contents of the bitstream. The tab *Design Summary* shows the reports generated by the implementation tools.

15.1.2.4 Software

The XPS menu *Project → Export Hardware* opens a dialog window to export the required files to SDK, as shown in Fig. 15.12. Select the option to export the BIT and BMM files to permit SDK to program the FPGA. It creates a new directory which is allocated in the XPS project folder.

SDK starts opening a dialog to set the workspace folder. Write the path *c:\edk13.1\led7seg\SDK\workspace* to create it into the SDK folder which was generated by XPS during the hardware exportation. The software development involves two stages:

(1) The BSP generation. Creates a set of headers and libraries to control the hardware from the microprocessor.
(2) The executable ELF file. It builds the file executed by the embedded microprocessor.

15.1.2.5 Board Support Package

The BSP provides the Application Programming Interface (API) for the target hardware and Operating System (OS). The BSP generates a set of libraries and header files that facilitates the development of software executables. The application C/C++ source files are compiled and linked using the API to build the binary ELF which is executed by the microprocessor.

The SDK menu *File → New → New Board Support Package Project* launches a wizard to create a BSP project (see Fig. 15.13). Choose the platform *hw_platform_0* which is the hardware exported from XPS. Then set the *standalone* OS [18] since it provides interrupts management and it does require a large memory capacity. Other

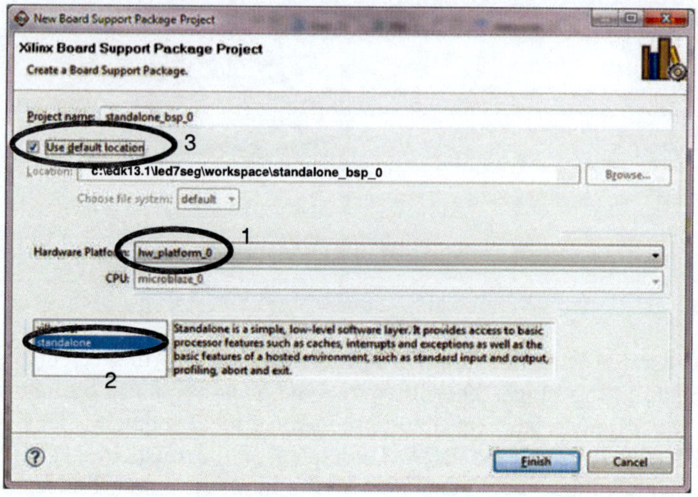

Fig. 15.13 Configuration of the BSP project

OS provide more advanced features, but they require external memory. The BSP project *standalone_bsp_0* is located by default in a folder contained in the SDK workspace.

The wizard generates the BSP project which is linked to a MSS file. The MSS [14] is a text file which list the drivers used by the peripherals and the OS for the microprocessor. The Libgen [13] tool reads the MSS file to generate the BSP. As with the MHS file, the MSS can be graphically or manually edited. Figure 15.14 shows the graphical view which configures the MSS. Change the standard input (*stdin*) and output (*stdout*) to the instance *rs232* in order to permit the console functions to use the UART peripheral.

The MSS file can also be manually edited, and it reflects the configuration changes done in the previous dialog.

```
BEGIN OS
 PARAMETER OS_NAME = standalone
 PARAMETER OS_VER = 3.01.a
 PARAMETER PROC_INSTANCE = microblaze_0
 PARAMETER STDIN = rs232
 PARAMETER STDOUT = rs232
END
```

The rest of the MSS file shows the drivers and peripherals. A peripheral driver is a collection of declarations and functions that can be used to control it from the executable. By default the BSP wizard sets a specific driver to every peripheral, but the user can change it to set a generic driver or no driver. The generic driver can control any peripheral, but the user must have a deeper knowledge about its internal architecture. The system's hardware provides two GPIO peripherals: the

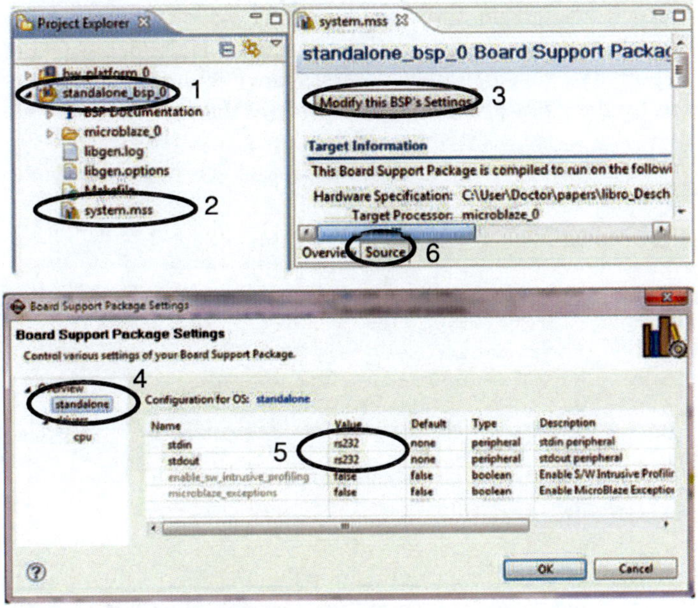

Fig. 15.14 Configuration on the BSP

switches and the *led7seg* instances. Select the generic driver for them in order to understand better the role of internal registers of peripherals.

```
BEGIN DRIVER
 PARAMETER DRIVER_NAME = generic
 PARAMETER DRIVER_VER = 1.00.a
 PARAMETER HW_INSTANCE = switches
END

BEGIN DRIVER
 PARAMETER DRIVER_NAME = generic
 PARAMETER DRIVER_VER = 1.00.a
 PARAMETER HW_INSTANCE = led7seg
END
```

The SDK automatically calls the Libgen tool [13] when the MHS is changed to build the BSP. The user may disable the *Build Automatically* behaviour in order to clean or build the BSP using the commands under the menu *Project*. The Libgen tool compiles the source files of the peripheral drivers and the OS, and it stores them into the A (archive) library files. It also generates the H (header) files that declare the functions contained in the libraries. The library and header files are stored in the folders *lib* and *include* of the instance *microblaze_0*. The SDK can display the contents of both folders and open the header files.

An important header file is the *xparameters.h* which declares a set of parameters about the hardware. Every peripheral has its own parameters that are obtained from the exported hardware, as the addresses range of the GPIOs. The declarations can be used by the C/C++ source files to control the peripherals.

```
/* Definitions for driver GPIO */
#define XPAR_XGPIO_NUM_INSTANCES 2

/* Definitions for peripheral LED7SEG */
#define XPAR_LED7SEG_BASEADDR 0x81400000
#define XPAR_LED7SEG_HIGHADDR 0x8140FFFF

/* Definitions for peripheral SWITCHES */
#define XPAR_SWITCHES_BASEADDR 0x82400000
#define XPAR_SWITCHES_HIGHADDR 0x8240FFFF
```

15.1.2.6 Executable

The SDK will build the ELF executable from the source C++ files that are compiled and linked with the functions stored in the BSP libraries. Click the menu *File → New → Xilinx New C++ Project* which opens a wizard to create a C++ project for the BSP. Change the default project name to *app1* and select the previously generated BSP *standalone_bsp_0*, as depicted in Fig. 15.15. The wizard creates a nested folder *app1/src* in the SDK workspace to store the source files that will be compiled.

Using the Windows Explorer delete the template file *main.cc* which was created by the wizard, and copy the new source files: *ledseg7.cc*, *led7seg.h* and *application.cc*. Go to SDK and click the menu *Refresh* of the contextual menu (right button of the mouse) of the project *app1*, in order to update the list of source files. The SDK can open and display the source files in its integrated editor (see Fig. 15.16).

The source files *led7seg.h* and *led7seg.cc* declare and implement a C++ class named *CLed7Seg* which controls the display through the GPIO. The EDK peripherals implement a set of 32-bit registers that are used to control them. The peripheral's registers are memory-mapped, therefore, MicroBlaze can access them when it executes read/write instructions to the content of a C/C++ pointer. The application can directly control the peripheral, although, it is necessary to have a deeper knowledge about the internal architecture. The first register of a GPIO [5] is the *GPIO_DATA* which is mapped at the base address of the peripheral. The register retrieves/sets the state of the input/output ports depending if the microprocessor reads/writes it.

The class constructor assigns the input argument to the integer (32-bit) pointer *GPIO_Data*. Any pointer used to access a peripheral's register should be declared *volatile*. If not, the compiler may optimize a set of sequential memory accesses through the pointer, changing the order or deleting some of them. The *GPIO*

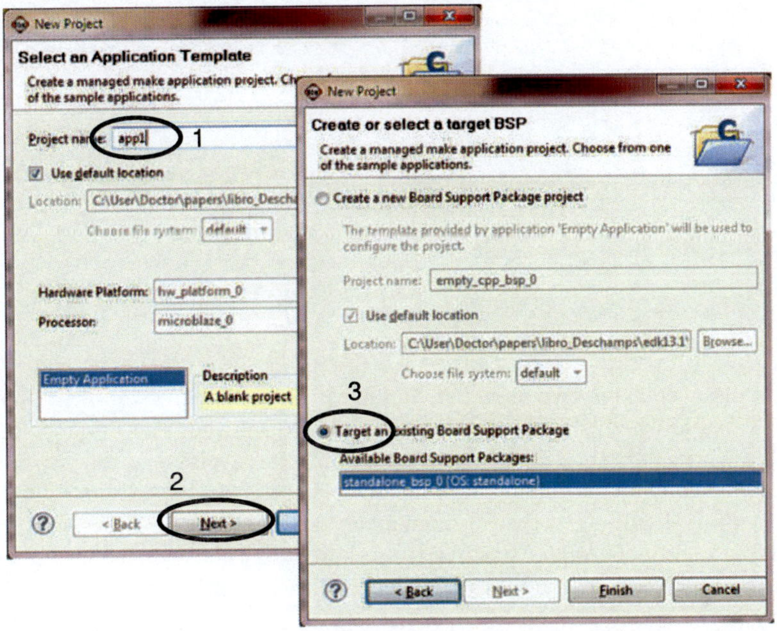

Fig. 15.15 Creating a new C++ project in SDK

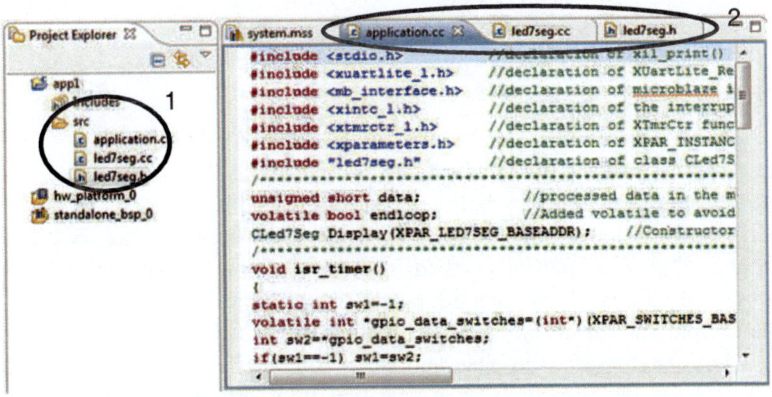

Fig. 15.16 Displaying the C++ source files

method concatenates the anodes and segments to write them into the *GPIO_DATA* register through its pointer. The header file of the class declares the two parameters that configure the active state of the anodes and segments of the display, therefore, they can be easily changed to adapt it to another prototyping board.

```
volatile int* GPIO_Data;        //Pointer to register GPIO_DATA
...
CLed7Seg::CLed7Seg(int gpio_baseaddr)     //Class constructor
{GPIO_Data=(int*)(gpio_baseaddr);}        //Pointer assignment
...

void CLed7Seg::GPIO(unsigned char anodes,
                    unsigned char segments)
{if(LED7SEG_SEGMENTS_ACTIVE_LOW) segments=(~segments)&0x7F;
 if(LED7SEG_ANODES_ACTIVE_LOW)   anodes=(~anodes)&0x0F;
 *GPIO_Data=(anodes<<7)|segments;}          //Writes GPIO_DATA
```

The class declares two member variables: *Data* and *Config*. The *Data* is a 16-bit variable which stores the number which is displayed. The *Config* is an 8-bit variable which uses the two LSBs to turn on/off the display and to show/hide the left-side zeros of the number. The method *Refresh* is periodically executed since the timer's ISR calls it. It reads the member variables and calls the *Digit* method to display one of the digits starting at the left side. The *Digit* method first computes the segments and anodes of a digit and then it calls the *GPIO* method to display it.

```
void CLed7Seg::Refresh()          //Called from the timer's ISR
{static char idx=3;               //Start at the left-side digit
 char off=(Config&0x02)>>1;       //Turn on/off the display
 char zeros=(Config&0x01);        //Show/hide the left-side zeros
 Digit(off,zeros,Data,idx);       //Display a single digit
 idx=(idx==0)? 3 : idx-1;}        //Next call will get next digit
```

The application C++ file is composed of 4 sections. The first section opens the required header files. The application controls the GPIOs directly, but the rest of the peripherals are controlled through their drivers. Therefore, it opens the header files that declare the functions stored in the BSP libraries. The file *xparameter.h* declares base addresses that are necessary to use the driver functions.

The second section initializes the object *Display* for the class *CLed7Seg*. The object's constructor gets the base address of the GPIO which drives the display. The section also declares the global variables *data* and *endloop* that are managed by the ISR of the UART.

The third section is composed of the two ISRs. The timer's ISR periodically reads the state of the two external switches and refreshes the display. First, it reads the register *GPIO_DATA* of the peripheral which attaches to the two external switches. Then, it compares the state of the switches against the previous call. A change in one of the switches will swap one of the two bits that configure the display, using the bitwise XOR operators. Finally, it refreshes the display. The other ISR is executed when the UART receives a new character which is read using its driver function [9]. Depending on the received character, it changes the data or the configuration of the display, or it quits the application.

```
#include <xparameters.h>    //declaration of MHS parameters
#include "led7seg.h"        //declaration of class CLed7Seg
...
CLed7Seg Display(XPAR_LED7SEG_BASEADDR);  //Object Display
unsigned short data; volatile bool endloop;
...
void isr_timer()                         //The timer ISR
{...
 volatile int *gpio_data_switches
      =(int*)(XPAR_SWITCHES_BASEADDR+0);//Pointer GPIO_DATA
 int sw2=*gpio_data_switches;             //Read switches
 ...
 Display.Config^=(sw1^sw2);       //Swap configuration bits
 Display.Data=data;               //Update the displayed data
 Display.Refresh();               //Refresh the display
 sw1=sw2;                         //Store the switches state
 ...}

void isr_rs232()                 //The UART ISR
{...
   char rs232_char=XUartLite_RecvByte(XPAR_RS232_BASEADDR);
   switch(rs232_char){           //Read received character
      case '+': data++; break;   //Increment displayed data
      case 'z': Display.Config^=LED7SEG_ZEROS; break;
      case 'x': endloop=true; break;   //Quit the app
      ...}
 ...}
```

The last section is the main function of the application. It configures and enables the interrupt sources, and then it executes a loop until the application quits. The loop can execute any computation without affecting the control of the display.

The timer peripheral [6] implements two registers to periodically generate an interrupt request: Timer Control/Status Register 0 (TCSR0) and Timer Load Register 0 (TLR0). The configuration of both registers asserts the interrupt signal every 5 ms (250,000 counts, 50 MHz clock). The constants and functions of the timer driver are declared in the header file *tmrctr_l.h* [10] which was generated by the BSP. They are low-level API functions since the programmer knows the functionality of the registers. These functions compute the address of the registers and write data into them through a pointer.

The driver of the interrupt controller provides functions [11] to register ISRs and to enable the interrupts sources. Finally the application enables the MicroBlaze's interrupt input through an OS function [18].

```
int main()
{...
 XTmrCtr_SetLoadReg(XPAR_TIMER_BASEADDR, 0,
   LED7SEG_REFRESH_COUNTS);      //Writes TLR0 register
 XTmrCtr_SetControlStatusReg(XPAR_TIMER_BASEADDR, 0,
   XTC_CSR_LOAD_MASK);           //Loads the counter register
 ...
 XIntc_RegisterHandler(XPAR_INT_CONTROL_0_BASEADDR,
   XPAR_INT_CONTROL_0_TIMER_INTERRUPT_INTR,
   (XInterruptHandler)isr_timer, NULL); //register Timer ISR
 ...
 XIntc_MasterEnable(XPAR_INT_CONTROL_0_BASEADDR);
 ...
 microblaze_enable_interrupts(); //standalone OS function
 ...}
```

By default, the wizard of the C++ project generates two targets: *Debug* and *Release*. They differ in the flags of the GNU compiler [13]. The target *Debug* compiles source files without optimizations and enabling debug symbols. The target *Release* compiles source files with optimizations to build smaller and faster code which is not suitable to debug. The targets configure other compiler flags that are derived from the MicroBlaze's configuration in order to use the optional machine instructions. The menu *Project → Build Project* builds the active target which can be changed anytime using the *Project → Build Configurations → Set Active*. Then, SDK runs the GNU tool chain which compiles the source files and it links the resulting object code with the BSP libraries. The executable ELF file is stored in a nested folder which is named as the target.

15.1.3 Programming and Debugging

The SDK menu *Xilinx Tools → Program FPGA* opens a dialog window which displays the BIT and BMM files that were imported from EDK, as shown in Fig. 15.17. Select the ELF file *app1.elf* of any of the two targets and press the button *Program*. It calls the Data2MEM tool [20] which generates a new bitstream *download.bit* from an existing bitstream and its annotated BMM file. The new bitstream configures the entire FPGA including the BRAM contents with the selected ELF file. Then, the new bitstream is programmed into the FPGA and the application begins to run.

The serial ports of the FPGA board and the PC are attached to test the application. The PC should execute a terminal configured with the same communication parameters as the embedded system. The terminal displays the received characters from the FPGA and sends the characters that are pressed on the keyboard. The SDK provides its own terminal which can be used for this

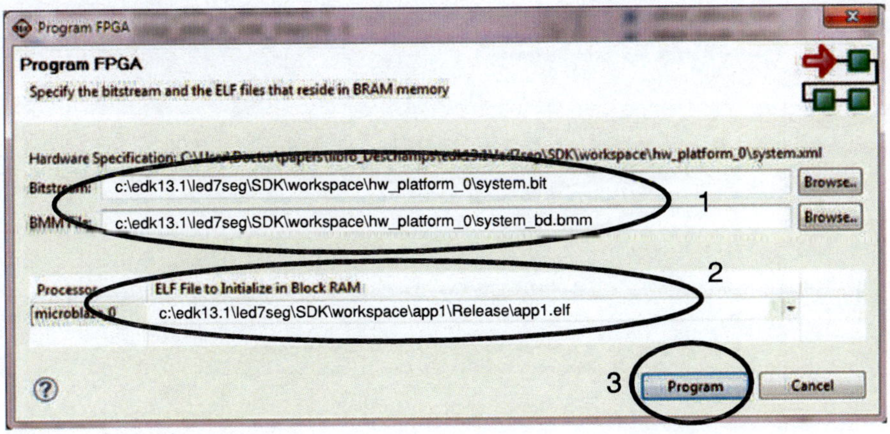

Fig. 15.17 Bitstream configuration to program the FPGA

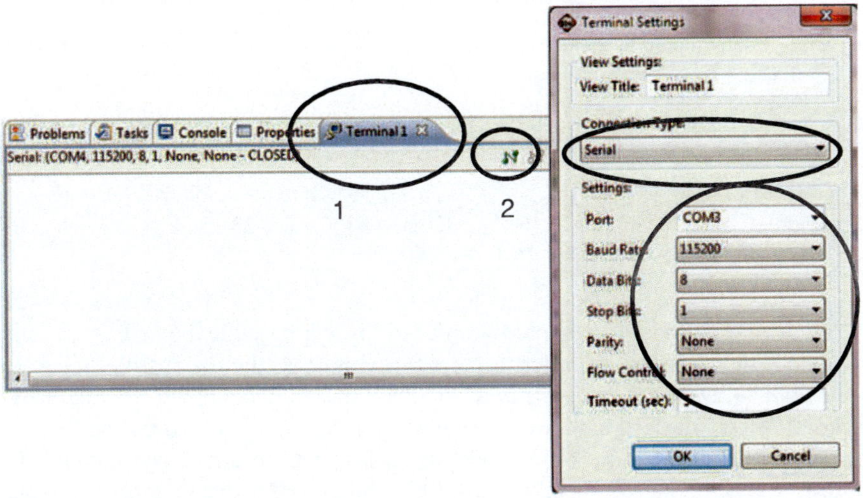

Fig. 15.18 Configuration of the terminal in SDK

purpose (see Fig. 15.18). Press the reset button of the FPGA board and play with the terminal.

The Xilinx Microprocessor Debugger (XMD) [13] is a low-level tool which manages the programming and debugging of the embedded system through the MDM peripheral and the JTAG programming cable. The user can interact with the XMD clicking the SDK menu *Xilinx Tools → XMD console*

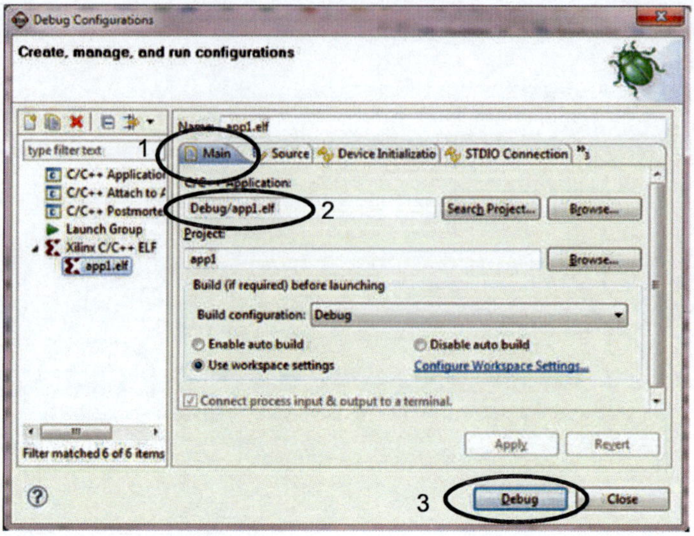

Fig. 15.19 Configuration of the debugger

```
XMD% fpga -f
  C:/edk13.1/led7seg/SDK/workspace/hw_platform_0/download.bit

Fpga Programming Progress
......10...20...30.......70....80...90....Done
Successfully downloaded bit file.
```

The debugging permits the programmer to inspect variables, insert breakpoints or execute step-by-step, in order to correct or improve the executable. Click the menu *Run* → *Debug Configurations* to open the dialog shown in Fig. 15.19. Ensure it selects the ELF under the target *Debug*. SDK will ask to switch to a new perspective which facilitates the debugging tasks, therefore it is recommended to confirm it. The debug perspective shows the source code, disassembly, variables, microprocessor's registers, memory, breakpoints, XMD console and more. The user can debug the executable, manually launching XMD commands which is quite uncomfortable. The SDK debugger relies on XMD to send or retrieve data through a graphical view. The debugger starts uploading the ELF into the BRAM and suspending it at the first executable line of the source file.

Set a breakpoint on a line of the timer's ISR and resume the application to observe the display refreshing (see Fig. 15.20). The tab *Variables* shows the local variables of the ISR that are updated when the user plays with the switches and resumes the application. The tab *Expressions* permits to display the object *Display* and the global variables *data* and *endloop*.

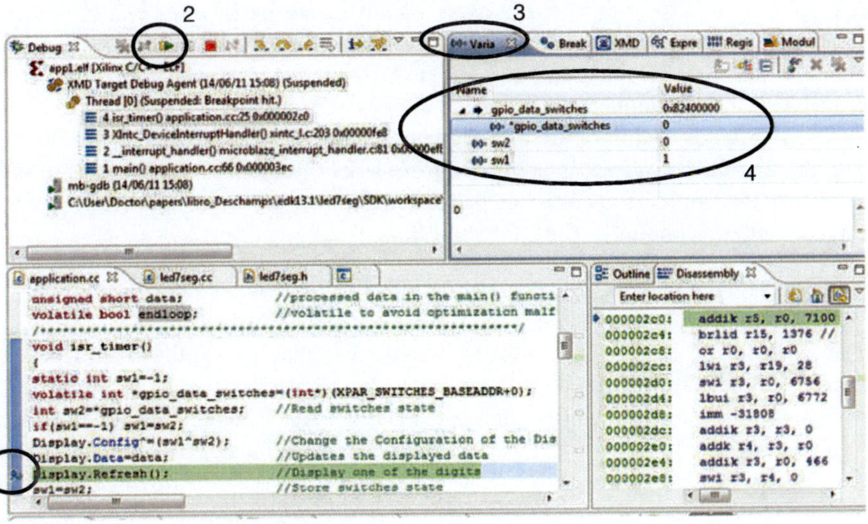

Fig. 15.20 Debugging the application

In a similar way the user can place another breakpoint at the assignment of the variable *rs232_char* in the ISR of the UART. The application will suspend when it receives a new character from the PC. Then, the ISR updates the data or the configuration of the display.

15.2 Case Study 1-B: Creating a Custom Peripheral

The previous embedded system devotes three peripherals (two GPIOs and a timer) to drive the display and to read the switches. The executable uses the timer's ISR to periodically read the switches and refresh the display. A dedicated peripheral can improve the design since it can replace the ISR and peripherals that are devoted to a specific task. The main disadvantage is the greater design effort since the designer must develop the source files of the hardware and the software driver. The designer will surely require simulating the source VHDL files, in order to verify and modify the hardware of the peripheral.

This case study starts from the previous one. It modifies the hardware design and the executable to incorporate the new peripheral. Copy the previous project folder and rename it to *led7seg_ip*. Then, open the XMP file to launch the XPS.

15.2.1 Design of a Custom Peripheral

A peripheral performs a specific task on hardware. Typically they are attached to the PLB bus as a slave in order to permit the microprocessor to access their internal registers to control them. More sophisticated peripherals can be attached as PLB masters in order to access memory themselves. These kinds of peripherals are quite harder to develop and they are much less common, therefore, they are not covered in this example.

The hardware of a peripheral is implemented from a set of source VHDL files. The driver of a peripheral is compiled from a set of source C and H files to build the BSP. The source files must be organized in a set of named folders in order to use the new peripheral in the EDK and SDK.

15.2.1.1 Hardware Design

The hardware is described in a set of VHDL files that are synthesized during the design flow. EDK requires two files in order to permit the integration of the peripheral to the embedded system: the MPD and PAO. The peripheral wizard from XPS generates the folders and a set of template files that must be modified in order to develop the desired IP. Open the wizard by clicking the menu *Hardware → Create or Import Peripheral.*

(1) Select the option *Create templates for a new peripheral.*
(2) Select the *XPS project repository* to create the template files into the local repository which is stored in the XPS directory. The folders *pcores* and *drivers* contain the local repository of the hardware and software.
(3) Set the name the peripheral to *plb_led7seg*, and the version to *1.00.a*. The wizard will create the folders *plb_led7seg_v1_00_a* in the local repositories.
(4) Select the PLB attachment.
(5) Unselect all the slave and master services of the IPIF (IP InterFace) to generate the simplest template files. The IPIFs are IPs deployed with EDK to facilitate the PLB interface.
(6) Continue with the default options of the next three dialogs.
(7) Select the option *Generate template driver file.* This option creates the template files for the driver in the software repository.

The MPD and PAO files are stored in the folder *data* which is nested in the local hardware repository *pcores.*

The Microprocessor Peripheral Description (MPD) file [14] is composed of four sections. The first section declares the description and implementation options. The second section sets the bus interface to Slave PLB (SPLB). The last two sections declare the parameters and ports of the SPLB. The data generated by the wizard in the MPD template must not be modified. However, the designer can add more parameters and ports to extend the peripheral. The parameters will be passed as VHDL generics and ports during the synthesis.

The new peripheral adds three parameters in order to configure the peripheral from XPS. The refresh period is an integer parameter which contains the number of microseconds, and its default value is set to *5,000* μs. The boolean data configure the active state of the ports that drive the display. The peripheral adds two output ports to

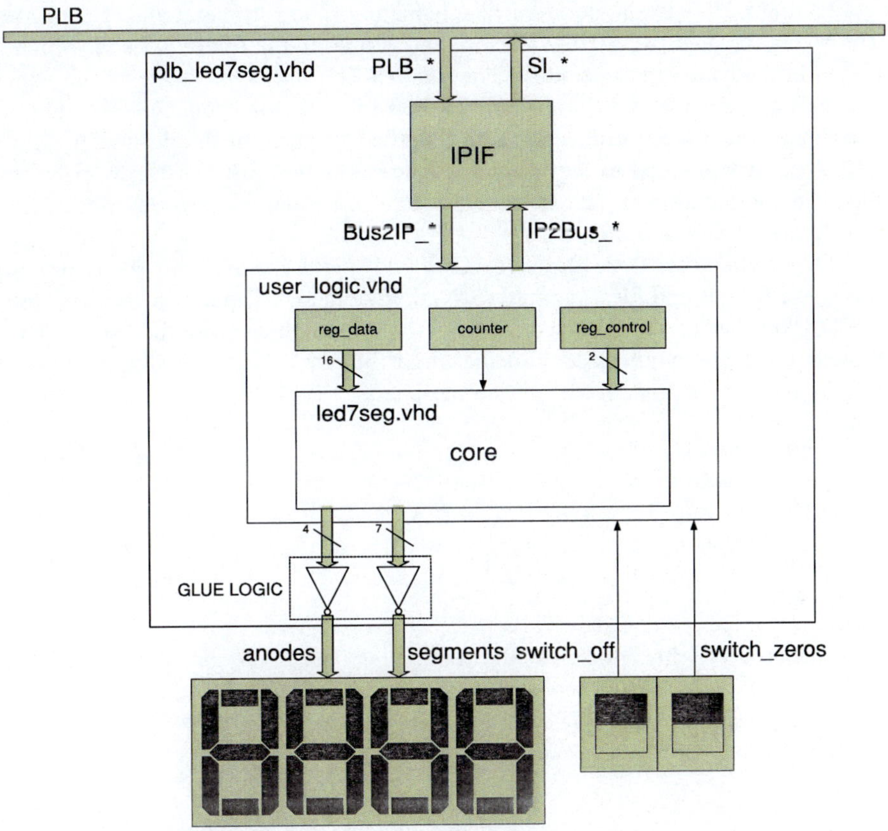

Fig. 15.21 Hierarchical schematic of the peripheral

drive the 7 segments and the 4 anodes of the display. Finally, it declares two input ports that attaches to the external switches that configure the display.

```
BEGIN plb_led7seg
...
## PERIPHERAL's SPECIFIC GENERICS
PARAMETER C_REFRESH_PERIOD_US = 5000, DT = integer
PARAMETER C_SEGMENTS_ACTIVE_LOW = false, DT = boolean
PARAMETER C_ANODES_ACTIVE_LOW = false, DT = boolean
## PERIPHERAL's SPECIFIC PORTS
PORT segments = "", DIR=O, VEC=[6:0]
PORT anodes = "", DIR=O, VEC=[3:0]
PORT switch_zeros= "", DIR=I
PORT switch_off= "", DIR=I
END
```

Figure 15.21 shows a hierarchical schematic of the peripheral and related files. The wizard created two VHDL files that are stored in the folder *hdl\vhdl* which is nested in the hardware repository. The *plb_led7seg.vhd* is the top-level file which connects an instance of the user logic to the PLB. The *user_logic.vhd* is a dummy peripheral, therefore, the file must be modified to perform the desired functionality. The tasks related to the timer's ISR are now described as hardware in this file. The computation related to the class *CLed7Seg* class is now described in the new hardware file *led7seg.vhd*.

The Peripheral Analyze Order (PAO) file [14] is the ordered list (bottom to top level) of libraries and files required to synthesize the IP. The first two entries refer to EDK libraries due to the selected IPIF. Then, it continues with the list of VHDL files that will be synthesized into the target library. The target library must be named as the repository folder of the peripheral.

```
lib proc_common_v3_00_a  all                          ##EDK library
lib plbv46_slave_single_v1_01_a  all                  ##EDK library
lib plb_led7seg_v1_00_a led7seg vhdl                  ##Added
lib plb_led7seg_v1_00_a user_logic vhdl               ##user logic
lib plb_led7seg_v1_00_a plb_led7seg vhdl              ##top-level
```

The top-level file declares the entity *plb_dec7seg* and its architecture. The template of the entity leaves space to add new generics and ports, therefore, the user must complete it. The VHDL generics and ports of the entity must be declared in the same way as in the MPD file.

```
ENTITY plb_led7seg IS
  GENERIC (
    -- ADD USER GENERICS BELOW THIS LINE ------------
    c_refresh_period_us:                 INTEGER:=5000;
    c_segments_active_low:               BOOLEAN:=FALSE;
    c_anodes_active_low:                 BOOLEAN:=FALSE;
    -- ADD USER GENERICS ABOVE THIS LINE ------------
    ...);
  PORT (
    -- ADD USER PORTS BELOW THIS LINE ---------------
    segments:        OUT STD_LOGIC_VECTOR(6 DOWNTO 0);
    anodes:          OUT STD_LOGIC_VECTOR(3 DOWNTO 0);
    switch_off:      IN STD_LOGIC;
    switch_zeros:    IN STD_LOGIC;
    -- ADD USER PORTS ABOVE THIS LINE ---------------
    ...);
```

The architecture declares an instance to the user logic. It leaves space to map the new generics and ports. The two switches are directly connected from the top-level input ports to the user logic. The glue logic attaches the output ports from the user logic to the top-level ports, in order to drive the display. The synthesizer will infer inverters between these ports if the configuration parameters are set to true. The user logic will provide its internal timer, therefore, the instance requires a new generic which configures the number of clock cycles to refresh the display. It is computed from the MPD parameters that define the refreshing period (microseconds) and the clock period (picoseconds). Finally, the user logic also requires a generic which configures the number of internal registers.

```
user_logic_i : ENTITY plb_led7seg_v1_00_a.user_logic
  GENERIC MAP (
    -- MAP USER GENERICS BELOW THIS LINE -------------
    c_refresh_counts => (c_refresh_period_us*1000)
                        /(c_splb_clk_period_ps/1000),
    -- MAP USER GENERICS ABOVE THIS LINE -------------
    c_num_reg => user_num_reg,
    ...)
  PORT MAP (
    -- MAP USER PORTS BELOW THIS LINE ----------------
    anodes=> user_anodes,
    segments=> user_segments,
    switch_off=> switch_off,
    switch_zeros=> switch_zeros,
    -- MAP USER PORTS ABOVE THIS LINE ----------------
    ...);
...
--Glue logic
segments <=NOT user_segments WHEN c_segments_active_low
           ELSE user_segments;
anodes   <=NOT user_anodes WHEN c_anodes_active_low
           ELSE user_anodes;
```

The architecture also contains the IPIF instance which eases the PLB connection. It adapts the PLB handshake to/from IP signals named *ipif_Bus2IP_*/ ipif_IP2Bus_**. The IPIF also decodes the PLB address bus to enable one of the peripheral's internal registers. The architecture declares the number of internal registers in the user logic, which is two in this case. This number affects the width of the signals *Bus2IP_RdCE* (Read Chip Enable) and *Bus2IP_WrCE* (Write Chip Enable) that arrives to the user logic. When MicroBlaze executes a read/write instruction to a peripheral's address range, the IPIF decoder sets one of the enable bits, in order to read/write one of the internal registers of the user logic. The IPIF maps registers on 32-bit boundaries from the peripheral's base address. The first register is mapped on *C_BASEADDR*, the second register on *C_BASEADDR+4*, and so on. In order to simplify the IPIF, the *C_HIGHADDR* parameter is usually configured much higher than necessary, but it leads to an incomplete address decoding. In this case, the peripheral's registers can be accessed from different addresses since the IFPF does not decode some bits of the address bus.

```
CONSTANT user_slv_num_reg : INTEGER := 2;
CONSTANT user_num_reg: INTEGER := user_slv_num_reg;
```

The template of the user logic is a dummy design which must be modified. The entity template declares an important generic which is the number of registers. The entity must be completed according to the top-level file. Therefore, the designer must add the new generics to configure the refreshing counts and the new ports to drive the display and to read the switches. The default values of the generics are overwritten since they are passed from the instance at the top-level file.

```
ENTITY user_logic IS
   GENERIC (
   -- ADD USER GENERICS BELOW THIS LINE -------------
   c_refresh_counts:                  INTEGER := 1000;
   -- ADD USER GENERICS ABOVE THIS LINE -------------
   c_num_reg:                         INTEGER := 1 );

   PORT (
   -- ADD USER PORTS BELOW THIS LINE ----------------
   segments:          OUT STD_LOGIC_VECTOR(6 DOWNTO 0);
   anodes:            OUT STD_LOGIC_VECTOR(3 DOWNTO 0);
   switch_zeros:      IN STD_LOGIC;
   switch_off:        IN STD_LOGIC;
   -- ADD USER PORTS ABOVE THIS LINE ----------------
   ...);
```

The architecture of the user logic declares two registers to control the display that are accessible from the PLB. The first one (*reg_data*) is a 16-bit data register which sets the 4 digits to display. The second one (*reg_control*) is a 2-bit register which controls the display configuration: one bit to turn on/off the display and the other bit to show/hide the left-hand zeros. The input ports *Bus2IP_RdCE* and *Bus2IP_WrCE* provide a single bit for each register to read or write them. The data comes from the port *Bus2IP_Data* during a write access to one of the registers. During a read access, one of the registers drives the port *IP2Bus_Data*. The architecture uses the signals *bus_data*/*ip_data* to get/set the data bus, since they are declared in descending bit order as the registers. The PLB must acknowledge the access completion before MicroBlaze can continue with a new access. The user logic asserts the ports *IP2Bus_RdAck*/*IP2Bus_WrAck* when it completes the access to the registers. In this case, the user logic does not require wait states to read/write the registers.

```
SIGNAL reg_control: STD_LOGIC_VECTOR (1 DOWNTO 0);
SIGNAL reg_data: STD_LOGIC_VECTOR (15 DOWNTO 0);
...
bus_data <= Bus2IP_Data;                       --Reads data bus
IP2Bus_Data  <= ip_data;                       --Writes data bus
IP2Bus_WrAck <= or_reduce(Bus2IP_WrCE);        --Acknoledges write
IP2Bus_RdAck <= or_reduce(Bus2IP_RdCE);        --Acknoledges read
IP2Bus_Error <= '0';

read_reg: PROCESS(Bus2IP_RdCE, reg_data, reg_control)
BEGIN
  ip_data<=(OTHERS=>'0');                      --default value
  CASE Bus2IP_RdCE(0 TO 1) IS                  --read a register
    WHEN "10"=>   ip_data(15 DOWNTO 0)<=reg_data;
    WHEN "01"=>   ip_data (1 DOWNTO 0)<=reg_control;
    WHEN OTHERS=> NULL;
  END CASE;
END PROCESS;

write_reg: PROCESS(Bus2IP_Reset, Bus2IP_Clk)
BEGIN
  IF Bus2IP_Reset='1' THEN                     --reset registers
    reg_data<=(OTHERS=>'0'); reg_control<=(OTHERS=>'0');
  ELSIF RISING_EDGE(Bus2IP_Clk) THEN
    IF Bus2IP_WrCE(0)='1' THEN                 --write reg_data
      reg_data<=bus_data(15 DOWNTO 0);
    END IF;
    IF Bus2IP_WrCE(1)='1' THEN                 --write reg_control
      reg_control<=bus_data(1 DOWNTO 0);
    ELSIF refresh='1' THEN                     --refresh reg_control
      reg_control<=reg_control XOR (sw1 XOR sw2);
    END IF;
  END IF;
END PROCESS;
```

The rest of the user logic performs the tasks executed in the timer's ISR in the software application. It implements a counter which periodically asserts the signal *refresh*. This signal is used to refresh the display and to capture the state of the switches in order to modify the control register. There is an instance, named *core*, of the IP *led7seg*, which drives the display from the registers contents.

```
core: ENTITY plb_led7seg_v1_00_a.led7seg(beh1) PORT MAP(
      rst=> Bus2IP_Reset,       --Reset input
      clk=> Bus2IP_Clk,         --Clock input
      refresh=> refresh,        --refresh input
      segments=> segments,      --Drive the display
      anodes=> anodes,          --Drive the display
      data=> reg_data,          --Input from data register
      zeros=> reg_control(0),   --Input from control register
      off=> reg_control(1) );   --Input from control register
```

The file *led7seg.vhd* implements the functionalities described in the C++ class *CLed7Seg* in the software application. It generates the signals that drive the display, updating the displayed digit when the port *refresh* is asserted. The architecture is composed of two processes. The first process updates the index of the digit to refresh, starting at the left side. The second process computes the display's ports based in the input ports *off*, *zeros* and *data* that are driven from the peripheral's registers.

15.2.1.2 Driver Design

A peripheral's driver is built from header and implementation files. The C files implement functions that control the peripheral. The header H files contain the declarations required to use the functions. The SDK requires two additional files, the MDD and the TCL, and a structured folder organization in a repository in order to build the BSP with the peripheral's driver. The Microprocessor Driver Definition (MDD) file [14] declares the supported peripherals, dependencies of the driver and files to copy. The TCL file is a script used to compile the source files and build the library. The XPS peripheral wizard generates the template files and folders. The designer usually does not modify the MDD and TCL files, but he must rewrite the H and C files to develop the driver.

The MSS file controls the peripheral drivers that are built in the BSP. The Libgen tool creates a folder named as the microprocessor and copies the H files into the folder *include*. Then it compiles the C files and adds the functions into the library *libxil.a* which is stored in the folder *lib* of the BSP.

The header file contains the declarations that are used to build the BSP and the executable. First, it defines the masks to control the display. Next, it declares C macros to read or write the peripheral's registers through a pointer. Finally, there are the declarations of the visible functions.

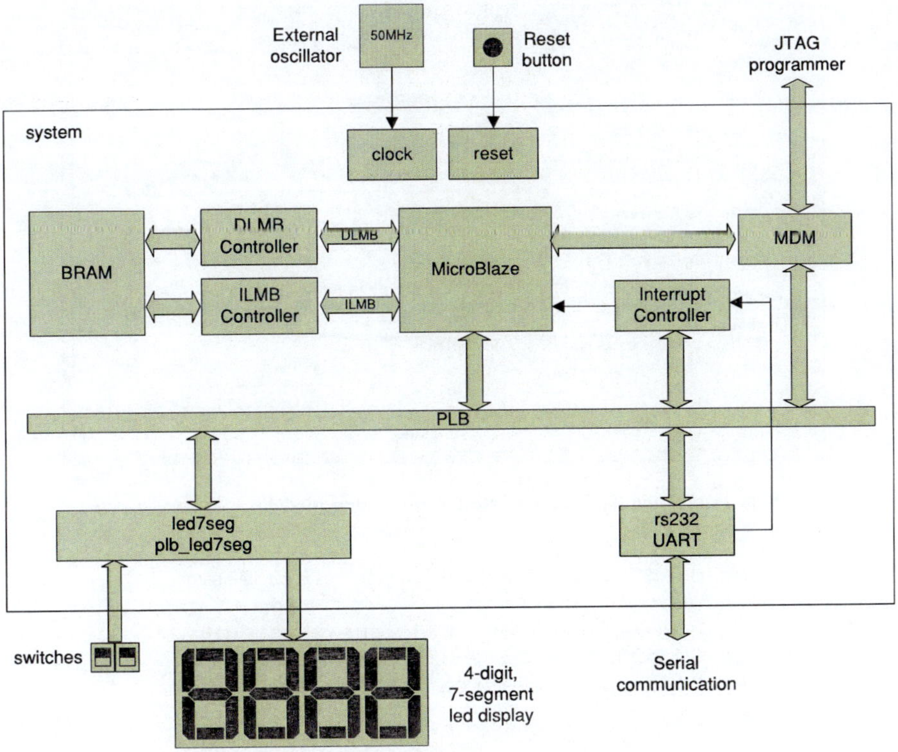

Fig. 15.22 Overview of the modified system architecture

```
#define LED7SEG_ZEROS_MASK (0x1)      //Display left-side zeros
#define LED7SEG_OFF_MASK (0x2)        //Turn-off display
...
#define Led7Seg_GetStatus(baseaddr)
   *((volatile int*)(baseaddr+4))                       //Read macro
#define Led7Seg_SetControl(baseaddr, config)
   *((volatile int*)(baseaddr+4))=(int)(config)  //Write macro
void Led7Seg_SwapOff(int baseaddr);                     //Function
void Led7Seg_SwapZeros(int baseaddr);                   //Function
```

The C file of the driver is compiled during the BSP generation. The library file stores the functions, but not the C macros. The library may store other internal functions that are not visible for the programmer since they are not declared in the H file. The function which swaps one of the bits of the control register executes two accesses to the register. First, MicroBlaze reads the register and changes one bit according to a mask. Next, it writes the resulting value into the same register.

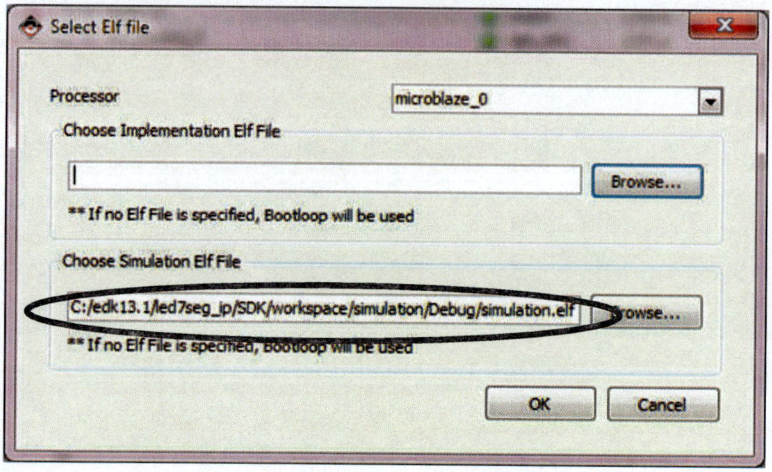

Fig. 15.23 Selection of the ELF to generate the simulation model

```
void Led7Seg_SwapControl(int baseaddr, char mask)
{Led7Seg_SetControl( baseaddr,
                     Led7Seg_GetStatus(baseaddr)^(mask) );}
void Led7Seg_SwapOff(int baseaddr)
{Led7Seg_SwapControl( baseaddr, LED7SEG_OFF_MASK );}
```

15.2.2 System's Hardware/Software Modification

The new peripheral replaces the timer and the GPIOs connected to the display and switches, as well as the software code related to them. Therefore, the hardware and software of the new system must be modified, accordingly.

15.2.2.1 Hardware Modification

The MHS file of the new system adds the instance *led7seg* of the PLB display controller *plb_led7seg*, and it removes the timer and GPIOs, as depicted in Fig. 15.22. The external ports connected to the display and switches are assigned to the display controller. The parameters that configure the refresh period and active state of the anodes and segments can be easily modified to adapt them to other FPGA board. The MHS parameters overwrite the default values defined in the peripheral's MPD file. The modified system requires just one interrupt source for the UART, therefore, the interrupt controller could be eliminated. However, the system maintains it since it does not increase significantly the area and it provides the possibility to add new peripherals to request interrupts.

```
PORT fpga_0_led7seg_segments_pin=led7seg_segments, DIR = O,
     VEC = [6:0]
PORT fpga_0_led7seg_anodes_pin=led7seg_anodes, DIR = O,
     VEC = [3:0]
PORT fpga_0_led7seg_zeros_pin=led7seg_switch_zeros, DIR = I
PORT fpga_0_led7seg_off_pin=led7seg_switch_off, DIR = I
...
BEGIN plb_led7seg
  PARAMETER INSTANCE = led7seg
  PARAMETER HW_VER = 1.00.a
  PARAMETER C_BASEADDR = 0x84C00000
  PARAMETER C_HIGHADDR = 0x84C0FFFF
  PARAMETER C_REFRESH_PERIOD_US = 4000
  PARAMETER C_SEGMENTS_ACTIVE_LOW = true
  PARAMETER C_ANODES_ACTIVE_LOW = true
  BUS_INTERFACE SPLB = mb_plb
  PORT segments = led7seg_segments
  PORT anodes = led7seg_anodes
  PORT switch_zeros = led7seg_switch_zeros
  PORT switch_off = led7seg_switch_off
END

BEGIN xps_intc
  ...
 PORT Intr = rs232_Interrupt
END
```

The names of the external FPGA ports, connected to the display and switches, have changed, therefore, the UCF must be updated.

```
Net fpga_0_led7seg_segments_pin<0>  LOC=E14; #segment-A
Net fpga_0_led7seg_anodes_pin<0>    LOC=D14; #anode-0
Net fpga_0_led7seg_zeros_pin        LOC=K13; #switch zeros
...
```

At this point, the designer can easily test the peripheral's hardware using the XMD tool. Using the XPS, implement the new hardware. Then, click *Device Configuration → Download Bitstream* to create the *download.bit* file and program the FPGA. If the XPS project has no ELF file, the bitstream configures the BRAM to store the default executable *bootloop* which runs a dummy loop. In order to program the FPGA, XPS calls the iMPACT tool with the commands of the batch file *download.cmd*. Check the number of the *–p* flag, which configures the FPGA position in the JTAG chain.

Once the FPGA is successfully programmed, click the *Debug* → *Launch XMD* to open the command shell. The XMD shell initially shows the microprocessor configuration when it connects to the MDM. Then, the user can test the peripheral when he runs the XMD commands: *mwr* (memory write) and *mrd* (memory read). Write the peripheral's base address (*0 × 84C00000*) to change the displayed number. Write the next register (*0 × 84C00004*) to change the configuration of the display. The peripheral's registers are accessible from different addresses due to the incomplete IPIF decoder.

```
...
MicroBlaze Processor Configuration :
-------------------------------------
Version............................8.10.a
Optimization.......................Area
Interconnect.......................PLB_v46
...
Connected to "mb" target. id = 0
Starting GDB server for "mb" target (id = 0)...

XMD% mwr 0x84c00000 0x0AF4         (Display number 0AF4)
XMD% mwr 0x84c00004 0x1            (Display left-side zeros)
XMD% mwr 0x84c00004 0x2            (Turn off display)
XMD% mrd 0x84c00000 4              (Read 4 registers. It shows)
84C00000:   00000AF4                     (The data register)
84C00004:   00000002                     (The control register)
84C00008:   00000AF4                     (The data register)
84C0000C:   00000002                     (The control register)
```

15.2.2.2 Software Modification

At the software level, the executable has to be also modified in order to delete the code related to the GPIOs and the timer's ISR. Export the new hardware from XPS and create a SDK workspace in the *c:\edk13.1\led7seg_ip\SDK\workspace* path. Check that the imported hardware presents the new display controller.

Create a new BSP project and select the standalone OS. By default, the MSS assigns the generic driver to the display controller *led7seg*. A more convenient way is to assign the specific driver to the peripheral in order to build the BSP.

```
BEGIN DRIVER
 PARAMETER DRIVER_NAME = plb_led7seg
 PARAMETER DRIVER_VER = 1.00.a
 PARAMETER HW_INSTANCE = led7seg
END
```

The Libgen tool generates the BSP according to the information described in the MSS file. It gets the source files from the folder *plb_led7seg_v1_00_a* which must be stored in the driver repositories. Then, it copies the header files and builds the libraries. The file *libgen.options* must be modified to configure the local repository to the XPS directory.

```
PROCESSOR=microblaze_0
REPOSITORIES=-lp ../../
HWSPEC=../hw_platform_0/system.xml
```

The C++ code related to the GPIOs and timer's ISR is deleted. Therefore, the files that declare and implement the class *CLed7Seg* are deleted in the project. The modified file *application.cc* calls the driver's functions to control the display.

```
#include <plb_led7seg.h>        //Driver's header file
...
void isr_rs232()
{...
   switch(rs232_char){          //Received key
     case '+':                  //Increments displayed number
       Led7Seg_SetData(XPAR_LED7SEG_BASEADDR,++data); break;
     case 'z':                  //Show/hide the left-side zeros
       Led7Seg_SwapZeros(XPAR_LED7SEG_BASEADDR); break;
   ...}
 ...}
```

The application can be programmed and debugged following the same steps as in the previous case study. Check that the BIT and BMM files are the imported ones from the current XPS project, before proceeding.

15.2.3 Functional Simulation

The source VHDL files should be simulated if the peripheral is not working as expected. The easiest way to test the VHDL files is by performing a functional simulation of the embedded system. The system can be simulated with the ISim or other supported simulator. The XPS can select the simulator when you click the *Edit → Preferences*. The Xilinx ISim has some limitations when it is compared with other third-party simulators. However, the ISim is easy to use, and it provides the simulation libraries.

Instead of simulating the application executable, which depends on the serial communication, it is preferable to develop a much simpler executable. Create a new Xilinx C++ project in SDK named *simulation*. Modify the C++ template file *main.cc* to add a sequence which just writes the peripheral's registers, as desired. The function *delay* is called to improve the observation on the simulation results. Build the target *Debug* to get the file *simulation.elf*.

```
Led7Seg_SetData(XPAR_LED7SEG_BASEADDR,0x001E);
Led7Seg_SetControl(XPAR_LED7SEG_BASEADDR,0x0);
delay();
Led7Seg_SetControl(XPAR_LED7SEG_BASEADDR,LED7SEG_ZEROS_MASK);
delay();
...
Led7Seg_SwapOff(XPAR_LED7SEG_BASEADDR);
delay();
...
```

The XPS project must import the ELF file in order to generate the simulation model. Click the menu *Project → Select Elf file* and choose the file *simulation.elf* which is stored in the SDK project (see Fig. 15.23)

XPS launches the Simgen tool [13] to generate the set of files required to simulate the embedded system running the selected executable. Click the XPS menu *Project → Project Options* which opens the dialog window shown in Fig. 15.24. Select the options: VHDL, generate testbench template and behavioural model. This is the recommended model since it simulates the peripheral's source files. The other models are more complicated, since they simulate the output files from the synthesis or the implementation stages. Click the XPS menu *Simulation → Generate Simulation HDL Files* to call the Simgen tool. The Simgen does not support some FPGA families, such as the Spartan-3, but there is a workaround. Choose any supported device since it does not affect the behavioural simulation of the peripheral, because it does not require the synthesizing of its source VHDL files. Finally, click the menu *Simulation → Launch HDL Simulator* which takes some time to compile the simulation files before starting the ISim [19].

Simgen creates the testbench file *system_tb.vhd*. It declares the instance *dut* (Device Under Test) of the system and the stimulus applied to its inputs. ISim shows the hierarchical view of the instances to simulate. Open the contextual menu of the top-level instance *system_tb* and click *Go To Source Code* in order to edit the testbench template, as shown in Fig. 15.25.

The template file drives the clock and reset signals, and it provides a user's section to write more stimulus. In order to simulate the external switches, change the state of the switch, which turns off/on the display, at 250 and 260 μs.

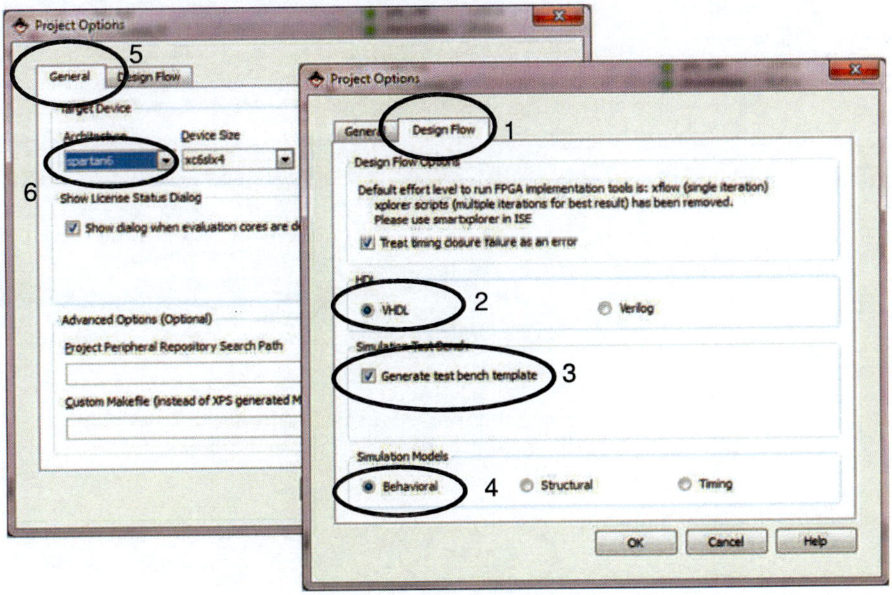

Fig. 15.24 Configuration of the simulation model

```
dut : system PORT MAP (
    fpga_0_clk_1_sys_clk_pin=>fpga_0_clk_1_sys_clk_pin,
    fpga_0_rst_1_sys_rst_pin=>fpga_0_rst_1_sys_rst_pin,
    ...
    fpga_0_led7seg_zeros_pin=>fpga_0_led7seg_zeros_pin,
    fpga_0_led7seg_off_pin =>fpga_0_led7seg_off_pin);
...
-- START USER CODE (Do not remove this line)
fpga_0_led7seg_zeros_pin<='0';
fpga_0_led7seg_off_pin<='0', '1' AFTER 250 us,
                             '0' AFTER 260 us;
-- END USER CODE (Do not remove this line)
```

The simulation will take a huge number of clocks to refresh the display. The display's ports are updated every 200,000 clock cycles when the refresh cycle is set to 4,000 μs (50 MHz clock frequency). Each instance of a peripheral is simulated from a wrapper file which configures the IP according the MHS file. Go through the hierarchical view, and select the instance *led7seg*. Open the wrapper file *led7seg_wrapper.vhd*, and change the generic *C_REFRESH_PERIOD_US* to shorten the simulation of the refresh. This change does not affect the synthesis or the implementation since it does not modify the MHS file.

Fig. 15.25 Hierarchical view of the instances to open the testbench file

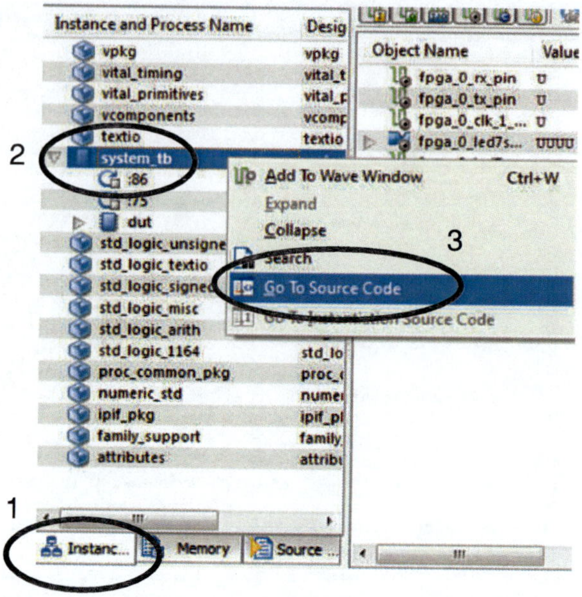

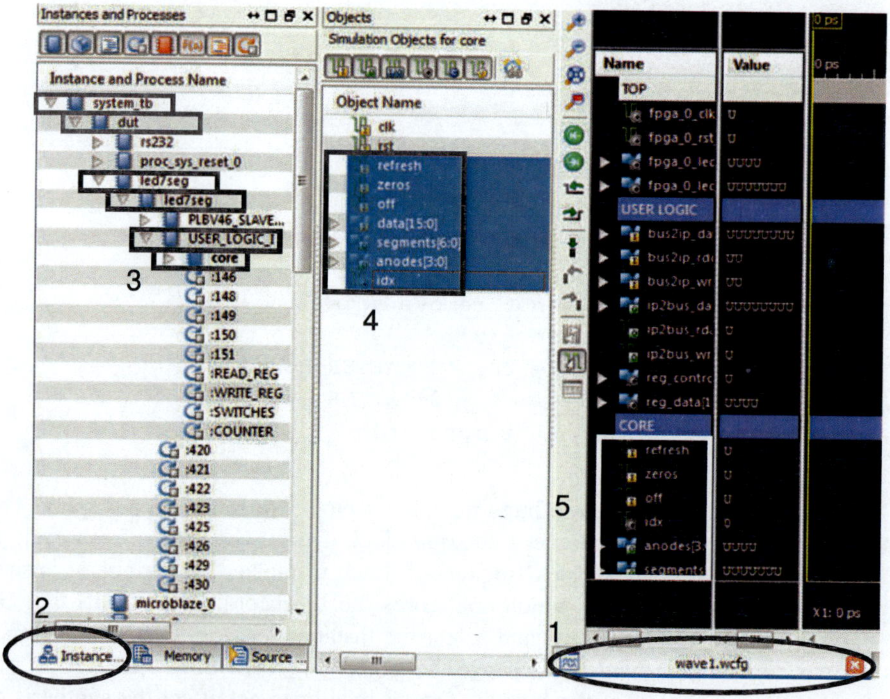

Fig. 15.26 Configuring the waveform window

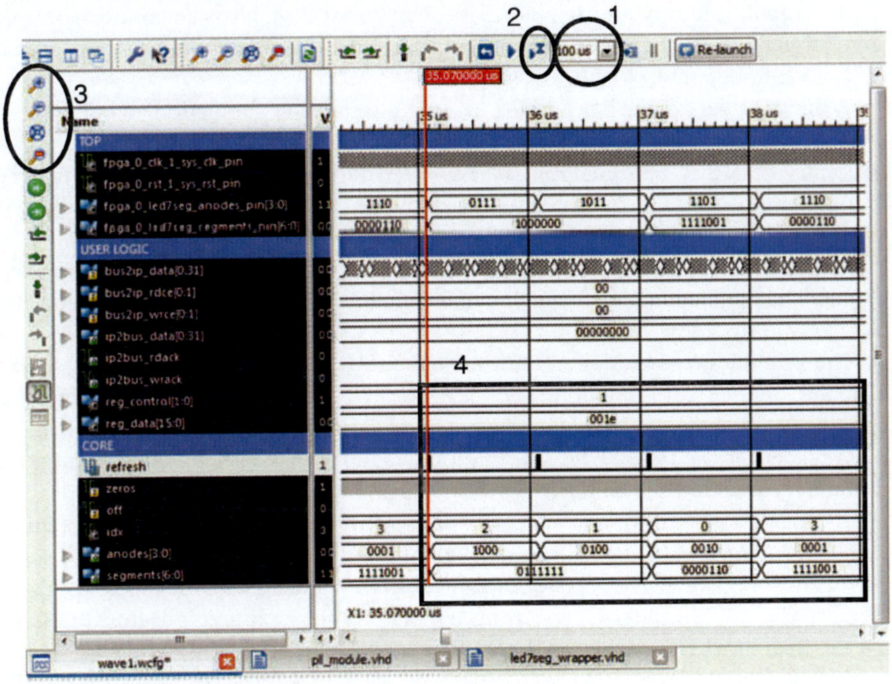

Fig. 15.27 Simulation waveform

```
led7seg : plb_led7seg
  GENERIC MAP (                          --Parameters from the MHS
    c_baseaddr => X"84C00000",
    c_highaddr => X"84C0FFFF",
    ...
    c_refresh_period_us => 1,            --Changed to simulate
    ...)
  PORT MAP (...);
```

Any change in the VHDL files requires the compilation of the model before starting the simulation. Therefore, click the ISim menu *Simulation* → *Relaunch*. In order to display the external ports of the system, select the instance *system_tb* and drag the desired signals into the waveform window. Add a divider, named *USER LOGIC*, in order to display a new group of waveforms separately from the previous ones. Go to the hierarchy of the instances *system_tb* → *dut* → *led7seg* → *led7seg* → *USER_LOGIC_I* and drag the IPIF signals and the peripheral's registers to the waveform. Repeat the previous steps to display the signals of the instance *core*, as shown in Fig. 15.26.

By default, signals are displayed in binary radix which is uncomfortable for some signals. Select the peripheral's registers and the IPIF busses to change the radix to hexadecimal.

Write 100 μs on the simulation time, and run the simulation, as shown in Fig. 15.27. The resulting waveform can be zoomed in/out to show the desired time interval. Check that the signal *refresh* is asserted during a clock cycle every 1 μs, due to the change done in the wrapper file. Display the peripheral's registers at 35 μs. At this point MicroBlaze has already executed the first four lines of the C++ file, therefore, the peripheral's registers (*reg_data*, *reg_control*) contain the expected values. The ports *anodes* and *segments* that drive the display are refreshed according to the value stored in those registers.

The simulation should continue another 100 μs to display how Microblaze executes the driver function *Led7Seg_SwapOff*. The function executes two consecutive accesses to the register *control_reg*: a read followed by a write access. The registers are accessed when the microprocessor executes read/write instructions to memory addresses within the peripheral's address range. The IPIF decodes the PLB address bus in order to assert one of the enabler bits of the signals *Bus2IP_RdCE*/*Bus2IP_WrCE* that arrive to the user logic, in order to read/write a peripheral's registers. The register *control_reg* is associated with the index 1 of these signals. The waveform window permits a search for the next/previous transition of a signal. Search a rising edge of the bit *Bus2IP_RdCE(1)*, as shown in Fig. 15.28. At this time, the MicroBlaze is reading the control register. The peripheral carries the register's content on the signal *IP2Bus_Data* and it asserts the *IP2Bus_RdAck* to complete the read access. The next access, MicroBlaze writes the control register. The peripheral gets the new content from the *Bus2IP_Data* and asserts the *IP2Bus_WrAck* to complete the write access.

Figure 15.29 shows the simulation of the user action on a switch at 250 μs and 260 μs to turn off/on the display, as described in the testbench file. The peripheral updates the control register when the signal *refresh* is asserted, after the user action.

If the C/C++ code of the simulation executable is changed it is necessary to close the ISim before executing Simgen. Save the waveform file before exiting ISim. Then, build the SDK project to get the new ELF file and generate the new simulation model with XPS. Finally, launch ISim, open the saved waveform file, and run a new simulation.

Once the simulation is finished, select the target FPGA in the project options and generate the bitstream. The system will be synthesized and implemented for the selected device.

15.3 Case Study 2: Implementation of a Custom Coprocessor

A coprocessor is a specific processor aimed to accelerate an algorithm. The hardware architecture is designed to efficiently speed-up a specific computation. The rest of computations and tasks are carried on the general-purpose

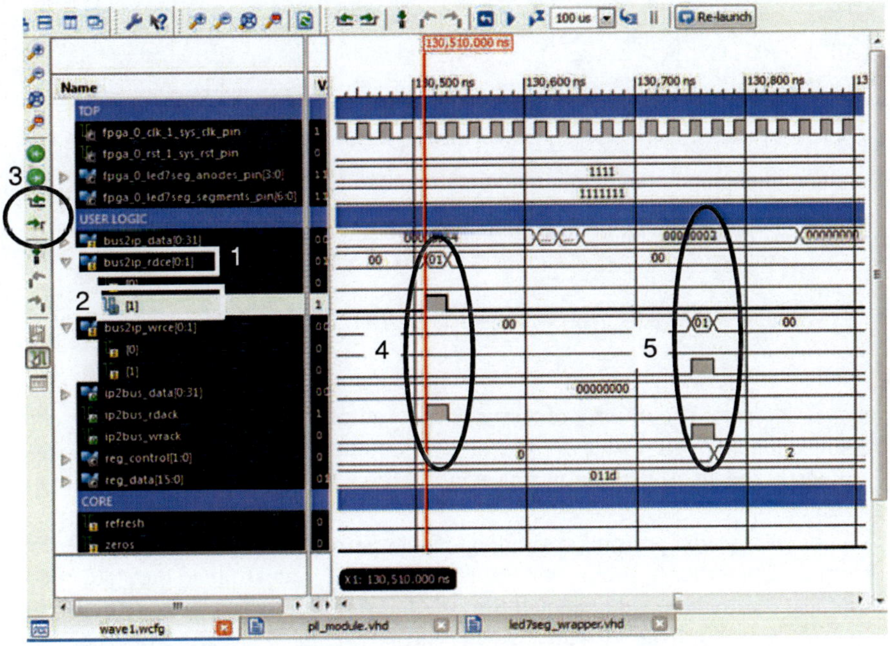

Fig. 15.28 Simulation of a read/write accesses to a peripheral's register

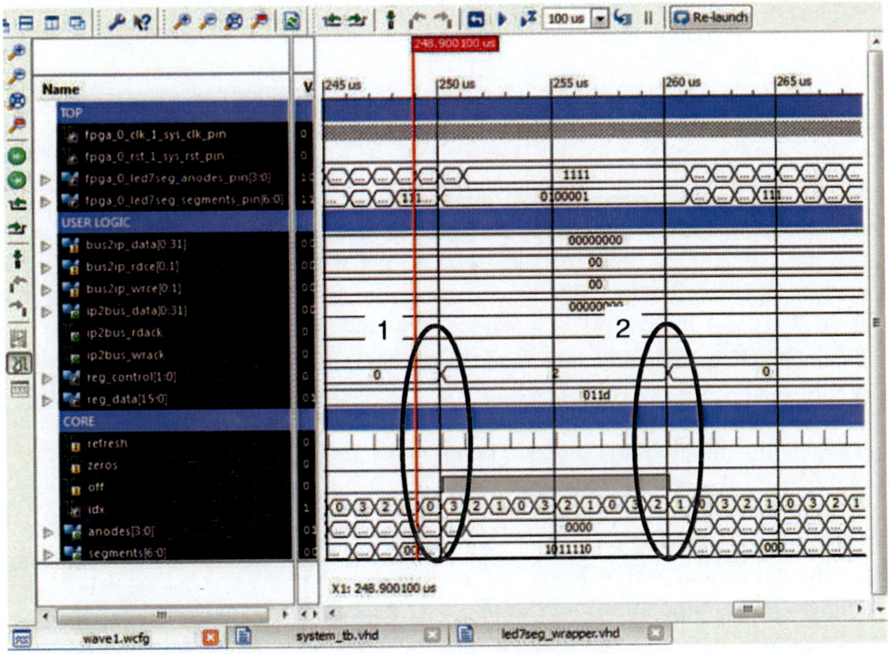

Fig. 15.29 Simulation of the user's action on a switch

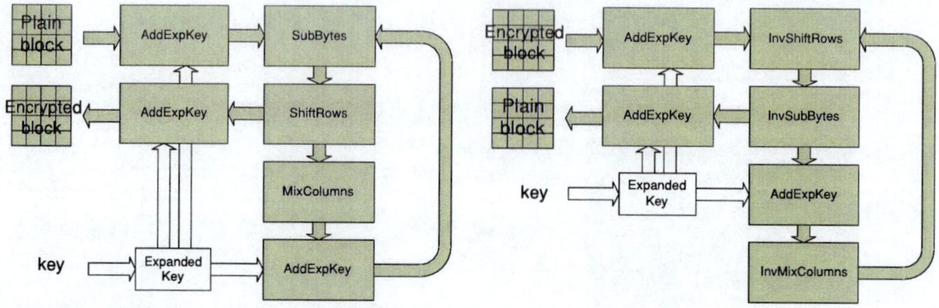

Fig. 15.30 Block encryption (*left*) and decryption (*right*) in the AES-128 cipher

microprocessor. The microprocessor writes commands and data to the coprocessor's registers in order to perform a computation. When the coprocessor completes, the microprocessor retrieves the computed data in order to continue the algorithm.

This case study presents an embedded system which communicates with an external PC in order to set the state of some led diodes and to read the state of some switches. The system is connected through the serial port, but the communications are ciphered using the Advanced Encryption Standard (AES-128). The system decrypts the commands received from the PC and encrypts the answer messages. The MicroBlaze cannot decrypt the messages at the desired speed, therefore, a coprocessor is developed to accelerate the AES-128 computation.

15.3.1 A Brief Introduction to the AES-128 Cipher

The AES [1] standard comprises three symmetric-key block ciphers: AES-128, AES-192 and AES-256, where the number denotes the key size. The cipher encrypts a fixed-size block of plain text to get a block of ciphered text. It also decrypts ciphered blocks in the reverse way using the same key. The block size of the AES ciphers is 128 bits independently of the key size.

All the steps executed during AES encryption/decryption are done using a variable named *state*. It is a 128-bit data arranged on an array of 4-rows x 4-columns of 8-bit elements (bytes). During the AES-128 encryption the *state* is initially written with a block of plain text. Then it is processed 9 rounds, and each round is composed of 4 steps, as shown in Fig. 15.30. The initial and the final rounds are slightly different since some steps are skipped. The AES *state* stores intermediary computations and the block of ciphered text at the end of the final round. In a similar way the decryption initializes the *state* with the ciphered block which is processed with inverted computations and steps to get the block of plain text.

One of the AES steps applies the expanded key which must be computed from the cipher key. The expansion of the key can be computed online or offline. The

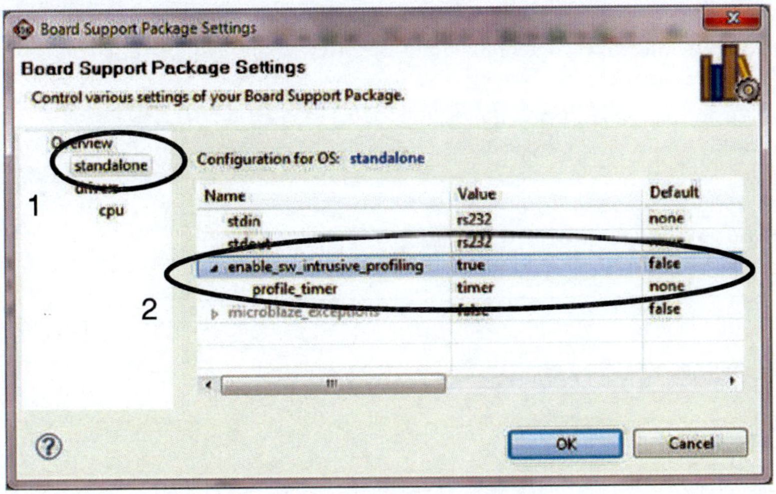

Fig. 15.31 Enabling and configuring the profiling in the BSP

offline mode computes the expanded key before starting the rounds, therefore, it is computed only when the cipher key changes.

The transmitter divides a message into blocks that are encrypted and transmitted. The receiver decrypts the received blocks to build the message. There are several operation modes and padding schemes to permit block ciphers to work with messages of any length [22]. This case study chooses the Electronic Code-Book (ECB) mode and null padding, due to its simplicity. The other modes require the IV (Initialization Vector) generator and a different block partitioning, but this fact does not affect the AES block cipher.

15.3.2 Software Implementation of the AES-128 Cipher

In order to study the applicability of a coprocessor on this application, the AES-128 cipher is implemented as a C++ class named *CAES128*. The system receives/ transmits ciphered messages from/to the external PC and it uses the C++ class to decrypt/encrypt blocks according to the cipher key. Microblaze reacts to the messages to set the leds or to report the state of the switches.

The hardware specification is quite similar to the first case study, but it pro-vides a single GPIO peripheral to read two switches and to set two leds. Copy the folder which allocates the XPS project of the first case study and rename it as *co-processor*. Edit the MHS file to add the 4-bit width GPIO and connect its bidirectional port to an external FPGA port. The timer will not only measure the time taken to encrypt/decrypt blocks, but it will also be used for the cipher's profiling. The profiler requires the timer to be able to interrupt the microprocessor, therefore, the MHS connects their interrupt ports. The system removes the interrupt controller since there is a single interrupt source.

```
PORT fpga_0_gpio_pin=gpio_GPIO_IO, DIR=IO, VEC=[3:0]
...
BEGIN xps_gpio
 PARAMETER INSTANCE = gpio
 PARAMETER C_GPIO_WIDTH = 4
 PORT GPIO_IO = gpio_GPIO_IO
 ...
END

BEGIN microblaze
 PARAMETER INSTANCE = microblaze_0
...
 PORT INTERRUPT = timer_Interrupt
END

BEGIN xps_timer
 PARAMETER INSTANCE = timer
...
 PORT Interrupt = timer_Interrupt
END
```

The UCF file is modified to attach the external port to the switches and leds according the FPGA board. Implement the hardware and export it to SDK.

```
Net fpga_0_gpio_pin<0> LOC=K12;          #led-0
Net fpga_0_gpio_pin<1> LOC=P14;          #led-1
Net fpga_0_gpio_pin<2> LOC=K13;          #switch-0
Net fpga_0_gpio_pin<3> LOC=K14;          #switch-1
```

Open the SDK workspace *c:\edk13.1\coprocessor\SDK\workspace*. Modify the BSP settings to enable the profiling and configure the profile timer (see Fig. 15.31). Clean and build the new BSP which includes a new library which is required when the application is profiled.

Delete the previous C++ projects and create a new project named *server* which targets the new BSP. The server application is composed of four files.

The files *caes128.h* and *caes128.cc* are the source files of the C++ class *CAES128* which implements the cipher. It is a straightforward implementation in order to facilitate the next steps. The class can be improved, in terms of speed and security, but it requires a deeper knowledge of the AES. The class provides the method *SetKey* to compute the expanded key at the beginning of the application. Then, it provides methods to *Encrypt* or *Decrypt* a single 128-bit block.

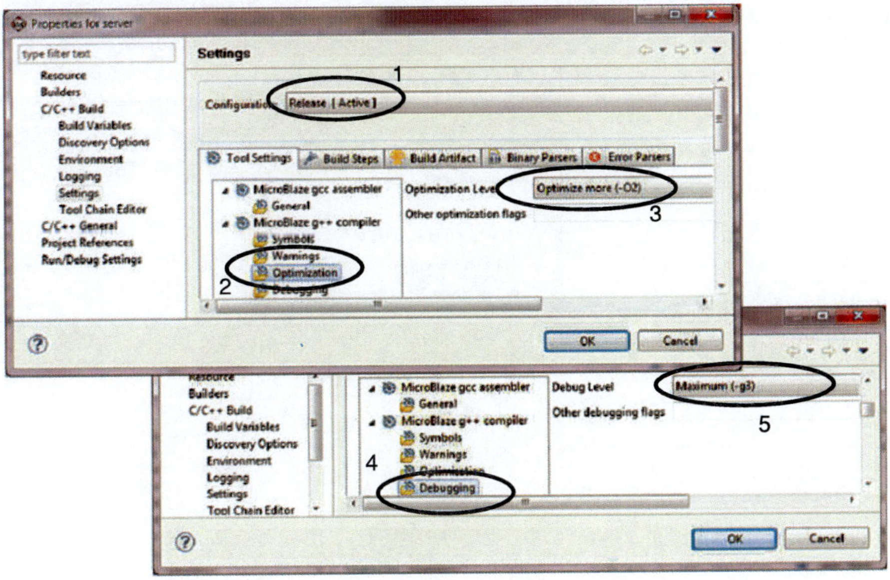

Fig. 15.32 Settings to build the release target

```
#define AES128_BLOCK_SIZE (128)          //128 bits
...
class CAES128
  {
  private:
    unsigned char State[4][4];           //The state variable
    ...
  public:
    void SetKey(unsigned char key[AES128_KEY_SIZE/8]);
    void Encrypt(unsigned char in[AES128_BLOCK_SIZE/8],
                 unsigned char out[AES128_BLOCK_SIZE/8]);
    void Decrypt(unsigned char in[AES128_BLOCK_SIZE/8],
                 unsigned char out[AES128_BLOCK_SIZE/8]);
  };
```

The file *app.cc* implements the application. It waits for commands launched from a PC in order to set the leds or to report the state of the switches. Any command produces an answer message which is transmitted to the PC. Commands and answer messages are encrypted during their transmission on the serial interface. The function *send_rs232_cipher* implements the ECB operation mode and null padding. It divides an answer message into 128-bit blocks that are individually encrypted and transmitted. In a similar way, the function *get_rs232_cipher* builds a command message from the received and decrypted blocks.

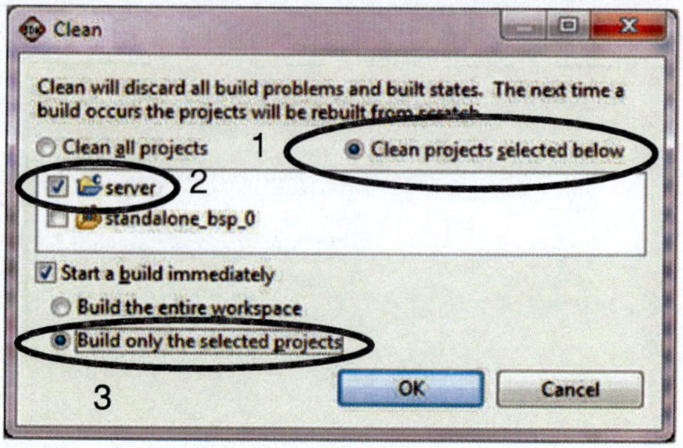

Fig. 15.33 Project cleaning and building

```
CAES128 Cipher;                    //Object to the AES-128 C++ class
...
void send_rs232_cipher(char *string)
{unsigned char block[AES128_BLOCK_SIZE/8];
 char c=0xFF;
 do{
   for(unsigned char j=0; j<AES128_BLOCK_SIZE/8; j++)
     {if(c) c=string[k++];              //null padding
     block[j]=c;}
   ...
   Cipher.Encrypt(block,block);         //ECB mode
   ...
   for(unsigned char j=0; j<AES128_BLOCK_SIZE/8; j++)
     XUartLite_SendByte(XPAR_RS232_BASEADDR,block[j]);
   } while(c);             //Repeat until the end of the string
   }
  ...
  int main()
  {unsigned char key[AES128_KEY_SIZE/8]={'0','1',...,'f'};
   Cipher.SetKey(key);            //Computes the expanded key
   ...
   while(!quit)
     {send_rs232_cipher("Enter CMD: ");
      XUartLite_SetControlReg(...);//Reset the receiving buffer
      get_rs232_cipher(input,INPUT_NCHARS);
      if(check_rs232())         //Check overrun or framing errors
        command(input);}        //Parse the received command
   ...}
```

The application stores into the variables *EncryptMaxCycles/DecryptMaxCycles* the maximum number of clock cycles taken to encrypt/decrypt a block. It relies on the file *cchronometer.h* which implements the C++ class to get the number of clock cycles from the timer. The compiler can skip the code related to the chronometer when the *CIPHER_CHRONOMETER* declaration is commented. The application also declares the *CIPHER_DISABLE* which can be used to skip the cipher to facilitate debugging.

Click the menu *Project* → *Properties* to edit the settings of the build of the target *Release*, as shown in Fig. 15.32. The compiler can optimize the code to improve the execution speed, by changing the order of instructions or the unrolling of loops. The ELF may not fit in the BRAM memory if the optimization level at the maximum ($-O3$). In order to display the measured encryption/decryption time, lower the optimization level ($-O2$) and set the debug to the maximum level ($-g3$).

Set the *Release* as the active target. Then, click the menu *Project* → *Clean* to clean the C++ project in order to build it from scratch (see Fig. 15.33).

In order to test or debug the application, the PC must also use the AES-128 cipher. There are many PC programs that permit the ciphering of messages through a communication channel, but they usually employ sophisticated protocols (SSH, SSL, etc.) to authenticate users before transmitting the cipher keys. The presented application is much simpler, since the PC and the embedded system are programmed with the same cipher key. In order to test the application, the PC executes a Linux program to encrypt/decrypt messages from/to a console which redirects the input/output to the serial port. The folder *linux_aes128* stores the executable files for Linux or Cygwin. Cygwin permits the execution of applications written for Linux on Windows PC. Cygwin cannot execute the Linux binaries, but it requires the building of the executable from the Linux source files.

Install Cygwin or a Linux virtual machine and open a shell to launch a set of commands. The first command changes to the directory which contains the path of the executable (it may change). The next one configures the communication parameters on the serial port (*/dev/com1* may change). The third command maps a file descriptor to the serial port. The last two commands decrypt or encrypt messages between the console and the serial port. The cipher key is set to the same value as the embedded system. Finally, program the FPGA and test the application. The Linux console shows unreadable text if the cipher keys are mismatched.

```
$ cd /cygdrive/c/edk13.1/linux_128
$ stty -F /dev/com1 115200 cs8 -parenb -cstopb -crtscts -isig
$ exec 9<>/dev/com1
$ ./aes128 -d -c -k "0123456789abcdef" <&9 &
$ ./aes128 -e -c -k "0123456789abcdef" >&9
```

In order to display the time taken to encrypt/decrypt blocks, debug the ELF of the target *Release* using SDK. Set a breakpoint at the last instruction of the C++ file to pause the execution when the user launches the quit command. Then resume the application and launch several commands to the embedded system through the Linux console. The variables *EncryptMaxCycles* and *DecryptMaxCycles* contain the maximum number of clock cycles required to encrypt and decrypt a block. The decryption of a block takes 1.76 ms (87,338 clock cycles, 50 MHz clock frequency), but the serial communication requires 1.39 ms to transmit it (115,200 bps). The FIFO of the UART may overrun during the reception of large messages since the MicroBlaze cannot decrypt blocks at the required speed. To solve the problem, the transmission speed can be lowered or the system can implement a flow control. A better solution is to accelerate the most time-spending computations.

15.3.3 Profiling

The profiler [12] is an intrusive tool which is used to test the application performances. It reports the number of calls and the execution time of every function. The profiling requires a dedicated timer able to interrupt the microprocessor in order to sample the program counter. The source files of the application must be compiled to add profiling ISR and code (compiler switch—*pg*). The linker attaches to the profiling library to build the executable.

The server application is not adequate to profile the since most of the execution time is devoted to wait for user messages from the serial port. A better approach is to execute a new application which continuously encrypts and decrypts messages. Therefore, create a new C++ project named *profiling*. The application file, named *profiling_app.cc*, uses the class *CAES128* to encrypt/decrypt messages in a loop. Change the compiler switches of the target *Release* to enable the profiler and the same optimization level as the server application (switches *-pg -O2*). Next, clean the project in order to build the application from scratch.

The profiler collects data and stores it in memory during the execution. Once the application completes, the collected data is downloaded to the PC in order to analyze it. The SDK must set the profiler memory which cannot be overlapped to the application memory. Use the SDK to program the FPGA with the imported bitstream and BMM, and set the ELF to *bootloop*. Open the XMD console to launch some commands. The first command changes the working directory to the application's folder. The second command establishes the connection to the MicroBlaze's MDM. The last command tries to download the ELF file into memory. It fails since the profiler memory is not defined, but it reports the allocated memory of the ELF. The hardware implements the local memory from the address *0 × 0000* to *0 × 3FFF,* and the XMD reported there is available free memory from address *0 × 2248.*

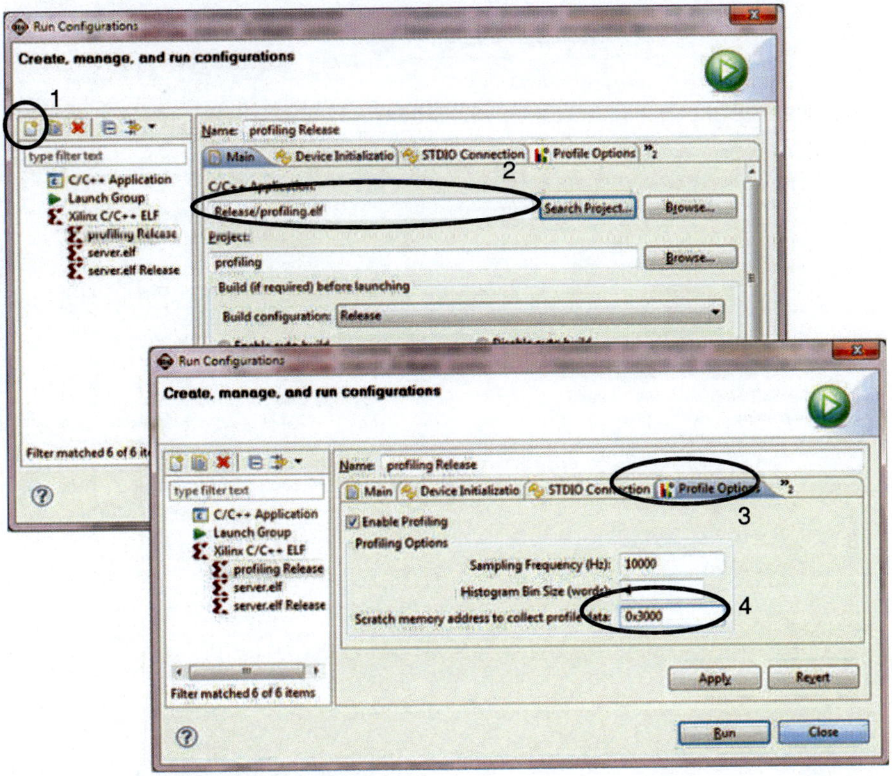

Fig. 15.34 Run configuration to profile

```
XMD% cd c:/edk13.1/coprocessor/SDK/workspace/profiling
XMD% connect mb mdm
XMD% dow Release/profiling.elf

Downloading Program -- Release/profiling.elf
        section, .vectors.reset: 0x00000000-0x00000003
        ...
        section, .stack: 0x00001e44-0x00002247

ERROR: Failed to download ELF file
        Memory needed for Profiling not defined
```

The profiling data can be stored into any free space of memory. Click the menu *Run* → *Run Configurations* to open the dialog depicted in Fig. 15.34. Then add the application to profile and set the profile memory from the *0* × *3000* address. Finally, run the application and wait until the application ends. The PC downloads

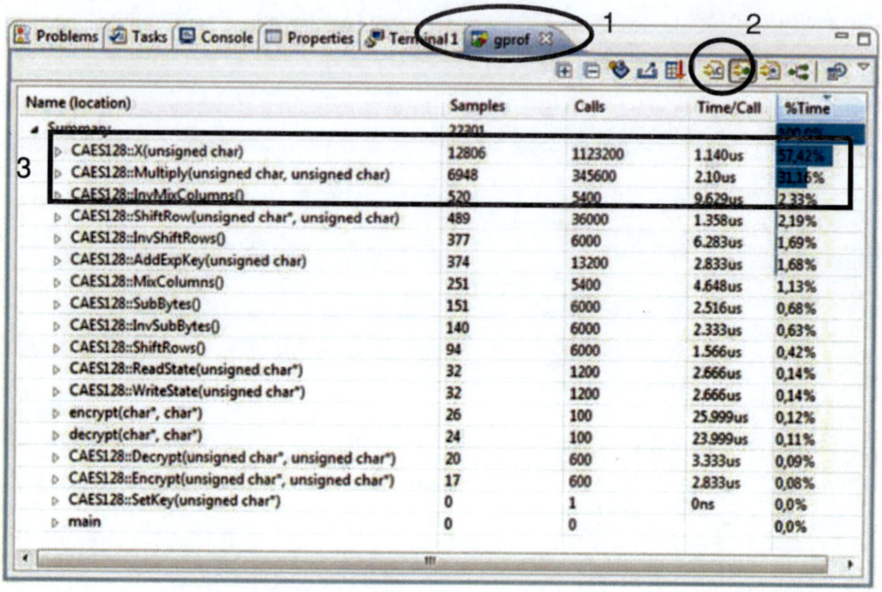

Name (location)	Samples	Calls	Time/Call	%Time
Summary	22301			100.0%
CAES128::X(unsigned char)	12806	1123200	1.140us	57.42%
CAES128::Multiply(unsigned char, unsigned char)	6948	345600	2.10us	31.16%
CAES128::InvMixColumns()	520	5400	9.629us	2.33%
CAES128::ShiftRow(unsigned char*, unsigned char)	489	36000	1.358us	2,19%
CAES128::InvShiftRows()	377	6000	6.283us	1,69%
CAES128::AddExpKey(unsigned char)	374	13200	2.833us	1,68%
CAES128::MixColumns()	251	5400	4.648us	1,13%
CAES128::SubBytes()	151	6000	2.516us	0,68%
CAES128::InvSubBytes()	140	6000	2.333us	0,63%
CAES128::ShiftRows()	94	6000	1.566us	0,42%
CAES128::ReadState(unsigned char*)	32	1200	2.666us	0,14%
CAES128::WriteState(unsigned char*)	32	1200	2.666us	0,14%
encrypt(char*, char*)	26	100	25.999us	0,12%
decrypt(char*, char*)	24	100	23.999us	0,11%
CAES128::Decrypt(unsigned char*, unsigned char*)	20	600	3.333us	0,09%
CAES128::Encrypt(unsigned char*, unsigned char*)	17	600	2.833us	0,08%
CAES128::SetKey(unsigned char*)	0	1	0ns	0,0%
main	0	0		0,0%

Fig. 15.35 Collected data from the application profiling

the collected data which is stored in the file *gmon.out*. It is a binary file which is interpreted by the GNU gprof tool.

Double click on the file *gmon.out* to display the results (see Fig. 15.35). The SDK shows a graphical view of the collected data which can be arranged in several ways, as the percentage of execution time devoted to each function. Two methods of the *CAES128* class take the 88% of the processing time: *X* and *Multiply*. They are child functions called from the function *InvMixColumn* which is one of the steps executed during the decryption. The child function *X* is also called from the step *MixColumn* during the encryption.

The profiling information can be used to re-implement the most time-spending functions to improve the execution time. However, a coprocessor can greatly accelerate a specific computation.

15.3.4 Coprocessor Design

The coprocessor aims to accelerate the step *InvMixColumn* and its child functions. The step *MixColumn* is quite similar, therefore, the hardware design of both steps does not significantly increase the effort. The coprocessor will accelerate the computations that take the 92% of the processing time obtained from the profiler.

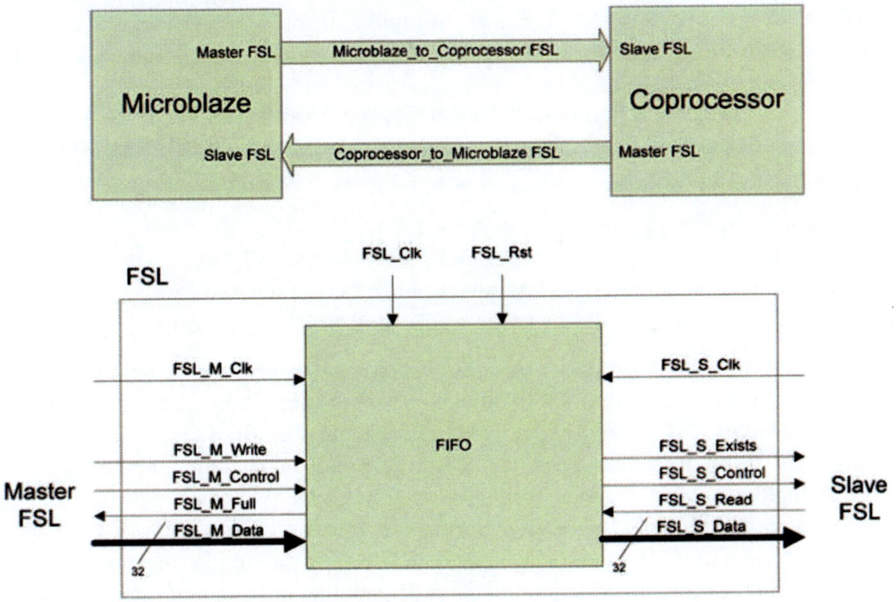

Fig. 15.36 Attachment of the MicroBlaze and coprocessor (*top*), and FSL schematic (*bottom*)

Copy the previous XPS folder and rename it as *coprocessor_ip* in order to develop the coprocessor, its driver, and to update the hardware and software of the application.

15.3.4.1 Hardware Design

Coprocessors are attached to MicroBlaze through Fast Simplex Link (FSLs), as depicted in Fig. 15.36 (top). A MicroBlaze's master-FSL connects to the coprocessor's slave-FSL in order to write the coprocessor's registers. In the reverse way, the MicroBlaze's slave-FSL permits the reading of the coprocessor's registers.

The FSL [21] is a point-to-point link which permits a low latency and fast communication due to its simple handshake. The FSL does not provide an address bus, as seen in Fig. 15.36 (bottom). Therefore, the coprocessor's registers must be sequentially accessed in a predetermined order. Each FSL provides, by default, a 16-depth FIFO which can temporally store data written from the master, when the slave is not ready to read it. The FSL provides a single 32-bit width bus (*FLS_M_Data*, *FSL_S_Data*) which can carry data or commands, depending on the control signal (*FSL_M_Control*, *FSL_S_Control*). In order to read data, the FIFO signals when data is available to read (*FSL_S_Exists*), and the slave acknowledges when data is retrieved (*FSL_S_Read*). Similarly, in order to write data, the FIFO signals when there is no free space (*FSL_M_Full*), and the master requests to write data (*FSL_M_Write*).

XPS can create the template files for coprocessors through the same wizard used for peripherals. Launch the wizard by clicking the menu *Hardware → Create or Import Peripheral*:

(1) Choose create templates in the XPS directory
(2) Set the name to *fsl_mixcolumns* and the version to *v1.00.a*
(3) Select FSL attachment and go ahead with the default settings
(4) Select to implement the driver template and finish the wizard

The coprocessor does not require additional parameters, ports or VHDL files, therefore, the MPD and PAO files are not modified. The template file *fsl_mixcolumns.vhd* is modified to design the coprocessor. It implements a 4x4 matrix of 8-bit registers (*reg_state*) to compute and store the AES *state*. Additionally, a 1-bit register (*reg_mode*) configures the computation mode as *MixColumns* or *InvMixColumns*. MicroBlaze will execute a control-type write instruction to the coprocessor's slave FSL to set the mode register. Next, it follows four data-type write instructions to set the registers of the AES *state*. Matrices are stored in rows in the microprocessor's memory, and each 32-bit data sets the 4 registers of a row in the coprocessor. The coprocessor starts the computation when the registers *reg_state* and *reg_mode* have been written. The coprocessor must acknowledge the read of data to the slave-FSL in order to delete it from the FIFO.

The coprocessor computes and stores the result in the registers *reg_state*. The coprocessor writes the resulting data to its master-FSL in a quite similar way. It writes a new row on the FSL when the computation is completed and the FIFO is not full. MicroBlaze executes four read instructions to the FSL in order to retrieve the resulting AES *state*.

A key idea to accelerate a computation is in parallel processing. A full-parallel implementation could compute the entire array of the AES *state* in a single clock cycle, although it may occupy a large area. However, MicroBlaze would not completely take profit of this architecture since it takes several more clock cycles to execute the FSL instructions to set and retrieve the AES *state*. A semi-parallel architecture offers a good trade-off between speed and area. The coprocessor computes the left-side column of the *reg_state* in a clock cycle. The state register shifts one column and the computation is repeated for the next 3 columns. Therefore, it takes 4 clock cycles to complete.

```
SUBTYPE t_byte IS STD_LOGIC_VECTOR(7 DOWNTO 0);
TYPE t_state IS ARRAY(0 TO 3,0 TO 3) OF t_byte;
TYPE t_col IS ARRAY(0 TO 3) OF t_byte;
TYPE t_row IS ARRAY(0 TO 3) OF t_byte;
...
SIGNAL reg_mode: STD_LOGIC;
SIGNAL reg_state, state: t_state;
SIGNAL row,resrow: t_row;
...
p_read_sfsl: PROCESS(...)       --Slave-FSL to write registers
VARIABLE cnt_mode: INTEGER RANGE 0 TO 1;
VARIABLE cnt_row:  INTEGER RANGE 0 TO 4;
BEGIN
  IF RISING_EDGE(fsl_clk) THEN  --Counts number of FSL writes
    IF fsl_rst='1' OR start='1' THEN
      cnt_row:=0; cnt_mode:=0;
    ELSE
      IF wr_mode='1' THEN cnt_mode:=cnt_mode+1; END IF;
      IF wr_row='1'  THEN cnt_row:=cnt_row+1;   END IF;
    END IF;
  END IF;

  IF fsl_s_exists='1' AND fsl_s_control='0' AND
     NOT(cnt_row=4) THEN wr_row<=ready;   --enables write
     ELSE wr_row<='0'; END IF;            --to state register
  IF fsl_s_exists='1' AND fsl_s_control='1' AND
     NOT(cnt_mode=1) THEN wr_mode<=ready; --enables write
     ELSE wr_mode<='0'; END IF;            --to mode register

  fsl_s_read<=wr_mode OR wr_row;        --Acknowledge slave-FSL
  mode<=fsl_s_data(fsl_s_data'RIGHT); --The LSB sets the mode
  row<=(fsl_s_data(0 TO 7), fsl_s_data(8 TO 15),
        fsl_s_data(16 TO 23), fsl_s_data(24 TO 31));
  idx_row<=CONV_UNSIGNED(cnt_row, idx_row'LENGTH);--row index

  IF cnt_mode=1 AND cnt_row+1=4 AND wr_row='1'  --starts the
    THEN start<='1'; ELSE start<='0'; END IF;   --coprocessor
END PROCESS;

p_state: PROCESS
...
BEGIN
  WAIT UNTIL RISING_EDGE(fsl_clk);
  IF fsl_rst='1' THEN
    ready<='1';
  ELSIF ready='1' THEN
    IF wr_mode='1' THEN    --Writes the mode register
      reg_mode<=mode;
    END IF;
    IF wr_row='1' THEN     --Writes a row in state register
      reg_state<=f_set_row(reg_state, idx_row, row);
    END IF;
    ...
END PROCESS;
```

```
col0<=f_get_col(reg_state,0);           --Get column#0
ncol0<=f_mix_col(col0) WHEN reg_mode='0' ELSE
       f_invmix_col(col0);              --Computes column#0
state<=f_set_col(reg_state,0,ncol0);    --Stores column#0

p_state: PROCESS
  VARIABLE cnt_col: INTEGER RANGE 0 TO 4;
  VARIABLE ncol: t_col;
  BEGIN
    WAIT UNTIL RISING_EDGE(fsl_clk);
    IF fsl_rst='1' THEN
      ready<='1';
    ELSIF ready='1' THEN
      ...
      IF start='1' THEN
        ready<='0'; cnt_col:=0;         --Starts computation
      END IF;
    ELSE
      reg_state<=f_lshift_col(state);--Shifts state register
      cnt_col:=cnt_col+1;
      IF cnt_col=4 THEN
        ready<='1';                     --Computation completes
      END IF;
    END IF;
END PROCESS;
```

15.3.4.2 Driver Design

The XPS wizard generates the H and C templates for the coprocessor's driver. The driver provides a single function which sends the AES *state* and gets the result through FSLs. It executes a control-type write instruction followed by four data-type instructions to set the computation mode and the AES *state*. The *state* is transmitted in rows to avoid rearrangement of data from memory to the FSL. The last instructions read the resulting *state*.

The function uses macros to set/get data to/from FSLs. During the BSP generation, the compiler replaces them by the MicroBlaze instructions that read or write an FSL slot. MicroBlaze provides 16 master-FSLs and 16 slave-FSLs. The input slot is the FSL index from which the coprocessor reads the input data (coprocessor's slave-FSL). The output slot is the FSL index which connects to the coprocessor's master-FSL. The slot names must match with the name of the coprocessor's instance (*fsl_mixcolumns_0*) which will be declared in the MHS file, otherwise the BSP will fail.

```
#include <xparameters.h>
     //parameters XPAR_FSL_INSTANCENAME_INPUT/OUTPUT_SLOT_ID
     //INSTANCE_NAME must match with the MHS file
#include <fsl.h>   // cputfsl, putfsl, getfsl macros

#define FSL_INPUT_SLOT
      XPAR_FSL_FSL_MIXCOLUMNS_0_INPUT_SLOT_ID
#define FSL_OUTPUT_SLOT
      XPAR_FSL_FSL_MIXCOLUMNS_0_OUTPUT_SLOT_ID
void fsl_mixcolumns (unsigned char mode,
                     unsigned char state[4][4])
{
 unsigned int *p1=(unsigned int*)state[0];
 unsigned int *p2=(unsigned int*)state[0];
 cputfsl(mode,FSL_INPUT_SLOT);    //writes mode register
 putfsl(*p1++,FSL_INPUT_SLOT);    //writes row#0 to FSL
 putfsl(*p1++,FSL_INPUT_SLOT);    //        row#1
 putfsl(*p1++,FSL_INPUT_SLOT);    //        row#2
 putfsl(*p1  ,FSL_INPUT_SLOT);    //        row#3
 getfsl(*p2++,FSL_OUTPUT_SLOT);   //reads row#0 from FSL
 getfsl(*p2++,FSL_OUTPUT_SLOT);   //      row#1
 getfsl(*p2++,FSL_OUTPUT_SLOT);   //      row#2
 getfsl(*p2  ,FSL_OUTPUT_SLOT);   //      row#3
}
```

15.3.5 Modification of the Embedded System

The hardware must be modified to attach MicroBlaze to the coprocessor through FSLs. The class *CAES128* must accelerate the computation of the steps *MixColumns* and *InvMixColumns* using the coprocessor's driver.

15.3.5.1 Hardware Modification

The XPS offers a wizard (see Fig. 15.37) to connect a coprocessor when clicking the menu *Hardware → Configure Coprocessor*. It modifies the MHS file to attach the coprocessor's instance (*fsl_mixcolumns_0*) to MicroBlaze through two FSLs.

```
BEGIN microblaze
 PARAMETER INSTANCE = microblaze_0
 ...
 PARAMETER C_FSL_LINKS = 1
 BUS_INTERFACE SFSL0 = fsl_mixcolumns_0_to_microblaze_0
 BUS_INTERFACE MFSL0 = microblaze_0_to_fsl_mixcolumns_0
END

BEGIN fsl_mixcolumns
 PARAMETER INSTANCE = fsl_mixcolumns_0
 ...
 BUS_INTERFACE MFSL = fsl_mixcolumns_0_to_microblaze_0
 BUS_INTERFACE SFSL = microblaze_0_to_fsl_mixcolumns_0
END

BEGIN fsl_v20
 PARAMETER INSTANCE = fsl_mixcolumns_0_to_microblaze_0
 ...
END

BEGIN fsl_v20
 PARAMETER INSTANCE = microblaze_0_to_fsl_mixcolumns_0
 ...
END
```

15.3.5.2 Software Modification

Export the new hardware to SDK and set the workspace to the path *c:\edk13.1\coprocessor_ip\SDK\workspace* before continuing. By default, the MSS associates a generic driver to the coprocessor. The BSP generation with the coprocessor's driver is similar to the case of the peripheral. First, edit the file *libgen.options* to add the local repository which contains the driver's source files. Then, edit the MSS file to change the driver associated to the coprocessor.

```
BEGIN DRIVER
 PARAMETER DRIVER_NAME = fsl_mixcolumns
 PARAMETER DRIVER_VER = 1.00.a
 PARAMETER HW_INSTANCE = fsl_mixcolumns_0
END
```

Clean and build the BSP project to generate the new BSP from scratch. The software applications use the C++ class *CAES128* which implements the block cipher. The member methods *MixColumns* and *InvMixColumns* are modified in order that the coprocessor computes these steps. The class also provides conditional compilation to permit the computation by software for testing purposes.

```
#include "caes128.h"              //declares CAES128 class
#ifdef ENABLE_COPROCESSOR_MIXCOLUMNS
   #include <fsl_mixcolumns.h>    //coprocessor's driver
#endif
...
#ifdef ENABLE_COPROCESSOR_MIXCOLUMNS
   //Hardware computation on the coprocessor
   void CAES128::MixColumns()
      {fsl_mixcolumns(MIXCOLUMNS,State);}
   void CAES128::InvMixColumns()
      {fsl_mixcolumns(INVMIXCOLUMNS,State);}
#endif

#ifndef ENABLE_COPROCESSOR_MIXCOLUMNS
   //Software computation on the microprocessor
   ...
#endif
```

The rest of the source files are not modified, since the class *CAES128* carries out the encryption/decryption of blocks. Build the C++ projects to generate the new executables.

15.3.6 Simulation

The coprocessor can be simulated in a very similar way as described for the peripheral. The application *profiling* is used to build the simulation model since it continuously encrypts/decrypts blocks using the coprocessor.

Search when the coprocessor asserts the signal *start* which launches the computation, as shown in Fig. 15.38. The waveform previously shows the execution of five FSL write instructions by MicroBlaze. The coprocessor reads the data from its slave-FSL. The first FSL instruction is a control-type access which writes the mode register (reg_mode). The next four FSL instructions are data-type accesses to write the rows of the state register (*reg_state*). The coprocessor starts the computation after the completion of the five write accesses.

Figure 15.39 shows that the coprocessor takes 4 clock cycles to compute the state register. Then, it writes the four rows of the resulting *reg_state* to its master-FSL, in order the MicroBlaze can read them. The coprocessor does not have to wait for the MicroBlaze to read the resulting data, since the FIFO of the FSL is not full.

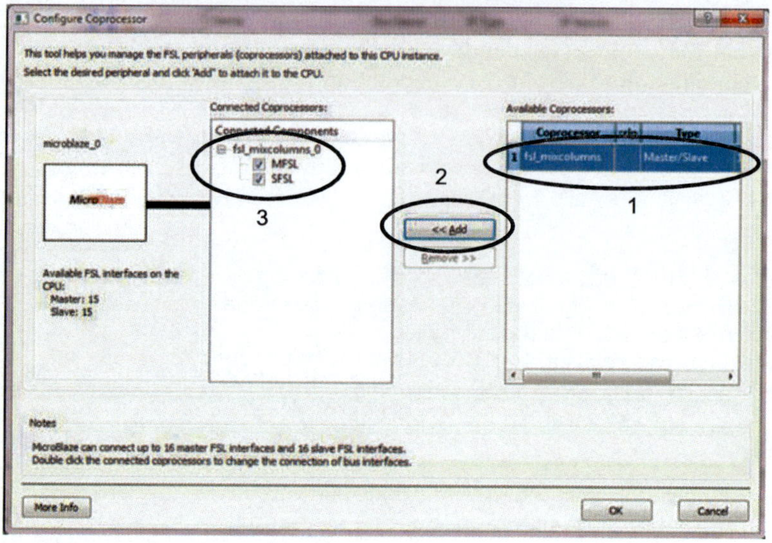

Fig. 15.37 EDK wizard to connect a coprocessor

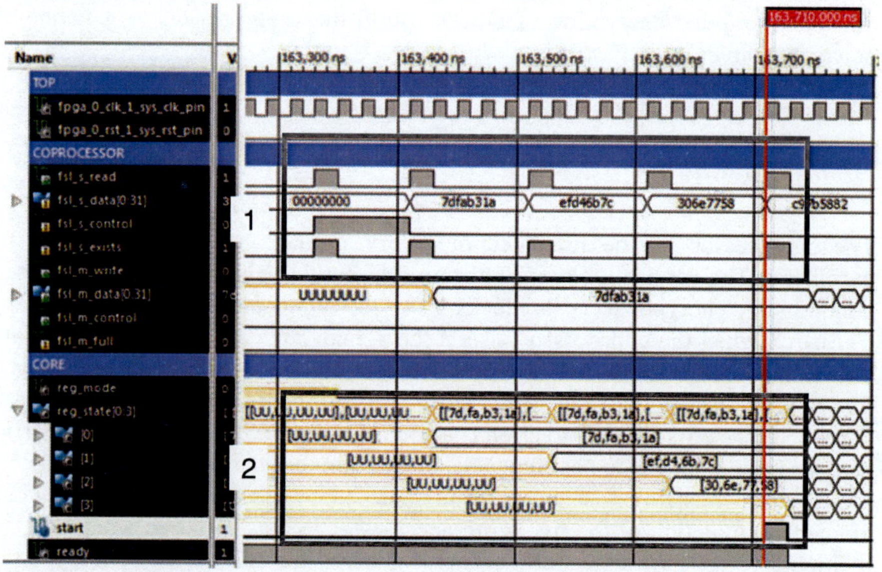

Fig. 15.38 Simulation of the coprocessor's slave FSL

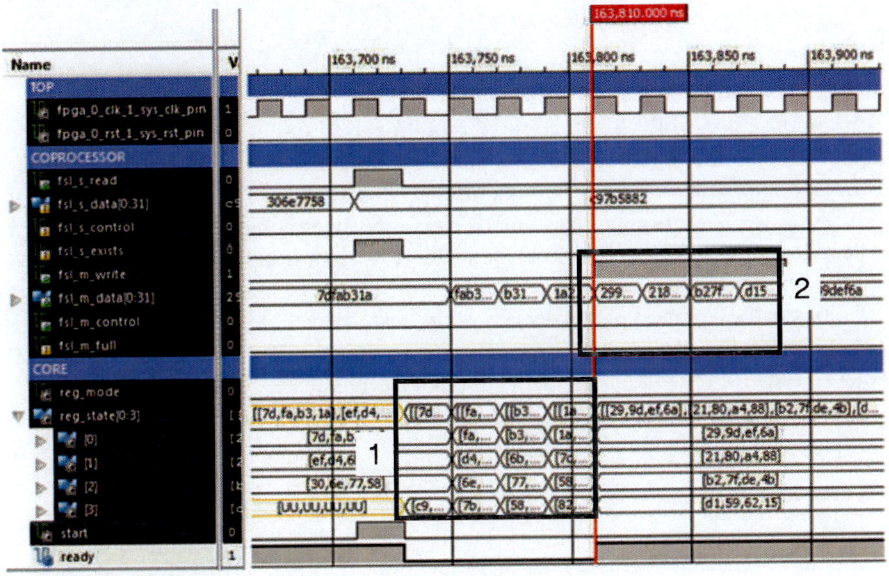

Fig. 15.39 Simulation of the coprocessor computation and the master FSL

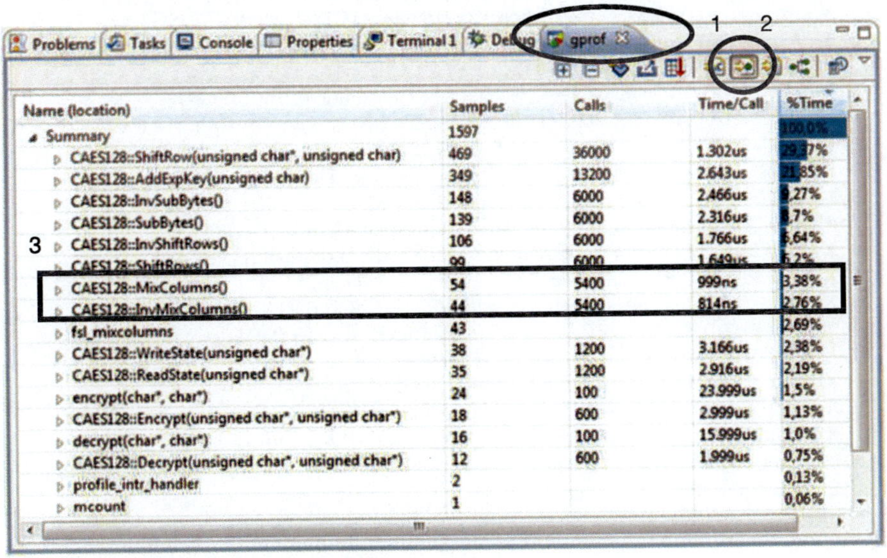

Fig. 15.40 Profiling data with the coprocessor

15.3.7 Experimental Results

Follow the steps previously described to collect the new profiling data. Before programming the FPGA, check the directories used to get the BIT, BMM and ELF files are in the current SDK workspace. Figure 15.40 shows the functions *MixColumns* and *InvMixColumns* currently represent only the 6.14% of the execution time. The time in these functions is mainly devoted to the transmitting and receiving of the *state* variable through the FSL instructions.

The application *server* can be tested as described before. The measured number of clock cycles to encrypt and decrypt a block are now 5,189 and 5,150, respectively. Therefore the decryption time is greatly improved from 1.76 ms to 103 μs (50 MHz clock frequency), which represents about 17 times faster.

References

1. Nist (2002) NIST Advanced Encryption Standard (AES) FIPS PUB 197
2. Xilinx (2005) Spartan-3 Starter Kit Board User Guide (UG130)
3. Xilinx (2010a) XST User Guide for Virttex-4, Virtex-5, Spartan-3, and Newer CPLD Devices (UG627)
4. Xilinx (2010b) LogiCORE IP Processor Local Bus PLB v4.6 (DS531)
5. Xilinx (2010c) XPS General Purpose Input/Output (GPIO) v2.00.a (DS569)
6. Xilinx (2010d) LogiCORE IP XPS Timer/Counter v1.02.a (DS573)
7. Xilinx (2010e) LogiCORE UART Lite v1.01.a (DS571)
8. Xilinx (2010f), LogiCORE IP XPS Interrupt Controller v2.01.a (DS572)
9. Xilinx (2010g) Xilinx Processor IP Library. Software Drivers. uartlite v2.00.a
10. Xilinx (2010h) Xilinx Processor IP Library. Software Drivers. tmrctr v2.03.a
11. Xilinx (2010i) Xilinx Processor IP Library. Software Drivers. intc v2.02.a
12. Xilinx (2010j) EDK Profiling Use Guide (UG448)
13. Xilinx (2011a) Embedded System Tools Reference Manual (UG111)
14. Xilinx (2011b) Platform Specification Format Reference Manual (UG642)
15. Xilinx (2011c) Command Line Tools User Guide (UG628)
16. Xilinx (2011d) MicroBlaze Processor Reference Guide (UG081)
17. Xilinx (2011e) LogicCORE Local Memory Bus LMB v10 (DS445)
18. Xilinx (2011f) Standalone (v.3.0.1.a) (UG647)
19. Xilinx (2011g) ISim User Guide v13.1 (UG660)
20. Xilinx (2011h) Data2MEM User Guide (UG658)
21. Xilinx (2011i) LogiCORE IP Fast Simplex Link (FSL) V20 Bus v2.11d (DS449)
22. Menezes AJ, Oorschot PC, Vanstone SA (1996) Handbook of applied cryptography. CRC Press, Boca Raton

Chapter 16
Partial Reconfiguration on Xilinx FPGAs

Partial Reconfiguration (PR) is the ability to change a portion (the reconfigurable partition) of the device without disturbing the normal operation of the rest (the static partition). A typical PR application is a reconfigurable coprocessor which switches the configuration of the reconfigurable partition at run-time when required by the application. The main advantage is the ability to map different coprocessor configurations in the reconfigurable partition in a time-multiplexed way, reducing the required area. The main drawbacks are the storage of partial bitstreams, the reconfiguration time and the increased complexity of the design flow. Reconfigurable systems must be implemented in a reconfigurable device, such as FPGAs. Most FPGAs are not partially reconfigurable since the internal architecture of the device must provide features to support it as well as the design flow tools. Moreover, the steps required to implement a reconfigurable system on a FPGA is highly dependent of the device architecture and cannot be easily adapted to other FPGA vendors. Virtex-4/5/6 are the Xilinx families of devices that are supported by the ISE 13.1 to perform partial reconfiguration [1].

16.1 Partial Reconfiguration on Xilinx FPGAs

Xilinx FPGAs that supports partial reconfiguration feature a set of common features at the hardware level:

- ICAP (Internal Configuration Access Port). It is an internal version of the SelectMAP programming interface. It permits to write or read-back the FPGA configuration from an embedded microprocessor or other internal circuit. The ICAP is controlled through commands that are embedded in the bitstream file.
- The configuration granularity is the minimum set of bit required to update FPGA resources which is a frame in Virtex-4/5/6 devices.

J.-P. Deschamps et al., *Guide to FPGA Implementation of Arithmetic Functions*, Lecture Notes in Electrical Engineering 149, DOI: 10.1007/978-94-007-2987-2_16,

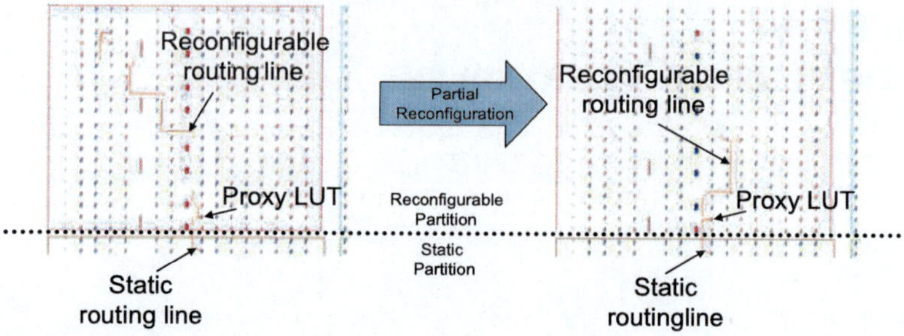

Fig. 16.1 FPGA layout displaying the proxy logic

- Configuration frame. The configuration memory of the FPGA is arranged in frames that are addressable memory spaces. Each frame is composed of a fixed number of 32-bit configuration words. There are several types of frames, such as the CLB or BRAM. A Virtex-5 frame [2] configures a single row of 20 CLBs, 4 BRAMs, etc.
- Partial bitstream. Contains the ICAP commands and configuration frames required to reconfigure a portion of the FPGA hardware resources. The configuration bits of the rest resources of the FPGA are not affected.
- Glicthless reconfiguration. It guaranties that if a configuration bit has the same value before and after the reconfiguration, the hardware resource controlled by the bit does not undergo any discontinuities in its operation. Therefore a portion of the static partition may be placed in the reconfigurable area and not be affected by the reconfiguration.

At the design flow level, the PR terminology is:

- HWICAP. It is an EDK peripheral which contains the ICAP plus FIFO memories and the PLB interface, in order to attach it to an embedded system.
- Static partition or top-level logic. It is composed of an embedded processor which drives the HWICAP with a partial bitstream. It cannot be affected during the partial reconfiguration, in order to complete it.
- Reconfigurable Partition (RP). It is composed of a subset of FPGA resources that can be reconfigured at run-time from the static partition. A RP implements several reconfigurable modules in a time-multiplexed way.
- Reconfigurable Module (RM). It is an implemented module in the reconfigurable partition.
- Partition pin. It connects a 1-bit net across the static and the dynamic partitions. The proxy logic transfers the bit value of the partition pin.
- Proxy logic. It is the interface logic between the static and the dynamic partitions. Each partition pin devotes a LUT placed in the RP which connects a reconfigurable routing line to a static routing line (see Fig. 16.1). The LUT and the reconfigurable routing line must be entirely contained in the RP. The reconfiguration does not affect the static routing line, therefore, the communication is reestablished when the partial reconfiguration is completed.

- Decoupling logic. Hardware resources that are changing their configuration bits can drive unexpected transient values on crossing nets that may cause a malfunction in the static partition. The static partition can deassert the nets from the proxy logic during the reconfiguration. Another approach is to reset the affected circuits of the static partition when the reconfiguration completes.

16.2 Design Flow for Partial Reconfiguration

The Project Navigator tool, which is frequently used to implement designs on Xilinx FPGA, does not support PR. The PR design flow is based on a technique named modular design which can be run by using the command-line ISE tools as well in the PlanAhead graphical tool. The modular design flow starts from a set of module netlists to implement (map, place and route) the layout of modules separately, merging them to build the complete design layout. A previous phase to modular design is to synthesize HDL sources to get the set of module netlists, where the reconfigurable modules have to meet a set of requirements in order to build a successful design.

16.3 Case Study

The case study exposes the design and PR flow of a self-reconfigurable embedded system. It summarizes some steps related to the design flow of the embedded system since they are detailed in the previous chapter. The case study is developed for the AVNET Virtex-5 LX Evaluation Kit [3, 4], but it can be easily adapted to other Virtex-4/5/6 boards.

Figure 16.2 depicts the embedded MicroBlaze which attaches to a reconfigurable coprocessor through FSLs. The coprocessor can map several configurations sharing a devoted set of FPGA resources in a time-multiplexed way. The MicroBlaze switches the coprocessor's configuration at run-time when required by the application. It reads a partial bitstream to control the HWICAP accordingly to perform the partial reconfiguration.

An important issue is the size of partial bitstreams which is usually larger than the internal BRAM capacity. Therefore, the system architecture must retrieve bitstreams from an external host or a non-volatile memory. An external host could store and transmit bitstreams when requested. The approach is slow and it depends on the communication channel. A better approach is to store the set of bitstreams in a non-volatile memory since many prototyping boards provide some FLASH which can be used for the required purpose. The case study provides autonomous reconfiguration since it reads the partial bitstreams from a FLASH memory to reconfigure the coprocessor.

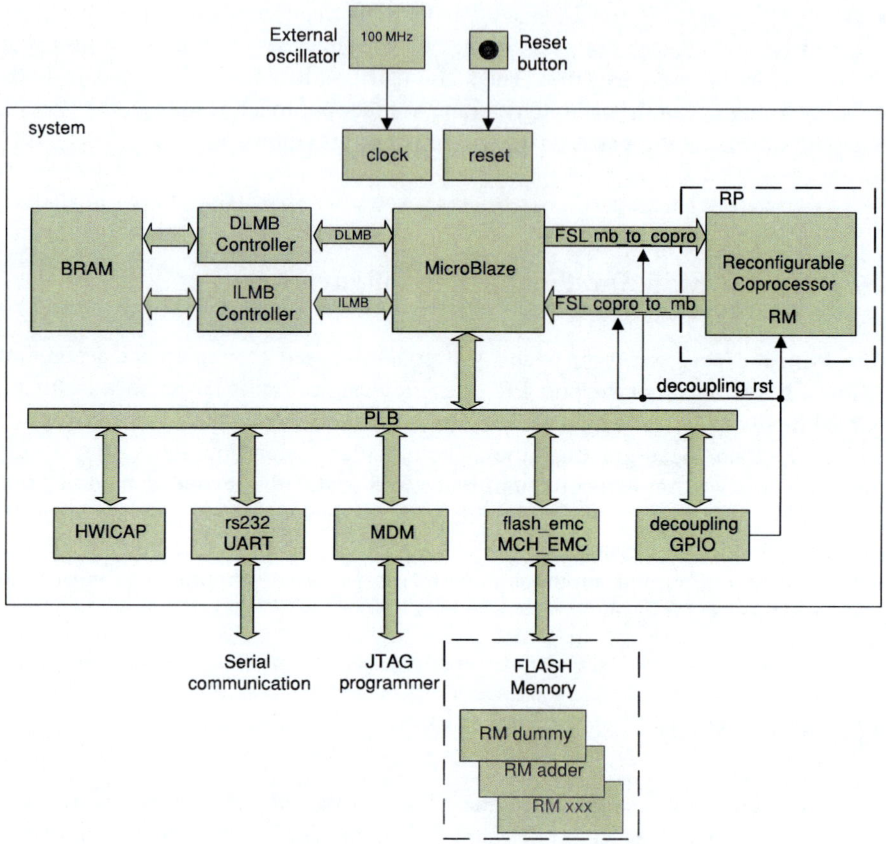

Fig. 16.2 Overview of the system architecture

The reconfigurable coprocessor is defined as the reconfigurable partition (RP) since it can allocate different reconfigurable modules (RM). The static partition is composed of the rest of the system. Each RM performs a different computation on 3×3 matrices: addition, multiplication, scalar multiplication and the determinant. There is also a dummy RM which does not perform any computation, but it is used in the design flow, as explained later.

The followed steps are the hardware design on XPS and synthesis of the RMs, the PR flow and the software development on SDK. The steps related to XPS and SDK are similar to the cases of the previous chapter, but they meet a set of requirements to permit the partial reconfiguration.

1. The hardware design starts with XPS to synthesize the embedded system in order to get the set of netlist files. The next step synthesizes the different configurations of the coprocessor to get the netlist files of the RMs.
2. PlanAhead runs the PR flow which starts declaring the reconfigurable partition and modules from the set of netlist files. The PR flow computes the layout of

the static partition and the RMs on the reconfigurable partition. The output is the set of partial bitstreams required to perform the partial reconfiguration.
3. The software development on SDK. The partial bitstreams are programmed into a FLASH memory. The application provides a C++ class which performs the partial reconfiguration controlling the HWICAP according to the bitstreams retrieved from the FLASH.

16.3.1 Hardware Design

The PR flow will start from a set of netlist files that are obtained from the synthesis of the hardware design of the embedded system and the RMs. The XPS facilitates the design and synthesis of the embedded system. The coprocessor will map different circuit configurations, therefore, it is also necessary to synthesize the RMs.

16.3.1.1 Design of the Embedded System on XPS

The architecture of the embedded system includes a MicroBlaze microprocessor attached to internal BRAMs through the data and instruction LMBs and memory controllers. The PLB attaches MicroBlaze to the peripherals: RS232 serial interface, the MDM and the HWICAP. The system includes the FSL reconfigurable coprocessor and the associated decoupling logic. The set of partial bitstreams will be retrieved from an external FLASH memory, therefore, the system includes an EMC (External Memory Controller) peripheral in order to permit the MicroBlaze reading them to perform the reconfiguration.

In order to design the hardware of the embedded system, create a new XPS project with the BSB wizard, as described in the previous chapter.

1. Choose the folder *c:\edk13.1\reconfig\XPS* to store the project file *system.xmp*.
2. Select the PLB, since the EDK does not support AXI for Virtex-5 devices.
3. Create a new design and select the *xc5vlx50-ff676-1* device which is embedded in the Virtex-5 prototyping board.
4. Select a single processor and continue to the end with the default options.

Edit the MHS file to change the polarity of reset button and the frequencies of the reference clock and the system's clock to the appropriate values of the prototyping board. Set the frequency of the system's clock to 100 MHz as it will be detailed later. The name of the system's clock is changed to *sys_clk*, therefore, it is necessary that the clock inputs reference to the new name, including the LMBs and the PLB. Finally, change the parameters *C_HIGHADDR* of the instruction and data BRAM controllers to increase the processor's local memory to 64 KB.

```
PORT fpga_0_clk_1_sys_clk_pin=CLK_S, DIR = I, SIGIS = CLK,
     CLK_FREQ = 100000000
PORT fpga_0_rst_1_sys_rst_pin=sys_rst_s, DIR = I, SIGIS=RST,
     RST_POLARITY = 1

BEGIN proc_sys_reset
 PARAMETER INSTANCE = proc_sys_reset_0
 PARAMETER C_EXT_RESET_HIGH = 1
 PORT Slowest_sync_clk = sys_clk
 ...
END

BEGIN clock_generator
 PARAMETER INSTANCE = clock_generator_0
 PARAMETER C_CLKIN_FREQ = 100000000
 PARAMETER C_CLKOUT0_FREQ = 100000000
 PARAMETER C_EXT_RESET_HIGH = 1
 PORT CLKOUT0 = sys_clk
 ...
END
...
BEGIN lmb_bram_if_cntlr
 PARAMETER INSTANCE = dlmb_cntlr
 PARAMETER C_HIGHADDR = 0x0000ffff
 ...
END

BEGIN lmb_bram_if_cntlr
 PARAMETER INSTANCE = ilmb_cntlr
 PARAMETER C_HIGHADDR = 0x0000ffff
 ...
END
```

The system declares the instance *rcopro* which is the reconfigurable coprocessor. The source files of the coprocessor *fsl_rcopro* are stored in the default local repository *pcores* created by the XPS wizard. The coprocessor's instance attaches to MicroBlaze through FSLs. There is a decoupling logic which can reset the FSLs and the coprocessor. MicroBlaze will control a GPIO (instance *decoupling_gpio*) in order to assert the 1-bit signal *decoupling_rst* when the reconfiguration is completed. If not, the FIFOs of the FSLs would store unexpected data, and the state machine of the coprocessor would start in an undesired state.

```
BEGIN microblaze
 PARAMETER INSTANCE = microblaze_0
 ...
 PARAMETER C_FSL_LINKS = 1
 BUS_INTERFACE MFSL0 = mb_to_copro
 BUS_INTERFACE SFSL0 = copro_to_mb
END

BEGIN fsl_v20
 PARAMETER INSTANCE = mb_to_copro
 PORT SYS_Rst = decoupling_rst
 ...
END

BEGIN fsl_v20
 PARAMETER INSTANCE = copro_to_mb
 PORT SYS_Rst = decoupling_rst
 ...
END

BEGIN fsl_rcopro
 PARAMETER INSTANCE = rcopro
 PARAMETER C_CONFIG_IDX = 0            #The dummy configuration
 BUS_INTERFACE MFSL = copro_to_mb
 BUS_INTERFACE SFSL = mb_to_copro
 PORT LMB_Rst = decoupling_rst
 ...
END

BEGIN xps_gpio
 PARAMETER INSTANCE = decoupling_gpio
 PORT GPIO_IO_O = decoupling_rst
 ...
END
```

The reconfigurable coprocessor can map several configurations. The parameter *C_CONFIG_IDX* in the MHS file specifies the initial configuration which is synthesized and implemented during the EDK flow. The value *0* sets the module *dummy* as the initial configuration of the coprocessor.

MicroBlaze will control the reconfiguration through the HWICAP peripheral [5]. The peripheral embeds the ICAP which is connected to read/write FIFOs and a slave PLB interface. The maximum frequency of the ICAP's input clock is 100 MHz, therefore, the system's clock accomplishes this constraint. The system will provide several clock domains if the ICAP sets a different clock frequency to the system's clock; however, it will complicate the design.

```
BEGIN xps_hwicap
 PARAMETER INSTANCE = hwicap
 PARAMETER HW_VER = 5.01.a
 PARAMETER C_BASEADDR = 0x86800000
 PARAMETER C_HIGHADDR = 0x8680ffff
 BUS_INTERFACE SPLB = mb_plb
 PORT ICAP_Clk = sys_clk
END
```

MicroBlaze must read a partial bitstream from the external FLASH to feed the ICAP, in order to reconfigure the coprocessor. EDK provides an external memory controller (EMC) [6] which permits access to asynchronous memories such as SRAM and FLASH. It can control several banks of different memories that share the same input/output ports. Each bank is accessible from a configurable address range and provides configuration parameters to set the width of the data bus and the timing specifications. The EMC permits access to 8-bit/16-bit data width memories, and it can match them to the 32-bit data bus of the PLB. The displayed configuration is valid for the 28F128J3D [7] device, the 16 MB FLASH embedded in the Virtex-5 development board [3]. The MicroBlaze can access to the FLASH from the address *0xA0000000* to the *0xA0FFFFFF*. The configuration can be complicated but, the board designer usually provides a reference design which can be used to get the parameters. The input/output ports of the EMC are externally connected to the address, data and control busses of the FLASH. The width of address bus of the FLASH is 24-bit, but the EMC generates 32-bit addresses. Therefore, the MHS slices the 24 lower bits that are externally connected.

```
#external FGPA ports connected to the FLASH
PORT fpga_0_flash_A_pin = flash_A, DIR = O, VEC = [23:0]
PORT fpga_0_flash_DQ_pin = flash_DQ, DIR = IO, VEC = [7:0]
PORT fpga_0_flash_CEN_pin = flash_CEN, DIR = O
PORT fpga_0_flash_OEN_pin = flash_OEN, DIR = O
PORT fpga_0_flash_WEN_pin = flash_WEN, DIR = O
PORT fpga_0_flash_RSTN_pin = net_vcc, DIR = O
...
BEGIN xps_mch_emc                          #The EMC from EDK
 PARAMETER INSTANCE = flash_emc
 ...
```

```
 PARAMETER C_MEM0_BASEADDR = 0xA0000000 #bank0 address range
 PARAMETER C_MEM0_HIGHADDR = 0xA0FFFFFF #      (16 MB)
 PARAMETER C_MEM0_WIDTH = 8             #bank0 8-bit data-bus
 PARAMETER C_INCLUDE_DATAWIDTH_MATCHING_0 = 1   #match 32-bit
 PARAMETER C_TCEDV_PS_MEM_0 = 75000     #bank0 timing
 PARAMETER C_TAVDV_PS_MEM_0 = 75000
 ...
 PORT Mem_A = 0b00000000 & flash_A #address-bus(24-bit slice)
 PORT Mem_DQ = flash_DQ            #data-bus (8-bit)
 PORT Mem_CEN = flash_CEN          #chip enable
 PORT Mem_OEN = flash_OEN          #output enable
 PORT Mem_WEN = flash_WEN          #write enable
END
```

The system must be synthesized in order to generate the netlist files required by the PR flow. The system netlist contains the module *dummy* as the initial configuration of the coprocessor. The module *dummy* just reads the slave FSL and writes the master FSL with zeros, but it does not perform any computation. This way, MicroBlaze will not hang up when it executes any read/write instructions to the coprocessor's FSLs.

XPS can continue to implement the FPGA layout, generating the bitstream *system.bit*. Previously, the UCF file must be modified to map the external ports depending on the development board. The implemented system is not valid to perform the PR but it can be exported to SDK in order to build the software. Moreover, the system provides an initial configuration for the coprocessor which can be tested and simulated, as described in the previous chapter. Finally, the system is able to program the set of partial bitstreams into the FLASH, as is detailed later.

16.3.1.2 Synthesis of the RMs

The coprocessor will map several reconfigurable modules. All of them are connected to MicroBlaze through the same FSL ports. The master and the slave FSL connections are quite similar to the case study exposed in the previous chapter. The VHDL source files that contain the different configurations of the coprocessor must be synthesized to obtain their netlist files. There are several ways, as with the Xilinx Project Navigator or with a third-party synthesizer.

This case study executes the XST synthesizer [8] from a command shell. The directory *c:\edk13.1\reconfig\XST* allocates the files required by XST. Open a Xilinx console clicking the XPS menu *Project→Launch Xilinx Shell*, and type the next commands:

```
> cd ..\XST                    Changes the work directory
> xst -ifn fsl_rcopro.scr      Executes XST with a batch file
```

The batch file *fsl_rcopro.scr* contains the XST commands to synthesize a set of source VHDL files. The NGC file stores the netlist of a single module for the target device.

```
run
-opt_mode speed
-netlist_hierarchy as_optimized
-opt_level 1
-top fsl_rcopro                    #top-level entity
-ifmt MIXED
-ifn fsl_rcopro.prj                #list of source VHDL files
-hierarchy_separator /
-iobuf NO
-p xc5vlx50ff676-1                 #target FPGA device
-ofn fsl_rcopro_dummy.ngc          #output NGC filename
```

The batch file relies on the file *fsl_rcopro.prj* which lists the files that are synthesized, as in the PAO file. The VHDL file *fsl_rcopro.vhd* declares the top-level entity of the coprocessor, and it contains the instance of the RM which is synthesized. The rest of the VHDL files (*fsl_rcopro_dummy.vhd*, *fsl_rcopro_adder.vhd*, etc.) are the source files of the different RMs.

```
ENTITY fsl_rcopro IS GENERIC (     --Top-level entity
   c_config_idx: INTEGER:=0        --Changes the synthesized RM
...);

ARCHITECTURE beh OF fsl_rcopro IS
BEGIN
   g0: IF c_config_idx=0 GENERATE  --instance to the dummy RM
      inst: ENTITY WORK.fsl_rcopro_dummy port map(...);
   END GENERATE;
   g1: IF c_config_idx=1 GENERATE  --instance to the adder RM
      inst: ENTITY WORK.fsl_rcopro_adder port map(...);
   END GENERATE;
   ...
END ARCHITECTURE;
```

The XST must be executed for every RM in order to get the set of netlist files. Change the parameter *C_CONFIG_IDX* and the name of the NGC output file to synthesize a new RM. The XST folder will store the netlist files such as *fsl_rcopro_dummy.ngc*, *fsl_rcopro_adder.ngc*, and so on.

16.3.2 Partial Reconfiguration Flow on PlanAhead

PlanAhead is a complex tool to perform the design floorplanning on Xilinx FPGAs, in order to improve clock frequency and delays, the placement of IOBs, and more. PlanAhead can run in a PR flow mode in order to facilitate the implementation of RMs on a reconfigurable partition. The PR flow requires an additional license feature. The main steps to follow are:

1. Set the PR flow and the design files. The flow starts from the top-level netlist file.
2. Set the reconfigurable partition and assign it the netlist of the RMs.
3. First layout implementation. It computes the FPGA layout of the static and the reconfigurable partition with the initial RM. The static partition is promoted to be imported during the next step.
4. Layout implementation for the rest of RMs. The flow continues importing the promoted static partition, but implementing a new layout in the reconfigurable partition for each new RM. The result is a set of FPGA layouts, where the layout of the static partition remains invariable.
5. Bitstream generation. The last step is the bitstream generation phase to compute the complete and partial bitstream of each implemented layout.

16.3.2.1 Set the PR Flow and the Design Files

Launch the PlanAhead and click the menu *Create a New Project* to open the wizard (see Fig. 16.3).

1. Select the root folder *c:\edk13.1\reconfig*, and set the project name to *PlanAhead.*
2. Choose the options *Specify synthesized (EDIF or NGC) netlist* and *Set PR Project*. The PR license must be correctly installed in order to set the PR project.
3. Select the top-level netlist file *system.ngc* which is stored in the folder *c:\edk13.1\reconfig\XPS\implementation.* It is the synthesized embedded system designed in XPS. It relies on the rest of the netlist files, therefore, add the same path to the *Netlist directories*. Do not select the copy option.
4. Continue with the default options to complete the wizard.

The BMM file *system.bmm* is required during the implementation of embedded systems, in order to generate the initial bitstream which configures the FPGA and the BRAM contents, as commented in the previous chapter. The BMM file is stored in the same location as the set of netlist files. Therefore, copy the file *system.bmm* to the folder *c:\edk13.1\reconfig\PlanAhead* which contains the PlanAhead project.

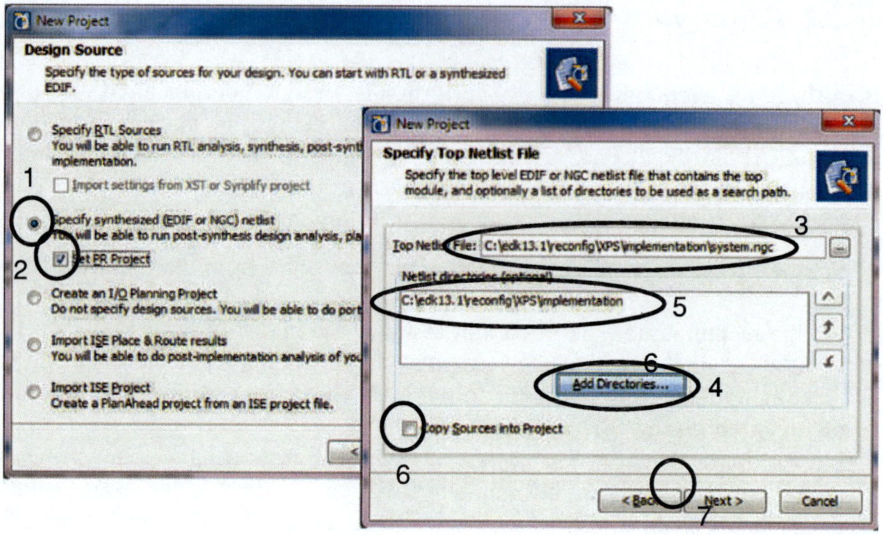

Fig. 16.3 Wizard to create a new PlanAhead PR project

16.3.2.2 Set the Reconfigurable Partition and the Reconfigurable Modules

Click the menu *Flow→Netlist Design* to display the list of netlist files which composes the system's hardware. The tab *Netlist* displays the instances of the embedded system. In order to set the reconfigurable partition (see Fig. 16.4):

1. Select the instance *rcopro* and open the contextual menu (right button of the mouse) and select *Set Partition*.
2. Select the option *is a reconfigurable Partition*, and change the name of the reconfigurable module to *module_dummy* since it is the initial configuration stored in the netlist files.

The tab *Netlist* now shows the *module_dummy* contained in the instance *rcopro*. In order to add a new RM to the RP (see Fig. 16.5):

1. Open the contextual menu of the *rcopro* and click *Add Reconfigurable Module*
2. Change the RM name to *module_adder*.
3. Select the netlist file *fsl_rcopro_adder.ngc* which is stored in the XST directory. Continue with the default settings to end the wizard.

The previous steps are repeated for every RM. PlanAhead displays all the RM modules in the tab *Netlist*. Finally, open the contextual menu of the configuration *module_dummy* to select the option *Set as Active Reconfigurable Module*.

The next step defines the FPGA area devoted as the reconfigurable partition. The RP must accomplish a set of design rules [1].

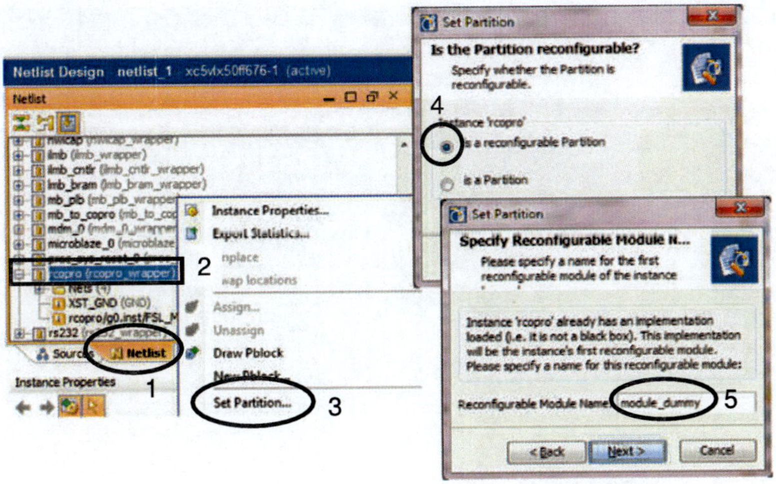

Fig. 16.4 Set the RP and the initial RM to the coprocessor's instance

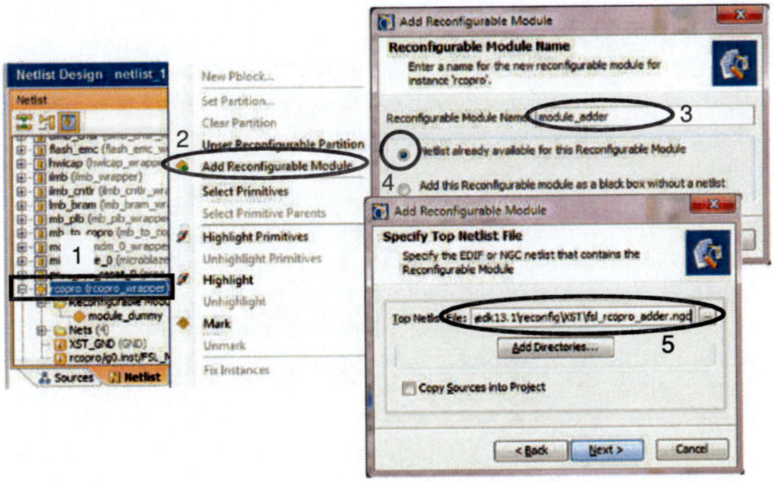

Fig. 16.5 Adding a RM to the RP

- The global clock logic must reside in the static partition
- Some types of hardware resources (global buffers, boundary-scan, etc.) should be placed in the static partition.
- Bidirectional partition pins are not permitted.
- In order to obtain a greater efficiency, the RP should align to the boundaries of the configuration frames.

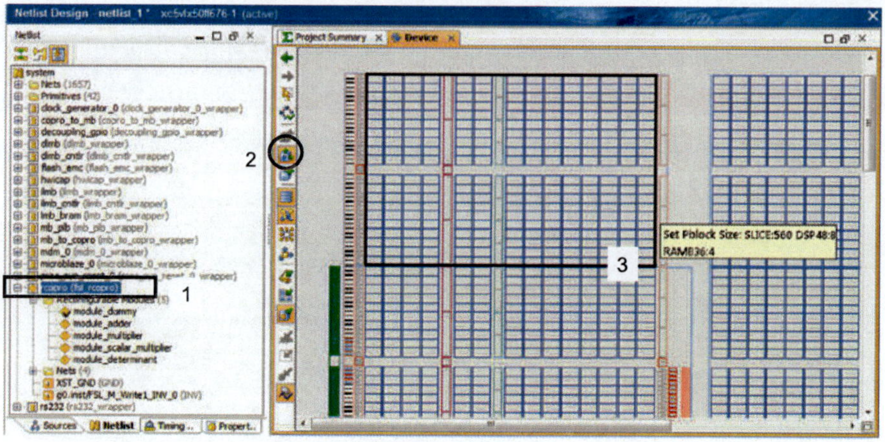

Fig. 16.6 Assigning the area of the RP

Select the instance *rcopro* and click the button *Set Pblock Size* in the device view, as shown in Fig. 16.6. The number of hardware resources in the RP must be higher than required by any of the synthesized RMs. This case study selects a region which starts from the top-left edge of the FPGA. Set the partition height to 20 CLBs (a Virtex-5 configuration frame) and the width to 28 CLB columns. The RP occupies the whole clock region of the FPGA which contains 560 SLICES, 4 RAMB36 and 8 DSP48 resources. Select all the resources in the next window to permit them to be reconfigured, with the exception of the RAMB36 since no RM requires them. The partial bitstream will reconfigure the hardware resources contained in the RP, with the exception of the BRAM contents. This way, the size of the partial bitstreams is reduced and the reconfiguration time improves.

Figure 16.7 displays the tab *Statistics* in the *Properties* of the Pblock *pblock_rcopro*. It shows the size of a partial bitstream and the detailed number of reconfigurable hardware resources. It also reports that the 100 % of the RP is contained in a single clock region, as desired. The tab *Attributes* shows the range of resources as *SLICE_X0Y100:SLICE_X27Y119* and *DSP48_X0Y40:DSP48_X0Y47*. The numbers denote the x–y coordinates starting from the bottom-left edge of the FPGA. Therefore, the SLICES are arranged in a 28 × 20 matrix and the DSP48 are in a single column. These parameters are annotated in a new UCF in order to implement the layout of the reconfigurable system.

It is recommended to check the design rules for the PR flow before continuing. Press *Run DRC* and unselect all rules except the related to the partial reconfiguration, as shown in Fig. 16.8. It should only report a warning for each RM since it is not still implemented by a configuration.

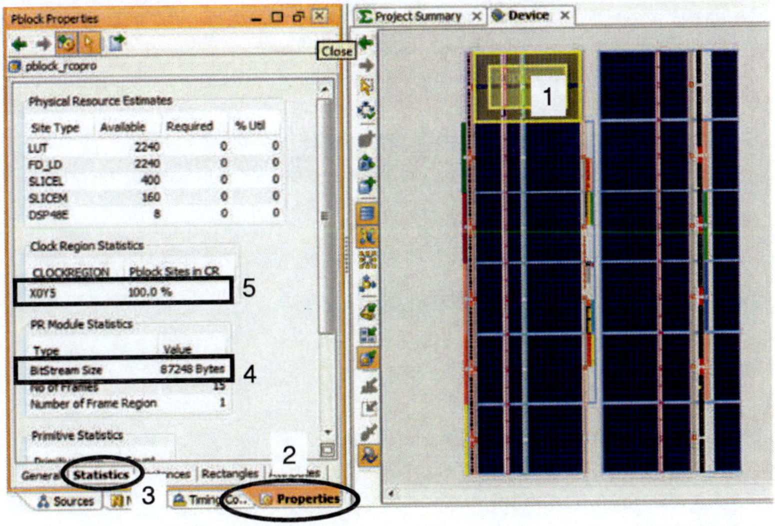

Fig. 16.7 Properties of the RP

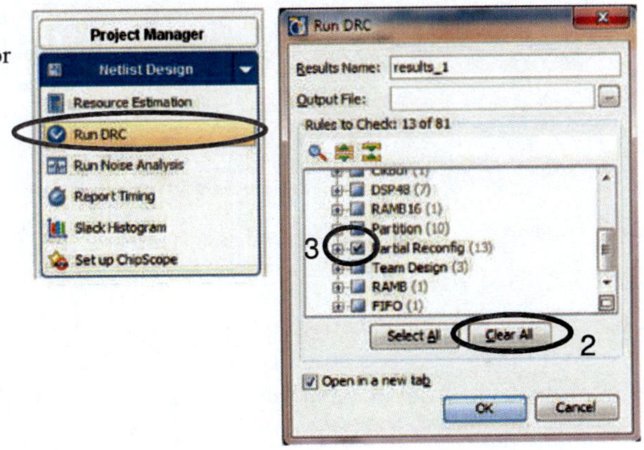

Fig. 16.8 Design rule check for the PR

16.3.2.3 First Layout Implementation

The tab *Design Runs* shows the settings to run the implementations (see Fig. 16.9). The default run created by PlanAhead will implement the layouts of the static partition and the dummy module into the reconfigurable partition. It creates a batch file which calls the ISE tools starting from the annotated constraints and netlist files.

1. Select the default run *config_1* created by PlanAhead
2. Display the properties of the partitions and check the coprocessor's RP will implement the RM *module_dummy*.

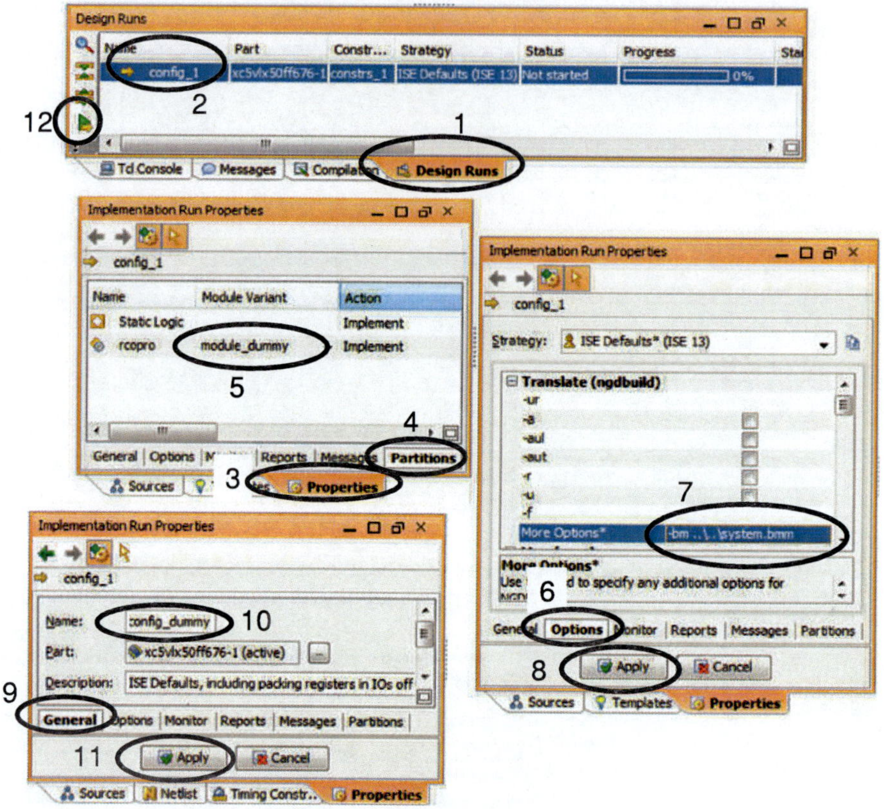

Fig. 16.9 Configuring the implementation run for the first layout

3. Display the implementation options to add *–bm..\.\system.bmm* to the NGD-Build tool. This option will load the BMM file which was copied from the XPS project, in order to generate the annotated BMM file which will be required later by the Data2MEM tool [9].
4. Display the general properties to change the run name to *config_dummy*.
5. Start the implementation run by pressing the button *Launch Selected Runs*.

The implementation of the layout takes some minutes to complete. PlanAhead opens a dialog window when it completes. Promote the implemented partitions since it is necessary in the next step.

16.3.2.4 Layout Implementation for the Next RMs

The next step implements the layouts for the rest of the reconfigurable modules. The PR flow starts importing the layout of the static partition from the first implementation. Then, it computes the layout of a new RM implemented into the RP.

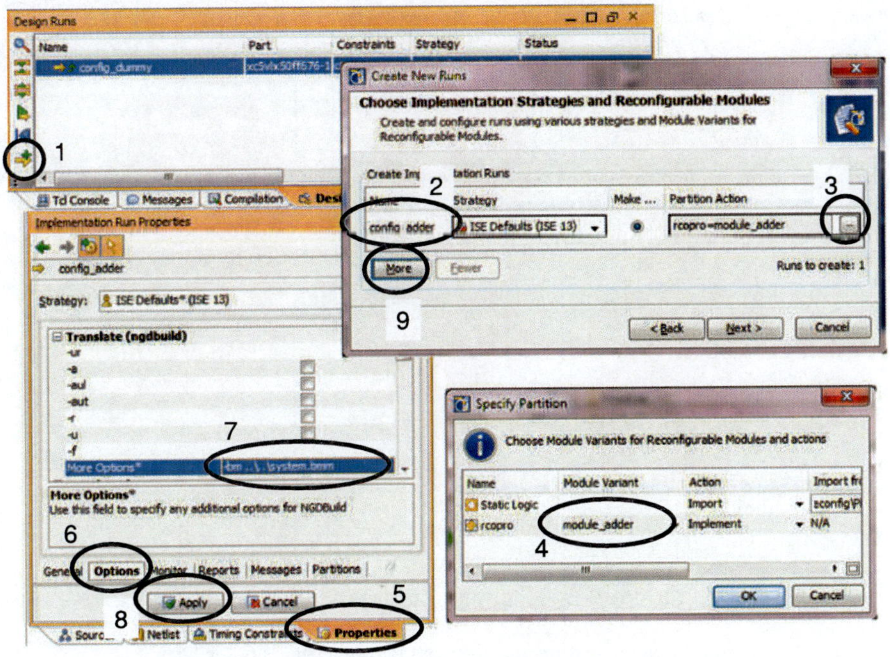

Fig. 16.10 Configuring a new implementation run for the adder RM

1. Click the button *Create New Runs* to configure a new implementation (see Fig. 16.10). Continue with the default netlist, constraints and device since they are the same as the first implementation.
2. Change the default name to *config_adder*. Configure the RP *rcopro* to implement the RM *module_adder*.
3. Add the BMM file in the same way as exposed in the first implementation run.
4. Press the button *More* to add the new runs to implement the rest of the RMs and repeat the previous two steps (but changing the name and the RM).
5. Finalize the wizard, but do not launch the run.

Select the new created runs to launch them. PlanAhead can take profit of a multi-core PC incrementing the number of jobs to compute several layouts in parallel. Each run imports the layout of the static partition and computes a new layout in the reconfigurable partition. After finishing the implementation, do not perform any action when asked by PlanAhead.

The ISE provides the FPGA Editor tool which displays an implemented layout. Open the file *config_dummy_routed.ncd* which is stored in the folder *c:\edk13.1\reconfig\PlanAhead\PlanAhead.runs\config_dummy* to display the FPGA first layout. In a similar way, open another implemented layout and check that the layouts are only different in the reconfigurable partition (see Fig. 16.11).

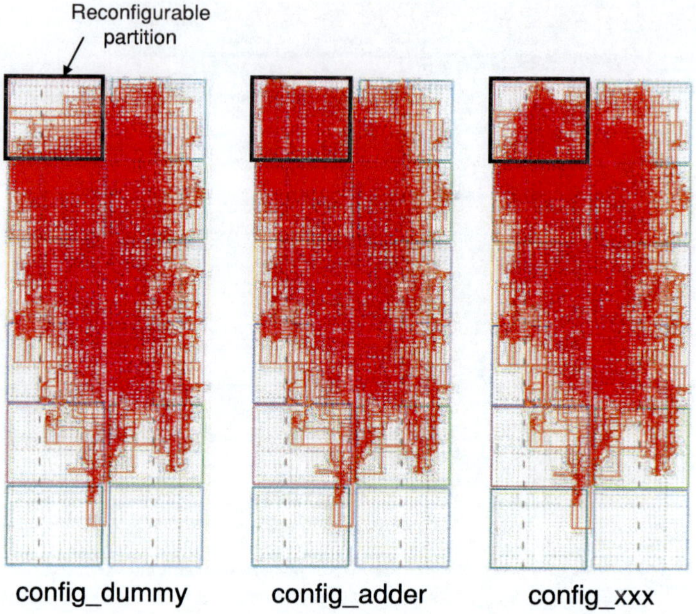

Fig. 16.11 The first implemented layout (*left*) and two other layouts for other RMs

16.3.2.5 Bitstreams Generation

It is the final step in the PR flow. Select all the implementation runs and open the contextual menu to generate the bitstreams with the default options. It launches the BitGen tool [10] from the implemented layouts. Each layout generates two bitstream BIT files: the total and the partial. The total bitstream configures all the hardware resources of the FPGA to program the reconfigurable embedded system. The partial bitstream only configures the resources of the reconfigurable partition to program the coprocessor. The BitGen also creates the file *system_bd.bmm* which annotates the source BMM file with the placement of the BRAM blocks. The annotated BMM file is required later to program the FPGA with an executable ELF file.

A change in the netlist files or in the PR settings requires updating the bitstream files. PlanAhead permits to reset the implemented runs in order to update the implemented layouts and bitstreams.

16.3.3 Software Development

Once the system hardware has been exported from XPS, the SDK can start, as detailed in the previous chapter. Create a new SDK workspace in the folder *c:\edk13.1\reconfig\XPS\SDK\workspace*. The imported bitstream file *system.bit* from XPS can program the FPGA with a static embedded system which cannot be reconfigured since it does not provide a reconfigurable partition. However, SDK can build the BSP and the ELF file for the imported system. Moreover, this static system can be tested, debugged and simulated, with the initial configuration for the coprocessor, in a similar way as in the previous chapter. The static system can be used also to program the external FLASH with the partial bitstreams. The reconfigurable system will read the partial bitstreams to program the coprocessor.

16.3.3.1 Flashing the Partial Bitstreams

The folder *bitstream* contains a simple batch file *bitstream.bat* which copies the required bitstream files from PlanAhead. Open a command shell by clicking the SDK menu *Xilinx Tools→Launch Shell*, in order to execute the batch file.

```
> cd ..\..\..\bitstream\
> bitstream
```

The total bitstream of the reconfigurable system is copied and renamed to *system_dummy.bit* since the module *dummy* is the initial coprocessor configuration. The partial bitstreams are named *partial_dummy.bit*, *partial_adder.bit* and so on. The file sizes of the partial bitstream are larger than those reported by PlanAhead since they include header data and the ICAP commands [2] to perform the partial reconfiguration. The batch also copies the BMM file *system_bd.bmm* which will be necessary later.

The NOR FLASH 28F128J3D [7] provides 16 MB of non-volatile memory arranged in 128 KB blocks. It provides the standard CFI (Common Flash Interface) which permits to erase and program the FLASH blocks. The application devotes the last 1 MB (8 blocks) to storing the partial bitstreams and each single block can allocate a partial bitstream file.

The easiest way to program the FLASH memory is from SDK. First, the FPGA must map an embedded system which provides a memory controller able to successfully access the external FLASH. Then, the SDK can load an executable into the embedded system which downloads a data file and executes the CFI queries to program the FLASH.

Therefore, program the FPGA with the bitstream *system.bit* imported from XPS, and then click the menu *Xilinx Tools→Program Flash* to open the dialog window shown in Fig. 16.12.

1. Set the file *partial_dummy.bit* which will be programmed into the FLASH.
2. Check that the EMC and bank are related to the external FLASH, and that the base address is correct (*0xA0000000*).
3. Set the address offset to 15 MB (*0xF00000*). The file is programmed in the resulting address of the addition the base address and the offset (*0xA0F00000*).
4. Program the FLASH. It will take some seconds to complete.

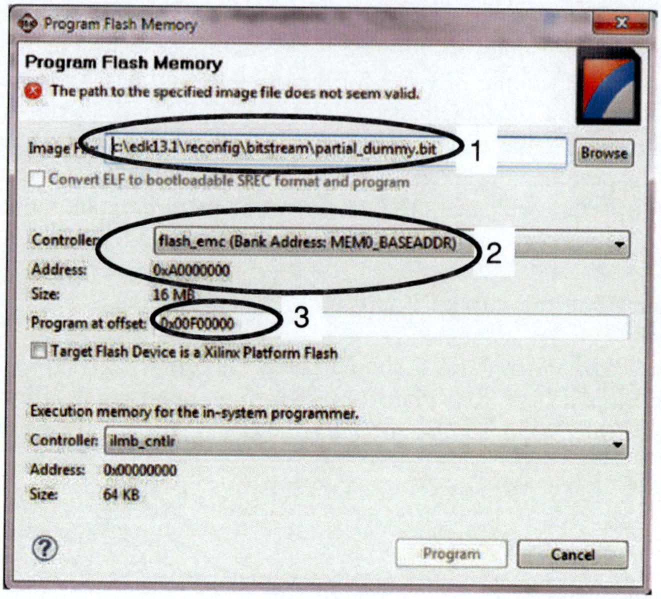

Fig. 16.12 Programming a partial bitstream into the FLASH

Repeat the process to program the rest of partial bitstreams into the next FLASH blocks (*0xA0F200000*, *0xA0F400000*, etc.). Therefore, the offset must be incremented 128 KB (*0x20000*) to point to the next block.

16.3.3.2 BSP Generation

The BSP generation is similar to the study cases of the previous chapter. The SDK provides a wizard to create the BSP project. The MSS file is modified to set the specific driver to the reconfigurable coprocessor in order to build the BSP with it.

```
BEGIN DRIVER
   PARAMETER DRIVER_NAME = fsl_rcopro
   PARAMETER DRIVER_VER = 1.00.a
   PARAMETER HW_INSTANCE = rcopro
END
```

Moreover, the file *libgen.options* must add the path of the local repository which contains the source files. They are stored in the default folder *drivers* allocated in the XPS root directory.

```
REPOSITORIES=-lp ../../../
```

The coprocessor driver is composed of a different function for each module to compute 3 × 3 matrices. They send data and retrieve results through the FSL slots, in the same way as described in the previous chapter. The driver function for the matrix addition and multiplication sends two matrices and it retrieves the resulting matrix. The scalar multiplication function sends an integer and a matrix and retrieves the computed matrix. The determinant function sends a matrix and it returns the integer result. Finally, the driver for the dummy module sends any data and it returns the data (should be zero) retrieved from the module.

```
void fsl_rcopro_putmatrix(int a[3][3])
{int i,j;
 for(i=0; i<3; i++) for(j=0; j<3; j++)          //3x3 loop
    putfsl(a[i][j],FSL_INPUT_SLOT); }           // writes FSL

void fsl_rcopro_getmatrix(int a[3][3])
{int i,j;
 for(i=0; i<3; i++) for(j=0; j<3; j++)          //3x3 loop
    getfsl(a[i][j],FSL_OUTPUT_SLOT); }          // reads FSL

void fsl_rcopro_adder(int a[3][3],int b[3][3],int o[3][3])
{fsl_rcopro_putmatrix(a);                  //sends first matrix
 fsl_rcopro_putmatrix(b);                  //sends second matrix
 fsl_rcopro_getmatrix(o); }                //gets resulting matrix
...
```

16.3.3.3 Executable

Create a new C++ project named *app1* to build the executable ELF file for the generated BSP. The source file *app1.cc* contains the main application functions, and it relies on the C++ class *CBitStream.* The class implements methods to read the bitstream headers from FLASH and to perform the partial reconfiguration. The application devotes an array of objects of the previous class which is initialized at the beginning of the application. Three parameters define the settings used to initialize the array reading the bitstream headers. Then, the application continues communicating with the user through the serial port in order to load the desired RM. It also performs a test of the active reconfigurable module which compares the resulting data from the coprocessor and the computed data from MicroBlaze. As commented previously, Microblaze must command the GPIO to reset the FSLs and the reconfigurable coprocessor after the partial reconfiguration completes.

```
#include "cbitstream.h"      //declaration of CBitStream class

#define NUM_BITSTREAMS (5)  //parameters for bitstreams
#define ADDR_BASE_BITSTREAMS
   (XPAR_FLASH_EMC_MEM0_BASEADDR+0xF00000)
#define ADDR_OFFSET_BITSTREAMS (0x20000)

CBitStream bitstream[NUM_BITSTREAMS];       //Array of objects

void rcopro_init_bitstreams()   //Initialize array of objects
{unsigned char *file=(unsigned char*)ADDR_BASE_BITSTREAMS;
 for(int j=0; j<NUM_BITSTREAMS; j++)
   {...
    bitstream[j].ReadHeader(file); //Reads a bitstream header
    file+=ADDR_OFFSET_BITSTREAMS; }//Computes next address
}

void rcopro_reconfig(char idx)      //Reconfigures a bitstream
{...
 if(bitstream[rcopro_idx].Reconfig())     //Calls a C++ method
   xil_printf("Succeed\r\n");
...
 rcopro_reset(true);       //GPIO asserts the decoupling_rst
 rcopro_reset(false);      //GPIO deasserts the decoupling_rst
}

int main()
{...
 rcopro_init_bitstreams(); //Initializes the array of objects
 ...
 while(!end)
   {rcopro_test();    //Tests the active reconfigurable module
    switch(XUartLite_RecvByte(XPAR_RS232_BASEADDR)) //Get key
      {case 27:  end=true; break;              //Quits when ESC
       case '0': rcopro_reconfig(0); break;  //Loads RM#0
       case '1': rcopro_reconfig(1); break;  //Loads RM#1
       ...}
 ...}
```

The class *CBitStream* contains the method *ReadHeader* to initialize its member data from a bitstream file. The member data will take invalid values if the header reading fails. The BIT files stored in the FLASH are composed of a header and raw data. The header contains several fields [11] that provide information about the layout: target FPGA, date, time and name of the NCD file. The last field contains the address and the size (in bytes) of the raw data.

The method *Reconfig* performs the partial reconfiguration after checking the data collected from the header. The raw data of a bitstream embeds the set of ICAP commands [2] that permit the partial reconfiguration of the FPGA. Therefore, the task performed by the C++ class during the reconfiguration is reduced to write into the HWICAP the 32-bit words retrieved from the raw data.

```
void CBitStream::ReadHeader(unsigned char *header)
{Header=header;                              //Address to the BIT file
 Addr=NULL; Size=0;                                   //Invalid values
 Fpga=NULL; Ncdfile=NULL; Date=NULL; Time=NULL;
 ...
 for(unsigned char key='a'; key<='e'; key++) //Header fields
    {...
     switch(key)
        {case 'a': Ncdfile=header; header+=length; break;
         case 'b': Fpga=header; header+=length; break;
         ...
         case 'e': Addr=header+2; Size=...; break;}
    }

bool CBitStream::Reconfig()
{if(Addr==NULL || Size==0)
    return false;                        //Quit if header is invalid
 unsigned char *addr=Addr;              //Start address of raw data
 unsigned char *max=Addr+Size;          //End address or raw data
 Hwicap_InitWrite();                    //Abort and reset the HWICAP
 while(addr<max)                        //Loop for the entire raw data
   {unsigned int word=0;                //32-bit configuration word
    for(unsigned char k=0; k<4; k++)    //Read 4 bytes
       word=(word<<8)|(*addr++);        //to set the conf. word
    if(!Hwicap_WordWrite(word))         //Write it to HWICAP
       return false;}                   //Quit if write fails
 return true;}                          //End successfully
```

The HWICAP peripheral [5] embeds the ICAP and provides two FIFOs and a set of registers to perform the partial reconfiguration or to read-back the configuration memory. The example uses three peripheral's registers to perform the partial reconfiguration: control, status, and write FIFO registers. MicroBlaze accesses to the registers through C/C++ pointers as commented in the previous chapter.

The C++ class implements the method *Hwicap_WordWrite* to write a configuration word to the HWICAP. First, the word is temporally stored in the write

FIFO. Then, it sets the control register to start the data transfer from the FIFO to the ICAP. The status register indicates whether the write to the HWICAP was completed, aborted or got an error. The method returns unsuccessfully in the last two cases. The method waits until the write of the configuration word completes before returning successfully.

The method *Hwicap_InitWrite* aborts and restarts the HWICAP before starting, in order to assure that the peripheral is empty of data from a previous reconfiguration.

```
#include <xparameters.h>    //declares XPAR_HWICAP_0_XXX
#include <xhwicap_l.h>      //declares HWICAP_XXX and XHI_XXX
#define HWICAP_ADDR (XPAR_HWICAP_0_BASEADDR)
...                         //Pointers to the HWICAP registers
volatile unsigned int *hwicap_control
   =(unsigned int*)(HWICAP_ADDR+XHI_CR_OFFSET); //Control reg
volatile unsigned int *hwicap_status
   =(unsigned int*)(HWICAP_ADDR+XHI_SR_OFFSET); //Status reg
volatile unsigned int *hwicap_writefifo
   =(unsigned int*)(HWICAP_ADDR+XHI_WF_OFFSET); //Write FIFO


inline bool CBitStream::Hwicap_WordWrite(unsigned int word)
{*hwicap_writefifo=word; //Write a configuration word to FIFO
 *hwicap_control=XHI_CR_WRITE_MASK; //Writes the FIFO to ICAP
 unsigned int status=*hwicap_status;//Reads status register
 if((status&XHI_SR_CFGERR_N_MASK==0)||
   (status&XHI_SR_IN_ABORT_N_MASK==0))
      return false;             //Aborted or got error
 if(status&XHI_SR_DONE_MASK==0)    //Waits the ICAP to finish
   while(*hwicap_status&XHI_SR_DONE_MASK==0);
 return true;}                  //Successfully completed
inline void CBitStream::Hwicap_InitWrite()
{//Aborts ICAP and FIFOs, and resets hwicap_interrupts
 *hwicap_control=(XHI_CR_SW_ABORT_MASK|XHI_CR_SW_RESET_MAS);}
```

The application would improve using an ISR associated to the HWICAP to avoid waiting to the completion of the write. The ISR would read a new configuration word when the HWICAP is ready to get a new one. Moreover, the ISR would write the FIFO with several configuration words before starting the data transfer to the ICAP. They were not implemented since they would obfuscate the basics about the partial reconfiguration.

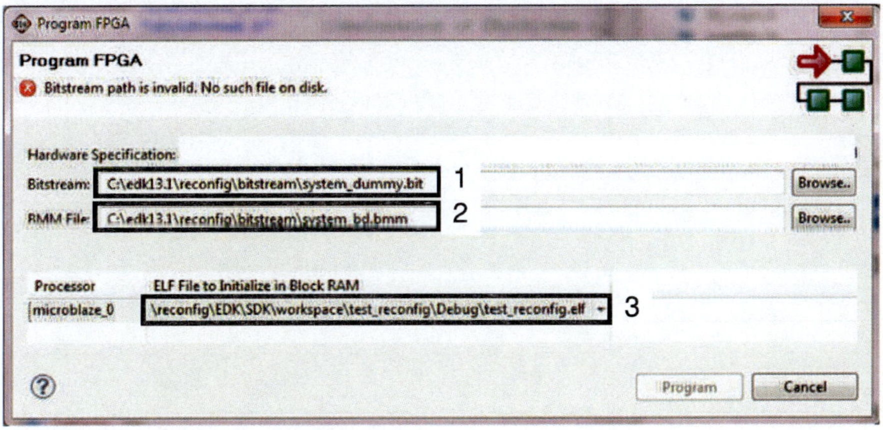

Fig. 16.13 Testing the reconfigurable system

16.3.3.4 Testing

In order to test the reconfigurable embedded system, the FPGA must be programmed with the bitstream generated by PlanAhead and the executable file. Therefore, select the bitstream *system_dummy.bit* and the BMM *system_bd.bmm* that were imported from PlanAhead (see Fig. 16.13). Then, select the executable ELF file *app1.elf*. SDK calls the Data2MEM [9] tool to generate the bitstream file *download.bit* which configures the FPGA and the BRAM contents.

The application can also be debugged in the reconfigurable system as described in the previous chapter.

References

1. Xilinx (2011a) Partial reconfiguration user's guide (UG702)
2. Xilinx (2010a) Virtex-5 FPGA configuration user guide (UG191)
3. Avnet (2009) Virtex-5 LX evaluation kit. User guide
4. Xilinx (2010b) Virtex-5 FPGA user guide (UG190)
5. Xilinx (2011b) LogiCORE IP XPS HWICAP (DS586)
6. Xilinx (2010c) LogiCORE IP XPS multi-channel external memory controller (DS575)
7. Numonyx (2007) Numonyx^{TM} embedded flash memory J3 v.D datasheet
8. Xilinx (2010d) XST user guide for virtex-4, virtex-5, spartan-3 and newer CPLD devices (UG627)
9. Xilinx (2011d) Data2MEM user guide (UG658)
10. Xilinx (2011c) Command line tools user guide (UG628)
11. PldTool (2010) Xilinx BIT bitstream files. http://www.pldtool.com/pdf/fmt_xilinxbit.pdf. Accessed July 2011

About the Authors

Jean-Pierre Deschamps received his MS degree in electrical engineering from the University of Louvain, Belgium, in 1967, his PhD in computer science from the Autonomous University of Barcelona, Spain, in 1983, and his second PhD degree in electrical engineering from the Polytechnic School of Lausanne, Switzerland, in 1984. He has worked in several companies and universities. He is currently a professor at the University Rovira i Virgili, Tarragona, Spain. His research interests include ASIC and FPGA design, digital arithmetic and cryptography. He is the author of nine books and of more than a hundred international papers.

Gustavo Sutter received his MS degree in Computer Science from State University UNCPBA of Tandil (Buenos Aires) Argentina, and his PhD degree from the Autonomous University of Madrid, Spain. He has been a professor at the UNCPBA, Argentina and is, currently, a professor at the Autonomous University of Madrid, Spain. His research interests include ASIC and FPGA design, digital arithmetic, and development of embedded systems. He is the author of three books and about fifty international papers and communications.

Enrique Cantó received his MS degree in electronic engineering (1995) and his PhD in electronic engineering (2001), both from the Polytechnic University of Barcelona, Spain. He has been a professor at the Polytechnic University of Barcelona, Spain and is, currently, a professor at the University Rovira i Virgili, Tarragona, Spain. His research interests include ASIC and FPGA design, development of embedded systems, and dynamic reconfiguration of programmable devices. He is the author of about fifty international papers and communications.

J.-P. Deschamps et al., *Guide to FPGA Implementation of Arithmetic Functions*, Lecture Notes in Electrical Engineering 149, DOI: 10.1007/978-94-007-2987-2,

Index

J.-P. Deschamps et al., *Guide to FPGA Implementation of Arithmetic Functions*, Lecture Notes in Electrical Engineering 149, DOI: 10.1007/978-94-007-2987-2,